Fungal Biology

Series Editors

Fungal biology has an integral role to play in the development of the biotechnology and biomedical sectors. It has become a subject of increasing importance as new fungi and their associated biomolecules are identified. The interaction between fungi and their environment is central to many natural processes that occur in the biosphere. The hosts and habitats of these eukaryotic microorganisms are very diverse; fungi are present in every ecosystem on Earth. The fungal kingdom is equally diverse, consisting of seven different known phyla. Yet detailed knowledge is limited to relatively few species. The relationship between fungi and humans has been characterized by the juxtaposed viewpoints of fungi as infectious agents of much dread and their exploitation as highly versatile systems for a range of economically important biotechnological applications. Understanding the biology of different fungi in diverse ecosystems as well as their interactions with living and non-living is essential to underpin effective and innovative technological developments.

This series will provide a detailed compendium of methods and information used to investigate different aspects of mycology, including fungal biology and biochemistry, genetics, phylogenetics, genomics, proteomics, molecular enzymology, and biotechnological applications in a manner that reflects the many recent developments of relevance to researchers and scientists investigating the Kingdom Fungi. Rapid screening techniques based on screening specific regions in the DNA of fungi have been used in species comparison and identification, and are now being extended across fungal phyla. The majorities of fungi are multicellular eukaryotic systems and therefore may be excellent model systems by which to answer fundamental biological questions.

A greater understanding of the cell biology of these versatile eukaryotes will underpin efforts to engineer certain fungal species to provide novel cell factories for production of proteins for pharmaceutical applications. Renewed interest in all aspects of the biology and biotechnology of fungi may also enable the development of "one pot" microbial cell factories to meet consumer energy needs in the 21st century. To realize this potential and to truly understand the diversity and biology of these eukaryotes, continued development of scientific tools and techniques is essential. As a professional reference, this series will be very helpful to all people who work with fungi and should be useful both to academic institutions and research teams, as well as to teachers, and graduate and postgraduate students with its information on the continuous developments in fungal biology with the publication of each volume.

Bhim Pratap Singh • Shekhar Agnihotri
Harinder Singh Oberoi
Editors

Fungal Additives and Bioactives in Food Processing Industries

Challenges and Prospects

Springer

Editors
Bhim Pratap Singh
Department of Agriculture and Environmental Sciences
National Institute of Food Technology Entrepreneurship and Management (NIFTEM), Kundli
Sonepat, Haryana, India

Shekhar Agnihotri
Department of Agriculture and Environmental Sciences
National Institute of Food Technology Entrepreneurship and Management (NIFTEM), Kundli
Sonepat, Haryana, India

Harinder Singh Oberoi
Department Interdisciplinary Sciences
National Institute of Food Technology Entrepreneurship and Management (NIFTEM), Kundli
Sonepat, Haryana, India

ISSN 2198-7777 ISSN 2198-7785 (electronic)
Fungal Biology
ISBN 978-3-032-04518-8 ISBN 978-3-032-04519-5 (eBook)
https://doi.org/10.1007/978-3-032-04519-5

This Springer imprint is published by the registered company Springer Nature Switzerland AG
The registered company address is: Gewerbestrasse 11, 6330 Cham, Switzerland

The book volume is dedicated to the soldiers of the Indian Armed Forces for their dedication and bravery shown during Operation SINDOOR.

Preface

The journey of compiling this book, *Fungal Additives and Bioactives in Food Processing Industries: Challenges and Prospects*, has been both exhilarating and challenging. Fungi, often overshadowed by their reputation as spoilage agents, hold untapped potential as game-changers in the food industry. This realization inspired us to compile cutting-edge research and innovative applications of fungal bioactives, bridging the gap between scientific discovery and industrial practice. Our goal is to present a comprehensive resource that empowers researchers and students exploring this emerging field while demonstrating to industry professionals how fungal-based solutions can enhance food security, quality, and sustainability.

This book delves into diverse facets of fungal applications, including novel extraction methods for fungal metabolites, the role of fungal endophytes as sources of bioactives, and using fungal metabolites as natural colorants and flavor enhancers. From metabolites that combat food-borne pathogens to enzymes that revolutionize food processing, we highlight fungi as a treasure trove of solutions aligned with the urgent need for eco-friendly alternatives. The later sections address regulatory frameworks and the promising synergy between fungal bioactives and nanotechnology. Each contribution reflects the collaborative efforts of experts across microbiology, biotechnology, and food science, showcasing the interdisciplinary nature of this field.

We extend our deepest gratitude to the contributors, reviewers, and publishers who made this endeavor possible. We hope this book not only sparks curiosity but also drives innovation, inspiring the next wave of research and industrial adoption of fungal-based solutions. Together, we can redefine fungi's role in the food industry, transforming challenges into opportunities for a healthier, more sustainable future.

Sonepat, Haryana, India

Bhim Pratap Singh
Shekhar Agnihotri
Harinder Singh Oberoi

Acknowledgments

The present book volume entitled *Fungal Additives and Bioactives in Food Processing Industries: Challenges and Prospects* is a collection of the latest developments and prospects of fungal bioactives in the food industry. The fungal bioactives have an overabundance of applications in the food industry that need the attention of researchers. For successful compilation of book chapters, first and foremost, we would like to extend our sincere gratitude to all the contributors for sparing time and extending their contribution in the form of a book review to this volume. Your knowledge and commitment made it possible to come up with this final volume in this shape. Equally, the support of the editorial and production team of Springer is also acknowledged. A special thanks to Arun Siva Shanmugam, Production Editor, for sparing significant time in providing the editorial comments as and when needed, which were crucial to bring up the volume in this shape.

We are also thankful to the Director, National Institute of Food Technology, Entrepreneurship and Management, for providing all administrative support during the compilation. At last, we extend our sincere appreciation to all the readers, and we hope that the book contains the information desired for their academic and research endeavors.

Bhim Pratap Singh
Shekhar Agnihotri
Harinder Singh Oberoi

Acknowledgements

Contents

About the Editors

Bhim Pratap Singh is a Professor and Head, Department of Agriculture and Environmental Sciences, National Institute of Food Technology Entrepreneurship and Management (NIFTEM), an Institute of National Importance (INI) under the Ministry of Food Processing Industries, Government of India. Dr. Singh has more than 17 years of teaching experience in the field of applied microbiology. He has authored more than 100 research papers and book chapters in journals and books of international repute. He has edited nine books published by Springer and Elsevier. He is a member of the Association of Microbiologists of India (AMI), the Asian PGPR Society for Sustainable Agriculture, the Mushroom Society of India (MSI), and the Indian Science Congress (ISC). He has received several prestigious awards from agencies such as DST, New Delhi, and other agencies in the field of Microbial Diversity. He has served as a guest editor in many journals like *Frontiers in Microbiology*, *Biology-MDPI*, *Journal of Fungi-MDPI*, *Frontiers in Molecular Biosciences*, etc. His major research areas are in postharvest disease management using natural agents and improving the shelf life of fresh commodities to reduce food loss happening during the food supply chain.

Shekhar Agnihotri is currently working as an Assistant Professor at the National Institute of Food Technology Entrepreneurship and Management, Sonepat (Haryana) under the Ministry of Food Processing Industries (MOFPI), Government of India. He holds a PhD in Nanotechnology from IIT Bombay, receiving the prestigious IIT-B Excellence Award for Best PhD thesis. Earlier, he received M.Tech in Biochemical Engineering jointly from IIT-BHU and IIT Kanpur. His prime expertise lies in the use and development of nano-enabled technologies in food bioprocessing, post-harvest processing, biotechnology, and the environment. Dr. Shekhar's research group focuses in bringing new innovations to generate value-added products from food/agro-waste, designing packaging materials, intelligent coatings and smart sensing technologies. He has over 35 publications in top tier journals, 04 International Books and 02 patents, to his credit. Dr. Shekhar is the recipient of DST young scientist award, Indo-Japan Young Fellowship Award, Gold access award (Royal Society of Chemistry, UK), and TIET Best Performer Award.

Harinder Singh Oberoi is currently holding the position of Director, National Institute of Food Technology Entrepreneurship and Management (NIFTEM), an Institute of National Importance (INI) under the Ministry of Food Processing Industries, Government of India. He completed his Doctorate in Microbiology from Punjab Agricultural University, Ludhiana, India, and Post-Doc in Bioprocessing and Value Addition from Kansas State University, Kansas, USA, with diverse exposure to different aspects of Post-Harvest Technology. He has over 28 years of experience working with the food and beverage industry, including big conglomerates, government R&D, and academia in post-harvest management, value addition, food processing, food safety, fermentation, and biovalorization. He is a recipient of several national and international awards. He is a fellow of the National Academy of Agricultural Sciences (NAAS) and the National Academy of Biological Sciences (NABS). He has published over 100 research articles in high-impact factor journals in the field of fermentation technology, bioprocessing, and food safety.

Chapter 1
Novel Extraction Methods for Bioactive Compounds from Fungi: An Overview

Charu Agnihotri, Shubham Singh Patel, Zothanpuia, and Bhim Pratap Singh

1 Introduction

Ascomycota and Basidiomycota are the largest phyla among macrofungi, including mushrooms. The phylum Ascomycota is divided into three subphyla: Taphrinomycotina, Saccharomycotina, and Pezizomycotina. Similarly, Basidiomycota is categorized into three subphyla: Ustilaginomycotina, Pucciniomycotina, and Agaricomycotina. Edible mushrooms are often produced by basidiospores, which are generated within certain subphyla of Basidiomycota (Ramos et al., 2019). The subphyla collectively encompass approximately 15,000 to 16,000 species, of which only around 700 are recognized as edible. Of these, approximately 60 species are cultivated worldwide, with just 10 species accounting for large-scale production, despite the remarkable diversity within these groups (Cerletti et al., 2021; Gong et al., 2020). Specific varieties of mushrooms, including *Agaricus bisporus* (button mushroom), *Pleurotus* species (oyster mushroom), *Lentinula edodes* (shitake), *Flammulina velutipes* (enoki mushroom), *Grifola frondose* (maitake), *Ganoderma species* (reshi), contribute high nutritional benefits (Leong et al., 2021; Taofiq et al., 2017). Recently, consumer demand for mushrooms has increased over the period of time as they are well-known biomolecules that are

---

The original version of the chapter has been revised. A correction to this chapter can be found at https://doi.org/10.1007/978-3-032-04519-5_16

C. Agnihotri · S. S. Patel · B. P. Singh (✉)
Department of Agriculture and Environmental Sciences, National Institute of Food Technology Entrepreneurship and Management (NIFTEM), Kundli, Sonepat, Haryana, India
e-mail: bpsingh@niftem.ac.in

Zothanpuia
Department of Biotechnology, Pachunga University College (PUC), Mizoram University, Aizawl, India

B. P. Singh et al. (eds.), *Fungal Additives and Bioactives in Food Processing Industries*, Fungal Biology, https://doi.org/10.1007/978-3-032-04519-5_1

described for their medical attributes. Polysaccharides, including β-glucans, phenolics, terpenoids, glycopeptides, peptides such as ribosomal inactivating proteins (RIPs), fungal immunomodulatory proteins (FIPs), and laccases are especially considered as potential bioactives (Agnihotri et al., 2024; Yadav & Negi, 2021). Additionally, extensive research over the past two decades has explored their bioactivities, including antioxidant, anticancer, antidiabetic, antimicrobial, and immunomodulatory effects (Kour et al., 2022). Recent studies showed the extraction of valuable compounds such as polysaccharides of mushrooms, especially from mushroom wastes, for further food and packaging applications (Table 1.1) (Leong et al., 2021). The extraction of targeted components from various mushrooms is affected by the nature of the compounds. For decades, conventional techniques such as hot water extraction, maceration, infusion, and decoction have been widely used for extracting bioactive compounds (Morales, 2024). However, these methods present significant limitations, including challenges in releasing bioactive compounds from the complex matrix, the need for large amounts of organic solvents, and high energy consumption (Chemat et al., 2019). These drawbacks can cause deterioration of the compounds and alterations in their structure. To handle these limitations, novel extraction techniques have been developed, such as microwave-assisted extraction, ultrasound-assisted extraction, supercritical fluid extraction, subcritical fluid extraction, pulsed electric field extraction, natural deep eutectic solvent extraction, and enzyme-assisted extraction (Aguiló-Aguayo et al., 2017; Barbosa et al., 2020; Marić et al., 2018; Parniakov et al., 2014). However, some novel techniques, such as microwave-assisted extraction, still require improvement, particularly in terms of solvent usage and temperature control (Marić et al., 2018). To overcome these issues, non-thermal extraction methods that operate at room temperature or under freeze conditions have emerged as promising alternatives (Marić et al., 2018). This chapter focuses on the influence of various extraction techniques on the recovery of bioactive compounds from mushrooms, highlighting their advantages and limitations.

2 Conventional Extraction Methods

Mushrooms are a good source for deriving biologically beneficial components or bioactive compounds, which require various extraction techniques to release them from the fungal matrix (Kour et al., 2022). Extraction operations allow for the driving out of such compounds of interest from a sample matrix (Farooq & Mir, 2022). Techniques like hot-water extraction, maceration, infusion, hydro distillation, and Soxhlet extraction have long been used to obtain target compounds from plants and other biological matrices (Table 1.2) (Subramanian & Anandharamakrishnan, 2023). These techniques are often termed conventional extraction techniques (CET) due to their complexity, lower operating costs, and easy installation. The selection and operation of a CET are mainly based on the polarity of the target bioactive compound and the solvent. Furthermore, solvent-based extraction methods are based on the selective extraction of bioactive compounds, with the various solvents being a controlling factor for the attained yield or composition of the isolated molecules (Kapadia et al., 2022). CET covers solvent-based extraction techniques (as mentioned before), which are discussed below.

Table 1.1 Mushroom bioactive compounds and their mechanisms

Mushrooms species	Bioactive compound	Properties	Mode of action	References
Amillariella mellea	AAMP	Anti-diabetic	It exerts a hypoglycemic effect and enhances insulin sensitivity, likely by promoting lipolysis	Yang et al. (2019)
Grifola frondosa	Polysaccharide GFP-N	Anti-diabetic	It enhances hypoglycemic effects by activating the insulin receptor substrate 1 (IRS-1), phosphatidylinositol-3-kinase/Akt (PI3K/Akt) signaling pathway, and glucose transporter 4 (GLUT-4) translocation, while suppressing c-Jun N-terminal kinase 1/2 (JNK1/2) activity	Chen et al. (2019)
Grifola frondosa	Ergosterol peroxide (EPO)	Anti-diabetic	It inhibits ROS production, upregulates glucose uptake, and modulates PI3K/Akt, MAPK, and GLUT4 signaling pathways	Wu et al. (2020)
Grifola frondosa	Polysaccharides	Anti-diabetic	It overexpresses cholesterol 7α-hydroxylase CYP7A1 and bile salt export pump BSEP, suggesting an enhancement of bile acid (BAs) synthesis and excretion in the liver	Guo et al. (2020)
Pleurotus pulmonarius	Hot water extracts (HWE)	Anti-diabetic	HWE achieved 99.23% α-amylase inhibition at 1000 μg/ml	Balaji et al. (2020)
Pleurotus pulmonarius	Catalase-like protein, glyceraldehyde-3-phosphate dehydrogenase-like protein	Anti-diabetic	It inhibited *Saccharomyces cerevisiae* α-glucosidase (24.18%) and porcine pancreatic α-amylase (41.80%) activities	Wahab et al. (2014)
P. ostreatus and *P. eryngii*	Glucan fractions,	Immunomodulatory	Prevented TNF-α-mediated inflammation in the dextran sulfate sodium (DSS)-induced inflamed intestine	Vetvicka et al. (2019)
Pleurotus eryngii	Water-soluble polysaccharide	Antitumor and immunomodulatory	Increased thymus and spleen indices, promoting lymphocyte proliferation, elevated TNF-α and IL-2 levels in Renca tumor-bearing mice	Yang et al. (2013)

(continued)

Table 1.1 (continued)

Mushrooms species	Bioactive compound	Properties	Mode of action	References
Pleurotus eryngii	Polysaccharides	Antioxidant activity, hepatoprotective	Captured reactive oxygen species (ROS)	Li and Shah (2016)
King Boletus mushroom	Protein hydrolysate (KBMPH)	Antihypertensive, antioxidant activity	Expressed angiotensin I-converting enzyme (ACE) inhibitory effect	Kaprasob et al. (2022)
Ganoderma lucidum	Peptides (DRVSIYGWG, ALLSISSF)	Antioxidant and anti-inflammatory	In RAW 264.7 murine macrophages, it downregulates the expression of proinflammatory mediators, including inducible nitric oxide synthase (iNOS), interleukin-6 (IL-6), tumor necrosis factor-alpha (TNF-α), and cyclooxygenase-2 (COX-2)	Aursuwanna et al. (2022)

2.1 *Hot-Water Extraction (HWE)*

HWE is a solvent-based method that requires water as the primary solvent and is carried out at high temperatures (usually at or higher than the boiling point of water) (Liu, 2015). HWE is based on the function of solubility, where a solute travels from one phase to another, usually from a solid to a liquid (typically water). It has been applied to extract phytochemicals like anthocyanins from blue pea flower, various bioactive compounds from papaya peel and pulp, prepare onion-stevia leaf extracts, extract galactoglucomannas from softwood and hemicellulose from by-products of food processing, and phenolic compounds from *Stevia rebaudiana* leaves (Halahlah et al., 2023; Sandra et al., 2020; Vidana Gamage & Choo, 2023; Wani & Uppaluri, 2022; Wolf et al., 2022; Yang et al., 2023). Likewise, the technique has also been applied to extract bioactive components from mushrooms. The technique was applied when crude polysaccharides were being derived from *Volvariella volvacea* mushrooms, which was performed at temperatures of 60 °C, 80 °C, and 100 °C (Tepsongkroh et al., 2019). Furthermore, it was also employed when bioactive compounds were extracted from *Inonotus obliqus* (chaga) mushrooms, which was done at 100 °C (Hwang et al., 2019). The procedure for conducting HWE for mushrooms typically involves pulverizing the dried mushroom samples or grinding fresh mushroom samples into a paste. A fixed amount is added to a determined volume of water, where the solid–liquid mixture is left to boil for a decided amount of time. The boiled mixture is then allowed to come to room temperature and subjected to filtration to separate the solid particles from the liquid portion. The filtrate containing the water-soluble components from the mushroom powder is retained, and the

Table 1.2 Conventional methods for the extraction of mushroom bioactives

Species	Bioactive extracted	Extraction technique	Extraction solvent	Parameters	Extraction yield	Reference
Volvariella volvacea (straw mushroom)	Crude polysaccharides	HWE	Water	Solid-liquid ratio: 1:25 (w/v) Temperature: 60 °C, 80 °C, and 100 °C Stirring speed: 200 rpm Time: 2 h	The highest yield obtained through HWE was 9.38% The combinations of HPP and HWE (600 MPa and 60 °C) produced a yield 2–12% better than HWE alone HPP can lead the charged groups to deprotonate and cause the salt bridges and hydrophobic interactions to disrupt, leading to an increase permeability	Tepsongkroh et al. (2023)
vInonotus obliqus (chaga mushroom)	Bioactive compounds	HWE	Water	Sample weight: 10 g Solvent volume: 400 mL Time: 2 h (water bath)	HWE produced the second lowest yield of α- and β-glucans, i.e., 2.66 ± 0.42% and 1.15 ± 0.12% (w/w, dry basis), respectively HTPE and EE produced better results for polysaccharide extraction and retention HWE lead to the second highest yield of 166 ± 8.6 mgGAE/g of total phenolic compounds in comparison to HTPE and EE	Hwang et al. (2019)
Ganoderma lucidum (reishi mushroom)	β-D-glucan	SHE	Water	Sample weight: 10 g (defatted) Solvent volume: 400 mL Reflux cycles: 60 Extraction time: 1–24 h	SHE resulted in the maximum yield of 107.13 mg of polysaccharides, but the lowest yield of β-D-glucan (25.03 ± 4.00 mg/g) SHE outperformed UAE and HWE SHE also had the highest energy consumption of 7.2 kWh in comparison to the utilized other techniques	
Ganoderma lucidum (reishi mushroom) G2 spores	GLS	SHE	Hexane (68–70 °C B.P.)	Sample weight: 30 g (broken form) Solvent volume: 150 mL	A yield of 22.51 ± 0.23% was obtained via SHE, greater than what obtained for MAE (20.26 ± 0.29%) However, oil obtained through MAE had lower lipid oxidation and higher antioxidant activity	Senphan et al. (2024)

(continued)

Table 1.2 (continued)

Species	Bioactive extracted	Extraction technique	Extraction solvent	Parameters	Extraction yield	Reference
Agrocybe aegerita (chestnut mushroom)	Crude polysaccharides	CRE	Distilled water	Solid-liquid ratio 1:15 (w/v) Time: 2 h Temperature: 100 °C	A yield of 3.2% was obtained through CRE However, the degradation of polysaccharides were observed due to factors like extended Concentration steps, time-consuming, and labor-intensive operations	Du et al. (2022)
Pleurotus sajor-caju (oyster mushroom)	Bioactive compounds	SHE	Ethanol and hexane	Sample weight: 5 g Solvent volume: 150 mL Solid-liquid ratio: 0.033 g/mL Time: 6 h Temperature: boiling temperature of the selected solvent(s)	SHE extraction with ethanol yielded the maximum yield of 15.6 ± 0.1% However, the highest antioxidant activity was showcased by PLE (10 MPa and 40 °C) with ethanol	Krümmel et al. (2022)
Leccinum scabrum (birch bolete mushroom)	Crude extract	Maceration	Ethyl acetate	Sample weight: 490 mg Solvent volume: 14.7 mL Solid-liquid ratio: 1:30 (w/v) Extraction time: 24 h Temperature: 25 °C	The yield obtained through maceration (4.8%) was among the lowest yields The highest yield reported was 37.2%, obtained via MAE with 70% ethanol However, the extract attained through maceration illustrated the maximum activity against α-glucosidase and α-amylase during the in vitro hypoglycemic study	Ferraro et al. (2024)
Pleurotus eous (pink oyster mushroom)	Crude extract	Maceration	95% methanol	Sample weight: 100 g (roughly ground) Solvent volume: 1 L Time: 7 days Agitation interval: Twice a day	The extraction yield was noted to be 11.2% TPC and TFC content were quantified to be 4.56 ± 0.93 mg/g and 0.34 ± 0.93 mg/g, respectively The obtained extract depicted significant antibacterial activity against *S. aureus*, as enumerated through turbidometric and viable cell count assays The extract also lowered TNF-α, IL-1β, and IL-6 levels in mice, exhibiting high potential in combating *S. aureus* infections	Dash et al. (2024)

Leucopaxillus gentianeus (bitter false funnelcap mushroom)	Cucurbitacin	SHE	Til oil Petroleum ether Chloroform Ethyl acetate Methanol Water	*SHE via Til oil*: Sample weight: 100 g Solvent volume: 1000 mL Temperature: 90–100 °C Reflux cycles: 15 Time: 4 days *SHE via petroleum ether*: Sample weight: 93 g Solven volume: 1000 mL Temperature: 90–100 °C Reflux cycles: 8 Time: 35 h *SHE via chloroform*: Sample weight: 88 g Solvent volume: 1000 mL Temperature: 90–100 °C Reflux cycles: 10 Time: 41 h *SHE via ethyl acetate*: Sample weight: 88 g Solvent volume: 1000 mL Temperature: 90–100 °C Reflux cycles: 10 Time: 41 h *SHE via methanol*: Sample weight: 79 g Solvent volume: 1000 mL Temperature: 90–100 °C Reflux cycles: 13 Time: 54 h *SHE via water*: Sample weight: 74 g Solvent volume: 1000 mL Temperature: 90–100 °C Reflux cycles: 12 Time: 48 h	The highest cucurbitacin content of 264 ng was obtained through SHE conducted with til oil Furthermore, the til oil extract depicted the highest progesterone and estrogen inhibition activity during the anti-breast cancer study	Kanani and Pandya (2023)

(continued)

Table 1.2 (continued)

Species	Bioactive extracted	Extraction technique	Extraction solvent	Parameters	Extraction yield	Reference
Agaricus blazei Murrill (almond mushroom)	Ergosterol	HAE	Ethanol	Sample weight: 600 mg Solvent volume: 20 mL Solid-liquid ratio: 30 g/L Stirring speed: 500 rpm Time: 10–150 min Temperature: 30–90 °C	The optimal conditions for HAE were identified as extraction time of 150 min and a temperature of 81.6 °C HAE produced the least amount of ergosterol in comparison to UAE and MAE, where UAE gave the best yield The optimal operating conditions for UAE and MAE were highlighted as 30 min and 400 W, and 25 min and 134.6 °C, respectively.	Taofiq et al. (2019)

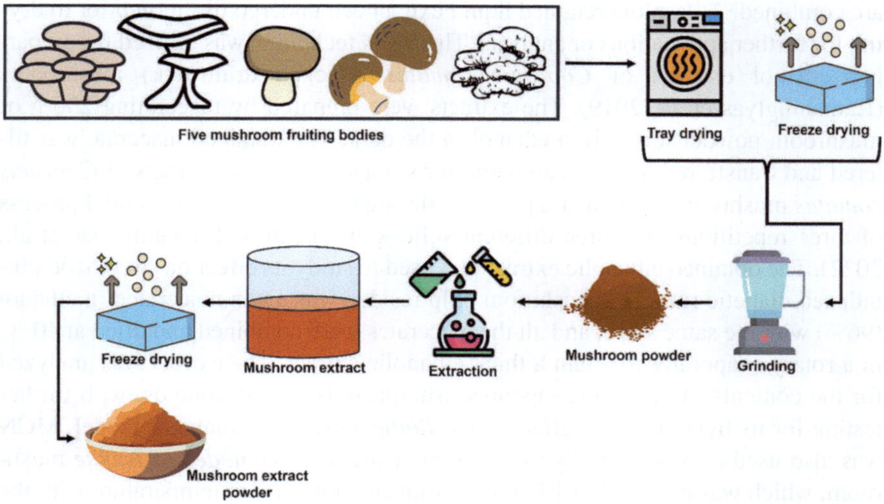

Fig. 1.1 Hot-water extraction for preparing extracts of five different mushrooms. (Based on the work by Tepsongkroh et al. (2019))

residue left behind on the filter paper can either be re-extracted with the same procedure or discarded (Tepsongkroh et al., 2019). The obtained filtrate (extract) can be dried to calculate how much of the mushroom powder got solubilized in water, which is the extract's yield. A schematic of the HWE process performed for five different freeze-dried and tray-dried mushrooms (*P. ostreatus*, *P. pulmonarius*, *V. volvacea*, *L. edodes*, and *S. commune*) to check their extracts' antioxidant properties and phenolic acid profile is depicted in Fig. 1.1.

2.2 *Maceration (MCN)*

MCN is an extraction technique where various solvents can be used, based on the chemical attributes of the compounds extracted from mushroom samples. One of the simplest extraction techniques involves soaking coarse or powdered sample material (like plants) in solvents like hexane, acetone, ethyl acetate, ethanol, and methanol (Farooq & Mir, 2022). The extraction process involves adding a fixed amount of powdered or coarsely ground samples to a closed container containing a defined volume of the selected solvent and leaving for at least 72 h (depending on the sample and efficiency of the solvent) at room temperature (Senapati & Behera, 2023). The process is aided by frequent agitation to facilitate optimal extraction from the solid particulates into the solvent by promoting diffusion and adding extra solvent to increase the extraction yield by discarding the concentrated solution from the sample surface. The liquid portion is removed, the solid-soaked sample is squeezed to obtain the maximum quantity of the extract, and both collected extracts are combined. At last, the retained liquid extract can undergo filtration prior to drying for further applications or analysis. The MCN technique was utilized for preparing ethanol extracts of *Coprinus comatus* (chicken drumstick) mushrooms (Ratnaningtyas et al., 2019). The extracts were prepared by macerating 200 g of mushroom powder for 24 h in ethanol in the dark. The obtained macerate was filtered and transferred into a clean bottle for storage. Ethanolic extracts of *Coprinus comatus* mushrooms were again prepared through MCN but in a sequential process of three repetitions and three different solid–solvent ratios (Ratnaningtyas et al., 2022). The obtained ethanolic extract was used to study its effect on streptozotocin-induced diabetic rats. The mushroom pulp residue was again macerated in ethanol (96%) with the same steps, and all the macerates were combined and dried at 40 °C in a rotary evaporator to attain a thick ethanolic extract. The extract was analyzed for the contents of alkaloids, saponins, triterpenoids, and flavonoids, with further testing for its hypoglycemic effect in the *Rattus norvegicus* diabetes model. MCN was also used in producing hydro-alcoholic extracts of *Ganoderma tsugae* mushroom, which was prepared in 190 proof grain alcohol at room temperature with the solution shaken twice per week, and the maceration period of the experiment was 1 month (Sharpe et al., 2021).

2.3 Infusion (IN) and Decoction (DCN)

IN is an extraction process where a sample material (like plants) is steeped in a boiling solvent (such as water), and the whole mixture is left to stand for a particular period of time in a stoppered container (Sharpe et al., 2021). It is based on preparing a dilute solution containing the easily soluble components of the sample material. The best instance of frequent utilization of IN is in tea, where the dried and crushed tea leaves are brewed at particular temperatures for a fixed amount of time. The technique has been employed to prepare extracts of *Auricularia delicata* and *Auricularia mesenterica* (wood ear mushrooms), which were assessed for their phytochemical constitution and antioxidant properties (Ghosh et al., 2021). IN was also applied to prepare an extract of *Flammulina velutipes* mushroom to determine its mycochemical composition and antioxidant properties (Sharma et al., 2022).

DCN, in contrast, follows the principle of HWE to extract water-soluble components from a sample material that are not susceptible to heat damage, with bioactive compounds typically extracted with water (Mathews et al., 2024). The procedure for carrying out DCN involves boiling the pulverized sample in a closed vessel filled with a defined amount of water for 30 min or until the solid–liquid mixture reduces to half the original volume (More et al., 2022). After the boiling reaches completion, the solids are separated from the liquid portion through a filter, which contains the water-soluble and thermos-stable bioactive compounds present in the sample powder. The major difference between IN and DCN is that the latter involves simmering the sample mixture rather than exposing it to harsh boiling (Suna et al., 2019). DCN is recommended for treating or extracting from samples with a high degree of physical hardness to derive thermally stable bioactive compounds (de Souza Ribeiro et al., 2022). The technology was utilized to prepare the DCN of *Cantharellus cibarius* mushroom (S. Novakovic et al., 2019). It involved heating the dry mushroom powder in Mili-Q water in a 1:10 ratio for 30 min at 100 °C. The prepared *Cantharellus cibarius* DCN was further analyzed with ferric radical reducing power (FRAP), cupric reducing antioxidant capacity (CUPRAC), lipid peroxidation, and DPPH free radical scavenging activity. Furthermore, the technique has been applied to prepare extracts of wild *Morchella esculenta* (L.) Pers., *Boletus edulis* (porcini mushroom), *Russula brevipes* Peck, *Auricularia auricula* (wood ear mushroom), and *Craterellus cornucopioides* (Mahtal et al., 2024; Novakovic, 2021; Novakovic et al., 2020, Ohiri & Odey, 2022; Sharma et al., 2019).

2.4 Hydrodistillation (HDN)

HDN is a common method utilized to extract bioactive compounds from plant materials, but mainly essential oils (Oreopoulou et al., 2019). HDN is valued for isolating aroma-active compounds contributing to mushrooms' characteristic scents and flavors. The HDN process involves packing the sample material into a still compartment containing sufficient water, which is brought to boiling temperatures. Direct steam is also injected into the sample material, as hot water and steam are the primary factors that liberate bioactive compounds from the sample material. A condenser is connected to the HDN assembly, where the vapors containing the oil and volatile bioactive compounds get condensed. Finally, the condensed mixture travels to a separator from the condenser to allow the essential oils and bioactive compounds to separate from the water phase. The method was applied to extract volatile oil from *Agaricus brasiliensis* mushroom for analyzing its antioxidant and antimicrobial activities (Mazzutti et al., 2012). A Clevenger-type apparatus (2 L capacity) was used, where 50 g of *Agaricus brasiliensis* grounded powder was added with 700 mL distilled water, and the entire operation was performed in duplicates for 6 h. The obtained volatile oil was dissolved in 2 mL of hexane in the decantation section of the Clevenger apparatus. HDN has been further employed in extracting essential oils and volatile bioactive components from fresh and dried *Stropharia rugosoannulata, Pleurotus eryngii* var. *ferulae*¸ *Pleurotus ostreatus*, *Pleurotus abalonus*, and *Boletopsis leucomelas* (Nosaka & Miyazawa, 2014; Usami et al., 2015; Usami et al., 2014; Wei et al., 2023).

2.5 Soxhlet Extraction (SHE)

SHE employs solvents at vaporization temperature and ambient pressure to selectively extract compounds of interest from solid samples, also known as a continuous solvent extraction method (Yu et al., 2023). It is also typically employed to extract biologically active lipids, sterols, and fatty acids. The SHE apparatus consists of a thimble, distillation flask, condenser, siphon, heating source (mantle), and a round round-bottom flask for collecting the solvent. A small sample amount is placed in the thimble fitted over a distillation flask filled with the extraction solvent. When the vapors from the boiling solvent return from the condenser to the thimble, the bioactive compounds are extracted with every drop of the solvent falling on the sample (Vibala et al., 2020). After reaching a specific level of submersion, the siphon starts to run, and all the solvent containing the sample solutes returns to the round-bottom flask. The whole process is continuously run for a decided period of time. The sample containing solvent can further undergo rotary evaporation to concentrate the isolated volatile compounds. SHE has been utilized to obtain bioactive extracts from *Pleurotus pulmonarius* mycelium, with acetone and water as extraction solvents (Amirullah et al., 2023). It was also implemented to extract bioactive compounds from *Pleurotus pulmonaris* (Fr.) Quél mushroom, where the mushroom's powder was subjected to boiling in 95% ethanol for 8 hrs (Amirullah et al., 2021). The obtained extract was filtered for further analysis, such as total phenolic content, DPPH radical scavenging activity, and oxidation inhibition of human plasma low-density lipoprotein through conjugated dienes and the thiobarbituric acid reactive species assay. The method was also used in extracting essential compounds and water-soluble polysaccharides from the fruiting bodies of *Grifola frondosa* grown on lignocellulosic substrates (Minakov et al., 2021). Water-soluble polysaccharides were also derived from *Ganoderma lucidum* mushroom, where 10 g of the defatted mushroom fiber was placed inside the thimble along with distilled water (400 mL) stored in the solvent reservoir (Alzorqi, Sudheer, et al., 2017b). Furthermore, ethanolic extracts of *Hericium erinaceus* mushrooms were prepared using SHE, where multiple temperature, time, and extraction condition combinations were created to optimize the extraction process (Sevindik et al., 2024). The optimization was carried out using an artificial neural network with the Levenberg–Marquardt algorithm.

2.6 Limitations of CET Approaches

The discussed CETs are known for being cost-effective and simple while delivering competitive results when used in extraction operations. However, their uses have various disadvantages, making their utilization problematic. The limitations in their usage include multiple extraction steps, substantial amounts of solvents, and prolonged extraction time (Bitwell et al., 2023). Furthermore, exposure to high temperatures during boiling and other operations might cause a significant amount of phytochemicals to degrade or decompose while being extracted from samples. Also, the yields of the extracts or the isolated bioactive compounds are usually mediocre in comparison to novel extraction techniques like microwave-assisted extraction (MAE), ultrasonication-assisted extraction (UAE), and supercritical fluid extraction (SFE) (El Maaiden et al., 2022). Moreover, the high volumes of organic solvents (like acetone, hexane, and petroleum ether) are typically toxic in nature and detrimental to the environment, requiring proper disposal facilities to contain and treat them to avoid water and soil contamination (Putra Nicky et al., 2023). Therefore, it would be beneficial if the employment of CETs in extraction procedures were minimized. Instead, the focus is shifted to unconventional or novel techniques such as UAE, MAE, and SFE to mitigate the drawbacks posed by CETs for better yield output and more efficient utilization of resources like energy and organic solvents to extract bioactive compounds from mushrooms.

3 Novel Extraction Techniques for Mushroom Bioactives

As discussed earlier, traditionally, solvent-assisted extraction methods have been employed to recover valuable compounds from mushrooms. Water extraction, maceration, percolation, pressurized liquid extraction, Soxhlet extraction, and hydroalcoholic extraction are widely used due to their cost-effectiveness and simplicity, although they require high temperature conditions and extended treatment times, leading to potential degradation of thermolabile compounds. Improved extraction techniques such as ultrasonic extraction (UAE), microwave extraction, subcritical water extraction, and enzyme-assisted extractions are employed for better extraction yield and bioactivity (Table 1.3; Fig. 1.2).

3.1 Ultrasonic-Assisted Extraction (UAE)

Ultrasonic-assisted extraction (UAE) employs thermal, mechanical, and cavitation effects to extract bioactive components. Ultrasonic waves are mechanical vibrations that propagate through solid, liquid, and gaseous media when their frequency falls within the ultrasonic range (Gong et al., 2020). Recent studies proposed various mechanisms underlying UAE, highlighting physical processes such as fragmentation, erosion, the acoustic capillary effect, membrane permeabilization, local shear stress, and structural disruption (Shen et al., 2023). Mushroom extracts were less influenced by UAE compared to potato and onion peels. Succinyl adenosine was effectively extracted after 1-hour agitation but degraded with prolonged extraction. Tryptophan extraction was minimally affected by sonication, while 5(S),8(R)-dihydroxy monocarboxylic acid and lysophosphatidylethanolamine degraded significantly at 25 kHz. Lower sonication intensity (45 kHz) preserved flavonoid glycosides, while 25 kHz favored aglycone flavonoids. Long agitation improved the extraction of membrane-bound compounds. Overall, UAE effectively extracted bioactives like flavonoids, amino acids, and carboxylic acids, with cytoplasmic compounds being more accessible (Alves Filho et al., 2021). Oyster mushrooms (*Pleurotus citrinopileatus*) are rich in phenolic compounds. In a study, optimized

Table 1.3 Novel Extraction methods for the extraction of mushroom bioactives

Species	Bioactive extracted	Novel extraction technique	Extraction solvent	Parameters	Extraction yield	Reference
Agaricus bisporus (button mushrooms)	Ergosterol (Vitamin D)	DESE	HDES consisted of menthol, thymol, or tetrabutyl ammonium hydrogen bromide (HBAs) and various long chain acids (HBDs)	Solvent volume Stirring time Sonication time Ergosterol extracted (μh/g of DW of mushroom powder)	Highest extraction yield of 6995.30 ± 2.23 μg/g of dry weight of mushroom powder obtained with HDSE containing DL-menthol-pyruvic acid Menthol contains asymmetric carbon atoms similar to ergosterol, which enhances its extraction efficiency	Khare et al. (2021)
Hericium erinaceus (lion's mane mushroom) biomass	Erinacine A	Ultrasonication	Ethanol and water	Solvent concentration (%) Extraction time (min) Solvent-material ratio (mL/g)	A maximum yield of 17.50% was obtained Extraction time was 45 min, and the solvent–material ratio was 20 mL/g	Valu et al. (2020)
Lentinula edodes (shiitake mushrooms)	Immunostimulatory PS	Autoclaving-ultrasonication extraction	Distilled water	*Autoclaving*: Macerated mushroom in distilled water (1:30, w/v) Temperature (121 °C) Pressure (15 psi) Time: 60 min *Ultrasonication*: Amplitude (60%) Treatment time: 30 min	Maximum PS mass of 64.4% obtained Ethanolic precipitation performed with 40% ethanol solution	Li et al. (2023)

Boletus bicolor (bolete mushroom)	Polyphenols	Ultrasonication	40% ethanol	Volume fraction of ethanol (%) Liquid-solid ratio (mL/g) Ultrasonication time (min) Ultrasonication temperature (°C)	Maximum yield of 13.69 ± 0.13 mg/g Polyphenolic compounds identified: syringic acid, chlorogenic acid, gallic acid, rosmarinic acid, protocatechuic acid, catechin, caffeic acid, and quercetin (via UPLC-MS)	Hu et al. (2023)
Agaricus bisporus (button mushrooms)	Chitin and chitosan	DESE and MAE	DES systems consisting of choline chloride (HBAs) Acetic acid, lactic acid, and glycerol (HBDs) DESs prepared in different molar ratios	Solid-solvent ratio (1:10) Microwave irradiation time (3 min, in 5 s pulses for each minute)	Highest deproteinization efficiency for obtaining chitin was 84.25% with DES containing choline chloride and acetic acid (1:2) Deacytlation of chitin (via MAE) to obtain chitosan was 69%	Ozel and Elibol (2024)
Pleutorus ostreatus (oyster mushroom)	Phenolics and flavonoids	Microwave hydrodiffusion and gravity	No solvents were used, as the sample moisture was enough to drive out the analytes	*Two stages*: Microwave firing at higher levels (510 W for 2 min and then 340 W for 2 min, with 2 min of cooling period) Microwave firing at low power levels (170 W for 2 min, with 2 min cooling period till extraction was complete) Extract collected via the action of gravity	A phenolic and flavonoid yield 6200 ± 167 μgGAEeq/g of the dried extract and 733 ± 27 mgQEeq of the dried extract was obtained at 26 min treatment of microwave hydrodiffusion and gravity Produced a 55.7% ± 2.8 and 82.3% ± 3.7 yield for phenolics and flavonoids than the conventional Soxhlet extraction	Singh Chouhan et al. (2020)

(continued)

Table 1.3 (continued)

Species	Bioactive extracted	Novel extraction technique	Extraction solvent	Parameters	Extraction yield	Reference
Pleurotus pulmonarius (lung oyster mushroom)	β-glucans	PEF	Water	PEF field strength: 5, 7.5, and 10 kV/cm Frequency: 15 Hz Number of pulses: 12,000	The yield of β-glucans from the caps and stems of *Pleurotus pulmonarius* mushroom was 34.03% and 50.22%, respectively More yield obtained in comparison to conventional hot water extraction	Thikham et al. (2024)
Agaricus bisporus (button mushrooms)	TPC	PEF	Distilled water:Tap water (1:1)	Sample weight: 220 g Field strength: 2–3 kV/cm Specific energy: 50–200 kJ/kg Time (h): 0–6 h	An increased yield of TPC of 96.86% was observed with PEF treatment, in comparison to conventional hot water extraction The best optimized parameters were: 2.5 kV/cm, 50 kJ/kg, and 6 h time	Calleja-Gómez et al. (2022)
Flammulina velutipes (golden needle mushroom)	Ergosterol	SEE and UASE	*SEE*: Saturated steam *UASE*: 95% ethanol containing 2 mol/L potassium hydroxide	*SEE*: Steaming pressure: 0.5 MPa, 1.0 MPa, and 1.5 MPa Explosion temperature: 151.87 °C, 184.10 °C, and 198.33 °C SES (IgR): 2.23, 3.32, and 3.74 *UASE*: Temperature: 25 °C Time: 30 min Power: 500 W Frequency: 40 kHz	The theoretical maximum extraction concentration (C_{∞}) and diffusion coefficient obtained for ergosterol were 34.10% and 78.04% SEE altered the structural properties of the mushroom, leading to an enhanced extraction of ergosterol via UASE	Liu et al. (2024)

Pleurotus citrinopileatus (golden oyster mushroom)	Ergothioneine	HHPE	Distilled water 20% ethanol 40% ethanol 60% ethanol 80% ethanol	Pressure: 0.1–400 MPa Time: 20–60 min Liquid-solid ratio: 1:1 to 1:20 (w/v)	The highest yield attained for Ergothioneine was 4.03 ± 0.01 mg/g (dry weight) The best optimized extraction conditions were: 250 MPa pressure, 52 min of extraction time, distilled water, and 1:10 liquid-solid ratio	Zhang et al. (2024)
Hericium erinaceous (lion's mane mushroom)	Bioactive compounds	SCFE-CO_2	CO_2 Ethanol	Temperature: 40–60 °C Pressure: 100–300 bar Ethanol flow rate: 0–1 mL/min CO_2 flow rate: 3 L/min Time: 60 min	Maximum yield of 3.74 mg/g (dry weigh) was obtained The optimal SCFE-CO_2 conditions were: 46.38 °C extraction temperature, 100 bar pressure, and 0.99 mL/min ethanol flow rate SCFE-CO_2 produced comparable results to the conventional maceration extraction	Joradon et al. (2024)

(continued)

Table 1.3 (continued)

Species	Bioactive extracted	Novel extraction technique	Extraction solvent	Parameters	Extraction yield	Reference
Inonotus obliquus (chaga mushroom)	α- and β-glucans	MAE, UAE, and SWE	*SWE*: Water *MAE*: 96% ethanol 50% ethanol Water *UAE*: 96% ethanol 50% ethanol Water	*SWE*: Solid-liquid ratio: 1:10 g/mL Temperature: 120 °C and 200 °C Pressure: 30 bar Time: 20 min *UAE*: Frequency: 40 kHz Temperature: 30 °C Power: 60 W/L Time: 60 min *MAE*: Irradiation power: 470 W Time: 30 min Frequency: 50 Hz	SWE gave the highest yield of α-glucans (3.46 ± 0.05 g/100 g and 2.50 ± 0.01 g/100 g) and β-glucans (19.00 ± 1.12 g/100 g and 19.61 ± 0.8 g/100 g) for Serbian and Mongolian chaga mushrooms, respectively	Lazić et al. (2024)
Lentinula edodes (shiitake mushroom)	Lentinan	NDES	HBAs (choline chloride, betaine, carnitine, and proline) and HBDs (urea, propionic acid, levulinic acid, ethylene glycol, glycerol, urea, oxalic acid, and tartaric acid)	Extraction time: 120 min Stirring speed: 700 rpm Temperature: 90 ± 0.5 °C Liquid-solid ratio: 30 mL/g	Highest yield of lentinan obtained was 14.69 ± 0.23 g/100 g The best performing NDES was the mixture of 15% carnitine, 40.8% urea, and 44.2% water on % weight basis ANN was utilized with genetic algorithm to optimize the extraction process	Wang et al. (2024)

Flammulina velutipes (wild enoki mushroom)	Polysaccharides	UAEE	Anhydrous ethanol	*Single-factor experiment*: Solid–liquid ratio: 1:10, 1:20, 1:25, and 1:30 g/mL Time: 30, 60, 90, 120, and 150 min Ultrasonic power: 100, 150, 200, 250, and 300 W Temperature: 35 °C, 45 °C, 55 °C, 65 °C, and 75 °C *RSM*: Time: 60 to 120 min Ultrasonic power: 100–200 W Temperature: 45–65 °C	The highest yield obtained was 109.54 ± 5.44 mg/g The optimal conditions for UAEE were temperature of 55 °C, ultrasonication power of 150 W, and extraction time of 90 min UAE treatment might have enhanced the activity of enzymes used in the extraction, which could have accelerated the cell wall membrane structure breakdown, and the exudation of the polysaccharides	Wang et al. (2024)
Lentinula edodes (shiitake mushroom)	Chitin-glucan complex (CGC)	NDES and bacterial fermentation	Urea:Glycerol (1:1) Urea:Glucose: Glycerol (1:1:2) Urea:Sorbital: Glycerol (2:1:1)	Liquid–solid ratio: 10:1 (v/w) Temperature: 70 °C (water bath) Time: 120 min	Maximum CGC yield of 27.86 ± 0.82% (w/w) obtained through fermentation with *Lactiplantibacillus plantarum* However, the maximum antioxidant activity was depicted by CGC obtained via NDES mixture of urea:Glycerol (1:1) CGC demonstrated good prebiotic activity with a good prebiotic score of 118–134%	Majumdar and Negi (2024)

(continued)

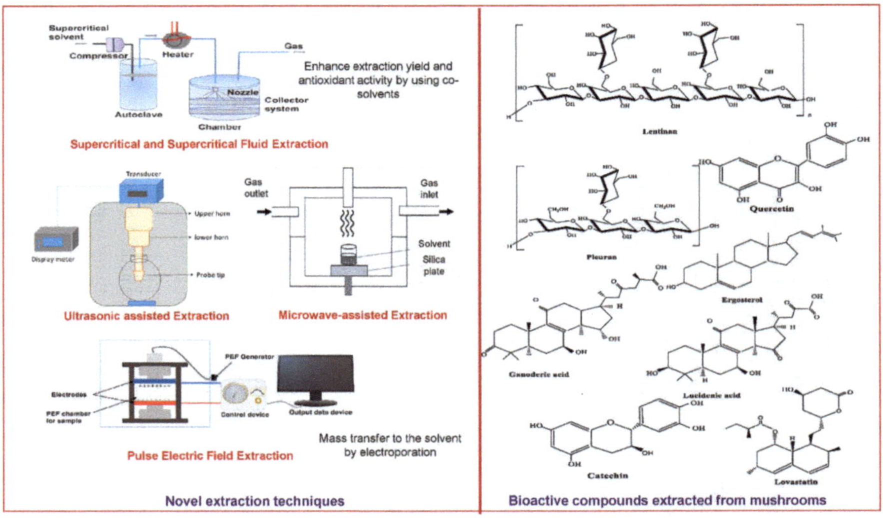

Fig. 1.2 Novel extraction techniques for bioactive compounds from mushrooms

ultrasound-assisted extraction (UAE) was achieved under optimal conditions of 44 °C, 14 min, and 20 mL/g for water extracts, and 39 °C, 13 min, and 20 mL/g for ethanol extracts. UAE provided higher polyphenol yield and antioxidant activity than conventional methods, with reduced time and energy consumption, highlighting its potential for health and food applications (Gogoi et al., 2019). UAE is used to extract biologically active metabolites such as β-glucans; however, the solvent concentration plays an important role in the extraction process. In an investigation, ergosterol was extracted from *A. bisporus* stalk efficiently in 70% ethanol concentration, and the residue achieved 12.2 ± 1.7 g/100 g dry matter polysaccharides (Umaña et al., 2020). Water-soluble polysaccharides (PS) were achieved from *Ganoderma lucidum* in optimized conditions at 590 W power, 58 min irradiation, and 81 °C, yielding 52.28 mg of PS by using UAE (Alzorqi, Singh, et al., 2017a). Furthermore, total PS content 37.05 ± 0.31 mg/g was extracted from *L. edodes* substrate (SLS) by using optimal conditions temperature of 50 °C, a liquid–solid ratio of 30 mL/g, and an ultrasonic power of 120 W (K. Wang et al., 2021). In an investigation, the extraction of bioactive peptides from mushroom foot was optimized using ultra-high-pressure processing (400 MPa, 10 min). Peptides with molecular weights of 0–3 kDa showed the highest activation of alcohol metabolic enzymes, increasing alcohol dehydrogenase (ADH) and aldehyde dehydrogenase (ALDH) activities by 70.79% and 71.35%, respectively (Zhao et al., 2017). Dextran gel chromatography identified two peptide fractions, with fraction p-I showing higher ADH and ALDH activation (72.00% and 73.43%) than p-II. Nine peptides were identified in p-I, and two synthesized peptides, IPLH and IPIVLL, were tested. IPLH demonstrated higher activation of ADH (42.7%) and ALDH (29.2%) compared to IPIVLL. All the above-mentioned studies depicted that UAE improves the extraction yield as well as their bioactivities.

3.2 Microwave-Assisted Extraction (MAE)

MAE is an advanced technology that improves the efficiency of extracting bioactive compounds. It operates on the fundamental principle of the interaction between polar molecules and microwaves, which induces molecular vibrations, and due to the collision of molecules, a sudden rise of heat occurs, facilitating cell rupture and mass transfer within the solvent (Destandau & Michel, 2022). Key factors influencing the bioactive molecules extraction yield are temperature, time, power supply, and solid/liquid ratio of the sample. Process power can negatively influence the extraction yield due to the structural disruption of bioactive molecules; however, the optimized prolonged extraction time may enhance the yield as it increases the disintegeration of biomolecules in the solvent (Xu et al., 2018). Polysaccharides were extracted from *L. edodes* using ethanol/ammonium sulfate two-phase aqueous extraction (MAATPE). In optimal conditions, with 26% ethanol, 19.58% ammonium sulfate, 78.7 °C, 19.55 min extraction time, and a 1:50 solid-to-solvent ratio, the polysaccharide extraction efficiency was enhanced up to 2.12% and 11.16% for top and bottom phases, respectively. Additionally, both phases showed strong antioxidant activity by using MAATPE (Lin et al., 2019). The traditional medicinal mushroom *Poria cocos* Wolf (*Wolfiporia cocos*) is known for its polysaccharides with significant health benefits. In a study, water-soluble polysaccharides were extracted using hot water extraction (PCP-H), ultrasonic-assisted extraction (PCP-U), enzyme-assisted extraction (PCP-E), and microwave-assisted extraction (PCP-M) and compared for their in vitro antioxidant activities, structural characteristics, and chemical composition (N. Wang et al., 2016). The results indicate that the extracts consisted of mannose, glucose, galactose, and arabinose, as confirmed by IR spectra. PCP-M showed the highest mannose and uronic acid content but a lower neutral sugar level. The optimized yield of polysaccharides reached 9.95%, consistent with predicted values. PCP-M demonstrated superior hydroxyl radical and DPPH radical scavenging activities. The physicochemical properties, functional characteristics, and digestion behavior of protein isolates (PIs) extracted from *Lactarius volemus* with microwave-assisted extraction (MAE) were systematically studied and compared with the conventional water bath-assisted extraction (WAE) method. The yield of PIs was significantly improved by 30.00% and 47.98% after 20 and 60 minutes, respectively (Shi et al., 2024). Furthermore, structural changes in PIs by unfolding their spatial conformation, transitioning from ordered to disordered structures, exposing hydrophobic groups, and increasing free sulfhydryl content were observed after the MAE treatment. The extracted PIs using MAE also depicted superior solubility, emulsifying, and foaming properties compared to those extracted from traditional methods. Additionally, the bioavailability and degree of hydrolysis of *Lactarius volemus* protein isolates were increased 6.06% and 19.78% respectively, in the small intestine. These studies suggest that MAE is a promising approach for efficiently producing bioactive molecules with enhanced functional and antioxidant properties.

3.3 Enzyme-Assisted Extraction (EAE)

The bioactive molecules enclosed within the fungal cell wall are efficiently released through enzymatic hydrolysis. The key advantages of EAE include its eco-friendliness, simplicity, and mild operational conditions such as low temperature, low energy consumption, and targeted action on specific sites (Leong et al., 2021). Additionally, it preserves the structural integrity of the biomolecules. However, despite its benefits, EAE has certain limitations. Enzyme activity is affected by various physical factors, including temperature, pH, solid–liquid ratio, and oxidative conditions (Marić et al., 2018). Nevertheless, optimizing these conditions can significantly enhance the efficiency of the process. EAE has been used to extract umami taste compounds and total free amino acids from six varieties of mushrooms, including shiitake (*L. edodes*), oyster (*P. ostreatus*), tea tree (*A. aegerita*), and white, brown, and portobello champignons (*A. bisporus*) (Poojary et al., 2017). The extraction process was optimized for pH levels ranging from 4.0 to 6.0, temperatures from 30 °C to 60 °C, enzyme concentrations varying from 0.1% to 5% (v/w), and extraction times ranging from 0.5 to 5 hours. The optimal conditions were identified as pH 7, 50 °C, and 5% enzyme concentration. EAE was used to extract polysaccharides from *Hericium erinaceus* fruit (Zhu et al., 2014). The process was optimized using the response surface methodology (Box–Behnken design). Optimal conditions (pH 5.71, 52.03 °C, 33.79 min) achieved the highest polysaccharide yield of 13.46 ± 0.37%. Enzyme-assisted extraction observed a 67.72% higher yield compared to conventional hot water extraction (HWE). Further characterization by FT-IR, SEM, CD, AFM, and GC revealed that the polysaccharides, composed of mannose, glucose, xylose, and galactose, had similar functional groups but distinct conformational differences compared to HWE. A study developed ACE-inhibitory peptides from shiitake mushroom protein hydrolysates using alkalase. Optimal hydrolysis conditions yielded the highest degree of hydrolysis (28.88%) at 50.2 °C, 3 hours, and an enzyme/substrate ratio of 1.16. Maximum ACE inhibitory activity (IC50 of 0.33 μg/mL) was achieved at 47 °C, 3 hours 28 minutes, and a ratio of 0.59 (Paisansak et al., 2021).

3.4 Pulsed Electric Field Assisted Extraction (PEFAE)

PEFAE (pulsed electric field-assisted extraction) employs electroporation through high-voltage pulses (10–100 kV/cm) of 100–1000 μs duration to increase cell wall porosity and membrane permeability by inducing localized disruptions and structural modifications. This method enables efficient extraction of intracellular components with low energy consumption (1–20 kJ/kg) and minimal heat application, making it ideal for temperature-sensitive compounds (Leong et al., 2021). Continuous PEF treatment significantly improves the extraction of compounds from a 9% w/w *Agaricus bisporus* mushroom suspension compared to traditional hot water extraction (HWE) methods. At 38.4 kV/cm for 272 μs and 85 °C, PEF achieved yields of 98% polysaccharides (7.9 mg/g), 51% polyphenols (1.6 mg GAE/g), and 49% proteins (2.7 mg/g). In contrast, conventional extraction at 98 °C for 1 hour yielded only 56% polysaccharides (6.4 mg/g), 25% polyphenols (1.3 mg GAE/g), and 45% proteins (2.6 mg/g). Minimal yields were observed with shorter conventional treatments (Xue & Farid, 2015). Another study demonstrated the extraction of β-glucan from *Pleurotus pulmonarius* by-products (caps and stems) using PEF and evaluated their prebiotic properties. PEF treatments (5–10 kV/cm, 15 Hz, 12,000 pulses) showed better results than conventional hot water extraction (100 °C, 6 h). Caps treated at 7.5 kV/cm yielded 4.18% with 14.10% β-glucan, while stems treated at 10 kV/cm yielded 2.34% with 21.39% β-glucan. Scanning electron micrographs (SEM) revealed significant structural damage caused by PEF, enhancing extraction efficiency. Both extracts promoted the growth of lactic acid bacteria and the production of short-chain fatty acids, highlighting their strong prebiotic potential (Thikham et al., 2024). Similarly, a study evaluated the effects of pulsed electric field (PEF), ultrasound (US), and their combination (PEF-US) on the drying efficiency, rehydration, and nutritional quality of *Lentinus edodes*. Pretreatments enhanced mass transfer during hot air drying at 65 °C, reducing drying time and energy consumption. Low-field NMR showed improved rehydration in pretreated samples, while SEM revealed significant structural changes, facilitating drying and rehydration. Combined PEF-US pre-treatment achieved the shortest drying time (4.7 h) and highest phenolics (224.17 μg/μL) and soluble sugars (3.90 mg/mL) (X. Li et al., 2021).

3.5 *Subcritical Water Extraction (SWE)*

Subcritical water (SW) is a solvent that maintains its state at a certain critical temperature between 100 °C and 374 °C and pressures of 1–22.1 MPa. As temperature increases, water's dielectric constant, viscosity, and surface tension decrease, while its diffusivity improves, resembling properties of organic solvents. For instance, at 250 °C and 25 bar, the dielectric constant drops to 25, facilitating the dissolution of medium- and low-polarity compounds (Essien et al., 2020). SWE enhances mass transfer via diffusion and convection, disrupts solute–matrix interactions by reducing activation energy, and uses elevated pressure to penetrate matrices, enabling efficient extraction. The extraction mechanism of subcritical water extraction (SWE) involves four sequential steps. Initially, the desorption of the solvent from the sample matrix occurs under high temperature and pressure conditions; followed by the diffusion of biomolecules into the matrix; following this, the solutes partition from the sample matrix into the extraction solvent; and finally, purified extract elutes from the extraction cell and is collected for chromatography (Zhang et al., 2020). Polysaccharides from *Inonotus obliquus* were extracted using SWE, optimized through the Box–Behnken Design (BBD). The optimal extraction conditions were determined as 200 °C, 13.24 min, and a solid-to-liquid ratio of 1:30 (w/v), yielding a crude polysaccharide extract with a recovery rate of 14.14 ± 0.17%. Following purification, the resulting polysaccharide fraction (IOP-P) exhibited a high total sugar content of 87.55% (Ma et al., 2024). The second most abundant polysaccharide, chitosan, found in mushroom by-products (*Pleurotus ostreatus*), was extracted and modified to chitin through SWE. Higher yield (6.26% ± 0.09%) of chitin was obtained under optimal conditions (120 °C and 10% (w/v) malic acid), which is 4.8 times greater than that achieved with CET (Natolino et al., 2025). Moreover, through the deacetylation of chitin, 20% of acetyl groups were removed, resulting in a less crystalline structure. Additionally, valuable compounds were obtained from the liquid residue through hydrolysis. A recent study demonstrated that β-glucan compounds were successfully extracted from *Ganoderma lucidum* and barley grains using SWE in a semi-batch process, with temperature ranges of 100–190 °C and controlled extraction times. Moreover, the incorporation of CO_2 into the system increased the β-glucan content by 3.5%. The extract was subsequently subjected to an atomization process, resulting in fine particles with diameters of less than 5 μm (Machmudah et al., 2023).

3.6 Supercritical Water Extraction (SCWE)

SCWE is attributed with the unique physicochemical qualities, including density, viscosity, diffusivity, and dielectric constant of supercritical fluids that are used for this technique. Lower viscosity and higher density enhance the extraction rate and diffusivity, which helps the fluid to penetrate the material easily (Rai et al., 2014). Moreover, supercritical fluids offer adjustable density controlled by pressure and temperature, enabling precise tuning of solvent strength. SFE uses GRAS (generally recognized as safe) solvents, ensures high efficiency with shorter processing times, and integrates easily with GC methods (López-Salas et al., 2024). Supercritical fluid extraction (SFE) combined with UV irradiation proved effective for obtaining ergosterol- and vitamin D2-enriched extracts from *Lentinula edodes*, offering potential for functional food development (Morales et al., 2017). Extraction yields and ergosterol content were more influenced by pressure than temperature, though higher temperatures (55–75 °C) promoted ergosterol formation. Post-extraction, ergosterol-enriched extracts dissolved in methanol or ethanol were irradiated (254 nm or full UV spectrum) to convert ergosterol to vitamin D2. The 254 nm irradiation was most efficient, with minimal vitamin D4 or lumisterol2 formation. Maximum transformation occurred within 1 hour (Morales et al., 2017). Supercritical CO_2 extraction (SCFE-CO_2) was used to extract bioactive compounds from lion's mane mushroom by-products. Optimal conditions (46.38 °C, 100 bar, 0.99 mL/min EtOH flow) yielded 11.64% mass, 3.74 mg/g ergosterol, 815.93 μg GAE/g TPC, and 42.97 μg TE/g antioxidant activity. EtOH flow rate significantly improved the yield and quality. SCFE-CO_2 performed comparably to maceration, showing no significant differences in results ($p \geq 0.05$) (Joradon et al., 2024). Additionally, a novel methodology was developed to extract antioxidant polysaccharides from *Pleurotus ostreatus* using a combination of H_2O + CO_2 supercritical fluid extraction (SFE). Total extract yield 30.69% was found at optimal conditions (25 MPa, 433.15 K, 20% H_2O) with 0.921 mg CHO_3. FTIR and NMR analyses revealed heteropolysaccharides, β-glucans, α-glucans, and oligosaccharides. The extract exhibited 80.83% antioxidant activity, no cytotoxicity, and improved cellular defence against oxidative damage caused by H_2O_2 (Barbosa et al., 2020). This SFE technique holds promise for use in pharmacological and food industries.

3.7 Natural Deep Eutectic Solvent Extraction (NDESE)

Deep eutectic solvents (DESs) include various anionic and/or cationic species. These species represent the eutectic mixture of Lewis/Bronsted acids and bases. The melting point of these mixtures is much lower than the individual compounds, making them a promising alternative (Vilková et al., 2020). Vitamin D deficiency can be addressed through fortified foods or supplements. DES (glycerol and choline chloride, 3:1) was used to extract vitamin D from *A. bisporus* mushrooms. Combined with ultrasonication, this method efficiently increased vitamin D2 content while reducing its precursor, ergosterol, through UV-C exposure. Most ergosterol was converted to vitamin D2, though a small amount of over-irradiated product was formed under UV-C light (Patil et al., 2018). In another study, various hydrophobic deep eutectic solvents were evaluated for ergosterol extraction from mushrooms. The process was optimized by RSM, optimal conditions yielded 6995.00 μg ergosterol/g dry weight with menthol:pyruvic acid (Khare et al., 2021). Additionally, the purified ergosterol was converted to ergocalciferol (vitamin D2) via UV radiation, achieving 2142.01 μg/g. The solvent was reused for six cycles, maintaining 28% efficiency. The chitin–glucan complex, a novel prebiotic, was extracted from *Lentinula edodes* (Shiitake mushrooms) using NDESE. Fermentation with *Lactiplantibacillus plantarum* was subsequently employed to achieve the highest yield of polysaccharides (Majumdar & Negi, 2024). Urea:glycerol (1:1) was used as DES for further analysis due to its antioxidant potential. CGC extracted was further characterized by FTIR, NMR, and XRD, resulting in higher prebiotic potential (118–134%) due to the release of short-chain fatty acids (produced by *L. plantarum*) and triple-helical structure (Majumdar & Negi, 2024). Similarly, CMCs were extracted from *A. bisporus* (white button mushrooms) through NDESE (Kim et al., 2021). Among five DESs tested, betaine-urea (BU) produced the highest yields with low protein and mineral content, acetylation degrees of 77.3% (BU_S) and 57.3% (BU_P), and thermal stability at 318 °C. Both showed low crystallinity (32.0% and 37.0%) and porous surfaces due to reduced H-bonds. This method demonstrates the potential of DESs for efficient biomaterial extraction (Kim et al., 2021). Recent interest has grown in extracting valuable compounds from mushroom by-products, particularly stems. A study found mushroom stems contain 23.02% protein and 63.01% carbohydrates. Alkaline extraction yielded higher protein content (45.90%)

but lower recovery (15.39%) compared to DES methods (22.04–21.47%) (Karabulut et al., 2024). Moreover, extracts prepared by all the extraction methods were characterized by FTIR and SEM; NDESE extract produced uniform particles and better preserved emulsion, foaming properties, and protein solubility (72–75%) than alkaline methods (Karabulut et al., 2024).

4 Conclusion and Future Perspectives

The demand for natural bioactive compound molecules, including polysaccharides (e.g., β-glucans), phenolics, terpenoids, and peptides, is increasing due to the adverse effects of synthetic compounds. Mushrooms are among the natural bioresources having a plethora of bioactive compounds having potential applications in various sectors, including health and food industries. Nowadays, researchers are focusing on exploring the potential applications of by-products of mushrooms like stipes and mushroom spent for the extraction of bioactive compounds. For getting the compound of interest, the extraction method plays an important role, and there is a need to standardize the best bioactive compound extraction method, including nonthermal technologies. The traditional methods used for the extraction of bioactive compounds should be replaced with the latest methods without impacting the cost of extraction. The methods should be environmentally friendly and should be economic as well.

Acknowledgment The authors are thankful to the National Institute of Food Technology, Entrepreneurship, and Management (NIFTEM), Kundli, for providing facilities to perform this work.

References

Agnihotri, C., Aarzoo, Agnihotri, S., Kamal, S., & Singh, B. P. (2024). Mushroom bioactives: Traditional resources with nutraceutical importance. In *Traditional resources and tools for modern drug discovery: Ethnomedicine and pharmacology* (pp. 617–639). Springer.

Aguiló-Aguayo, I., Walton, J., Viñas, I., & Tiwari, B. K. (2017). Ultrasound assisted extraction of polysaccharides from mushroom by-products. *LWT, 77*, 92–99. https://doi.org/10.1016/j.lwt.2016.11.043

Alves Filho, E. G., Lima, M., Silva, L., Ribeiro, P., Tiwari, B. K., Fernandes, F. N., & Brito, E. S. (2021). Green ultrasound-assisted extraction of bioactive compounds from button mushrooms, potatoes, and onion peels. *ACS Food Science & Technology, 1*(7), 1274–1284. https://doi.org/10.1021/acsfoodscitech.1c00153

Alzorqi, I., Singh, A., Manickam, S., & Al-Qrimli, H. F. (2017a). Optimization of ultrasound assisted extraction (UAE) of β-d-glucan polysaccharides from Ganoderma lucidum for prospective scale-up. *Resource-Efficient Technologies, 3*(1), 46–54. https://doi.org/10.1016/j.reffit.2016.12.006

Alzorqi, I., Sudheer, S., Lu, T.-J., & Manickam, S. (2017b). Ultrasonically extracted β-d-glucan from artificially cultivated mushroom, characteristic properties and antioxidant activity. *Ultrasonics Sonochemistry, 35*, 531–540. https://doi.org/10.1016/j.ultsonch.2016.04.017

Amirullah, N. A., Zainal Abidin, N., Abdullah, N., & Manickam, S. (2021). Application of ultrasound towards improving the composition of phenolic compounds and enhancing in vitro bioactivities of Pleurotus pulmonarius (Fr.) Quél extracts. *Biocatalysis and Agricultural BiotechnologyBiocatal Agric Biotech, 31*, 101881. https://doi.org/10.1016/j.bcab.2020.101881

Amirullah, N. A., Abdullah, E., Zainal Abidin, N., Abdullah, N., & Manickam, S. (2023). Influence of extraction technologies on the therapeutic properties of Pleurotus spp. (oyster mushrooms)—A critical review. *Food Bioscience, 56*, 103352. https://doi.org/10.1016/j.fbio.2023.103352

Aursuwanna, T., Noitang, S., Sangtanoo, P., Srimongkol, P., Saisavoey, T., Puthong, S., Reamtong, O., & Karnchanatat, A. (2022). Investigating the cellular antioxidant and anti-inflammatory effects of the novel peptides in lingzhi mushrooms. *Heliyon, 8*(10). https://doi.org/10.1016/j.heliyon.2022.e11067

Balaji, P., Madhanraj, R., Rameshkumar, K., Veeramanikandan, V., Eyini, M., Arun, A., Thulasinathan, B., Al Farraj, D. A., Elshikh, M. S., Alokda, A. M., Mahmoud, A. H., Tack, J. C., & Kim, H. J. (2020). Evaluation of antidiabetic activity of Pleurotus pulmonarius against streptozotocin-nicotinamide induced diabetic wistar albino rats. *Saudi Journal of Biological Sciences, 27*(3), 913–924. https://doi.org/10.1016/j.sjbs.2020.01.027

Barbosa, J. R., Freitas, M. M. S., Oliveira, L. C., Martins, L. H. S., Almada-Vilhena, A. O., Oliveira, R. M., Pieczarka, J. C., Brasil, D. S. B., & Carvalho Junior, R. N. (2020). Obtaining extracts rich in antioxidant polysaccharides from the edible mushroom Pleurotus ostreatus using binary system with hot water and supercritical CO2. *Food Chemistry, 330*, 127173. https://doi.org/10.1016/j.foodchem.2020.127173

Bitwell, C., Indra, S. S., Luke, C., & Kakoma, M. K. (2023). A review of modern and conventional extraction techniques and their applications for extracting phytochemicals from plants. *Scientific African, 19*, e01585. https://doi.org/10.1016/j.sciaf.2023.e01585

Calleja-Gómez, M., Castagnini, J. M., Carbó, E., Ferrer, E., Berrada, H., & Barba, F. J. (2022). Evaluation of pulsed electric field-assisted extraction on the microstructure and recovery of nutrients and bioactive compounds from mushroom (Agaricus bisporus). *Separations, 9*(10), 302. Retrieved from https://www.mdpi.com/2297-8739/9/10/302

Cerletti, C., Esposito, S., & Iacoviello, L. (2021). Edible mushrooms and Beta-Glucans: Impact on human health. *Nutrients, 13*(7). https://doi.org/10.3390/nu13072195

Chemat, F., Abert-Vian, M., Fabiano-Tixier, A. S., Strube, J., Uhlenbrock, L., Gunjevic, V., & Cravotto, G. (2019). Green extraction of natural products. Origins, current status, and future challenges. TrAC. *Trends in Analytical Chemistry, 118*, 248–263. https://doi.org/10.1016/j.trac.2019.05.037

Chen, Y., Liu, D., Wang, D., Lai, S., Zhong, R., Liu, Y., Yang, C., Liu, B., Sarker, M. R., & Zhao, C. (2019). Hypoglycemic activity and gut microbiota regulation of a novel polysaccharide from Grifola frondosa in type 2 diabetic mice. *Food and Chemical Toxicology, 126*, 295–302. https://doi.org/10.1016/j.fct.2019.02.034

Dash, P., Kar, B., Gochhi, M., Ghosh, G., Rai, V. K., Das, C., Pradhan, D., Rajwar, T. K., Halder, J., Dubey, D., Manoharadas, S., & Rath, G. (2024). Antimicrobial properties of the edible pink oyster mushroom, Pleurotus eous: In-vivo and in-vitro studies. *Microbial Pathogenesis, 196*, 106915. https://doi.org/10.1016/j.micpath.2024.106915

de Souza Ribeiro, M. M., dos Santos, L. C., de Novais, N. S., Viganó, J., & Veggi, P. C. (2022). An evaluative review on Stryphnodendron adstringens extract composition: Current and future perspectives on extraction and application. *Industrial Crops and Products, 187*, 115325. https://doi.org/10.1016/j.indcrop.2022.115325

Destandau, E., & Michel, T. (2022). Microwave-assisted extraction. In J. Prado & M. Rostagno (Eds.), *Natural product extraction: Principles and applications*. The Royal Society of Chemistry.

Du, B., Peng, F., Xie, Y., Wang, H., Wu, J., Liu, C., & Yang, Y. (2022). Optimization extraction and antioxidant activity of crude polysaccharide from chestnut mushroom (Agrocybe aegerita) by accelerated solvent extraction combined with response surface methodology (ASE-RSM). *Molecules, 27*(8). https://doi.org/10.3390/molecules27082380

El Maaiden, E., Bouzroud, S., Nasser, B., Moustaid, K., El Mouttaqi, A., Ibourki, M., Boukcim, H., Hirich, A., Kouisni, L., & El Kharrassi, Y. (2022). A comparative study between conventional and advanced extraction techniques: Pharmaceutical and cosmetic properties of plant extracts. *Molecules, 27*(7). https://doi.org/10.3390/molecules27072074

Essien, S. O., Young, B., & Baroutian, S. (2020). Recent advances in subcritical water and supercritical carbon dioxide extraction of bioactive compounds from plant materials. *Trends in Food Science and Technology, 97*, 156–169. https://doi.org/10.1016/j.tifs.2020.01.014

Farooq, S., & Mir, S. (2022). Chapter 2-extraction techniques. In S. A. Mir & A. Manickavasagan (Eds.), *MABT-PEA in the FI Shah*. Academic Press. https://doi.org/10.1016/B978-0-12-822475-5

Ferraro, V., Spagnoletta, A., Rotondo, N. P., Marsano, R. M., Miniero, D. V., Balenzano, G., De Palma, A., Colletti, A., Gargano, M. L., Lentini, G., & Cavalluzzi, M. M. (2024). Hypoglycemic properties of Leccinum scabrum extracts—An in vitro study on α-glucosidase and α-amylase inhibition and metabolic profile determination. *Journal of Fungi, 10*(10). https://doi.org/10.3390/jof10100718

Ghosh, S., Sett, S., Saha, R., Roy, A., & Acharya, K. (2021). Comparative phytochemical screening and antioxidant properties of infusion, decoction and hydroalcoholic extracts of wood ear mushrooms; Auricularia delicata and Auricularia mesenterica. *Indian Phytopathology, 74*(1), 113–121. https://doi.org/10.1007/s42360-020-00301-3

Gogoi, P., Chutia, P., Singh, P., & Mahanta, C. L. (2019). Effect of optimized ultrasound-assisted aqueous and ethanolic extraction of Pleurotus citrinopileatus mushroom on total phenol, flavonoids and antioxidant properties. *Journal of Food Process Engineering, 42*(6). https://doi.org/10.1111/jfpe.13172

Gong, P., Wang, S., Liu, M., Chen, F., Yang, W., Chang, X., Liu, N., Zhao, Y., Wang, J., & Chen, X. (2020). Extraction methods, chemical characterizations and biological activities of mushroom polysaccharides: A mini-review. *Carbohydrate Research, 494*, 108037. https://doi.org/10.1016/j.carres.2020.108037

Guo, W.-L., Deng, J.-C., Pan, Y.-Y., Xu, J.-X., Hong, J.-L., Shi, F.-F., Liu, G.-L., Qian, M., Bai, W.-D., Zhang, W., Liu, B., Zhang, Y.-Y., Luo, P.-J., Ni, L., Rao, P.-F., & Lv, X.-C. (2020). Hypoglycemic and hypolipidemic activities of Grifola frondosa polysaccharides and their relationships with the modulation of intestinal microflora in diabetic mice induced by high-fat diet and streptozotocin. *International Journal of Biological Macromolecules, 153*, 1231–1240. https://doi.org/10.1016/j.ijbiomac.2019.10.253

Halahlah, A., Abik, F., Lahtinen, M. H., Kemppinen, A., Kaipanen, K., Kilpeläinen, P. O., Granato, D., Ho, T. M., & Mikkonen, K. S. (2023). Effects of pH and temperature of ultrafiltration on the composition and physicochemical properties of hot-water-extracted softwood galactoglucomannans. *Industrial Crops and Products, 198*, 116656. https://doi.org/10.1016/j.indcrop.2023.116656

Hu, D.-B., Xue, R., Zhuang, X.-C., Zhang, X.-S., & Shi, S.-L. (2023). Ultrasound-assisted extraction optimization of polyphenols from boletus bicolor and evaluation of its antioxidant activity. *Frontiers in Nutrition, 10*. https://doi.org/10.3389/fnut.2023.1135712

Hwang, A. Y., Yang, S. C., Kim, J., Lim, T., Cho, H., & Hwang, K. T. (2019). Effects of non-traditional extraction methods on extracting bioactive compounds from chaga mushroom (Inonotus obliquus) compared with hot water extraction. *LWT, 110*, 80–84. https://doi.org/10.1016/j.lwt.2019.04.073

Joradon, P., Rungsardthong, V., Ruktanonchai, U., Suttisintong, K., Thumthanaruk, B., Vatanyoopaisarn, S., Uttapap, D., & Mendes, A. C. (2024). Extraction of bioactive compounds from lion's mane mushroom by-product using supercritical CO2 extraction. *Journal of Supercritical Fluids, 206*, 106162. https://doi.org/10.1016/j.supflu.2023.106162

Kanani, S. H., & Pandya, D. J. (2023). The phytochemical screening, total cucurbitacin content, and in vitro anti-breast cancer activity of Leucopaxillus gentianeus mushroom. *Future Journal of Pharmaceutical Sciences, 9*(1), 49. https://doi.org/10.1186/s43094-023-00496-y

Kapadia, P., Newell, A. S., Cunningham, J., Roberts, M. R., & Hardy, J. G. (2022). Extraction of high-value chemicals from plants for technical and medical applications. *International Journal of Molecular Sciences, 23*(18). https://doi.org/10.3390/ijms231810334

Kaprasob, R., Khongdetch, J., Laohakunjit, N., Selamassakul, O., & Kaisangsri, N. (2022). Isolation and characterization, antioxidant, and antihypertensive activity of novel bioactive peptides derived from hydrolysis of King Boletus mushroom. *LWT, 160*, 113287. https://doi.org/10.1016/j.lwt.2022.113287

Karabulut, G., Köroğlu, D. G., Feng, H., & Karabulut, Z. (2024). Sustainable fungi-based protein extraction from agro-waste mushroom stem using deep eutectic solvents. *Food Chemistry X, 24*, 101931. https://doi.org/10.1016/j.fochx.2024.101931

Khare, L., Karve, T., Jain, R., & Dandekar, P. (2021). Menthol based hydrophobic deep eutectic solvent for extraction and purification of ergosterol using response surface methodology. *Food Chemistry, 340*, 127979. https://doi.org/10.1016/j.foodchem.2020.127979

Kim, H., Kang, S., Li, K., Jung, D., Park, K., & Lee, J. (2021). Preparation and characterization of various chitin-glucan complexes derived from white button mushroom using a deep eutectic solvent-based ecofriendly method. *International Journal of Biological Macromolecules, 169*, 122–129. https://doi.org/10.1016/j.ijbiomac.2020.12.081

Kour, H., Kour, D., Kour, S., Singh, S., Jawad Hashmi, S. A., Yadav, A. N., Kumar, K., Sharma, Y. P., & Ahluwalia, A. S. (2022). Bioactive compounds from mushrooms: Emerging bioresources of food and nutraceuticals. *Food Bioscience, 50*, 102124. https://doi.org/10.1016/j.fbio.2022.102124

Krümmel, A., Gonçalves Rodrigues, L. G., Vitali, L., & Ferreira, S. R. S. (2022). Bioactive compounds from Pleurotus sajor-caju mushroom recovered by sustainable high-pressure methods. *LWT, 160*, 113316. https://doi.org/10.1016/j.lwt.2022.113316

Lazić, V., Klaus, A., Kozarski, M., Doroški, A., Tosti, T., Simić, S., & Vunduk, J. (2024). The effect of green extraction technologies on the chemical composition of medicinal Chaga mushroom extracts. *Journal of Fungi, 10*(3), 225. Retrieved from https://www.mdpi.com/2309-608X/10/3/225

Leong, Y. K., Yang, F.-C., & Chang, J.-S. (2021). Extraction of polysaccharides from edible mushrooms: Emerging technologies and recent advances. *Carbohydrate Polymers, 251*, 117006. https://doi.org/10.1016/j.carbpol.2020.117006

Li, S., & Shah, N. P. (2016). Characterization, antioxidative and bifidogenic effects of polysaccharides from Pleurotus eryngii after heat treatments. *Food Chemistry, 197*, 240–249. https://doi.org/10.1016/j.foodchem.2015.10.113

Li, X., Li, J., Wang, R., Rahaman, A., Zeng, X.-A., & Brennan, C. S. (2021). Combined effects of pulsed electric field and ultrasound pretreatments on mass transfer and quality of mushrooms. *LWT, 150*, 112008. https://doi.org/10.1016/j.lwt.2021.112008

Li, J. H., Zhu, Y. Y., Gu, F. T., & Wu, J. Y. (2023). Efficient isolation of immunostimulatory polysaccharides from Lentinula edodes by autoclaving-ultrasonication extraction and fractional precipitation. *International Journal of Biological Macromolecules, 237*, 124216. https://doi.org/10.1016/j.ijbiomac.2023.124216

Lin, Y., Zeng, H., Wang, K., Lin, H., Li, P., Huang, Y., Zhou, S., Zhang, W., Chen, C., & Fan, H. (2019). Microwave-assisted aqueous two-phase extraction of diverse polysaccharides from Lentinus edodes: Process optimization, structure characterization and antioxidant activity. *International Journal of Biological Macromolecules, 136*, 305–315. https://doi.org/10.1016/j.ijbiomac.2019.06.064

Liu, S. (2015). A synergetic pretreatment technology for woody biomass conversion. *Applied Energy, 144*, 114–128. https://doi.org/10.1016/j.apenergy.2015.02.021

Liu, W., Niu, J., Han, F., Zhong, K., Li, R., Sui, W., Ma, C., & Wu, M. (2024). Steam explosion-assisted extraction of Ergosterol and polysaccharides from Flammulina velutipes (Golden needle mushroom) root waste. *Food, 13*(12), 1860. Retrieved from https://www.mdpi.com/2304-8158/13/12/1860

López-Salas, L., Expósito-Almellón, X., Borrás-Linares, I., Lozano-Sánchez, J., & Segura-Carretero, A. (2024). Design of experiments for green and GRAS solvent extraction of phenolic compounds from food industry by-products – a systematic review. *TrAC, Trends in Analytical Chemistry, 171*, 117536. https://doi.org/10.1016/j.trac.2024.117536

Ma, Y., Zheng, X., Chu, Z., Nan, W., Zhao, Y., Bai, Y., Ma, R., Jia, Y., Lü, X., & Wang, X. (2024). Polysaccharides from Inonotus obliquus employing subcritical water extraction: Extraction optimization, physiochemical properties and bioactivities analysis. *Industrial Crops and Products, 222*, 119638. https://doi.org/10.1016/j.indcrop.2024.119638

Machmudah, S., Wahyudiono, Adschiri, T., & Goto, M. (2023). Hydrothermal extraction and micronization in a one-step process for enhancement of β–glucan concentrate at subcritical water conditions. *South African Journal of Chemical Engineering, 46*, 72–87. https://doi.org/10.1016/j.sajce.2023.07.009

Mahtal, A., Lamchouri, F., & Toufik, H. (2024). Mineral composition, phenolic content and antibacterial activity of aqueous and organic extracts of wild Morchella esculenta (L.) Pers. from the province of Taza, Morocco. *Biochemistry Research International, 2024*(1), 6769243. https://doi.org/10.1155/bri/6769243

Majumdar, S., & Negi, P. S. (2024). Extraction of chitin-glucan complex from shiitake (Lentinula edodes) fruiting bodies using natural deep eutectic solvents and its prebiotic potential. *International Journal of Biological Macromolecules, 273*, 133046. https://doi.org/10.1016/j.ijbiomac.2024.133046

Marić, M., Grassino, A. N., Zhu, Z., Barba, F. J., Brnčić, M., & Rimac Brnčić, S. (2018). An overview of the traditional and innovative approaches for pectin extraction from plant food wastes and by-products: Ultrasound-, microwaves-, and enzyme-assisted extraction. *Trends in Food Science & Technology, 76*, 28–37. https://doi.org/10.1016/j.tifs.2018.03.022

Mathews, A., Arbal, A. V., Kaarunya, A., Jha, P. K., Le-Bail, A., & Rawson, A. (2024). Chapter five – conventional vs modern extraction techniques in the food industry. In S. M. Jafari & S. Akhavan-Mahdavi (Eds.), *Extraction processes in the food industry* (pp. 97–146). Woodhead Publishing.

Mazzutti, S., Ferreira, S. R. S., Riehl, C. A. S., Smania, A., Smania, F. A., & Martínez, J. (2012). Supercritical fluid extraction of Agaricus brasiliensis: Antioxidant and antimicrobial activities. *Journal of Supercritical Fluids, 70*, 48–56. https://doi.org/10.1016/j.supflu.2012.06.010

Minakov, D. V., Morozhenko, Y. V., & Bazarnova, N. G. (2021). Content of extractive substances and polysaccharides in fruit bodies of Grifola frondosa cultivated on lignocellulose substrates depending on extraction methods. *Russian Journal of Bioorganic Chemistry, 47*(7), 1389–1394. https://doi.org/10.1134/S1068162021070116

Morales, D. (2024). Fomes fomentarius: An underexplored mushroom as source of bioactive compounds. *Food Bioscience, 61*, 104781. https://doi.org/10.1016/j.fbio.2024.104781

Morales, D., Gil-Ramirez, A., Smiderle, F. R., Piris, A. J., Ruiz-Rodriguez, A., & Soler-Rivas, C. (2017). Vitamin D-enriched extracts obtained from shiitake mushrooms (Lentinula edodes) by supercritical fluid extraction and UV-irradiation. *IFSET, 41*, 330–336. https://doi.org/10.1016/j.ifset.2017.04.008

More, P. R., Jambrak, A. R., & Arya, S. S. (2022). Green, environment-friendly and sustainable techniques for extraction of food bioactive compounds and waste valorization. *Trends in Food Science and Technology, 128*, 296–315. https://doi.org/10.1016/j.tifs.2022.08.016

Natolino, A., Voce, S., Alò, E., & Comuzzo, P. (2025). Exploring an eco-friendlier strategy for chitosan production and valuable compounds recovery from mushroom by-products with modified subcritical water. *IFSET, 100*, 103923. https://doi.org/10.1016/j.ifset.2025.103923

Nosaka, S., & Miyazawa, M. (2014). Characterization of volatile components and odor-active compounds in the oil of edible mushroom Boletopsis leucomelas. *Journal of Oleo Science, 63*(6), 577–583.

Novakovic, S. (2021). The potential of the application of boletus edulis, Cantharellus cibarius and Craterellus cornucopioides in frankfurters: A review. *IOP Conference Series: Earth and Environmental Science, 854*(1), 012068. https://doi.org/10.1088/1755-1315/854/1/012068

Novakovic, S., Djekic, I., Vunduk, J., Klaus, A., Lorenzo, J. M., Barba, F., & Tomasevic, I. (2019). An insight into in vitro antioxidant activity of Cantharellus cibarius hot water extract for the potential application in meat products. *IOP Conference Series: Earth and Environmental Science, 333*(1), 012089. https://doi.org/10.1088/1755-1315/333/1/012089

Novakovic, S., Djekic, I., Klaus, A., Vunduk, J., Đorđević, V., Tomovic, V., Kоćić-Tanackov, S., Lorenzo, J. M., Barba, F. J., & Tomasevic, I. (2020). Application of porcini mushroom (boletus edulis) to improve the quality of frankfurters. *Journal of Food Processing & Preservation, 44*(8), e14556. https://doi.org/10.1111/jfpp.14556

Ohiri, R. C., & Odey, O. P. (2022). Human hemoglobin S erythrocyte-stabilizing and Antisickling potential of extract of wood ear mushroom, Auricularia auricula (Agaricomycetes). *International Journal of Medicinal Mushrooms, 24*(3), 25–34. https://doi.org/10.1615/IntJMedMushrooms.2022042977

Oreopoulou, A., Tsimogiannis, D., & Oreopoulou, V. (2019). Chapter 15 - extraction of polyphenols from aromatic and medicinal plants: An overview of the methods and the effect of extraction parameters. In R. R. Watson (Ed.), *Polyphenols in plants* (2nd ed., pp. 243–259). Academic Press.

Ozel, N., & Elibol, M. (2024). Chitin and chitosan from mushroom (Agaricus bisporus) using deep eutectic solvents. *International Journal of Biological Macromolecules, 262*, 130110. https://doi.org/10.1016/j.ijbiomac.2024.130110

Paisansak, S., Sangtanoo, P., Srimongkol, P., Saisavoey, T., Reamtong, O., Choowongkomon, K., & Karnchanatat, A. (2021). Angiotensin-I converting enzyme inhibitory peptide derived from the shiitake mushroom (Lentinula edodes). *Journal of Food Science and Technology, 58*(1), 85–97. https://doi.org/10.1007/s13197-020-04517-z

Parniakov, O., Lebovka, N. I., Van Hecke, E., & Vorobiev, E. (2014). Pulsed electric field assisted pressure extraction and solvent extraction from mushroom (Agaricus Bisporus). *Food and Bioprocess Technology, 7*(1), 174–183. https://doi.org/10.1007/s11947-013-1059-y

Patil, J., Ghodke, S., Jain, R., & Dandekar, P. (2018). Extraction of vitamin D from button mushroom (Agaricus bisporus) using deep eutectic solvent and Ultrasonication. *ACS Sustainable Chemistry & Engineering, 6*(8), 10578–10586. https://doi.org/10.1021/acssuschemeng.8b01915

Poojary, M. M., Orlien, V., Passamonti, P., & Olsen, K. (2017). Enzyme-assisted extraction enhancing the umami taste amino acids recovery from several cultivated mushrooms. *Food Chemistry, 234*, 236–244. https://doi.org/10.1016/j.foodchem.2017.04.157

Putra Nicky, R., Yustisia, Y., Heryanto, R. B., Asmaliyah, A., Miswarti, M., Rizkiyah Dwila, N., Yunus Mohd Azizi, C., Irianto, I., Qomariyah, L., & Rohman Gus Ali, N. (2023). Advancements and challenges in green extraction techniques for Indonesian natural products: A review. *South African Journal of Chemical Engineering, 46*(1), 88–98. https://doi.org/10.1016/j.sajce.2023.08.002

Rai, A., Punase, K. D., Mohanty, B., & Bhargava, R. (2014). Evaluation of models for supercritical fluid extraction. *International Journal of Heat and Mass Transfer, 72*, 274–287. https://doi.org/10.1016/j.ijheatmasstransfer.2014.01.011

Ramos, M., Burgos, N., Barnard, A., Evans, G., Preece, J., Graz, M., Ruthes, A. C., Jiménez-Quero, A., Martínez-Abad, A., Vilaplana, F., Ngoc, L. P., Brouwer, A., van der Burg, B., del Carmen Garrigós, M., & Jiménez, A. (2019). Agaricus bisporus and its by-products as a source of valuable extracts and bioactive compounds. *Food Chemistry, 292*, 176–187. https://doi.org/10.1016/j.foodchem.2019.04.035

Ratnaningtyas, N. I., Hernayanti, Ekowati, N., Sukmawati, D., & Widianti, H. (2019). Chicken drumstick mushroom (Coprinus comatus) ethanol extract exerts a hypoglycaemic effect in the Rattus norvegicus model of diabetes. *Biocatalysis and Agricultural Biotechnology, 19*, 101050. https://doi.org/10.1016/j.bcab.2019.101050

Ratnaningtyas, N. I., Hernayanti, H., Ekowati, N., & Husen, F. (2022). Ethanol extract of the mushroom Coprinus comatus exhibits antidiabetic and antioxidant activities in streptozotocin-induced diabetic rats. *Pharmaceutical Biology, 60*(1), 1126–1136. https://doi.org/10.1080/13880209.2022.2074054

Sandra, P., Zorić, Z., Bursać Kovačević, D., Elez Garofulić, I., & Dragović-Uzelac, V. (2020). Pressurized hot water extraction of phenolic compounds from leaves of Stevia rebaudiana: An UPLC-ESI-MSMS study. *Journal of Food Process Engineering, 43*(2), e13319. https://doi.org/10.1111/jfpe.13319

Senapati, M. R., & Behera, P. C. (2023). Chapter 3 – novel extraction conditions for phytochemicals. In S. Pati, T. Sarkar, & D. Lahiri (Eds.), *Recent Frontiers of phytochemicals* (pp. 27–61). Elsevier.

Senphan, T., Benjakul, S., Sukketsiri, W., Chotphruethipong, L., & Sriket, C. (2024). Comparative studies on characterizations and cytotoxicity of oil extracted from Lingzhi (Ganoderma lucidum) G2 spore using Soxhlet extraction and microwave-assisted extraction. *Applied Food Research, 4*(2), 100483. https://doi.org/10.1016/j.afres.2024.100483

Sevindik, M., Gürgen, A., Khassanov, V. T., & Bal, C. (2024). Biological activities of ethanol extracts of Hericium erinaceus obtained as a result of optimization analysis. *Food, 13*(10). https://doi.org/10.3390/foods13101560

Sharma, Y. P., Sharma, R., Khatua, S., & Acharya, K. (2019). Morphotaxonomy and comparative mycochemical study and antioxidant activity of hydromethanol, infusion and decoction extracts from Russula brevipes peck. *Indian Phytopathology, 72*(3), 445–452. https://doi.org/10.1007/s42360-019-00173-2

Sharma, R., Khatua, S., Acharya, K., & Sharma, Y. P. (2022). Mycochemical composition and antioxidant activity of Flammulina velutipes: A comparative study on hydromethanol, decoction and infusion extracts. *Vegetos, 35*(3), 607–613. https://doi.org/10.1007/s42535-022-00362-x

Sharpe, E., Farragher-Gnadt, A. P., Igbanugo, M., Huber, T., Michelotti, J. C., Milenkowic, A., Ludlam, S., Walker, M., Hanes, D., Bradley, R., & Bou-Abdallah, F. (2021). Comparison of antioxidant activity and extraction techniques for commercially and laboratory prepared extracts from six mushroom species. *Journal of Agriculture and Food Research, 4*, 100130. https://doi.org/10.1016/j.jafr.2021.100130

Shen, L., Pang, S., Zhong, M., Sun, Y., Qayum, A., Liu, Y., Rashid, A., Xu, B., Liang, Q., Ma, H., & Ren, X. (2023). A comprehensive review of ultrasonic assisted extraction (UAE) for bioactive components: Principles, advantages, equipment, and combined technologies. *Ultrasonics Sonochemistry, 101*, 106646. https://doi.org/10.1016/j.ultsonch.2023.106646

Shi, R., He, Y., Wang, Q., Cai, J., Gantumur, M.-A., & Jiang, Z. (2024). Insight into the physicochemical characteristics, functionalities and digestion behavior of protein isolate derived from Lactarius volemus (L.volemus): Impacts of microwave-assisted extraction. *Food Chemistry, 431*, 137070. https://doi.org/10.1016/j.foodchem.2023.137070

Singh Chouhan, K. B., Tandey, R., Sen, K. K., Mehta, R., & Mandal, V. (2020). Microwave hydrodiffusion and gravity model with a blend of high and low power microwave firing for improved yield of phenolics and flavonoids from oyster mushroom. *Sustainable Chemistry and Pharmacy, 17*, 100311. https://doi.org/10.1016/j.scp.2020.100311

Subramanian, P., & Anandharamakrishnan, C. (2023). Chapter two—Extraction of bioactive compounds. In C. Anandharamakrishnan & P. Subramanian (Eds.), *Industrial application of functional foods, ingredients and nutraceuticals* (pp. 45–87). Academic Press.

Suna, S., Tamer, C. E., & Özcan-Sinir, G. (2019). 13 – trends and possibilities of the usage of medicinal herbal extracts in beverage production. In A. M. Grumezescu & A. M. Holban (Eds.), *Natural beverages* (pp. 361–398). Academic Press.

Taofiq, O., Heleno, S. A., Calhelha, R. C., Alves, M. J., Barros, L., González-Paramás, A. M., Barreiro, M. F., & Ferreira, I. C. F. R. (2017). The potential of Ganoderma lucidum extracts as bioactive ingredients in topical formulations, beyond its nutritional benefits. *Food and Chemical Toxicology, 108*, 139–147. https://doi.org/10.1016/j.fct.2017.07.051

Taofiq, O., Corrêa, R. C. G., Barros, L., Prieto, M. A., Bracht, A., Peralta, R. M., González-Paramás, A. M., Barreiro, M. F., & Ferreira, I. C. F. R. (2019). A comparative study between conventional and non-conventional extraction techniques for the recovery of ergosterol from Agaricus blazei Murrill. *Food Research International, 125*, 108541. https://doi.org/10.1016/j.foodres.2019.108541

Tepsongkroh, B., Jangchud, K., & Trakoontivakorn, G. (2019). Antioxidant properties and selected phenolic acids of five different tray-dried and freeze-dried mushrooms using methanol and hot water extraction. *Journal of Food Measurement and Characterization, 13*(4), 3097–3105. https://doi.org/10.1007/s11694-019-00232-2

Tepsongkroh, B., Thaihuttakij, C., Supawong, S., & Jangchud, K. (2023). Impact of high pressure pre-treatment and hot water extraction on chemical properties of crude polysaccharide extract obtained from mushroom (Volvariella volvacea). *Food Chemistry X, 19*, 100864. https://doi.org/10.1016/j.fochx.2023.100864

Thikham, S., Jeenpitak, T., Shoji, K., Phongthai, S., Therdtatha, P., Yawootti, A., & Klangpetch, W. (2024). Pulsed electric field-assisted extraction of mushroom ß-glucan from Pleurotus pulmonarius by-product and study of prebiotic properties. *IJFST, 59*(6), 3939–3949. https://doi.org/10.1111/ijfs.17144

Umaña, M., Eim, V., Garau, C., Rosselló, C., & Simal, S. (2020). Ultrasound-assisted extraction of ergosterol and antioxidant components from mushroom by-products and the attainment of a β-glucan rich residue. *Food Chemistry, 332*, 127390. https://doi.org/10.1016/j.foodchem.2020.127390

Usami, A., Ono, T., Kashima, Y., Nakahashi, H., Marumoto, S., Nosaka, S., Watanabe, S., & Miyazawa, M. (2014). Comparison of agitake (Pleurotus eryngii var. ferulae) volatile components with characteristic odors extracted by hydrodistillation and solvent-assisted flavor evaporation. *Journal of Oleo Science, 63*(1), 83–92.

Usami, A., Motooka, R., Nakahashi, H., Marumoto, S., & Miyazawa, M. (2015). Chemical composition and character impact odorants in volatile oils from edible mushrooms. *Chemistry & Biodiversity, 12*(11), 1734–1745. https://doi.org/10.1002/cbdv.201400395

Valu, M.-V., Soare, L. C., Sutan, N. A., Ducu, C., Moga, S., Hritcu, L., Boiangiu, R. S., & Carradori, S. (2020). Optimization of ultrasonic extraction to obtain Erinacine a and polyphenols with antioxidant activity from the fungal biomass of Hericium erinaceus. *Food, 9*(12). https://doi.org/10.3390/foods9121889

Vetvicka, V., Gover, O., Karpovsky, M., Hayby, H., Danay, O., Ezov, N., Hadar, Y., & Schwartz, B. (2019). Immune-modulating activities of glucans extracted from Pleurotus ostreatus and Pleurotus eryngii. *Journal of Functional Foods, 54*, 81–91. https://doi.org/10.1016/j.jff.2018.12.034

Vibala, B. V., Praseetha, P. K., & Vijayakumar, S. (2020). Evaluating new strategies for anticancer molecules from ethnic medicinal plants through in silico and biological approach – a review. *Gene Reports, 18*, 100553. https://doi.org/10.1016/j.genrep.2019.100553

Vidana Gamage, G. C., & Choo, W. S. (2023). Hot water extraction, ultrasound, microwave and pectinase-assisted extraction of anthocyanins from blue pea flower. *Food Chemistry Advances, 2*, 100209. https://doi.org/10.1016/j.focha.2023.100209

Vilková, M., Płotka-Wasylka, J., & Andruch, V. (2020). The role of water in deep eutectic solvent-base extraction. *Journal of Molecular Liquids, 304*, 112747. https://doi.org/10.1016/j.molliq.2020.112747

Wahab, N. A. A., Abdullah, N., & Aminudin, N. (2014). Characterisation of potential antidiabetic-related proteins from Pleurotus pulmonarius (Fr.) Quél. (Grey oyster mushroom) by MALDI-TOF/TOF mass spectrometry. *BioMed Research International, 2014*(1), 131607. https://doi.org/10.1155/2014/131607

Wang, N., Zhang, Y., Wang, X., Huang, X., Fei, Y., Yu, Y., & Shou, D. (2016). Antioxidant property of water-soluble polysaccharides from Poria cocos Wolf using different extraction methods. *International Journal of Biological Macromolecules, 83*, 103–110. https://doi.org/10.1016/j.ijbiomac.2015.11.032

Wang, K., Guo, J., Cheng, J., Zhao, X., Ma, B., Yang, X., & Shao, H. (2021). Ultrasound-assisted extraction of polysaccharide from spent Lentinus edodes substrate: Process optimization, precipitation, structural characterization and antioxidant activity. *International Journal of Biological Macromolecules, 191*, 1038–1045. https://doi.org/10.1016/j.ijbiomac.2021.09.174

Wang, D., Zhang, M., Law, C. L., & Zhang, L. (2024). Natural deep eutectic solvents for the extraction of lentinan from shiitake mushroom: COSMO-RS screening and ANN-GA optimizing conditions. *Food Chemistry, 430*, 136990. https://doi.org/10.1016/j.foodchem.2023.136990

Wani, K. M., & Uppaluri, R. V. S. (2022). Pulsed ultrasound-assisted extraction of bioactive compounds from papaya pulp and papaya peel using response surface methodology: Optimization and comparison with hot water extraction. *Applied Food Research, 2*(2), 100178. https://doi.org/10.1016/j.afres.2022.100178

Wei, L., Wang, W., Hou, Y., Xie, X., Li, X., Chen, F., Wang, Z., Zhou, Y., & Li, F. (2023). Jing B (2023) chemical composition, antibacterial test, and antioxidant activity of essential oils from fresh and dried Stropharia rugosoannulata. *Journal of Chemistry, 1*, 6965755. https://doi.org/10.1155/2023/6965755

Wolf, M., Berger, F., Hanstein, S., Weidenkaff, A., Endreß, H.-U., Oestreich, A. M., Ebrahimi, M., & Czermak, P. (2022). Hot-water hemicellulose extraction from fruit processing residues. *ACS Omega, 7*(16), 13436–13447. https://doi.org/10.1021/acsomega.1c06055

Wu, S.-J., Tung, Y.-J., & Ng, L.-T. (2020). Anti-diabetic effects of Grifola frondosa bioactive compound and its related molecular signaling pathways in palmitate-induced C2C12 cells. *Journal of Ethnopharmacology, 260*, 112962. https://doi.org/10.1016/j.jep.2020.112962

Xu, N., Sun, Y.-H., Guo, X.-L., Liu, C., Mao, Q., & Hou, J.-M. (2018). Optimization of ultrasonic-microwave synergistic extraction of polysaccharides from Morchella conica. *Journal of Food Processing & Preservation, 42*(2), e13423. https://doi.org/10.1111/jfpp.13423

Xue, D., & Farid, M. M. (2015). Pulsed electric field extraction of valuable compounds from white button mushroom (Agaricus bisporus). *Innovative Food Science & Emerging Technologies, 29*, 178–186. https://doi.org/10.1016/j.ifset.2015.03.012

Yadav, D., & Negi, P. S. (2021). Bioactive components of mushrooms: Processing effects and health benefits. *Food Research International, 148*, 110599. https://doi.org/10.1016/j.foodres.2021.110599

Yang, Z., Xu, J., Fu, Q., Fu, X., Shu, T., Bi, Y., & Song, B. (2013). Antitumor activity of a polysaccharide from Pleurotus eryngii on mice bearing renal cancer. *Carbohydrate Polymers, 95*(2), 615–620. https://doi.org/10.1016/j.carbpol.2013.03.024

Yang, S., Meng, Y., Yan, J., Wang, N., Xue, Z., Zhang, H., & Fan, Y. (2019). Polysaccharide-enriched fraction from Amillariella Mellea fruiting body improves insulin resistance. *Molecules, 24*(1). https://doi.org/10.3390/molecules24010046

Yang, H.-I., Ameer, K., Chung, Y. B., Min, S.-G., & Eun, J.-B. (2023). Optimization of spray drying process for recovery of onion–stevia leaf hot water extract powder using response surface methodology. *FSNT, 11*(4), 1770–1784. https://doi.org/10.1002/fsn3.3207

Yu, X., Tu, X., Tao, L., Daddam, J., Li, S., & Hu, F. (2023). Royal Jelly Fatty Acids: Chemical composition, extraction, biological activity, and Prospect. *Journal of Functional Foods, 111*, 105868. https://doi.org/10.1016/j.jff.2023.105868

Zhang, J., Wen, C., Zhang, H., Duan, Y., & Ma, H. (2020). Recent advances in the extraction of bioactive compounds with subcritical water: A review. *Trends Food Sci. Tecnol., 95*, 183–195. https://doi.org/10.1016/j.tifs.2019.11.018

Zhang, C., Xie, Y., Liu, D., Liu, R., & Han, J. (2024). Effects of drying process and high hydrostatic pressure on extraction of antioxidant Ergothioneine from Pleurotus citrinopileatus singer. *Food, 13*(6), 878. Retrieved from https://www.mdpi.com/2304-8158/13/6/878

Zhao, R.-J., Huo, C.-Y., Qian, Y., Ren, D.-F., & Lu, J. (2017). Ultra-high-pressure processing improves proteolysis and release of bioactive peptides with activation activities on alcohol metabolic enzymes in vitro from mushroom foot protein. *Food Chemistry, 231*, 25–32. https://doi.org/10.1016/j.foodchem.2017.03.058

Zhu, Y., Li, Q., Mao, G., Zou, Y., Feng, W., Zheng, D., Wang, W., Zhou, L., Zhang, T., Yang, J., Yang, L., & Wu, X. (2014). Optimization of enzyme-assisted extraction and characterization of polysaccharides from Hericium erinaceus. *Carbohydrate Polymers, 101*, 606–613. https://doi.org/10.1016/j.carbpol.2013.09.099

Chapter 2
Fungal Endophytes as a 'Super House' of Novel Bioactive Compounds

Devendra Sillu, Navneet Kaur Dhiman, Pranshu Saxena, Nandini Sareen, and Gurleen Kaur Sodhi

1 Introduction

Since ancient times, plants have served as a remarkable source for diverse bioactive compounds. These naturally derived substances, extracted from plants, exhibit a broad spectrum of applications, extending their use across the pharmaceutical, cosmeceutical, and food industries (Roy et al., 2022). The literature suggests an increased demand for safe and sustainable food ingredients that can provide health benefits as well. However, gathering these bioactive molecules from plant sources is a rigorous process (Jha & Sit, 2022). Plants harbour different concentrations of these compounds depending upon the species. Additionally, growth cycle, geographical location, and seasonal variations also impact the quality of bioactive compounds. While traditional methods of extraction, such as hydro distillation and Soxhlet extraction, are time-consuming, large-scale extraction requires substantial biomass, making it a cost-intensive process as well (Lefebvre et al., 2021; Daud et al., 2022). Over time, there has been growing research on a niche of microorganisms residing within the plants known as endophytes (Rai et al., 2021; Hashem et al., 2023). These endophytic microorganisms have coevolved with plants. This close relationship led to the occurrence of horizontal gene transfer, enabling

D. Sillu
Department of Environmental Science and Engineering, Guangdong Technion—Israel Institute of Technology, Shantou, People's Republic of China

N. K. Dhiman · P. Saxena · N. Sareen
Department of Biotechnology, Thapar Institute of Engineering and Technology, Patiala, Punjab, India

G. K. Sodhi (✉)
University Institute of Biotechnology, Chandigarh University, Mohali, Punjab, India
e-mail: gurleen.e17702@cumail.in

B. P. Singh et al. (eds.), *Fungal Additives and Bioactives in Food Processing Industries*, Fungal Biology, https://doi.org/10.1007/978-3-032-04519-5_2

endophytes to replicate and synthesise diverse bioactive compounds (Tiwari & Bae, 2020).

Endophytic fungi produce different compounds capable of exhibiting antioxidant, antimicrobial, and other beneficial properties, making them valuable for the development of functional foods (Rai et al., 2021; Wen et al., 2022; Vishwakarma et al., 2024). Since spoilage of food is a primary challenge in the food industry, utilisation of fungal endophytes for the extraction of anti-microbial compounds offers a natural solution (Caruso et al., 2022; Eshboev et al., 2024). By inhibiting the bacterial and fungal pathogens, these compounds enhance the shelf life. Compounds like alkaloids have been reported to inhibit foodborne pathogens such as *Escherichia* and *Salmonella* (Daley & Cordell, 2021; Khalkho et al., 2024). Similarly, terpenoids are also recognised for their antimicrobial activity and can be employed to enhance the shelf life. Other compounds like flavonoids are known for their anti-inflammatory potential. It provides a health-conscious, green solution as opposed to traditional and synthetic alternatives. Besides anti-inflammation, flavonoids also exhibit antioxidant potential, which reduces the damage caused by oxidation of lipids in food products (Adeleke & Babalola, 2021; Preethi et al., 2021; Hashem et al., 2023). This untapped reservoir of compounds with diverse food processing applications presents a unique avenue for future research.

The change and shift towards greener and sustainable solutions have driven the researchers to explore alternative solutions. And natural compounds derived from endophytic fungi have garnered much attention for their rich and unique metabolic diversity (Singh et al., 2021; Vishwakarma et al., 2024). Incorporation of endophyte-derived bioactive compounds can lead to the production of improved functional foods. Furthermore, to meet the increasing demand, emerging approaches such as genome editing and synthetic biology can be explored. In addition, optimisation of bioprocess and fermentation processes can aid in the development of economical and scalable production units for commercialisation. With this, the chapter will shed light on the symbiotic association of endophytic fungi and plants. Diversity of bioactive compounds and their applications in food science are also discussed. Besides highlighting the use, the chapter also discusses challenges related to safety, scalability, and regulatory compliance, which can pave the way for future studies.

2 Endophytic Fungi: Diversity and Bioactive Compound Production

The term 'Endophyte' was first introduced by De Bary in 1886. Endophytes are microorganisms that thrive and establish themselves within the internal tissues of host plants. These adaptable and widespread organisms are found in plants across nearly all geographic and climatic conditions, making them a ubiquitous presence in nature (Baron & Rigobelo, 2022; Sodhi & Saxena, 2023a, b). Historically, endophytic fungi have been classified into two primary groups: clavicipitaceous endophytes (C-endophytes) and nonclavicipitaceous endophytes (NC-endophytes). It is not easy to grow C-endophytes in laboratory conditions, and they are typically found in certain cool- and warm-season grasses. These fungi live inside the plant, mainly in the shoots, where they spread through the spaces between plant cells and are transmitted by vertically passing fungi to offspring via seed infection. Clay and Schardl classified clavicipitaceous endophytes into three types: Type I (pathogenic, causing visible symptoms), Type II (mixed interactions, both beneficial and harmful), and Type III (asymptomatic, causing no harm). NC-endophytes were gathered from asymptomatic matter of avascular plants, ferns, allies, conifers, and Angiosperms (Khiralla et al., 2016; Aamir et al., 2020; Bhunjun et al., 2024).

Fungal endophytes have been successfully extracted from a diverse array of plant species, including crops, invasive plants, and woody trees, with a particular focus on medicinal plants, mosses, ferns, and lichens. *Alternaria, Colletotrichum, Fusarium, Gibberella, Glomerella, Guignardia, Leptospaerulina, Nigrospora, Phoma, Phomopsis,* and *Xylaria* are the genera most commonly isolated (Gautam et al., 2013; Maulana et al., 2018; Kuo et al., 2021; Sodhi & Saxena, 2023a, b). It is believed that biodiversity occurs in tropical and temperate regions. Arnold and Lutzoni conducted a comparative study of endophytic fungal communities, ranging from the Canadian Arctic to the lowland tropical forests of central Panama. Their research, which examined 21 plant species across six locations, revealed a linear decline in endophytic fungal diversity from tropical regions to northern boreal forests. Over 80 genera of endophytes were identified in tropical and subtropical plant regions (Arnold & Lutzoni, 2007). Recently, researchers obtained 343 endophytes from mangroves representing at least 34 distinct genera, with the most prevalent being *Diaporthe* sp., *Colletotrichum* sp., *Fusarium* sp., *Trichoderma* sp., and *Xylaria* sp. (Sebastianes et al., 2017). This shows that mangroves possess a huge biodiversity of endophytic fungi. In another study, researchers isolated more than 200 genera of endophytic fungi from mangroves (Manathunga et al., 2024).

Through co-evolution over time, a mutually beneficial association steadily formed between endophytic fungi and their host plants. The plant supplies nutrients and a habitat for the continuance of its endophytes; in turn, the endophytes produce various bioactive compounds that help the plant withstand external stresses (Baron & Rigobelo, 2022; Sodhi & Saxena, 2023a, b). In the past five years, more than 300 endophytes capable of producing bioactive molecules have been successfully isolated and cultivated in laboratory settings. Endophytes generate a wide range of bioactive metabolites with distinct structures, including alkaloids, benzopyranones, chinones, flavonoids, phenolic acids, quinones, steroids, terpenoids, tetralones, xanthones, and other compounds (Manganyi & Ateba, 2020; Mishra et al., 2021; Agarwal et al., 2024). Certain fungal endophytes have evolved the capacity to yield identical or structurally similar bioactive compounds to those derived from plants, for example, Paclitaxel (Taxol), Campothecin, and Azadirachtin (Table 2.1).

Table 2.1 Bioactive compounds commercially produced by endophytic fungi

Endophytic fungi	Host plant	Chemical compound	Biological activities	Reference
Alternaria sp.	*Catharanthus roseus*	Vinblastine	Anti-cancer agent	Saxena et al. (2024)
Taxus sp.	*Taxus brevifolia (Pacific Yew)*	Paclitaxel (Taxol)	Anti-cancer agent for prostate, ovarian, breast and lung cancer	Adhikari et al. (2023)
Phomopsis sp.	*Plumeria acutifolia*	Terpenoid	Antimicrobial agent	Kumar et al. (2017)
Chaetomium chiversii C5-36- 62	*Ephedra fasciculate*	Radicicol	Cytotoxic agent	Kumar et al. (2017)
Alternaria sp., *Fusarium oxysporum, Penicillium* sp., *Monilia* sp.	*Sinopodophyllum, Diphylleia, Dysosma, Juniperus*	Podophyllotoxin	Anti-cancer, antiviral, antioxidant, immunostimulant, and anti-rheumatic agents	Saxena et al. (2024)
Botryosphaeria parva, Fusarium sp., *Nodulisporium* sp.	*Nothapodytes* sp., *Apodytes dimidiata*	Camptothecin	Inhibits intra-nuclear enzyme, Topoisomerase-1, which is required for DNA replication and transcription	Saxena et al. (2024)

3 Applications of Fungal-Derived Bioactive Compounds in the Food Industry

3.1 Anti-microbial

Fungal endophytes serve as a valuable reservoir of low-molecular-weight organic bioactive compounds with distinct structures and antimicrobial potential. They are synthesised by endophytes as an adaptive response to various environmental factors, serving specific functional roles. They can be extracted and prepared in laboratory settings for an array of antimicrobial applications (Usman et al., 2024). The utilisation of such antibiotic metabolites extends beyond pharmaceuticals and can be applied to alleviate foodborne diseases caused by phytopathogens at pre- or post-harvest stages, thereby enhancing food and agriculture security (Basappa et al., 2023; Manathunga et al., 2024). The bioactive compounds simultaneously exhibit strong antagonistic capabilities against fungal or bacterial pathogens and promote plant growth by improving nutrient absorption (X. Qin et al., 2024). These aids are owed to active regulation, enabling fungi to coexist with plants asymptomatically while offering various advantages. In a report, *Phoma* sp., isolated from *Arisaema erubescens*, has been revealed to produce beta-sitosterol, a significant medicinal steroid. This compound has demonstrated effective antifungal and antibacterial properties against a wide variety of plant pathogenic fungi and bacteria (Jha, 2015).

In another report, chlorinated benzophenone derivatives have demonstrated inhibitory effects against plant pathogenic fungi *Verticillium albo-atrum*, *Gibberella zeae*, and *Fusarium culmorum* with an MIC activity of 7.2, 144.4, and 114.4 mM, respectively (Haryani et al., 2019). Rashid (2021) isolated three fungal endophytes, namely, *Diaporthe phaseolorum*, *Aspergillus fumigatus*, and *Aspergillus versicolor* from tomato plant. Bioactive compounds from these fungi, molecularly identified as acetol, hexanoic acid, and acetic acid, displayed strong antibacterial activity (ZoI: ≤4.14 cm) against *Xanthomonas vesicatoria* and can be effective for biocontrol of tomato bacterial spot disease. Likewise, an extract from *Penicillium* sp., obtained from *Carapa guianensis*, demonstrated a concentration-dependant inhibition of *Enterococcus faecalis* (de Miranda et al., 2019).

These bioactive compounds can also be utilised as safe food preservatives to combat spoilage in perishable foods, which pose significant challenges to the global food supply chain (Deshmukh et al., 2022; Rani et al., 2021; Vega-Portalatino et al., 2023). The bioactive compound '7-amino-4-methylcoumarin', isolated from *Ginkgo biloba*, was utilised as a natural preservative on account of its microorganism inhibitory activity, including *Escherichia coli*, *Staphylococcus aureus*, *Vibrio anguillarum*, *Shigella* sp., *Yersinia* sp., *Salmonella typhi*, *Salmonella enteritidis*, *Salmonella typhimurium*, *Vibrio parahaemolyticus*, and *Aeromonas hydrophila* (Liu et al., 2008; Qin et al., 2009; Vishwakarma et al., 2024).

Similarly, inhibition of bacteria such as *Bacillus subtilis* and *Staphylococcus aureus* has been reported from the fungal endophyte extracts of *Alternaria alternata*, isolated from *Picrorhiza kurroa* (Chandra et al., 2021). *Penicillium* is particularly significant in the creation of potent antimicrobials (Castronovo et al., 2021). For instance, penicolinates A–E, classified as polyketides, are derived from *Penicillium* sp. BCC16054. Among these, penicolinates B and C have demonstrated notable antimicrobial potential against *Candida albicans* and *Bacillus cereus* (Zhang et al., 2022). *Penicillium citrinum* associated with *Ocimum tenuifolium* has been described to produce novel alkaloids and polyketides. Among these, citrinin, alternariol, and perinadine A have shown mild antimicrobial activity versus the common food poisoning bacterium *Staphylococcus aureus* (Kaul et al., 2017).

Recently, Bouarab-Chibane et al. (2019)) recorded strain-dependent antibacterial activity of 35 fungal-derived polyphenols (1 g L^{-1} conc.) against food-spoiling bacterial strains. Chitosan, a cationic polysaccharide, has been well explored for its antimicrobial, biocompatible, and biodegradable characteristics in the food industry (Gutiérrez, 2017). Fungal-derived chitosan with properties such as high fibre content, lower viscosity, medium-low molecular weight, and a high degree of deacetylation is well suited for application in food as well as biomedicine owing to its solubility (Ghormade et al., 2017; Setiawan, 2024). A group of researchers obtained nano chitosan from mangrove fungal endophyte *Fusarium* sp., 20CB07, which demonstrated antibacterial potential on *Pseudomonas aeruginosa* and *Staphylococcus aureus* with 20 mm and 8 mm ZoI, respectively, at 50 μg mL^{-1}. Similarly, Z. Wang et al. (2020)) found that GCP, a polysaccharide produced by the endophytic fungus *Chaetomium globosum* CGMCC 6882, exhibited antibacterial activity against *Staphylococcus aureus* by disrupting cell permeability.

3.2 Anti-oxidant

Antioxidants are chemical compounds that mitigate oxidative stress/damage induced by free radicals and have been reported to suppress deleterious effects arising from oxidative chain reactions (Usman et al., 2024). Antioxidants as additives are widely employed to preserve food by preventing oxidation and neutralising free radicals during processing and storage. Characterised by their high stability and low volatility, antioxidants play a crucial role in maintaining overall consumer appeal (Darwish, 2019). Numerous research efforts have been put forth to find naturally derived antioxidant agents (Sharaf et al., 2022). Fungal endophytes have grabbed attention as promising sources of antioxidants, especially since the discovery of antioxidant agents produced by *Pestalotiopsis microspora*. Ever since this advancement, endophytic fungal extracts derived from various plants, including medicinal, rainforest, and halophyte species, have been reported with similar potential (Zhou et al., 2018).

Synthetic phenolic antioxidants like butylated hydroanisole (BHA), butylated hydrotoluene (BHT), and tert-butylated hydroquinone (TBHQ) are extensively employed in packaged food products to minimise lipid rancidity (Yang et al., 2002). Natural phenolic compounds have shown antioxidant activity comparable to synthetic BHT. The progressive growth of antioxidants has facilitated the development of industrial technologies for producing oxidation-resistant edible flaxseed oil (Chandra et al., 2020; Shadyro et al., 2017). Gurgel et al. (2024)) extracted and optimised phenolic compounds production using *Cophinforma mamane CF2-13*, an isolate from *Fridericia chica*. The extract presented caffeic acid as a phenolic with significant antioxidant capacity. Moreover, they reported the olive oil oxidation protection ability, emphasising its potential use as an antioxidant. Lipid oxidation and microbial-induced deterioration represent common challenges in the meat industry. Active packaging involving the development of film or coating using biopolymers and natural antioxidants can prolong the products' shelf life (Domínguez et al., 2018).

Łopusiewicz et al. (2018)) reported gelatin coatings enriched with fungal melanin (0.1%, 0.5%, 1%) effectively reduced oxidative rancidity in pork lard after 7, 14, and 21 days of storage, highlighting their potential for preventing lard oxidation. Another study utilised postbiotics from *Saccharomyces cerevisiae* var. *boulardii* ATCC MYA-796 (PSB) and *Malva sylvestris* seed polysaccharide mucilage (MSM) to develop a protective film for the preservation of meat. It displayed superior radical scavenging activity and antibacterial action towards foodborne pathogens. The PSB-MSM film efficiently increased meat shelf life while maintaining the sensory aspect (Abbasi et al., 2023). A study by Davis et al. (2024)) presented a cost-effective approach for producing chitosan from an endophytic fungus cultivated on a pineapple peel-based medium. The fungal chitosan demonstrated significant antioxidant activity (IC_{50} = 0.22 mg L^{-1}) and was used to create a biodegradable film, which degraded by 28.92% over 50 days. Rezzani et al. (2022)) focused on developing films using the high-pressure homogenised yeast cell wall (YCW-H) and naphtho-γ-pyrone (CL-NGP) extract. CL-NGP exhibited significant antioxidant activity (IC_{50} = 17.78 mg L^{-1}), which remained effective after incorporation into YCW-H films. Additionally, in food simulation migration studies, CL-NGP migration was observed in fatty food simulants while retaining antioxidant activity, indicating the potential of these bioactive films for sustainable food packaging applications.

Recently, endophytes have seized attention as prolific makers of exopolysaccharides (EPSs) with significant antioxidant potential. Tao et al. (2024)) extracted three EPSs from *Monascus purpureus*, *Penicillium citrinum*, and *Aspergillus versicolor*, growing inside medicinal plant *Cyclocarya paliurus*, which presented antioxidant activity in vitro. Y. Wang et al. (2019)) assessed the antioxidant potential of EPS-producing *Alternaria tenuissima* F1 inhabiting *Angelica sinensis* and found that the fungal extract contains compounds with a non-reducing sugar and glycoprotein with a pyranoid ring, which aid in exhibiting strong scavenging activity. A study isolated a novel exopolysaccharide produced by *Fusarium merismoides* A6 (FM-EPS). Fractionated FM-EPS1 exhibited stronger antioxidant activity compared to FM-EPS2, while FM-EPS2 demonstrated significant antiproliferative effects on HeLa and HepG2 cells. These findings suggest that FM-EPSs have potential applications as antioxidant and anticancer agents in industries such as food, feed, and nutraceuticals (Chen et al., 2023).

3.3 *Nutraceuticals*

Nutraceuticals are dietary supplement products with pharmaceutical properties. From an industry perspective, nutraceuticals are defined as 'non-toxic food components with potential health benefits' (Al-Obaidi, Alobaidi, et al., 2021a; Das et al., 2012). They may be consumed as standalone supplements or functional foods without the need for therapeutic prescriptions. These products are crucial in promoting human health and longevity, offering benefits such as disease prevention, symptom reduction, improved wellness, and immune system enhancement. These are derived from plants, animals, and fungi as bioactive molecules. Endophytes have arisen as a novel source of bioactive molecules per pharmacological aspect, making them a promising frontier for nutraceutical innovation (Al-Obaidi et al., 2021b; Devi et al., 2020; Wadhwa et al., 2024).

β-Glucans, extensively employed in the nutraceutical industry, are natural polysaccharides in plant cell walls, like oats and barley, as well as in various fungal species and yeast (Ciecierska et al., 2019; Devi et al., 2020). β-Glucans influenced by their molecular weight, bond frequency, and structural variations have been found to be immunomodulating, hypocholesterolemic, anticancer, antioxidant, antidiabetic, hypolipidaemic, and neuroprotective (AbdelRazek et al., 2023; Agrawal et al., 2022; De Araújo et al., 2017; Jayasekara et al., 2022; Kobayashi et al., 2005; Mengdai et al., 2020; Rondanelli et al., 2009; Vetvicka, 2011; Xu et al., 2020). Glucans from *Lentinula edodes* have been utilised as a substitute for wheat flour in baked goods, making low-calorie, fibre-rich foods without compromising the product's air-holding capacity or texture (Kim et al., 2011). Similar report by Kim et al. (2009)), incorporating *L. edodes* β-glucans into noodles as a replacement for wheat flour, enhanced quality characteristics while providing antioxidant and hypocholesterolemic effects.

More recently, Jayasekara et al. (2022)) produced high levels of β-glucan on poor malt-peptone extract media from *Xylaria* sp. BCC 1067. The bioactive displayed additional anti-bacterial aspect against *Propionibacterium acnes*, and thus, the authors suggested its utilisation in the nutraceutical industry as a health-promoting substance. Zeng et al. (2019)) obtained polysaccharide DG2 from *Fusarium solani* DO7, the endophytic fungus of *Dendrobium officinale*. The metabolite was deduced as a galactoglucan through structural characterisation. The investigation of immunological activity revealed that DG2 increased the TNF-α, IL-6, and iNOs mRNA expression levels in RAW264.7 cells, thereby stimulating their production. These findings indicated that DG2 can boost macrophage-mediated immune responses, making it a promising functional food supplement candidate.

Tang et al. (2021)) produced vitexin with a yield of 18.72 mg L^{-1} from *Fusarium solani* G6 of *Cajanus cajan*, which showed osteoblastic proliferation-stimulating action and could potentially be utilised as a therapeutic agent for osteoporosis. Phytoseterol was isolated from three endophytic fungi *Colletotrichum gloeosporioides*, *Colletotrichum siamense* and *Phoma* sp. of *Cissus quadrangularis* plant. The metabolite extract due to the presence of β-carotene displayed antioxidant and radical scavenging activity with IC_{50} = 5.748–19.937 μg mL^{-1} (Awari et al., 2024). Fu et al. (2024)) extracted multiple bioactives from endophyte *Talaromyces neorugulosus* R-209 of pigeon pea.

Birat et al. (2022)) isolated *Nigrospora zimmermanii* (Apiosporaceae) from *Catharanthus roseus,* capable of producing vincristine, an anticancer vinca alkaloid. The studies showed that parent fungi produced maximum vincristine at 5.344 μg mL^{-1} under laboratory conditions. Tawfike et al. (2019)) identified anticancer and anti-trypanosome bioactive compounds from the *Aspergillus flocculus* endophyte from the plant *Markhamia platycalyx*. The active metabolites, namely, cis-4-hydroxymellein, 5-hydroxymellein, diorcinol, botryoisocoumarin A, and mellein at 30 μM conc., subdued the growth of K562. 3-Hydroxymellein and diorcinol showed 56% and 97% growth reduction, respectively, against *Trypanosoma brucei brucei*, which causes sleeping sickness. Moreover, the anti-trypanosomal activity was credited to steroidal compounds. Ergosterol, a precursor to vitamin D2, has also been used in fortified foods to address vitamin D deficiencies (Nzekoue et al., 2022; Papoutsis et al., 2020).

t-Resveratrol (3,5,4′-trihydroxystilbene) and α-viniferin, metabolites in the class of polyphenol stilbenes, with their pharmacological, biological, and nutraceutical properties are gaining significant scientific attention as nutritional supplements and medicinal ingredients in the treatment of metabolic syndromes (Cherniack, 2011; Zhu, 2011; Zhu et al., 2011). In the first of its kind report, Roat and Saraf (2020) used fungal endophyte *Aspergillus stellifer* AB4 isolated from *Vitis vinifera*, for the optimisation of t-resveratrol and α-viniferin with maximum yields of 300 μg L^{-1} and 324 μg L^{-1}. Similarly, Dwibedi et al. (2021)) investigated resveratrol production from *Arcopilus aureus*. Optimisation leads to 1.49-fold higher resveratrol with a yield of 133.53 μg mL^{-1}. Fungal resveratrol displayed significant tyrosinase inhibition at IC_{50} of 2.654 μg mL^{-1}, which opens up avenues for its production in commercial nutraceutical and cosmeceutical applications.

3.4 Biofortification

One in every three people in the world suffers from hidden hunger caused by the lack of vitamins and minerals in their diet, which has negative consequences for their health. Over the past 40 years, there has been significant development in the agriculture sector of developing countries aimed at increasing cereal production (Iversen et al., 2023). Recently, however, the focus has shifted not only to reducing hunger but also to providing nutrient-rich food to combat hidden hunger. To combat this hidden hunger, biofortification is used to expand the nutritional value of crops by increasing their micronutrient (nickel, copper, zinc, and iron) content through biological means. Endophytic fungi release small molecules called siderophores (Ofori et al., 2022; Zulfiqar et al., 2024). The chemical composition of fungi's siderophores contains hydroxylated and acylated ornithine groups. These molecules are responsible for chelating iron and bioavailability of the iron molecule. They improve iron uptake while simultaneously reducing the availability of iron for harmful pathogens, thereby limiting their growth. Siderophores also exhibit antibacterial properties (Dhiman et al., 2023; Mushtaq et al., 2024).

Fungal endophytes colonise plant's roots, stems, and leaves. These endophytes can replace soil microbes to enhance the capacity to fix nitrogen in plants. In non-legume plants, *Metarhizium rebertsii* transfers nitrogen from dead insects to host plants, demonstrating a non-soil nitrogen source (Zheng et al., 2024). Selenium (Se) is another compound that is very important for human health by playing a key role in the immune system and metabolism. Deficiency of selenium in the soil leads to low selenium in crops, which causes diseases like Keshan and Kashin-Beck disease. To prevent this deficiency of selenium, we can use mycorrhizal fungi and root endophytic fungi (Ye et al., 2020; Yang et al., 2021). *Arbuscular Mycorrhizal Fungi* (AMF) help plants absorb important nutrients like phosphorus and sulphur. Selenium and sulphur belong to the same chemical group, so they share similar chemical properties, meaning selenium can follow a similar pathway to sulphur in plants, thus improving selenium intake. However, selenium uptake can be affected by other nutrients like phosphate and sulphate, which compete for absorption. When phosphate fertilisers are applied to the soil, they can reduce the ability of the plant to take up selenium as both ions are absorbed through similar pathways (Devi et al., 2022;

Sodhi & Saxena, 2023a, b). However, research on *Astragalus* species found that sulphur deficiency can actually increase selenium accumulation in plants. This is because a lack of sulphur stimulates the expression of sulphate transporters, which can then facilitate more selenium uptake as well (Montanari et al., 2023).

In lentils, *Trichoderma fungi* and their bioactive metabolites are utilised to enhance the micronutrient content, particularly iron (Fe) and zinc (Zn), which are essential for human health. It even provides protection to the plant. An experimental setup was done where bioactive metabolites from 11 different *Trichoderma* strains were supplied to lentil plants in both greenhouse and field conditions. The treatment involved applying the fungal strains either to the seed, the soil, or both. When the results were checked for greenhouse conditions, iron and zinc saw the largest increase in content, particularly *T. afroharzianum T22*, *T. harzianum TH1*, and *T. virens GV41*. For field conditions, it was observed that *T. asperellum KV906* and the *hydrophobin HYTLO1* had increased iron and zinc contents. The test displayed that the metabolite not only boosted mineral content but also improved seed germination rates and overall crop yield (Marra et al., 2021).

Endophytes have the capability to revolutionise agriculture by enhancing crop productivity in an eco-friendly manner. Many endophytic fungi produce bioactive metabolites with immunosuppression, anti-cancer activity, antimicrobial, antioxidant activity, and antibiotic properties (Conrado et al., 2022; Omomowo et al., 2023). Regardless of their potential, the full extent of bioactive natural molecules from endophytes remains largely unexplored. Furthermore, endophytes have significant applications in agriculture, industry, and medicine, and their bioactive compounds have contributed to handling ailments like cancer, diabetes, and various neurological disorders. Advances in biotechnology, including genetic engineering and microbial fermentation, can enhance the utilisation of these microbial resources for human benefit. In conclusion, further exploration of endophytic fungi can open many doors.

4 Conclusion and Future Perspectives

The research on endophytic fungi for the production of bioactive compounds has opened new avenues in offering sustainable solutions for problems associated with the food industry. The potential of diverse compounds such as terpenoids, flavonoids, and alkaloids demonstrates remarkable potential in food security and safety. The numerous health hazards linked with the synthetic food additives and preservatives call for a shift towards greener, cleaner, and environmentally conscious trends. Thus, the utilisation of bioactive compounds harnessed from fungal endophytes presents a sustainable alternative. As a green alternative in food processing, future studies should focus on the optimisation of fermentation processes, thereby decreasing the demand for synthetic chemicals and reducing the carbon footprint. Optimising strain-specific variants and culture conditions for large-scale production using biotechnological advances could pave the way for commercialisation.

Furthermore, utilising synthetic biology and omics tools could help researchers enhance the yield of desired bioactive compounds in the fermentation process. For delivery of the compounds, novel formulations like nanoencapsulation or microencapsulation could facilitate the incorporation of fungal-derived compounds into food processing. Analysis of mycotoxins and other toxicity assessments, as per the guidelines of regulatory agencies, could validate the stability and efficacy of the compounds. Additionally, consumer awareness and transparency in marketing of the products could pave the way for trust and consumer confidence. To achieve this, these interdisciplinary collaborations could help unlock the complete ability of harnessed bioactive compounds. This would help transition towards a sustainable and greener practice in food processing.

References

Aamir, M., Rai, K. K., Zehra, A., Kumar, S., Yadav, M., Shukla, V., & Upadhyay, R. S. (2020). Fungal endophytes: Classification, diversity, ecological role, and their relevance in sustainable agriculture. In *Microbial endophytes* (pp. 291–323). Woodhead Publishing. https://doi.org/10.1016/B978-0-12-818734-2.00014-9

Abbasi, A., Sabahi, S., Bazzaz, S., Tajani, A. G., Lahouty, M., Aslani, R., & Hosseini, H. (2023). An edible coating utilizing *Malva sylvestris* seed polysaccharide mucilage and postbiotic from *Saccharomyces cerevisiae* var. *boulardii* for the preservation of lamb meat. *International Journal of Biological Macromolecules, 246*, 1–25. https://doi.org/10.1016/j.ijbiomac.2023.125660

AbdelRazek, M. M., Elissawy, A. M., Mostafa, N. M., Moussa, A. Y., Elanany, M. A., Elshanawany, M. A., & Singab, A. N. B. (2023). Chemical and biological review of endophytic fungi associated with *Morus sp.*(Moraceae) and in silico study of their antidiabetic potential. *Molecules, 28*(4), 1–29. https://doi.org/10.3390/molecules28041718

Adeleke, B. S., & Babalola, O. O. (2021). Pharmacological potential of fungal endophytes associated with medicinal plants: A review. *Journal of Fungi, 7*(2), 147. https://doi.org/10.3390/jof7020147

Adhikari, P., Joshi, K., & Pandey, A. (2023). Taxus associated fungal endophytes: Anticancerous to other biological activities. *Fungal Biology Reviews, 45*, 100308. https://doi.org/10.1016/j.fbr.2023.100308

Agarwal, S., Sharma, G., & Mathur, V. (2024). Plant endophytes: A treasure house of antimicrobial compounds. In *Medicinal plants and antimicrobial therapies* (pp. 107–123). Springer Nature Singapore. https://doi.org/10.1007/978-981-99-7261-6_5

Agrawal, S., Samanta, S., & Deshmukh, S. K. (2022). The antidiabetic potential of endophytic fungi: Future prospects as therapeutic agents. *Biotechnology and Applied Biochemistry, 69*(3), 1159–1165. https://doi.org/10.1002/bab.2192

Al-Obaidi, J. R., Alobaidi, K. H., Al-Taie, B. S., Wee, D. H.-S., Hussain, H., Jambari, N. N., Ahmad-Kamil, E., & Ariffin, N. S. (2021a). Uncovering prospective role and applications of existing and new nutraceuticals from bacterial, fungal, algal and cyanobacterial, and plant sources. *Sustainability, 13*(7), 1–19. https://doi.org/10.3390/su13073671

Al-Obaidi, J. R., Jambari, N. N., & Ahmad-Kamil, E. (2021b). Mycopharmaceuticals and nutraceuticals: Promising agents to improve human well-being and life quality. *Journal of Fungi, 7*(7), 1–21. https://doi.org/10.3390/jof7070503

Arnold, A. E., & Lutzoni, F. (2007). Diversity and host range of foliar fungal endophytes: Are tropical leaves biodiversity hotspots? *Ecology, 88*(3), 541–549. https://doi.org/10.1890/05-1459

Awari, V. S., Barvkar, V. T., Ade, A. B., & Borde, M. Y. (2024). Endophytic fungi from *Cissus quadrangularis* plant a promising source of bioactive compounds. *Brazilian Journal of Microbiology*, 1–18. https://doi.org/10.1007/s42770-024-01500-0

Baron, N. C., & Rigobelo, E. C. (2022). Endophytic fungi: A tool for plant growth promotion and sustainable agriculture. *Mycology, 13*(1), 39–55. https://doi.org/10.1080/21501203.2021.2003178

Basappa, K. S., Raghava, S., & Umesha, S. (2023). Potentials of Endophytic fungus untapped, including novel bioactive compounds. *Pharmacognosy Communications, 13*(2), 1–14. https://doi.org/10.5530/pc.2023.2.10

Bhunjun, C. S., Phukhamsakda, C., Hyde, K. D., McKenzie, E. H., Saxena, R. K., & Li, Q. (2024). Do all fungi have ancestors with endophytic lifestyles? *Fungal Diversity, 125*(1), 73–98. https://doi.org/10.1007/s13225-023-00500-1

Birat, K., Binsuwaidan, R., Siddiqi, T. O., Mir, S. R., Alshammari, N., Adnan, M., Nazir, R., Ejaz, B., Malik, M. Q., & Dewangan, R. P. (2022). Report on vincristine-producing endophytic fungus *Nigrospora zimmermanii* from leaves of *Catharanthus roseus*. *Metabolites, 12*(11), 1–13. https://doi.org/10.3390/metabo12111119

Bouarab-Chibane, L., Forquet, V., Lantéri, P., Clément, Y., Léonard-Akkari, L., Oulahal, N., Degraeve, P., & Bordes, C. (2019). Antibacterial properties of polyphenols: Characterization and QSAR (quantitative structure–activity relationship) models. *Frontiers in Microbiology, 10*, 1–23. https://doi.org/10.3389/fmicb.2019.00829

Caruso, D. J., Palombo, E. A., Moulton, S. E., & Zaferanloo, B. (2022). Exploring the promise of endophytic fungi: A review of novel antimicrobial compounds. *Microorganisms, 10*(10), 1990. https://doi.org/10.3390/microorganisms10101990

Castronovo, L. M., Vassallo, A., Mengoni, A., Miceli, E., Bogani, P., Firenzuoli, F., Fani, R., & Maggini, V. (2021). Medicinal plants and their bacterial microbiota: A review on antimicrobial compounds production for plant and human health. *Pathogens, 10*(2), 1–17. https://doi.org/10.3390/pathogens10020106

Chandra, P., Sharma, R. K., & Arora, D. S. (2020). Antioxidant compounds from microbial sources: A review. *Food Research International, 129*, 1–76. https://doi.org/10.1016/j.foodres.2019.108849

Chandra, H., Kumari, P., Prasad, R., Gupta, S. C., & Yadav, S. (2021). Antioxidant and antimicrobial activity displayed by a fungal endophyte *Alternaria alternata* isolated from *Picrorhiza kurroa* from Garhwal Himalayas, India. *Biocatalysis and Agricultural Biotechnology, 33*, 1–7. https://doi.org/10.1016/j.bcab.2021.101955

Chen, G., Xu, Z., Wang, F., Liu, L., Wei, Y., Li, J., Zhang, L., Zheng, K., Wu, L., & Men, X. (2023). Extraction, characterization, and biological activities of exopolysaccharides from plant root soil fungus *Fusarium merismoides* A6. *Brazilian Journal of Microbiology, 54*(1), 199–211. https://doi.org/10.1007/s42770-022-00842-x

Cherniack, E. P. (2011). Polyphenols: Planting the seeds of treatment for the metabolic syndrome. *Nutrition, 27*(6), 617–623. https://doi.org/10.1016/j.nut.2010.10.013

Ciecierska, A., Drywien, M., Hamulka, J., & Sadkowski, T. (2019). Nutraceutical functions of beta-glucans in human nutrition. *Roczniki Państwowego Zakładu Higieny, 70*(4), 1–10. https://doi.org/10.32394/rpzh.2019.0082

Conrado, R., Gomes, T. C., Roque, G. S. C., & De Souza, A. O. (2022). Overview of bioactive fungal secondary metabolites: Cytotoxic and antimicrobial compounds. *Antibiotics, 11*(11), 1604. https://doi.org/10.3390/antibiotics11111604

Daley, S. K., & Cordell, G. A. (2021). Biologically significant and recently isolated alkaloids from endophytic fungi. *Journal of Natural Products, 84*(3), 871–897. https://doi.org/10.1021/acs.jnatprod.0c01212

Darwish, A. M. G. (2019). Fungal mycotoxins and natural antioxidants: Two sides of the same coin and significance in food safety. *Microbial Biosystems, 4*(1), 1–16. https://doi.org/10.21608/mb.2019.37468

Das, L., Bhaumik, E., Raychaudhuri, U., & Chakraborty, R. (2012). Role of nutraceuticals in human health. *Journal of Food Science and Technology, 49*, 173–183. https://doi.org/10.1007/s13197-011-0269-4

Daud, N. M., Putra, N. R., Jamaludin, R., Norodin, N. S. M., Sarkawi, N. S., Hamzah, M. H. S., & Salleh, L. M. (2022). Valorisation of plant seed as natural bioactive compounds by various extraction methods: A review. *Trends in Food Science & Technology, 119*, 201–214. https://doi.org/10.1016/j.tifs.2021.12.024

Davis, D., Umesh, M., Santhosh, A. S., Suresh, S., Shanmugam, S., & Kikas, T. (2024). Extraction of fungal chitosan by leveraging pineapple Peel substrate for sustainable biopolymer production. *Polymers, 16*(17), 1–14. https://doi.org/10.3390/polym16172455

De Araújo, T., Andrade, E., Lobato, R., Orlando, D., Gomes, N., De Sousa, R., Zangeronimo, M., & Pereira, L. (2017). Effects of beta-glucans ingestion (*Saccharomyces cerevisiae*) on metabolism of rats receiving high-fat diet. *Journal of Animal Physiology and Animal Nutrition, 101*(2), 349–358. https://doi.org/10.1111/jpn.12452

de Miranda, M. d. C. M., Carvalho, C. M., Faria, F. S., Nobreza, A., Pereira, T. M., Pivatto, K., da Costa, M. V., Guedes, O. A., de Araújo Estrela, C. R., & Borges, Á. H. (2019). Antibacterial activity of phytochemical extracts and endophytic fungi of *Carapa guianensis* against *Enterococcus faecalis* in endodontic infections an *in vitro* study. *The Open Dentistry Journal, 13*(1). https://doi.org/10.2174/1874210601913010249

Deshmukh, S. K., Dufossé, L., Chhipa, H., Saxena, S., Mahajan, G. B., & Gupta, M. K. (2022). Fungal endophytes: A potential source of antibacterial compounds. *Journal of Fungi, 8*(2), 1–94. https://doi.org/10.3390/jof8020164

Devi, R., Kaur, T., Guleria, G., Rana, K. L., Kour, D., Yadav, N., Yadav, A. N., & Saxena, A. K. (2020). *Fungal secondary metabolites and their biotechnological applications for human health new and future developments in microbial biotechnology and bioengineering* (pp. 147–161). Elsevier.

Devi, R., Kaur, T., Kour, D., Yadav, A., Yadav, A. N., Suman, A., & Saxena, A. K. (2022). Minerals solubilizing and mobilizing microbiomes: A sustainable approach for managing minerals' deficiency in agricultural soil. *Journal of Applied Microbiology, 133*(3), 1245–1272. https://doi.org/10.1111/jam.15627

Dhiman, K., Sharma, D., Kumari, R., & Tomar, P. (2023). Biofortification of crops using microbes–a promising sustainable agriculture strategy. *Journal of Plant Nutrition, 46*(12), 2912–2935. https://doi.org/10.1080/01904167.2022.2160755

Domínguez, R., Barba, F. J., Gómez, B., Putnik, P., Kovačević, D. B., Pateiro, M., Santos, E. M., & Lorenzo, J. M. (2018). Active packaging films with natural antioxidants to be used in meat industry: A review. *Food Research International, 113*, 93–101. https://doi.org/10.1016/j.foodres.2018.06.073

Dwibedi, V., Rath, S. K., Prakash, R., & Saxena, S. (2021). Response surface statistical optimization of fermentation parameters for resveratrol production by the endophytic fungus *Arcopilus aureus* and its tyrosinase inhibitory activity. *Biotechnology Letters, 43*, 627–644. https://doi.org/10.1007/s10529-020-03032-7

Eshboev, F., Mamadalieva, N., Nazarov, P. A., Hussain, H., Katanaev, V., Egamberdieva, D., & Azimova, S. (2024). Antimicrobial action mechanisms of natural compounds isolated from endophytic microorganisms. *Antibiotics, 13*(3), 271. https://doi.org/10.3390/antibiotics13030271

Fu, J.-X., Jiao, J., Gai, Q.-Y., Fu, Y.-J., Gao, J., Zhang, Z.-Y., Wang, Y., & Wang, X.-Q. (2024). Enhanced production of health-promoting phenolic compounds using a novel endophytic fungus *Talaromyces neorugulosus* R-209 isolated from pigeon pea in a natural habitat by l-phenylalanine feeding. *World Journal of Microbiology and Biotechnology, 40*(10), 1–15. https://doi.org/10.1007/s11274-024-04122-8

Gautam, A. K., Kant, M., & Thakur, Y. (2013). Isolation of endophytic fungi from Cannabis sativa and study their antifungal potential. *Archives of Phytopathology and Plant Protection, 46*(6), 627–635. https://doi.org/10.1080/03235408.2012.754055

Ghormade, V., Pathan, E., & Deshpande, M. (2017). Can fungi compete with marine sources for chitosan production? *International Journal of Biological Macromolecules, 104*, 1415–1421. https://doi.org/10.1016/j.ijbiomac.2017.01.112

Gurgel, R. S., Pereira, D. I., Martins, B., Falcão, L. S., Lacerda, C. D., Neves, C. M., Pinto, A. F., Jordão, A. M., & Albuquerque, P. M. (2024). Global phenolic composition and antioxidant capacity of extracts from the Endophytic fungi Cophinforma mamane with potential use in food systems: The effects of time, temperature, and solvent on the extraction process. *Applied Sciences, 14*(19), 1–16. https://doi.org/10.3390/app14198784

Gutiérrez, T. J. (2017). *Chitosan applications for the food industry* (pp. 183–232). Derivatives, composites and applications. https://doi.org/10.1002/9781119364849

Haryani, Y., Hilma, R., Delfira, N., Martalinda, T., Puspita, F., Friska, A., Juwita, D., Farniga, A., & Ardi, F. (2019). *Potential antibacterial activity of endophytic fungi Penicillium sp. and Trichoderma sp. derived from mangrove Ceriops tagal (Perr.) CB Robb and Bruguiera sp.* Paper presented at the Journal of Physics: Conference Series.

Hashem, A. H., Attia, M. S., Kandil, E. K., Fawzi, M. M., Abdelrahman, A. S., Khader, M. S., & Abdelaziz, A. M. (2023). Bioactive compounds and biomedical applications of endophytic fungi: A recent review. *Microbial Cell Factories, 22*(1), 107. https://doi.org/10.1186/s12934-023-02097-0

Iversen, T. O., Westengen, O. T., & Jerven, M. (2023). The history of hunger: Counting calories to make global food security legible. *World Development Perspectives, 30*, 100504. https://doi.org/10.1016/j.wdp.2023.100504

Jayasekara, L. C. B., Poonsawad, A., Watchaputi, K., Wattanachaisaereekul, S., & Soontorngun, N. (2022). Media optimization of antimicrobial activity production and beta-glucan content of endophytic fungi *Xylaria sp.* BCC 1067. *Biotechnology Reports, 35*, 1–13. https://doi.org/10.1016/j.btre.2022.e00742

Jha, D. (2015). Bioprospecting for microbial endophytes and their natural products. *NBU Journal of Plant Sciences, 9*(1), 1–13. https://ir.nbu.ac.in/handle/123456789/4509

Jha, A. K., & Sit, N. (2022). Extraction of bioactive compounds from plant materials using combination of various novel methods: A review. *Trends in Food Science & Technology, 119*, 579–591. https://doi.org/10.1016/j.tifs.2022.10.018

Kaul, S., Gupta, S., Sharma, S., & Dhar, M. K. (2017). The fungal endobiome of medicinal plants: A prospective source of bioactive metabolites. In *Medicinal plants and fungi: recent advances in research and development* (pp. 167–228). https://doi.org/10.1007/978-981-10-5978-0_7

Khalkho, J. P., Beck, A., Priyanka, Panda, B., & Chandra, R. (2024). Microbial allies: Exploring fungal endophytes for biosynthesis of terpenoid indole alkaloids. Archives of Microbiology, 206(8), 340. doi: https://doi.org/10.1007/s00203-024-03400-0.

Khiralla, A., Spina, R., Yagi, S., Mohamed, I., & Laurain-Mattar, D. (2016). Endophytic fungi: Occurrence, classification, function and natural products. In *Endophytic fungi: diversity, characterization and biocontrol* (pp. 1–19). https://doi.org/10.1007/978-3-319-29532-9_1

Kim, S.-Y., Chung, S.-I., Nam, S.-H., & Kang, M.-Y. (2009). Cholesterol lowering action and antioxidant status improving efficacy of noodles made from unmarketable oak mushroom (*Lentinus edodes*) in high cholesterol fed rats. *Journal of the Korean Society for Applied Biological Chemistry, 52*, 207–212. https://doi.org/10.3839/jksabc.2009.038

Kim, J., Lee, S. M., Bae, I. Y., Park, H. G., Gyu Lee, H., & Lee, S. (2011). (1–3)(1–6)-β-Glucan-enriched materials from *Lentinus edodes* mushroom as a high-fibre and low-calorie flour substitute for baked foods. *Journal of the Science of Food and Agriculture, 91*(10), 1915–1919. https://doi.org/10.1002/jsfa.4409

Kobayashi, H., Yoshida, R., Kanada, Y., Fukuda, Y., Yagyu, T., Inagaki, K., Kondo, T., Kurita, N., Suzuki, M., & Kanayama, N. (2005). Suppressing effects of daily oral supplementation of beta-glucan extracted from *Agaricus blazei* Murill on spontaneous and peritoneal disseminated metastasis in mouse model. *Journal of Cancer Research and Clinical Oncology, 131*, 527–538. https://doi.org/10.1007/s00432-005-0672-1

Kumar, G., Chandra, P., & Choudhary, M. (2017). Endophytic fungi: A potential source of bioactive compounds. *Chemical Science Review and Letters, 6*, 2373–2381.

Kuo, J., Chang, C. F., & Chi, W. C. (2021). Isolation of endophytic fungi with antimicrobial activity from medicinal plant Zanthoxylum simulans Hance. *Folia Microbiologica, 66*(3), 385–397. https://doi.org/10.1007/s12223-020-00841-0

Lefebvre, T., Destandau, E., & Lesellier, E. (2021). Selective extraction of bioactive compounds from plants using recent extraction techniques: A review. *Journal of Chromatography A, 1635*, 461770. https://doi.org/10.1016/j.chroma.2020.461770

Liu, X., Dong, M., Chen, X., Jiang, M., Lv, X., & Zhou, J. (2008). Antimicrobial activity of an endophytic *Xylaria sp.* YX-28 and identification of its antimicrobial compound 7-amino-4-methylcoumarin. *Applied Microbiology and Biotechnology, 78*, 241–247. https://doi.org/10.1007/s00253-007-1305-1

Łopusiewicz, Ł., Jędra, F., & Bartkowiak, A. (2018). The application of melanin modified gelatin coatings for packaging and the oxidative stability of pork lard. *World Scientific News, 101*, 108–119.

Manathunga, K. K., Gunasekara, N. W., Meegahakumbura, M. K., Ratnaweera, P. B., Faraj, T. K., & Wanasinghe, D. N. (2024). Exploring Endophytic fungi as natural antagonists against fungal pathogens of food crops. *Journal of Fungi, 10*(9), 1–33. https://doi.org/10.3390/jof10090606

Manganyi, M. C., & Ateba, C. N. (2020). Untapped potentials of endophytic fungi: A review of novel bioactive compounds with biological applications. *Microorganisms, 8*(12), 1934. https://doi.org/10.3390/microorganisms8121934

Marra, R., Lombardi, N., Piccolo, A., Bazghaleh, N., Prashar, P., Vandenberg, A., & Woo, S. (2021). Mineral biofortification and growth stimulation of lentil plants inoculated with Trichoderma strains and metabolites. *Microorganisms, 10*(1), 87. https://doi.org/10.3390/microorganisms10010087

Maulana, A. F., Turjaman, M., Sato, T., Hashimoto, Y., Cheng, W., & Tawaraya, K. (2018). Isolation of endophytic fungi from tropical forest in Indonesia. *Symbiosis, 76*, 151–162. https://doi.org/10.1007/s13199-018-0555-5

Mengdai, X., Xiaoxing, M., Hao, H., Liegang, L., & Wei, Y. (2020). Yeast beta-Glucan effects on gut microbiota for regulating insulin signaling system on alzheimer's disease. *Proceedings of the Nutrition Society, 79*(OCE2), 1. https://doi.org/10.1017/S0029665120002116

Mishra, R. C., Kalra, R., Dwivedi, N., & Goel, M. (2021). Exploring endophytes using "omics": An approach for sustainable production of bioactive metabolites. *Mycoremediation and Environmental Sustainability, 3*, 349–376. https://doi.org/10.1007/978-3-030-54422-5_15

Montanari, S., Salinitro, M., Simoni, A., Ciavatta, C., & Tassoni, A. (2023). Foraging for selenium: A comparison between hyperaccumulator and non-accumulator plant species. *Scientific Reports, 13*(1), 10661. https://doi.org/10.1038/s41598-023-37249-z

Mushtaq, Z., Liaquat, M., Jehan, S., Akram, M. T., Jaffar, M. T., & Alasmari, A. (2024). Microbial biofortification of iron in Solanum tuberosum L. through siderophore-producing rhizobacteria along with L-tryptophan and iron sulphate supply. *Potato Research*, 1–19. https://doi.org/10.1007/s11540-024-09785-2

Nzekoue, F. K., Sun, Y., Caprioli, G., Vittori, S., & Sagratini, G. (2022). Effect of the ultrasound-assisted extraction parameters on the determination of ergosterol and vitamin D2 in *Agaricus bisporus, A. bisporus* Portobello, and *Pleurotus ostreatus* mushrooms. *Journal of Food Composition and Analysis, 109*, 1–44. https://doi.org/10.1016/j.jfca.2022.104476

Ofori, K. F., Antoniello, S., English, M. M., & Aryee, A. N. (2022). Improving nutrition through biofortification–A systematic review. *Frontiers in Nutrition, 9*, 1043655. https://doi.org/10.3389/fnut.2022.1043655

Omomowo, I. O., Amao, J. A., Abubakar, A., Ogundola, A. F., Ezediuno, L. O., & Bamigboye, C. O. (2023). A review on the trends of endophytic fungi bioactivities. *Scientific African, 20*, e01594. https://doi.org/10.1016/j.sciaf.2023.e01594

Papoutsis, K., Grasso, S., Menon, A., Brunton, N. P., Lyng, J. G., Jacquier, J.-C., & Bhuyan, D. J. (2020). Recovery of ergosterol and vitamin D2 from mushroom waste-potential valorization by food and pharmaceutical industries. *Trends in Food Science & Technology, 99*, 351–366. https://doi.org/10.1016/j.tifs.2020.03.005

Preethi, K., Manon Mani, V., & Lavanya, N. (2021). Endophytic fungi: A potential source of bioactive compounds for commercial and therapeutic applications. In *Endophytes: Potential source of compounds of commercial and therapeutic applications* (pp. 247–272). https://doi.org/10.1007/978-3-030-82572-4_11.

Qin, J.-C., Zhang, Y.-M., Gao, J.-M., Bai, M.-S., Yang, S.-X., Laatsch, H., & Zhang, A.-L. (2009). Bioactive metabolites produced by *Chaetomium globosum*, an endophytic fungus isolated from *Ginkgo biloba*. *Bioorganic & Medicinal Chemistry Letters, 19*(6), 1572–1574. https://doi.org/10.1016/j.bmcl.2009.02.025

Qin, X., Xu, J., An, X., Yang, J., Wang, Y., Dou, M., Wang, M., Huang, J., & Fu, Y. (2024). Insight of endophytic fungi promoting the growth and development of woody plants. *Critical Reviews in Biotechnology, 44*(1), 78–99. https://doi.org/10.1080/07388551.2022.2129579

Rai, N., Kumari Keshri, P., Verma, A., Kamble, S. C., Mishra, P., Barik, S., & Gautam, V. (2021). Plant associated fungal endophytes as a source of natural bioactive compounds. *Mycology, 12*(3), 139–159. https://doi.org/10.1080/21501203.2021.1887716

Rani, A., Saini, K. C., Bast, F., Varjani, S., Mehariya, S., Bhatia, S. K., Sharma, N., & Funk, C. (2021). A review on microbial products and their perspective application as antimicrobial agents. *Biomolecules, 11*(12), 1–28. https://doi.org/10.3390/biom11121860

Rashid, T. S. (2021). Bioactive metabolites from tomato endophytic fungi with antibacterial activity against tomato bacterial spot disease. *Rhizosphere, 17*, 1–5. https://doi.org/10.1016/j.rhisph.2020.100292

Rezzani, G. D., Choque, E., Salvay, A. G., Mathieu, F., & Peltzer, M. A. (2022). New antioxidant active packaging films based on yeast cell wall and naphtho-γ-pyrone extract. *Polymers, 14*(10), 1–16. https://doi.org/10.3390/polym14102066

Roat, C., & Saraf, M. (2020). Isolation and characterization of t-resveratrol and α-viniferin, a bioactive secondary metabolite of an endophytic fungus *Aspergillus stellifer* ab4, from *Vitis vinifera*. *The Journal of Microbiology, Biotechnology and Food Sciences, 9*(4), 1–6. https://doi.org/10.15414/jmbfs.2020.9.4.708-713

Rondanelli, M., Opizzi, A., & Monteferrario, F. (2009). The biological activity of beta-glucans. *Minerva Medica, 100*(3), 237–245.

Roy, A., Khan, A., Ahmad, I., Alghamdi, S., Rajab, B. S., Babalghith, A. O., & Islam, M. R. (2022). Flavonoids a bioactive compound from medicinal plants and its therapeutic applications. *BioMed Research International, 2022*(1), 5445291. https://doi.org/10.1155/2022/5445291

Saxena, S., Dufossé, L., Deshmukh, S. K., Chhipa, H., & Gupta, M. K. (2024). Endophytic fungi: A treasure trove of antifungal metabolites. *Microorganisms, 12*(9), 1903. https://doi.org/10.3390/microorganisms12091903

Sebastianes, F. L. S., de Azevedo, J. L., & Lacava, P. T. (2017). Diversity and biotechnological potential of endophytic microorganisms associated with tropical mangrove forests. In *Diversity and benefits of microorganisms from the tropics* (pp. 37–56). https://doi.org/10.1007/978-3-319-55804-9_3

Setiawan, A. (2024). Antibacterial activity of nano chitosan derived from mangrove fungus endophyte, *Fusarium sp.* 20CB07. *AACL Bioflux, 17*(1), 1–15. http://www.bioflux.com.ro/docs/2024.449-463.pdf

Shadyro, O. I., Sosnovskaya, A. A., & Edimecheva, I. P. (2017). Flaxseed oil stabilization using natural and synthetic antioxidants. *European Journal of Lipid Science and Technology, 119*(10), 1–11. https://doi.org/10.1002/ejlt.201700079

Sharaf, M. H., Abdelaziz, A. M., Kalaba, M. H., Radwan, A. A., & Hashem, A. H. (2022). Antimicrobial, antioxidant, cytotoxic activities and phytochemical analysis of fungal endophytes isolated from *ocimum basilicum*. *Applied Biochemistry and Biotechnology*, 1–19. https://doi.org/10.1007/s12010-021-03702-w

Singh, A., Singh, D. K., Kharwar, R. N., White, J. F., & Gond, S. K. (2021). Fungal endophytes as efficient sources of plant-derived bioactive compounds and their prospective applications in natural product drug discovery: Insights, avenues, and challenges. *Microorganisms, 9*(1), 197. https://doi.org/10.3390/microorganisms9010197

Sodhi, G. K., & Saxena, S. (2023a). Plant growth promotion and abiotic stress mitigation in rice using endophytic fungi: Advances made in the last decade. *Environmental and Experimental Botany, 209*, 105312. https://doi.org/10.1016/j.envexpbot.2023.105312

Sodhi, G. K., & Saxena, S. (2023b). Promising drought and salinity tolerance features of Nigrospora species existing as endophytes in Oryza sativa. *3. Biotech, 13*(8), 262. https://doi.org/10.1007/s13205-023-03512-6

Tang, P.-J., Zhang, Z.-H., Niu, L.-L., Gu, C.-B., Zheng, W.-Y., Cui, H.-C., & Yuan, X.-H. (2021). *Fusarium solani* G6, a novel vitexin-producing endophytic fungus: Characterization, yield improvement and osteoblastic proliferation activity. *Biotechnology Letters, 43*, 1371–1383. https://doi.org/10.1007/s10529-021-03118-w

Tao, X., Chen, X., Dong, R., Wang, G., Xu, X., Yu, Q., Chen, Y., Wang, X., & Xie, J. (2024). Characterization and antioxidant properties of three exopolysaccharides produced by the *Cyclocarya paliurus* endophytic fungus. *International Journal of Biological Macromolecules*, 1–32. https://doi.org/10.1016/j.ijbiomac.2024.132110

Tawfike, A. F., Romli, M., Clements, C., Abbott, G., Young, L., Schumacher, M., Diederich, M., Farag, M., & Edrada-Ebel, R. (2019). Isolation of anticancer and anti-trypanosome secondary metabolites from the endophytic fungus *Aspergillus flocculus* via bioactivity guided isolation and MS based metabolomics. *Journal of Chromatography B, 1106*, 71–83. https://doi.org/10.1016/j.jchromb.2018.12.032

Tiwari, P., & Bae, H. (2020). Horizontal gene transfer and endophytes: An implication for the acquisition of novel traits. *Plants, 9*(3), 305. https://doi.org/10.3390/plants9030305

Usman, M., Shah, I. H., Sabir, I. A., Malik, M. S., Rehman, A., Murtaza, G., Azam, M., ur Rahman, S., Rehman, A., & Ashraf, G. A. (2024). Synergistic partnerships of endophytic fungi for bioactive compound production and biotic stress management in medicinal plants. *Plant Stress*, 1–12. https://doi.org/10.1016/j.stress.2024.100425

Vega-Portalatino, E. J., Rosales-Cuentas, M. M., Valdiviezo-Marcelo, J., Arana-Torres, N. M., Espinoza-Espinoza, L. A., Moreno-Quispe, L. A., & Cornelio-Santiago, H. P. (2023). Antimicrobial and production of hydrolytic enzymes potentials of bacteria and fungi associated with macroalgae and their applications: A review. *Frontiers in Marine Science, 10*, 1–15. https://doi.org/10.3389/fmars.2023.1174569

Vetvicka, V. (2011). Glucan-immunostimulant, adjuvant, potential drug. *World Journal of Clinical Oncology, 2*(2), 1–5. https://doi.org/10.5306/wjco.v2.i2.115

Vishwakarma, S., Chaudhry, V., Chand, S., Sagar, K., Gupta, K. K., Bhardwaj, N., Prasad, R., Kumar, P., & Chandra, H. (2024). The potential of fungal endophytes in plants: Sources of bioactive compounds. *Indian Journal of Microbiology*, 1–15. https://doi.org/10.1007/s12088-024-01406-3

Wadhwa, K., Kapoor, N., Kaur, H., Abu-Seer, E. A., Tariq, M., Siddiqui, S., Yadav, V. K., Niazi, P., Kumar, P., & Alghamdi, S. (2024). A comprehensive review of the diversity of fungal secondary metabolites and their emerging applications in healthcare and environment. *Mycobiology*, 1–53. https://doi.org/10.1080/12298093.2024.2416736

Wang, Y., Li, Y., Li, S., Li, Q., Fan, W., Kiatoukosin, L., & Chen, J. (2019). Extracellular polysaccharides of endophytic fungus *Alternaria tenuissima* F1 from *Angelica sinensis*: Production conditions, purification, and antioxidant properties. *International Journal of Biological Macromolecules, 133*, 172–183. https://doi.org/10.1016/j.ijbiomac.2019.03.246

Wang, Z., Zhu, J., Li, W., Li, R., Wang, X., Qiao, H., Sun, Q., & Zhang, H. (2020). Antibacterial mechanism of the polysaccharide produced by *Chaetomium globosum* CGMCC 6882 against *Staphylococcus aureus*. *International Journal of Biological Macromolecules, 159*, 231–235. https://doi.org/10.1016/j.ijbiomac.2020.04.269

Wen, J., Okyere, S. K., Wang, S., Wang, J., Xie, L., Ran, Y., & Hu, Y. (2022). Endophytic fungi: An effective alternative source of plant-derived bioactive compounds for pharmacological studies. *Journal of Fungi, 8*(2), 205. https://doi.org/10.3390/jof8020205

Xu, M., Mo, X., Huang, H., Chen, X., Liu, H., Peng, Z., Chen, L., Rong, S., Yang, W., & Xu, S. (2020). Yeast β-glucan alleviates cognitive deficit by regulating gut microbiota and metabolites in Aβ1–42-induced AD-like mice. *International Journal of Biological Macromolecules, 161*, 258–270. https://doi.org/10.1016/j.ijbiomac.2020.05.180

Yang, M.-H., Lin, H.-J., & Choong, Y.-M. (2002). A rapid gas chromatographic method for direct determination of BHA, BHT and TBHQ in edible oils and fats. *Food Research International, 35*(7), 627–633. https://doi.org/10.1016/S0963-9969(01)00164-8

Yang, D., Hu, C., Wang, X., Shi, G., Li, Y., Fei, Y., & Zhao, X. (2021). Microbes: A potential tool for selenium biofortification. *Metallomics, 13*(10), mfab054. https://doi.org/10.1093/mtomcs/mfab054

Ye, Y., Qu, J., Pu, Y., Rao, S., Xu, F., & Wu, C. (2020). Selenium biofortification of crop food by beneficial microorganisms. *Journal of Fungi, 6*(2), 59. https://doi.org/10.3390/jof6020059

Zeng, Y.-J., Yang, H.-R., Wang, H.-F., Zong, M.-H., & Lou, W.-Y. (2019). Immune enhancement activity of a novel polysaccharide produced by *Dendrobium officinale* endophytic fungus *Fusarium solani* DO7. *Journal of Functional Foods, 53*, 266–275. https://doi.org/10.1016/j.jff.2018.12.038

Zhang, X., Yin, Q., Li, X., Liu, X., Lei, H., & Wu, B. (2022). Structures and bioactivities of secondary metabolites from *Penicillium* genus since 2010. *Fitoterapia, 163*, 1–38. https://doi.org/10.1016/j.fitote.2022.105349

Zheng, J., Shi, J., & Wang, D. (2024). Diversity of soil fungi and entomopathogenic fungi in subtropical mountain forest in Southwest China. *Environmental Microbiology Reports, 16*(4), e13267. https://doi.org/10.1111/1758-2229.13267

Zhou, J., Diao, X., Wang, T., Chen, G., Lin, Q., Yang, X., & Xu, J. (2018). Phylogenetic diversity and antioxidant activities of culturable fungal endophytes associated with the mangrove species *Rhizophora stylosa* and *R. mucronata* in the South China Sea. *PLoS One, 13*(6), 1–18. https://doi.org/10.1371/journal.pone.0197359

Zhu, H.-L. (2011). Resveratrol and its analogues: Promising antitumor agents. *Anti-Cancer Agents in Medicinal Chemistry (Formerly Current Medicinal Chemistry-Anti-Cancer Agents), 11*(5), 479–490. https://doi.org/10.2174/187152011795677427.

Zhu, X., Liu, Q., Wang, M., Liang, M., Yang, X., Xu, X., Zou, H., & Qiu, J. (2011). Activation of Sirt1 by resveratrol inhibits TNF-α induced inflammation in fibroblasts. *PLoS One, 6*(11), 1–7. https://doi.org/10.1371/journal.pone.0027081

Zulfiqar, U., Khokhar, A., Maqsood, M. F., Shahbaz, M., Naz, N., Sara, M., & Ahmad, M. (2024). Genetic biofortification: Advancing crop nutrition to tackle hidden hunger. *Functional & Integrative Genomics, 24*(2), 34. https://doi.org/10.1007/s10142-024-01308-z

Chapter 3
Fungal Secondary Metabolites as Food Preservatives

Nisha Kumari Pandit, Pallavi Sharma, Anee Mohanty, Prangya Ranjan Rout, and Sumer Singh Meena

1 Introduction

Foods are natural substances derived from plants or animals, consumed primarily for nutritional value and to sustain life. They comprise organic and inorganic components, including moisture, proteins, fats, carbohydrates, and minerals, contributing to their quality and health benefits. However, food can also be a source of illness when contaminated (Joardder & Masud, 2019). Foodborne illnesses, often underreported, can range from mild discomfort to severe, life-threatening conditions. These illnesses primarily occur through two mechanisms: infection and intoxication. Infections arise from food contamination by viruses, bacteria, parasites, and moulds, while intoxications result from ingesting toxic substances like botulinum toxin and aflatoxin. Prominent foodborne pathogens, including *Vibrio parahaemolyticus, Listeria monocytogenes,* and *Escherichia coli*, contribute significantly to foodborne diseases worldwide.

Synthetic antimicrobials control these pathogens; however, their effectiveness is increasingly compromised due to antimicrobial resistance development (Copetti, 2019). This alarming trend and the economic burdens of foodborne illnesses underscore the urgent need for sustainable, natural alternatives to ensure food safety and preservation. Effective food preservation is key to increasing the shelf life of foods while retaining their desirable qualities. Food preservation involves strategies and practices designed to regulate the factors, both internal and external, that cause food to deteriorate. The core purpose is to enhance the shelf life while maintaining the food's nutritional properties, visual appeal, consistency, and flavour (Amit et al., 2017). Traditional food preservation methods include thermal processing

N. K. Pandit · P. Sharma · A. Mohanty · P. R. Rout · S. S. Meena (✉)
Department of Biotechnology, Dr B R Ambedkar National Institute of Technology, Jalandhar, Punjab, India
e-mail: meenass@nitj.ac.in

B. P. Singh et al. (eds.), *Fungal Additives and Bioactives in Food Processing Industries*, Fungal Biology, https://doi.org/10.1007/978-3-032-04519-5_3

(pasteurisation and canning), refrigeration, freezing, dehydration, and chemical preservatives. Synthetic chemical preservatives such as nitrates, sulfites, and benzoates help prevent food degradation by inhibiting the activity of microorganisms and biochemical reactions. These are widely used due to their cost-effectiveness and reliability. Non-chemical approaches like refrigeration and freezing slow microbial activity, while drying and vacuum packaging limit the availability of moisture and oxygen essential for microbial growth. More recently, advanced technologies like irradiation, high-pressure processing, and nanotechnology have been adopted. Despite their benefits, conventional methods have several limitations. Synthetic preservatives can pose health risks, such as allergic reactions or carcinogenic effects, and are subject to stringent regulatory scrutiny. Physical methods, while effective, may alter the texture, flavour, or nutritional value of foods. Additionally, the growing consumer preference for clean-label products and natural components highlights the need for alternatives. Antimicrobial resistance and environmental concerns related to synthetic preservatives further compound the issue, necessitating the exploration of natural, sustainable solutions (Joardder & Masud, 2019).

Various natural pigments, including β-carotene, lycopene, lutein, astaxanthin, and other carotenoids commonly used as food additives, are primarily derived from microalgae and plants. However, the yields of plant-based metabolites are largely influenced by seasonal variations. While microalgae offer the advantage of year-round cultivation, their production becomes costly in regions where natural sunlight must be replaced with artificial lighting to ensure consistent output. Alternatively, fungi have been extensively documented as producers of a wide range of food additives, though microbial production demands significant investments in advanced technology, infrastructure, and compliance with regulatory standards (Dufossé et al., 2021). Identifying the most suitable natural additives, such as stability, comes with challenges since natural pigments often degrade under neutral and alkaline pH conditions.

In contrast, synthetic dyes remain highly stable across a broad pH spectrum (De Mejia et al., 2020). Furthermore, ensuring food safety involves maintaining proper sanitation throughout the food supply chain and adhering to microbiological and chemical safety standards. Food products must be free from harmful additives and impurities (Fung et al., 2018). A significant concern in this context is the production of mycotoxins, such as citrinin, during fungal fermentation. Mycotoxins impact the integrity and safety of food, leading to significant losses in the crop production industry as fungi and moulds destroy up to one-third of the crop produced (Avery et al., 2019). Moreover, some fungal metabolites, like 1-octen-3-ol, pose potential health risks. For instance, exposure to 10 mg/m^3 of volatilised 1-octen-3-ol for 2 h has been associated with minor eye, nose, and throat irritation. At lower concentrations, this compound may exhibit neurotoxic and cytotoxic effects. Consequently, even naturally derived substances, such as fungal aromas or metabolites, require thorough safety evaluations to ensure their suitability as food additives. This is essential to balance their functional benefits with consumer health and safety needs (Bojke et al., 2018).

Despite several challenges, there has been an escalating demand for natural products from fungi by both consumers and industries due to their physiochemical properties and functional attributes. This chapter explains the fungal secondary metabolites used in food preservation, such as polyketides, terpenoids, alkaloids, and organic acids. It delves into their biosynthesis, including the pathways and steps involved, and examines their roles in food preservation. It also discusses how these metabolites prevent food spoilage by exhibiting antibacterial, antifungal, antiviral, and antioxidant properties. Their applications as food preservatives are well known, highlighting their roles as organic acids, antimicrobial agents, antioxidants, and natural colourants. Other uses, such as enzymes, coatings, and their role in food fermentation, are also covered. The chapter also addresses the limitations and challenges associated with fungal secondary metabolites, including high production costs, stability issues, and regulatory concerns, which can limit their broader use. By overcoming these challenges, fungal secondary metabolites could be key in advancing natural and sustainable food preservation methods.

2 Fungal Secondary Metabolites

Fungal secondary metabolites are bioactive compounds produced by fungal strains; however, they do not directly affect their growth, development, or reproduction. Most often, these metabolites perform ecological functions, such as protection against competitors, predators, and pathogens, or facilitate the adaptation of the fungus to its environment. However, due to their bioactivity, many fungal secondary metabolites are important in medicine, agriculture, and industry (Keller, 2019). Fungal secondary metabolites exhibit extreme diversity and range in chemical class from alkaloids, terpenoids, and polyketides to non-ribosomal peptides and phenolic compounds, among others, to mycotoxins (Bills & Gloer, 2016). Some have found their way into pharmacy as drugs, agricultural usage as pesticides, food or beverage conservation, and fermentations that find industrial utilisation. This section explores the classification of fungal secondary metabolites and their biosynthesis, providing an understanding of how they are synthesised, regulated, and utilised.

2.1 Classification of Fungal Secondary Metabolites

Fungal secondary metabolites can be classified based on their chemical structure, biosynthetic origin, and biological activity. The major classes (Table 3.1) include terpenoids, alkaloids, polyketides, non-ribosomal peptides, phenolic compounds, and other miscellaneous compounds (Conrado et al., 2022). Each of these classes has unique properties that serve specific ecological functions.

Terpenoids, or isoprenoids, are compounds formed from isoprene units ($C5H8$). They are the largest class of secondary metabolites and include various structures,

Table 3.1 Classification of fungal secondary metabolites

Class	Structure	Examples	Biological activities	References
Terpenoids	Derived from isoprene units	Ergosterol, fumonisin, aflatoxins	Antibiotic, anticancer, and anti-inflammatory properties	Amirzakariya and Shakeri (2022), Chen et al. (2023)
Alkaloids	Nitrogen-containing compounds	Psilocybin, ergot alkaloids, lysergic acid	Psychoactive, antimicrobial, and anticancer properties	Zhang et al. (2012), Zorrilla and Evidente (2022)
Polyketides	Synthesised by the polymerisation of acetyl-CoA or malonyl-CoA units	Aflatoxins, lovastatin, griseofulvin	Antifungal, cholesterol-lowering, and anticancer properties	Santiago et al. (2014)
Non-ribosomal peptides	Synthesised by non-ribosomal peptide synthetases (NRPS)	Penicillin, cyclosporine, griseofulvin	Antibiotic, immunosuppressive, and antifungal properties	Jegorov et al. (2006) and Niu et al. (2020)
Phenolic compounds	Aromatic compounds with a hydroxyl group (-OH)	Resveratrol, stilbenes, 6-methoxyphenol	Antioxidant, anti-inflammatory, and antimicrobial properties	Stasiuk and Kozubek (2010)
Miscellaneous	Various structures	Polysaccharides, volatile organic compounds	Antioxidant, immune-modulating, and anti-cancer activities	Keller et al. (2005)

including monoterpenes, sesquiterpenes, diterpenes, triterpenes, and tetraterpenes. Fungi synthesise terpenoids, such as ergosterol, a major component of fungal cell membranes, exhibiting antifungal activity. It is essential for maintaining the integrity of the membrane. Another example is fumonisin, a mycotoxin produced by *Fusarium* spp., which is toxic to animals and humans. Similarly, aflatoxins from *Aspergillus* species are classified as potent carcinogens and can cause human liver cancer (Amirzakariya & Shakeri, 2022; Chen et al., 2023). Alkaloids are organic compounds that contain nitrogen derived from amino acids. The metabolites of these compounds have different physiological effects and can be toxic, psychoactive, or medicinal. Fungi produce some alkaloids as a defence mechanism against herbivores and microbes. Psilocybin is a naturally occurring hallucinogen that is found in psychedelic mushrooms, and *Claviceps* species produce ergot alkaloids and have been used in medicine since ancient times. However, their overdose can cause poisoning (Meyer & Slot, 2023; Hulvová et al., 2013).

Polyketides are a vast group of secondary metabolites produced by condensing simple precursors like acetyl-CoA or malonyl-CoA. These compounds often undergo various modifications, such as methylation and hydroxylation, to yield complex molecules. These compounds exhibit various biological activities, including antibiotic, antifungal, and anticancer properties. Notable examples include aflatoxins, which are highly toxic and carcinogenic compounds produced by Aspergillus species that often contaminate crops like peanuts and corn, and lovastatin, a cholesterol-lowering drug derived from *Aspergillus terreus* (Awuchi et al., 2020.; Patil et al., 2011). Similarly, non-ribosomal peptides, synthesised by non-ribosomal peptide synthetases (NRPS), bypass the conventional ribosomal protein biosynthesis pathway. These peptides can have many biological activities, including antibacterial and antifungal.

Penicillin, an antibiotic produced by *Penicillium* species, inhibits bacterial cell wall synthesis. Similarly, griseofulvin, another antifungal derived from *Penicillium griseofulvum,* prevents fungal skin infections (Jumiati et al., 2021). Phenolic compounds are distinguished by the aromatic ring to which one or more hydroxyl groups are attached. Such compounds often possess antioxidant, antimicrobial, and anti-inflammatory activities and are generally produced in fungi as part of a stress or competitive response. Examples include resveratrol, an antioxidant with potential anticancer properties, produced by fungi such as *Botrytis cinerea*, and 6-methoxyphenol, a volatile compound known for its antimicrobial activity (Jumiati et al., 2021).

2.2 Biosynthesis of Secondary Metabolites in Fungi

Fungal secondary metabolites are synthesised through a chain of enzymatic pathways that are strictly regulated. These pathways are usually under the control of the gene clusters, and the chain of these pathways can be activated by internal and external factors, including nutrient availability, environmental stress, and competition with other organisms (Keller, 2019). Secondary metabolite biosynthesis involves multiple pathways tailored to the specific class of compounds being produced. In terpenoid biosynthesis, the precursors are isoprenoids (isopentenyl pyrophosphate (IPP) and dimethylallyl pyrophosphate (DMAPP)) generated via either the mevalonate (MVA) pathway or the non-mevalonate (MEP) pathway. These compounds originate from amino acids such as tryptophan, tyrosine, and phenylalanine for alkaloid biosynthesis. The biosynthetic pathway involves several decarboxylation, methylation, and oxidation steps mediated by various enzymes. In the case of polyketide biosynthesis, polyketides are formed through the stepwise condensation of malonyl-CoA and acetyl-CoA units, a process catalysed by type I or type II polyketide synthases (PKS). These PKS enzymes are large, multi-domain complexes that facilitate the assembly of complex polyketide structures. Similarly, non-ribosomal peptide biosynthesis involves adding amino acid precursors sequentially, catalysed by non-ribosomal peptide synthetases (NRPS). Although the NRPS enzyme complex shares similarities with the machinery of polyketide synthesis, its modular units are distinct, carrying different amino acids for peptide assembly (Jan et al., 2021; Reshi et al., 2023; Zhan et al., 2022).

The biosynthesis of secondary metabolites in fungi is a highly regulated, multi-step process involving various enzymes, pathways, and precursors. A key aspect of this process is the organisation and activation of biosynthetic gene clusters (BGCs). These clusters, which are typically large, encode the enzymes necessary for synthesising secondary metabolites. Activation of these clusters often occurs in response to specific stress conditions or during particular growth phases under which they are not typically expressed. For instance, polyketides are synthesised by gene clusters encoding polyketide synthases (PKSs), while non-ribosomal peptides are produced by clusters containing genes for non-ribosomal peptide synthetases (NRPSs) (Guerriero et al., 2018; Polturak & Osbourn, 2021). The biosynthesis of secondary metabolites begins with the formation of essential building blocks derived from precursor molecules. These precursors are typically sourced from central metabolic pathways such as glycolysis, the tricarboxylic acid (TCA) cycle, or amino acid metabolism. For example, isoprenoids are synthesised from acetyl-CoA through the mevalonate or methylerythritol phosphate (MEP) pathway, while amino acids serve as precursors for non-ribosomal peptides. Subsequently, the process involves modular enzyme synthesis, where large, multifunctional enzyme complexes play a crucial role in producing secondary metabolites. Notable examples include polyketide synthases (PKSs) and non-ribosomal peptide synthetases (NRPSs). PKS enzymes facilitate the condensation of acetyl-CoA and other precursors to form polyketides, whereas NRPS enzymes catalyse the assembly of amino acids into peptides without the involvement of ribosomes (Polturak & Osbourn, 2021).

The biosynthesis of secondary metabolites is tightly controlled through transcriptional and post-transcriptional mechanisms that respond to environmental and physiological signals such as nutrient availability, temperature, and stress factors. Transcription factors like BrlA in *Aspergillus* and LaeA in other fungi play key roles in activating or repressing secondary metabolite pathways. Additionally, epigenetic modifications and chromatin remodelling regulate the expression of biosynthetic gene clusters. Environmental factors such as nutrient starvation (carbon or nitrogen), oxidative stress, UV exposure, and cellular differentiation trigger secondary metabolite production, often providing survival or defensive advantages under adverse conditions. Following these regulatory mechanisms, metabolite biosynthesis involves transport and export processes. Specialised transporters ensure the efficient export of secondary metabolites to the extracellular environment and prevent intracellular accumulation and potential toxicity.

Furthermore, these metabolites are often modified through tailoring enzymes such as methyltransferases, glycosyltransferases, and hydrolases, which enhance their biological activity, solubility, or stability, ultimately contributing to their functional diversity (Khare et al., 2020; Patra et al., 2013). The final step in the biosynthesis of secondary metabolites involves post-synthetic modifications, where the initially formed precursor structures undergo additional enzymatic transformations. These modifications, such as methylation, oxidation, reduction, or glycosylation, significantly enhance the secondary metabolites' biological activity, stability, or solubility. These enzymatic reactions play a crucial role in imparting the metabolites with their unique and often potent biological properties, including toxic or defensive characteristics, which contribute to their ecological and functional roles (Kumar & Kumar, 2019).

The production of secondary metabolites in fungi is regulated at the molecular level through genetic and epigenetic mechanisms. Specific gene clusters or loci involved in secondary metabolite biosynthesis are controlled by global regulators such as LaeA, a key regulator of fungal secondary metabolism. External factors, including nutrient availability, light, and temperature, play critical roles in inducing or repressing the expression of these gene clusters. Environmental signals like nitrogen and carbon availability, pH, and oxygen levels also significantly influence secondary metabolite production. For example, nutrient scarcity often triggers fungi to upregulate secondary metabolite production, enhancing their competitiveness and defence mechanisms against predators. Fungi signal transduction pathways are sophisticated networks that regulate gene expression for secondary metabolite biosynthesis. These pathways are broadly categorised into the nitrogen-activated protein kinase (MAPK) and cyclic AMP (cAMP) pathways. Additionally, other stress-related pathways contribute to fine tuning the regulation, ensuring a coordinated response to environmental and physiological changes (Khare et al., 2020; Patra et al., 2013; Shwab & Keller, 2008).

Fungal secondary metabolites serve a wide range of ecological and biological functions. Many of these compounds, such as mycotoxins, play critical roles in defence and protecting fungi from predators, herbivores, and microbial competitors. Other exhibit antibiotic properties, with notable examples like penicillin and griseofulvin, which inhibit the growth of competing microorganisms. In addition to their antimicrobial functions, certain secondary metabolites function as signalling molecules, facilitating communication within fungal communities or between fungi and other organisms. These metabolites are also vital for helping fungi adapt to environmental stresses, with antioxidants enhancing tolerance to oxidative stress and other compounds offering protection against UV radiation and desiccation (Macías-Rubalcava & Sánchez-Fernández, 2017; Vining, 2007).

3 Mechanism of Action of Fungal Secondary Metabolites

Fungal secondary metabolites are recognised for their significant potential in pharmaceutical, agricultural, and industrial applications due to their remarkable bioactive properties. These compounds have evolved in fungi primarily as part of their defense mechanisms and possess antibacterial, antifungal, antiviral, antioxidant, and anticancer activities. The mechanisms by which they exert their effects may vary greatly depending on the class of secondary metabolites. However, their combined action typically involves inhibiting microbial growth, suppressing oxidative damage, and interfering with the biological processes of pathogens (Bennett & Wallsgrove, 1994; Macheleidt et al., 2016). The antibacterial, antifungal, antiviral, and antioxidant properties of fungal secondary metabolites, along with their mechanisms of action, especially in inhibiting microbial growth, will be explored in this section.

3.1 Antibacterial and Antifungal Properties of Fungal Secondary Metabolites

Fungi have evolved to produce various secondary metabolites with potent antimicrobial properties. These metabolites are used to suppress the growth of competing microorganisms, including bacteria and other fungi. Many of these secondary metabolites have been developed into therapeutics, such as antibiotics and antifungal drugs. Fungal secondary metabolites are known to carry out antibacterial properties, which involve inhibiting bacterial growth by targeting specific cellular structures or processes. The antibacterial activities of most fungal metabolites have been attributed to the inhibition of cell wall biosynthesis and protein synthesis coupled with DNA replication in bacteria. Penicillin, derived from *Penicillium* species, is a classic example. It works by binding to the enzyme transpeptidase, which plays a crucial role in crosslinking peptidoglycan chains, an essential part of the bacterial cell wall. By inhibiting this process, penicillin weakens the cell wall, leading to cell lysis and bacterial death; other fungal metabolites may target different aspects of bacterial physiology, including inhibiting protein synthesis or DNA replication, further contributing to their antibacterial properties (Park & Strominger, 1957). Griseofulvin, derived from the *Penicillium griseofulvum* fungus, is recognised for its potent antifungal properties and exhibits antibacterial effects. It hinders the division of fungal cells by interfering with the formation of microtubules, which holds significance for particular bacteria that depend on microtubules for their intracellular functions (Huber & Gottlieb, 1968). Fumonisin, produced by various *Fusarium* species, is a mycotoxin that interferes with sphingolipid metabolism in eukaryotic cells and can similarly impact bacteria by compromising their membrane integrity (Stockmann-Juvala & Savolainen, 2008).

Most of the fungal secondary metabolites are synthesised to inhibit the growth and spread of other fungi. The compounds mainly inhibit cell wall biosynthesis, compromise cell membrane integrity, or inhibit protein synthesis in pathogenic fungi. Echinocandins, for example, are antifungal secondary metabolites produced by *Glarea lozoyensis*. These lipopeptides inhibit the synthesis of 1,3-β-D-glucan, a key component of the fungal cell wall, which leads to cell wall disruption and the death of the fungal cells (Szymański et al., 2022). Similarly, besides its antibacterial activity, griseofulvin is an effective antifungal agent. It inhibits fungal cell division by disrupting the mitotic spindle, a crucial structure for fungal reproduction, particularly in dermatophytes (Huber & Gottlieb, 1968). Trichothecenes are potent inhibitors of protein synthesis in fungi produced by *Fusarium* species. They bind to the ribosomal subunits, preventing the elongation of polypeptides and leading to the death of the fungal cells (Ueno, 1978). Table 3.2 shows the antibacterial and antifungal mechanisms of various fungal secondary metabolites.

Table 3.2 Antibacterial and antifungal mechanisms of fungal secondary metabolites

Compound	Produced by fungi	Target pathgen	Mechanism of action	References
Penicillin	*Penicillium* sp.	Gram-positive bacteria	Inhibition of cell wall synthesis by binding to transpeptidase	Park and Strominger (1957)
Griseofulvin	*Penicillium griseofulvum*	Bacteria, fungi	Disruption of microtubule function, inhibition of cell division	Huber and Gottlieb (1968)
Fumonisin	*Fusarium* sp.	Bacteria, fungi	Disruption of sphingolipid metabolism, destabilisation of cell membranes	Stockmann-Juvala and Savolainen (2008)
Echinocandins	*Glarea lozoyensis*	Fungi	Inhibition of protein synthesis by binding to ribosomal subunits	Ueno (1978)

3.2 Antiviral and Antioxidant Properties of Fungal Secondary Metabolites

Fungal secondary metabolites are primarily recognised for their antibacterial and antifungal activities, but some also possess antiviral properties. These metabolites disrupt the viral replication cycle by targeting different stages, such as attachment, penetration, replication, or the assembly of new viral particles. A notable example is cordycepin, derived from *Cordyceps militaris.* Cordycepin inhibits viral RNA synthesis by incorporating it into the viral RNA, thereby preventing elongation. This compound has been shown to inhibit viruses such as the influenza virus and hepatitis C virus (Guarino, 1967; Radhi et al., 2021). Lentinan, a compound found in the shiitake mushroom *Lentinula edodes*, has shown promising antiviral activity, particularly against herpes simplex virus (HSV). It boosts immune responses that inhibit viral replication and has been studied as a potential antiviral treatment for HIV (Ren et al., 2018; Rincão et al., 2012). Griseofulvin, primarily recognised for its antifungal properties, also acts as an antiviral agent. It interferes with the viral replication process and disrupts the entry of viruses into host cells (Aris et al., 2022; Jin et al., 2008).

Fungal secondary metabolites are highly effective in their antioxidant activity and can neutralise reactive oxygen species (ROS) and free radicals, thereby reducing oxidative stress. Oxidative stress is linked to the development of ageing, inflammation, and various diseases, including cancer and neurodegenerative disorders. An example of an antioxidant fungal secondary metabolite is resveratrol, a polyphenolic compound found in fungi such as *Botrytis cinerea.* Resveratrol scavenges free radicals and reduces oxidative damage, making it invaluable in protecting cells from oxidative stress and inflammation (de la Lastra & Villegas, 2007). Ganoderic acids, derived from the medicinal fungus *Ganoderma lucidum*, exhibit potent antioxidant properties that protect cells from oxidative damage. In addition, these compounds demonstrate modulatory effects on the immune system (Zhu et al., 2019). On the

Table 3.3 Antiviral and antioxidant mechanisms of fungal secondary metabolites

Compound	Produced by fungi	Target pathogen/ condition	Mechanism of action	References
Cordyecepin	*Cordyceps militaris*	Viruses (Influenza, Hepatitis C)	Inhibition of RNA synthesis, interference with viral replication	Guarino (1967) and Radhi et al. (2021)
Lentinan	*Lentinula edodes*	Herpes simplex virus (HSV)	Immune modulation, inhibition of viral replication	Ren et al. (2018) and Rincão et al. (2012)
Griseofulvin	*Penicillium griseofulvum*	Various viruses	Disruption of viral replication and entry into host cells	Aris et al. (2022) and Jin et al. (2008)
Resveratrol	*Botrytis cinerea*	Oxidative stress, inflammation	Scavenging of free radicals, reduction of oxidative damage	de la Lastra and Villegas (2007)
Ganoderic acids	*Ganoderma lucidum*	Oxidative stress	Antioxidant activity, reduction of oxidative damage	Zhu et al. (2019)
Psilocybin	*Psilocybe* sp.	Oxidative stress, inflammation	Scavenging of free radicals, protection against oxidative damage	Nkadimeng et al. (2020)

other hand, despite its psychoactive effect, psilocybin, produced by the *Psilocybe* mushroom, has been linked to antioxidant effects and could serve as an intriguing treatment for oxidative stress (Nkadimeng et al., 2020). Table 3.3 shows the antiviral and antioxidant mechanisms of various fungal secondary metabolites.

4 Application of Fungal Secondary Metabolites in Food Preservation

4.1 Organic Acids as Fungal Secondary Metabolites in Food Preservation

Organic acids are perhaps one of the most commonly utilised secondary metabolites in fungi as food preservatives. Natural production through various fungi fermentation produces these organic acids, such as *Aspergillus*, *Lactobacillus*, and *Penicillium*. The organic acids help decrease food pH, creating a hostile environment to prevent the development of spoilage bacteria, yeast, and mould growth (Anyasi et al., 2017). Citric acid is one of the most widely used organic acids and is produced mainly by *Aspergillus niger* through fermentation. It plays an important role in the food industry as an acidulant, preservative, and flavour enhancer. Citric acid extends the shelf life of food products by lowering their pH, thereby inhibiting the growth of spoilage microorganisms and pathogens. It is commonly used to preserve canned fruits, vegetables, and juices by maintaining acidity and preventing microbial spoilage.

Additionally, it is utilised in producing soft drinks and fruit juices to adjust pH levels and enhance flavour. Citric acid is also used to produce cheese and yoghurt, contributing to the fermentation process and helping preserve these dairy items. Citric acid lowers the pH of food. It is inhibitory to most bacteria and moulds because most require neutral pH for optimum growth, which is the basic mechanism of action; citric acid even chelates metal ions. Thus, the growth of microorganisms that rely on these ions for cellular processes is inhibited (Anyasi et al., 2017; Książek, 2023; Redondo-Blanco et al., 2020).

Lactic acid is another key organic acid produced through fermentation by fungi. Some species of *Lactobacillus* produce lactic acid primarily. It serves an important function in food products, preventing microbial spoilage, especially at a lower pH, and prevents harmful microbes from growing. Lactic acid is part of the processing for yoghurt, cheese, and pickles, even at the industrial scale. It is also used in meat products to reduce the pH and inhibit the growth of pathogens such as *Salmonella* and *Escherichia coli* to reduce spoilage. Lactic acid fermentation preserves vegetables, such as cabbage for sauerkraut and cucumbers for pickles, as it creates an acidic pH for the product and, therefore, is not conducive to the growth of many bacteria, moulds, and yeasts. It also presents antimicrobial action by preventing spoilage microorganisms from growing, thereby increasing the shelf life of perishable food (Rathod et al., 2021; Zapaśnik et al., 2022).

4.2 Natural Preservatives

Fungi are a significant source of diverse biological and chemical compounds, many of which have applications in food preservation. Secondary metabolites synthesised by filamentous fungi are particularly valuable, providing compounds such as organic acids, antimicrobial agents, and enzymes that enhance food safety and shelf life (Hyde et al., 2019). For instance, some fungi are widely utilised to convert multiple carbon sources into organic acids like itaconic, lactic, fumaric, malic, tartaric, succinic, oxalic, and citric acids, which are widely used in food systems to inhibit spoilage microorganisms and pathogens. Among these, citric acid stands out as the most versatile for food applications, with more than 95% of its global production achieved through the fermentation of glucose or sucrose by *Aspergillus niger*, and additional contributions from fungi like *Aspergillus wentii* and *Penicillium citrinum* (Copetti, 2019). Citric acid acts as an acidulant and pH regulator, creating an environment that suppresses microbial growth while its chelating properties prevent oxidative spoilage. It also serves as a stabiliser, antioxidant, and flavour enhancer, preserving food quality and sensory attributes during storage (Sahasrabudhe & Sankpal, 2001).

Another significant fungal secondary metabolite is kojic acid, predominantly produced by *Aspergillus oryzae*. Kojic acid has received much attention in the food industry as a precursor for flavouring agents such as maltol and ethyl maltol and as an antioxidant to prolong the shelf life of fruits and vegetables. It also acts as a colour preservative for seafood and meats. Its applications include forming composite films for antimicrobial packaging that aid in increasing the shelf life of stored foods (Liu et al., 2020). Additionally, the azaphilone pigments, including monascin and ankaflavin, derived from fungi of the *Monascus* genus, have been used for centuries as food colours, spices, and preservatives. By forming complexes with proteins and lipid surfaces, these yellow-pigmented molecules enhance the visual appeal and provide substantial food preservation benefits through their biological activities. They can also effectively preserve food quality and possess health-promoting properties such as prevention or reduced risk posed by high lipid levels, diabetes, inflammation, and fatty liver diseases (Lin & Xu, 2020).

Furthermore, extracellular enzymes from the lignocellulosic-degrading and polysaccharide-secreting fungi, including lignin-degrading peroxidases, multicopper oxidases, and glycoside hydrolases, are incredibly vital in degrading complex substrates into simpler compounds. These enzymes, which can be isolated from white-and brown-rot fungi, can create natural biocides or improve the effectiveness of the food processing cycle, thus increasing shelf-life and decreasing food spoilage. Combining all these diverse functionalities, fungal secondary metabolites are the environmentally friendly, effective solutions to modern food preservation needs that meet the growing demand for safer and longer shelf-life foods (Jain et al., 2020; Saini et al., 2020).

4.3 Antibacterial

Fungal secondary metabolites have been extensively studied to be used as natural antimicrobial agents in food preservation. These metabolites, such as alkaloids, terpenoids, phenolic compounds, and flavonoids, exhibit antimicrobial, antibacterial, and antifungal characteristics, making them highly effective in controlling microbial spoilage in food. Fungal metabolites may function by reducing the growth rate of pathogenic microorganisms, thereby increasing the shelf life of food (Devi et al., 2020). Fungi such as *Alternaria brassicicola, Curvularia* species, and *Penicillium aurantiogriseum* have been reported to suppress the growth of several foodborne pathogens, for instance, *E.coli and Listeria monocytogenes.* In addition, *Curvularia* species metabolite, ergosterol, and altechromone A from *E.coli* provide microbial and antioxidant characteristics, making them more effective in preserving food quality (Gu, 2009). These fungal metabolites have antimicrobial properties against spoilage and pathogenic bacteria, viruses, and moulds that cause foodborne illness. These natural antimicrobial agents present an alternative to synthetic chemicals, often associated with resistance issues. With growing resistance to microbial resistance, fungal metabolites provide a solution to safer methods of food preservation (Kaaniche et al., 2019). Recent studies on biologically derived antimicrobial peptides from fungal metabolites are promising because of the diverse microorganisms present and yet to be exploited due to their structural and functional characteristics.

Compounds such as polyketides and sulfonyl metabolites derived from fungi like *Penicillium* and *Exserohilum rostratum* effectively inhibit the growth of bacteria, fungi, and yeasts that cause food spoilage and contamination. For example, arvoredol, produced by *Penicillium* sp. F37, prevents biofilm formation by *Staphylococcus epidermidis,* which is particularly beneficial in maintaining hygiene in food processing environments (Scopel et al., 2017). Sulfonyl metabolites like pensulfonamide from *P.aculeatum* have demonstrated activity against *E.coli* and *C. albicans* (Hawas et al., 2022), while oxalicin C and penicierythritol A from *P.chrysogenum* target phytopathogens like *Ralstonia solanacearum,* offering the potential for protecting plant-based food systems (Xu et al., 2020). Additionally, Andrastin derivatives from *P.vulpinum* exhibit antimicrobial effects against common foodborne pathogens such as *Bacillus megaterium, Enterobacter aerogenes,* and *E. coli* (Qin et al., 2020). Metabolites like monocerin and ravenelin from *Exserohilum rostratum,* isolated from Amazonian plants, also show moderate activity against *Salmonella typhimurium* and *B. subtilis* (Pina et al., 2021). Sesquiterpenoids emericellins A and B, derived from *Emericells* sp. associated with *Panax notoginseng,* are effective against *Bacillus cereus* and *Salmonella typhi,* further emphasising their antibacterial potential (Pang et al., 2018). Additionally, alterperylenol and stemphyperylenol from *Alternaria* sp. have been shown to inhibit phytopathogenic bacteria such as *Clavibacter michiganensis,* reducing plant-based food spoilage (Zhao et al., 2018). These studies demonstrate the promise of fungal secondary metabolites as natural, effective, and sustainable antibacterial agents for food preservation.

4.4 Antifungal

Antifungal activity is crucial in preserving food by preventing fungal contamination spoilage and extending shelf life. Fungal secondary metabolites, due to their bioactivity, serve as natural antifungal agents, offering a sustainable and efficient solution to food preservation challenges. These metabolites inhibit or destroy fungal growth, safeguarding food quality and safety. Several secondary metabolites have demonstrated potent antifungal activity. Benzomalvin C, isolated from *Penicillium raistrick*, effectively inhibits fungal growth (Stierle & Stierle, 2000). Similarly, tenuazonic acid, altertoxin, alternariol, and alternariol monomethyl ether from *Alternaria* sp. exhibit antifungal and cytotoxic effects, making them valuable in controlling fungal contamination in food systems (Qin et al., 2009). Herbarin-A, extracted from *Candida albicans*, is another example of an antifungal metabolite that can combat foodborne fungi (Gu, 2009). Indole alkaloids from *Aspergillus flavus* and *Fusarium solani* have also shown remarkable antifungal efficacy, further underlining the potential of fungal secondary metabolites in food preservation (Abo-Kadoum et al., 2013). Flavonoids extracted from *Helminthosporium* sp. and β-1,3-glucanolytic compounds from *Fusarium solani* offer additional avenues for natural antifungal agents in food safety applications (Ammar et al., 2013; Awad et al., 2018). Advanced investigations have also revealed the antifungal potential of sesterterpenoids from *Aspergillus insuetus* and naphthoquinones from *Colletotrichum acutatum* and *E. gomezpompae*. These metabolites exhibit dual functionalities, including antifungal and anticancer properties, highlighting their diverse applications (Ammar et al., 2013; Kildgaard et al., 2014; Macías-Rubalcava & Sánchez-Fernández, 2017).

Metabolites such as emericelactones A-D from *Emericella* sp. have demonstrated potent antifungal activity against phytopathogens like *Verticillium dahlia* and fungi such as *Helminthosporium maydis* (Pang et al., 2018). Additionally, metabolites from *Fusarium oxysporum* CFE177 and *Aspergillus oryzae* CFE108 show antifungal activity against clinical pathogens like *Candida* and *Cryptococcus,* which contribute to food spoilage (Santos et al., 2012). Irpexlacte derivatives from *Irpex lacteus* offer antifungal and antioxidant properties, explicitly targeting oxidative stress and fungal pathogens such as *Pseudomonas aeruginosa* (Duan et al., 2019). Other promising antifungal agents include alterperylenol and stemphyperylenol from *Alternaria* sp., which inhibit fungal pathogens like *Alternaria brassicicola* and mitigate spoilage of plant-based foods (Zhao et al., 2018). Furthermore, sulfonyl metabolites like pensulfonamide display antifungal effects on *Candida albicans,* showcasing their dual antimicrobial and antifungal potential (Hawas et al., 2022). Fungal secondary metabolites are integral to developing natural antifungal treatments, ensuring the longevity and safety of food products. These compounds enhance food quality and reduce reliance on synthetic preservatives by mitigating fungal spoilage and contamination, contributing to sustainable food preservation strategies.

4.5 Antioxidants

Fungal secondary metabolites serve as potent antioxidants, critical in preserving food by preventing oxidative damage, leading to spoilage, rancidity, and discolouration. Berovic and Podgornik studied the submerged culture of *Pleurotus ostreatus* and identified key secondary metabolites such as oleic acid, palmitic acid, linoleic acid, stearic acid, and their methyl esters (Papaspyridi et al., 2012). These compounds, along with some other bioactive metabolites present in other species of the genus *Pleurotus*, such as polysaccharides FII, eryngiolode A, β-glucan, ubiquinone-9, and terpenoids like pleuroton A and B, show high antioxidant potential (Wisbeck et al., 2017). Other examples of secondary metabolites are phenolic compounds that are released through the shikimate pathway among some fungi and plants. This group includes folic acid and salicylic acid, known for their antioxidant properties and role in free radical scavenging. Phenolic compounds, including phenylpropanoids, significantly reduce oxidative stress in food systems (Singh et al., 2021).

Gallic acid attracts significant attention among phenolics due to its increased antioxidant capability and versatile use. It has been observed in the food industry that gallic acid enhances tannase production under submerged fermentation. It is important for the instantaneous tea and a clarifying substance in wine, fruit juices, and other beverages (Belmares et al., 2004). Further, gallic acid also plays a role in the preservation of foods by increasing product stability and preventing oxidative damage, thereby enhancing the shelf life of food. Propyl gallate, a gallic acid derivative, is widely used as an antioxidant for pharmaceutical products and the food industry. Its application in functional foods extends beyond preservation, as it also offers health benefits such as antidiabetic, anti-inflammatory, anticoagulant, and antitumor effects (Behdarvand-Margha et al., 2021).

Kojic acid, a major fungal secondary metabolite primarily found in *Aspergillus oryzae,* is involved in food preservation due to its antioxidative activities. It acts as a precursor for flavour enhancers like maltol and ethyl maltol, is an antioxidant in fruits and vegetables, preserves the colour of seafood and meats, and is utilised to manufacture antimicrobial packaging materials (Liu et al., 2020). Similarly, citric acid, a secondary metabolite produced primarily by *Aspergillus niger*, is a versatile antioxidant, with about 70% of its global production used in the food industry. It functions as an acidulant, chelating agent, stabiliser, pH regulator, and preservative, effectively delaying spoilage, preventing oxidative damage, and maintaining the freshness and quality of food products (Sahasrabudhe & Sankpal, 2001).

Additionally, carotenoids, a class of fungal secondary metabolites, are gaining popularity for their incorporation into functional foods due to their extensive antioxidant benefits. These metabolites are used in various bakery products, and carotenoid-rich seafood, such as the seaweed *Undaria pinnatifida* (wakame), is incorporated into pasta to enhance bio-functional and nutritional values (Prabhasankar et al., 2009). Nutraceutical formulations containing carotenoids focus on their antioxidant properties, with β-carotene being a prominent example. Enrichment with β-carotene has been shown to decrease the quality degradation of extra virgin olive oil exposed to microwave heating and UV light while enhancing its shelf life and nutritional value (Murillo-Cruz et al., 2021).

Quinone and azaphilones, two notable classes of polyketide-derived natural pigments produced by fungi, offer significant potential in the food industry. These compounds are natural additives for colour, flavour, and fragrance, while their antioxidant and antimicrobial properties help extend shelf life by preventing microbial spoilage. The diverse applications of fungal secondary metabolites emphasise their pivotal role in contemporary food preservation approaches (Dulo et al., 2021). Furthermore, L-carnitine reduces oxidative stress, showcasing the versatility of fungal secondary metabolites in food preservation (Rousta et al., 2021). The potent antioxidant properties of these metabolites enhance food quality and shelf life while contributing significantly to the development of sustainable and health-promoting food preservation techniques.

4.6 Packaging and Coatings

Active packaging is an advanced food packaging technique that incorporates multifunctional systems intended to enhance the shelf life of food products. Secondary metabolites produced by fungi have numerous bioactivities that can effectively create enhanced food coatings and packaging materials. These metabolites have potent antifungal, antimicrobial, and antioxidant properties that enhance the quality and safety of foods. They are increasingly recognised for their potential to develop advanced packaging and coating technologies aimed at food preservation. The fungal enzymatic capacity of breaking down polymeric materials is utilised to develop bagasse, paper, and other sustainable materials for packaging. These materials act as effective barriers against environmental factors like moisture and oxygen, major contributors to food spoilage (Prasad & Kochhar, 2014). In addition, food safety and shelf life are improved by the presence of β-glucans and other fungal polysaccharides, which contribute to the structural integrity of these materials. Fungal-based edible coatings could be applied directly on the food surface, thus providing a strong barrier against microbes and maintaining the freshness of food. Active packaging involves fungal metabolites with antimicrobial activity, assuring safe food consumption against foodborne pathogens such as *Listeria monocytogenes* and *E.coli.* For example, fungal species such as *Aspergillus niger* and *Aspergillus oryzae* are used in industries to produce enzymes like glucoamylase, which help maintain food quality during storage by breaking down complex sugars and preventing spoilage (Archer, 2000; Copetti, 2019). Figure 3.1 shows the application of fungal secondary metabolites in active packaging and coating.

In addition to enhancing food preservation, these fungal-based metabolites are biodegradable, reducing the environmental impact associated with conventional synthetic packaging. Advancements in biotechnology have further streamlined the production of fungal metabolites, enabling large-scale application and customisation of coatings and packaging to meet specific industry needs. These innovations highlight the dual benefits of fungal secondary metabolites in ensuring food safety and sustainability while addressing the increasing demand for environmentally friendly solutions.

4.7 Enzymes in Food Preservation

Fungal enzymes are widely used as food additives because they help to improve and maintain quality, texture, and shelf life. These enzymes, produced as secondary metabolites by fungi, can be employed to preserve food, enhance its nutritional value, and protect it from microbial spoilage. Amylases, cellulases, proteases, lipases, and pectinases obtained from fungi have been widely used in the food industry for processing, preservation, and other related uses. Fungi such as *Aspergillus, Penicillium,* and *Rhizopus* are capable of secreting enzymes that cause

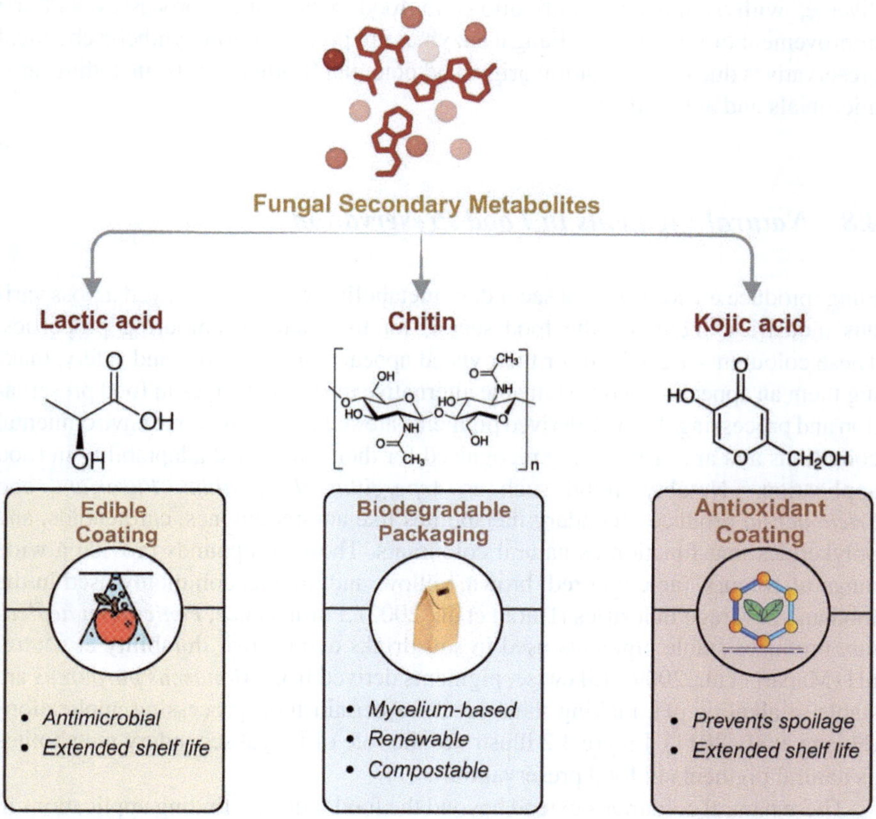

Fig. 3.1 Applications of fungal secondary metabolites in active packaging and coating: lactic acid for antimicrobial edible coatings, chitin for renewable, biodegradable packaging, and kojic acid for antioxidant coatings to prevent spoilage and extend shelf life

the breakdown of starches, celluloses, and proteins. For instance, amylases derived from *Aspergillus* species are commonly used to break down starch into simpler sugars, which are crucial for fermentation and the production of food ingredients like glucose and alcohol. These enzymes also play a role in improving the texture and taste of various food products (Carlsen & Nielsen, 2001).

Besides enhancing the enzymatic reactions, the fungal enzymes are involved in the regulation of microbial load in food. Fungi like *Trichoderma* and *Aspergillus* species produce proteases and lipases, which break down proteins and fats in food, thereby minimising the risk of spoilage and rancidity (Durán et al., 2002). They enhance the food preservation process and also help reduce pathogenic microbes that may pose health risks when consumed. Further, fungal enzymes are used in the bioconversion of agricultural by-products, converting lignocellulosic biomass to biofuel, thus contributing to waste reduction and sustainability (Harnpicharnchai et al., 2009). Thus, the utilisation of these fungal enzymes in food enzymes is diverse, with significant contributions to food preservation processes and the improvement of food quality. Fungal enzymes are preferred over synthetic chemical preservatives due to their natural origin and potential health benefits, including antimicrobials and antioxidants.

4.8 Natural Pigments in Food Preservation

Fungi produce a wide range of secondary metabolites that are employed across various industries, including the food sector, for their natural colouring properties. These colourants are valued for their visual appeal, natural origin, and safety, making them an appealing and sustainable alternative to synthetic dyes in food preservation and processing. Fungal-derived pigments are stable under diverse environmental conditions and are increasingly recognised for their safety and adaptability in food applications. Notably, fungi such as *Aspergillus, Penicillium, Fusarium,* and *Trichoderma* produce secondary metabolites like anthraquinones, carotenoids, and polyketides that function as natural colourants. These compounds provide a wide range of colours, including red, brown, yellow, and orange, commonly used in the food and beverage industries (Durán et al., 2002). For instance, *Penicillium aculeatum* produces stable pigments used in soft drinks due to their durability at neutral pH (Mapari et al., 2009). Likewise, pigments derived from *Monascus purpureus* are stable at alkaline pH, making them ideal for certain food processing applications (Huang et al., 2011). Figure 3.2 illustrates the role of fungal secondary metabolites as natural pigments in food preservation.

These natural colourants extend beyond the food industry, finding applications in cosmetics, pharmaceuticals, and textiles. The stability of fungal pigments under various processing conditions, including heat, light, and pH, makes them ideal for essential colour retention products. Furthermore, their safety and biodegradability enhance their appeal in food preservation, as they are non-toxic and environmentally friendly alternatives to synthetic dyes. Notably, fungi such as *Drechslera* species produce hydroxythraquinones, which serve as colourants offering vibrant red, bronze, and maroon hues, meeting the demands of various industrial applications (Margalith, 1993). These pigments are especially advantageous in food preservation, as they function as colourants and possess antioxidant properties that can enhance the shelf life of food products.

In conclusion, fungal secondary metabolites present a sustainable and safe alternative to synthetic colourants in the food industry. Their diverse colour offerings, stability, and antioxidant benefits make them a valuable asset for food preservation. The increasing demand for natural products in food processing underscores the need for continued exploration and utilisation of fungal-derived colourants in the food sector.

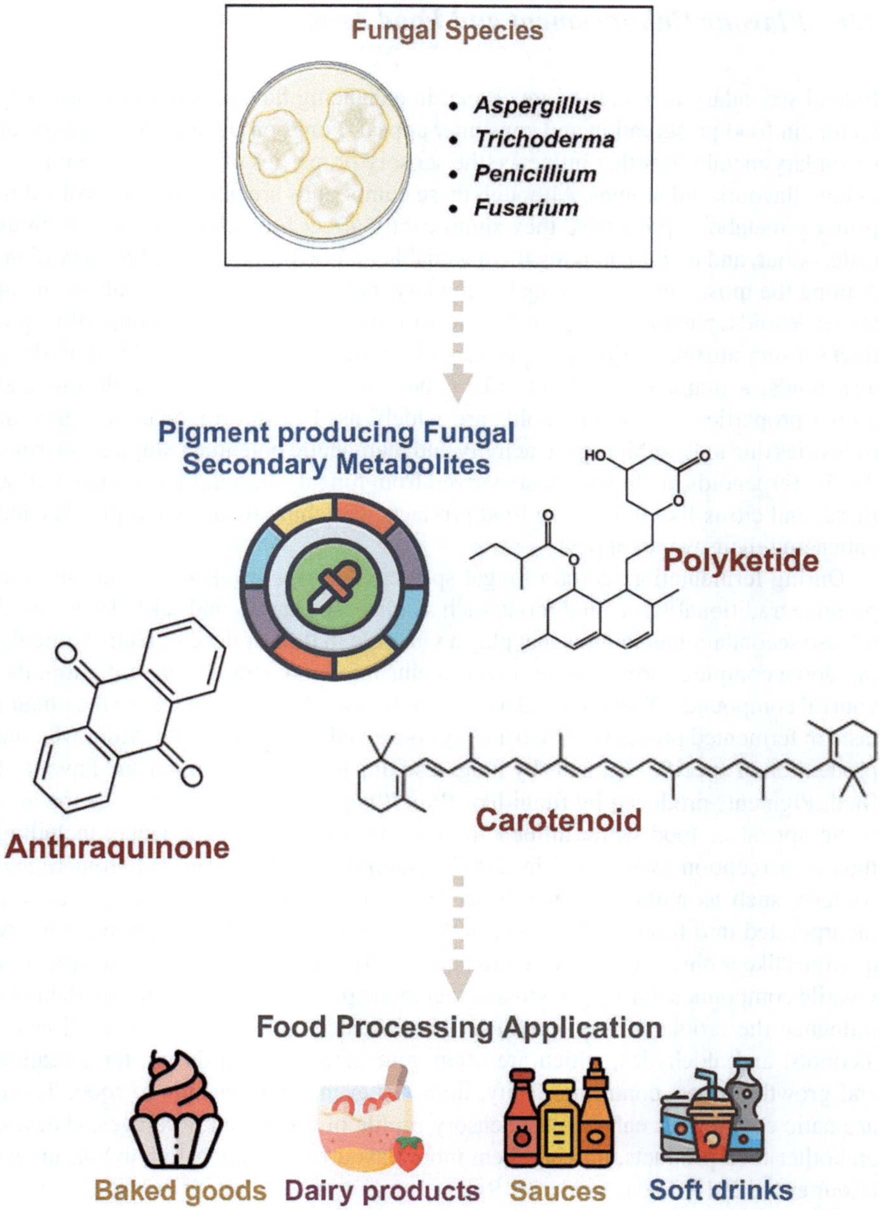

Fig. 3.2 Fungal species such as *Aspergillus, Trichoderma, Penicillium*, and *Fusarium* produce pigment-based secondary metabolites, including anthraquinones, polyketides, and carotenoids. These pigments are applied in food processing industries for colouring baked goods, dairy products, sauces, and soft drinks

4.9 Flavour Enhancement and Food Aroma

Fungal secondary metabolites are crucial in enhancing flavour and aroma, two key factors in food preservation and consumer appeal. Fungi produce a diverse range of secondary metabolites that influence the sensory properties of food, contributing to unique flavours and aromas. Although these compounds are not directly involved in primary metabolic processes, they significantly affect food characteristics such as taste, odour, and texture, making them valuable in food processing and preservation. Among the most significant fungal secondary metabolites for flavour enhancement are terpenoids, phenolic compounds, and various organic acids, each imparting distinct sensory attributes. For example, fungi like *Aspergillus* and *Penicillium* produce terpenoids, a major class of secondary metabolites known for their flavour and aroma properties. These terpenoids are widely used in the fragrance and flavour industries due to their biological activity and therapeutic potential (Singh & Sharma, 2015). Terpenoids, including those derived from fungal communities, impart earthy, floral, and citrus-like flavours to food products, enriching their flavour profiles and enhancing their overall appeal.

During fermentation, certain fungal species like *Aspergillus oryzae* are used to produce traditional fermented foods such as miso, soy sauce, and sake. These fungi release secondary metabolites that play a vital role in flavour development by breaking down complex carbohydrates and proteins into more straightforward, more flavourful compounds. These secondary metabolites are key contributors to the umami taste in fermented products, which are highly valued in Asian cuisine. Moreover, the production of specific pigments by fungi also impacts both the aroma and flavour of food. Pigments produced by fungi like *Penicillium* and *Fusarium* enhance the aesthetic appeal of food while influencing the overall sensory experience, including flavour perception (Wang & Lin, 2007). Natural colourants derived from fungal sources, such as anthraquinones from *Aspergillus* and *Trichoderma* species, are incorporated into food products to enhance the visual appeal and improve sensory qualities like aroma and flavour (Durán et al., 2002). Fungi also produce a variety of volatile compounds during growth and metabolic processes, which can significantly influence the aroma of food products. For instance, compounds like ethyl ester, alcohols, and aldehydes, which are often generated by fungi during fermentation and growth phases, contribute fruity, floral, or even spicy flavours to food. These aromatic compounds enhance the sensory profile of fermented beverages, cheeses, and other food products, making them more flavourful and appealing to consumers (Kour et al., 2019; Yadav et al., 2019).

4.10 Food Fermentation

Fungal secondary metabolites are vital to food fermentation, enhancing flavour, texture, aroma, and preservation in a variety of food products. Fungi play a key role in fermentation processes used to produce staples like bread, soy sauce, miso, and alcoholic beverages. These metabolites, generated during fungal growth and metabolism, significantly influence fermented foods' sensory qualities and shelf life. Among the most extensively studied fungal secondary metabolites in fermentation are organic acids, which aid in preservation by reducing pH and inhibiting spoilage microorganisms. Organic acids such as lactic acid, citric acid, and acetic acid, produced by fungi like *Aspergillus* and *Penicillium,* act as natural preservatives while contributing to the tangy and sour flavour characteristics of many fermented foods. These acids enhance taste and play a crucial role in extending the shelf life of fermented products (Khan et al., 2014; Rana et al., 2020).

In addition to organic acids, fungi produce a variety of enzymes that play a vital role in breaking down complex compounds during fermentation. These enzymes, including amylases, proteases, cellulases, and lipases, break down carbohydrates, proteins, and fats in raw materials into simpler, more digestible, and flavourful compounds. For example, enzymes produced by *Aspergillus oryzae* during soybean fermentation are key to degrading starches and proteins, resulting in the distinctive umami flavour characteristics of soy sauce and miso. The combined action of these enzymes and fungal secondary metabolites is essential for developing the texture and flavour profiles of the final fermented products (Devi et al., 2020; Dufossé et al., 2021).

Fungal secondary metabolites also contribute to the production of volatile compounds during fermentation, which directly influences the aroma of the food. These volatiles, including esters, alcohols, and aldehydes, are produced as by-products of metabolic processes and play a significant role in fermented foods' flavour and aroma characteristics. For example, the volatile compounds produced by *Saccharomyces cerevisiae* in brewing are essential in developing the distinct flavours and aromas of beer and wine. Similarly, other fungal species used in the fermentation of dairy products, such as *Penicillium* species in cheese-making, produce secondary metabolites that enhance cheese's flavour profile by contributing unique aromatic compounds. Furthermore, fungi produce bioactive secondary metabolites such as terpenoids, which are used in food flavouring and preservation. Terpenoids, produced by fungal communities, not only provide characteristic aromas but also have antimicrobial properties, which help inhibit the growth of spoilage microorganisms during fermentation. These metabolites contribute to the longevity and stability of fermented products, improving their shelf life and reducing the need for additional preservatives (Bills & Gloer, 2016; Devi et al., 2020).

5 Challenges and Limitations

Fungal secondary metabolites are increasingly being considered for their various uses in food preservation. While primary metabolites such as enzymes, organic acids, and proteins have been widely used in conventional food products, secondary metabolites present advantages beyond traditional applications. These compounds are essential in food preservation, as they enhance natural flavours, develop vibrant colours, extend shelf life, and contribute health-promoting properties, ensuring the safety and quality of food products. As the demand for natural ingredients in food preservation increases, fungal secondary metabolites are expected to be crucial in enhancing food preservation, but this application has some drawbacks and limitations. One of the main challenges is the high costs of production. Enzymes, colourants, and flavour compounds produced by fungi are known to be produced through costly processes such as fermentation, which necessitate complex equipment, strictly controlled environment conditions, and high-quality inputs. These factors can greatly enhance the operation costs. Another difficulty in the process of obtaining fungal metabolites at a commercial scale or in transition from laboratory to industrial scale is the required optimisation of the process of fermentation, which presents an additional expensive challenge. Further, some of these metabolites may require expensive raw materials, which could increase production costs and reduce the process's viability for large-scale applications. Another important constraint is fungal metabolite stability. Many of these molecules, including pigment and flavour compounds, are susceptible to environmental conditions such as temperature, pH, and light. Their chemical instability may limit their application in food preservation since they may lose effectiveness or degrade with time, lowering the overall quality of the stored food. For example, fungal pigments may lose colour with extended storage, particularly if subjected to temperature and pH changes. Additionally, the shelf life of fungal metabolites may be reduced, especially in packaged foods where the breakdown of enzymes or bioactive substances may result in less effective preservation. The microbial contamination risks associated with fungal culture management can also compromise the stability and safety of the final product if not adequately controlled.

Regulatory concerns present a significant challenge to the widespread use of fungal secondary metabolites. Food additive approval processes are frequently delayed due to strict guidelines and a lack of standard safety protocols. The potential toxicity or allergenic reactions associated with some fungal metabolites require extensive testing to ensure their safety for human consumption. Additionally, sustainability concerns arise, particularly regarding the environmental impact of large-scale production. While fungal metabolites are often considered more eco-friendly than synthetic preservatives, their production still requires significant resources, and waste management, including unused biomass and by-products, poses environmental challenges.

Another limitation is the variability in the composition of fungal secondary metabolites. The production of fungal metabolites can vary greatly depending on factors such as the fungal strain used, fermentation conditions, and the type of substrate employed. This variability can result in inconsistencies in the quality and effectiveness of the metabolites, making it difficult to standardise their use in food preservation. For example, the concentration of pigments or enzymes produced by the same fungal species can differ depending on environmental conditions, leading to variations in the quality of the final product. The genetic diversity of fungal strains also contributes to this issue, and strain optimisation or genetic engineering may be required to ensure consistent and high-quality yields of metabolites, further adding to the complexity and cost of production. Lastly, the consumer acceptance of fungal metabolites in food preservation can be a barrier to their widespread adoption. Despite the increasing demand for natural ingredients, some consumers may be hesitant to accept products containing fungal-derived substances, associating them with undesirable qualities such as mould or contamination. Overcoming this perception will require effective public education and transparent communication regarding fungal metabolites' safety, benefits, and natural origins. Addressing these consumer concerns is crucial for promoting the acceptance of fungal-based food preservatives in the marketplace.

6 Concluding Remarks and Perspectives

Fungal secondary metabolites display various biological activities in food systems, making them natural food additives with potential advantages, including rich flavours, natural colours, antioxidative effects, and health effects. However, their adoption is limited because of challenges, including high production costs, stability, regulatory barriers, and consumer acceptance. Overcoming these limitations is critical for realising the potential of fungal metabolites as sustainable alternatives to synthetic preservatives. Considering the above challenges posed by fungal secondary metabolites, there is a need for future studies to cover these aspects and work on optimising production processes, enhancing stability, and minimising costs. Recent developments in fermentation methods, genetic modification, and strain optimisation have consequences for the consistency and efficiency of production, with increasingly large-scale production becoming possible. Standardised safety protocols and regulations need to be addressed for further food preservation using fungal metabolites to be widely adopted. Further, educating consumers about the safety and benefits of fungal-derived preservatives will also play a key role in increasing acceptance. By overcoming these barriers, fungal secondary metabolites have the potential to revolutionise the food preservation industry, offering more sustainable, eco-friendly alternatives to traditional preservatives.

References

Abo-Kadoum, M. A., Abo-Dahab, N. F., Awad, M. F., & Abdel-Hadi, A. M. (2013). Marine-derived fungus, *Penicillium aurantiogriseum* AUMC 9757: A producer of bioactive secondary metabolites. *Journal of Basic and Applied Mycology – Egypt, 4*, 77–83.

Amirzakariya, B. Z., & Shakeri, A. (2022). Bioactive terpenoids derived from plant endophytic fungi: An updated review (2011–2020). *Phytochemistry, 197*, 113130. https://doi.org/10.1016/j.phytochem.2022.113130

Amit, S. K., Uddin, M. M., Rahman, R., Islam, S. M. R., & Khan, M. S. (2017). A review on mechanisms and commercial aspects of food preservation and processing. *Agriculture & Food Security, 6*(1), 51. https://doi.org/10.1186/s40066-017-0130-8

Ammar, M. I., Nenaah, G. E., & Mohamed, A. H. H. (2013). Antifungal activity of prenylated flavonoids isolated from *Tephrosia apollinea* L. against four phytopathogenic fungi. *Crop Protection, 49*, 21–25. https://doi.org/10.1016/j.cropro.2013.02.012

Anyasi, T. A., Edokpyi, J. N., & Anokwuru, C. (2017). Application of organic acids in food preservation. Biochem Res Trends. In C. Vargas (Ed.), *Organic acids: Characteristics, properties and synthesis* (pp. 1–47). Nova Science Publishers Inc.

Archer, D. B. (2000). Filamentous fungi as microbial cell factories for food use. *Current Opinion in Biotechnology, 11*(5), 478–483. https://doi.org/10.1016/S0958-1669(00)00129-4

Aris, P., Wei, Y., Mohamadzadeh, M., & Xia, X. (2022). Griseofulvin: An updated overview of old and current knowledge. *Molecules, 27*(20), 7034. https://doi.org/10.3390/molecules27207034

Avery, S. V., Singleton, I., Magan, N., & Goldman, G. H. (2019). The fungal threat to global food security. *Fungal Biology, 123*(8), 555–557. https://doi.org/10.1016/j.funbio.2019.03.006

Awad, N. E., Kassem, H. A., Hamed, M. A., El-Feky, A. M., Elnaggar, M. A. A., Mahmoud, K., & Ali, M. A. (2018). Isolation and characterisation of the bioactive metabolites from the soil derived fungus *Trichoderma viride*. *Mycology, 9*(1), 70–80. https://doi.org/10.1080/21501203.2017.1423126

Awuchi, C. G., Amagwula, I. O., Priya, P., Kumar, R., Yezdani, U., & Khan, M. G. (2020). Aflatoxins in foods and feeds: A review on health implications, detection, and control. *Bulletin of Environment, Pharmacology and Life Sciences, 9*, 149–155.

Behdarvand-Margha, Z., Ahangarpour, A., Shahraki, M., Komeili, G., & Khorsandi, L. (2021). The effects of gallic acid and metformin on male reproductive dysfunction in diabetic mice induced by methylglyoxal: An experimental study. *International Journal of Reproductive BioMedicine (IJRM)*, 715–724. https://doi.org/10.18502/ijrm.v19i8.9619

Belmares, R., Contreras-Esquivel, J. C., Rodríguez-Herrera, R., Coronel, A. R., & Aguilar, C. N. (2004). Microbial production of tannase: An enzyme with potential use in food industry. *Food Science and Technology, 37*(8), 857–864. https://doi.org/10.1016/j.lwt.2004.04.002

Bennett, R. N., & Wallsgrove, R. M. (1994). Secondary metabolites in plant defence mechanisms. *New Phytologist, 127*(4), 617–633. https://doi.org/10.1111/j.1469-8137.1994.tb02968.x

Bills, G. F., & Gloer, J. B. (2016). Biologically active secondary metabolites from the fungi. *Microbiology Spectrum, 4*(6), 4.6.01. https://doi.org/10.1128/microbiolspec.FUNK-0009-2016

Bojke, A., Tkaczuk, C., Stepnowski, P., & Gołębiowski, M. (2018). Comparison of volatile compounds released by entomopathogenic fungi. *Microbiological Research, 214*, 129–136. https://doi.org/10.1016/j.micres.2018.06.011

Carlsen, M., & Nielsen, J. (2001). Influence of carbon source on α-amylase production by *Aspergillus oryzae*. *Applied Microbiology and Biotechnology, 57*(3), 346–349. https://doi.org/10.1007/s002530100772

Chen, H. Y., Lei, J. Y., Li, S. L., Guo, L. Q., Lin, J. F., Wu, G. H., Lu, J., & Ye, Z. W. (2023). Progress in biological activities and biosynthesis of edible fungi terpenoids. *Food Science & Nutrition, 63*(25), 7288–7310. https://doi.org/10.1080/10408398.2022.2045559

Conrado, R., Gomes, T. C., Roque, G. S. C., & De Souza, A. O. (2022). Overview of bioactive fungal secondary metabolites: Cytotoxic and antimicrobial compounds. *Antibiotics, 11*(11), 1604. https://doi.org/10.3390/antibiotics11111604

Copetti, M. V. (2019). Fungi as industrial producers of food ingredients. *Current Opinion in Food Science, 25*, 52–56. https://doi.org/10.1016/j.cofs.2019.02.006

de la Lastra, C. A., & Villegas, I. (2007). Resveratrol as an antioxidant and pro-oxidant agent: Mechanisms and clinical implications. *Biochemical Society Transactions, 35*(5), 1156–1160. https://doi.org/10.1042/BST0351156

De Mejia, E. G., Zhang, Q., Penta, K., Eroglu, A., & Lila, M. A. (2020). The colours of health: Chemistry, bioactivity, and market demand for colourful foods and natural food sources of colourants. *Food Science and Technology, 11*(1), 45–182. https://doi.org/10.1146/annurev-food-032519-051729

Devi, R., Kaur, T., Guleria, G., Rana, K. L., Kour, D., Yadav, N., Yadav, A. N., & Saxena, A. K. (2020). Fungal secondary metabolites and their biotechnological applications for human health. In *New and future developments in microbial biotechnology and bioengineering* (pp. 147–161). https://doi.org/10.1016/B978-0-12-820528-0.00010-7

Duan, X. X., Qin, D., Song, H. C., Gao, T. C., Zuo, S. H., Yan, X., Wang, J. Q., Ding, X., Di, Y. T., & Dong, J. Y. (2019). Irpexlacte A-D, four new bioactive metabolites of endophytic fungus *Irpex lacteus* DR10-1 from the waterlogging tolerant plant *Distylium chinense*. *Phytochemistry Letters, 32*, 151–156. https://doi.org/10.1016/j.phytol.2019.06.001

Dufossé, L., Fouillaud, M., & Caro, Y. (2021). Fungi and fungal metabolites for the improvement of human and animal nutrition and health. *Journal of Fungi, 7*(4), 274. https://doi.org/10.3390/jof7040274

Dulo, B., Phan, K., Githaiga, J., Raes, K., & De Meester, S. (2021). Natural quinone dyes: A review on structure, extraction techniques, analysis and application potential. *Waste and Biomass Valorization, 12*(12), 6339–6374. https://doi.org/10.1007/s12649-021-01443-9

Durán, N., Teixeira, M. F. S., De Conti, R., & Esposito, E. (2002). Ecological-friendly pigments from fungi. *Food Science & Nutrition, 42*(1), 53–66. https://doi.org/10.1080/10408690290825457

Fung, F., Wang, H. S., & Menon, S. (2018). Food safety in the 21st century. *Biomedical Journal, 41*(2), 88–95. https://doi.org/10.1016/j.bj.2018.03.003

Gu, W. (2009). Bioactive metabolites from *Alternaria brassicicola* ML-P08, an endophytic fungus residing in *Malus halliana*. *World Journal of Microbiology and Biotechnology, 25*(9), 1677–1683. https://doi.org/10.1007/s11274-009-0062-y

Guarino, A. J. (1967). Cordycepin. In D. Gottlieb & P. D. Shaw (Eds.), *Mechanism of action* (pp. 468–480). Springer. https://doi.org/10.1007/978-3-642-46051-7_38

Guerriero, G., Berni, R., Muñoz-Sanchez, J. A., Apone, F., Abdel-Salam, E. M., Qahtan, A. A., Alatar, A. A., Cantini, C., Cai, G., Hausman, J. F., Siddiqui, K. S., Hernández-Sotomayor, S. M. T., & Faisal, M. (2018). Production of plant secondary metabolites: Examples, tips and suggestions for biotechnologists. *Genes, 9*(6), 309. https://doi.org/10.3390/genes9060309

Harnpicharnchai, P., Champreda, V., Sornlake, W., & Eurwilaichitr, L. (2009). A thermotolerant β-glucosidase isolated from an endophytic fungi, *Periconia* sp., with a possible use for biomass conversion to sugars. *Protein Expression and Purification, 67*(2), 61–69. https://doi.org/10.1016/j.pep.2008.05.022

Hawas, U. W., El-Kassem, L. T. A., Ahmed, E. F., & Alghamdi, R. A. (2022). Bioactive sulfonyl metabolites from the Red Sea endophytic fungus *Penicillium aculeatum*. *Natural Product Research, 36*(11), 2713–2721. https://doi.org/10.1080/14786419.2021.1917571

Huang, C. H., Pan, J. H., Chen, B., Yu, M., Huang, H. B., Zhu, X., Lu, Y. J., She, Z. G., & Lin, Y. C. (2011). Three bianthraquinone derivatives from the mangrove endophytic fungus *Alternaria* sp. ZJ9-6B from the South China Sea. *Marine Drugs, 9*(5), 832–843. https://doi.org/10.3390/md9050832

Huber, F. M., & Gottlieb, D. (1968). The mechanism of action of griseofulvin. *Canadian Journal of Microbiology, 14*(2), 111–118. https://doi.org/10.1139/m68-019

Hulvová, H., Galuszka, P., Frébortová, J., & Frébort, I. (2013). Parasitic fungus *Claviceps* as a source for biotechnological production of ergot alkaloids. *Biotechnology Advances, 31*(1), 79–89. https://doi.org/10.1016/j.biotechadv.2012.01.005

Hyde, K. D., Xu, J., Rapior, S., Jeewon, R., Lumyong, S., Niego, A. G. T., Abeywickrama, P. D., Aluthmuhandiram, J. V. S., Brahamanage, R. S., Brooks, S., Chaiyasen, A., Chethana, K. W. T., Chomnunti, P., Chepkirui, C., Chuankid, B., De Silva, N. I., Doilom, M., Faulds, C., Gentekaki, E., & Stadler, M. (2019). The amazing potential of fungi: 50 ways we can exploit fungi industrially. *Fungal Diversity, 97*(1), 1–136. https://doi.org/10.1007/s13225-019-00430-9

Jain, K. K., Kumar, A., Shankar, A., Pandey, D., Chaudhary, B., & Sharma, K. K. (2020). De novo transcriptome assembly and protein profiling of copper-induced lignocellulolytic fungus *Ganoderma lucidum* MDU-7 reveals genes involved in lignocellulose degradation and terpenoid biosynthetic pathways. *Genomics, 112*(1), 184–198. https://doi.org/10.1016/j.ygeno.2019.01.012

Jan, R., Asaf, S., Numan, M., Lubna, & Kim, K. M. (2021). Plant secondary metabolite biosynthesis and transcriptional regulation in response to biotic and abiotic stress conditions. *Agronomy, 11*(5), 968. https://doi.org/10.3390/agronomy11050968

Jegorov, A., Hajduch, M., Sulc, M., & Havlicek, V. (2006). Nonribosomal cyclic peptides: Specific markers of fungal infections. *Journal of Mass Spectrometry, 41*(5), 563–576. https://doi.org/10.1002/jms.1042

Jin, H., Yamashita, A., Maekawa, S., Yang, P., He, L., Takayanagi, S., Wakita, T., Sakamoto, N., Enomoto, N., & Ito, M. (2008). Griseofulvin, an oral antifungal agent, suppresses hepatitis C virus replication in vitro. *Hepatology Research, 38*(9), 909–918. https://doi.org/10.1111/j.1872-034X.2008.00352.x

Joardder, M. U. H., & Masud, M. H. (2019). A brief history of food preservation. In M. U. H. Joardder & M. H. Masud (Eds.), *Food preservation in developing countries: Challenges and solutions* (pp. 57–66). Springer. https://doi.org/10.1007/978-3-030-11530-2_3

Jumiati, R., Chandra, B., & Asra, D. R. (2021). Drug discovery of griseofulvin: A review. *Asian Journal of Pharmaceutical Research and Development, 9*(4), 101–107. https://doi.org/10.22270/ajprd.v9i4.993

Kaaniche, F., Hamed, A., Abdel-Razek, A. S., Wibberg, D., Abdissa, N., El Euch, I. Z., Allouche, N., Mellouli, L., Shaaban, M., & Sewald, N. (2019). Bioactive secondary metabolites from new endophytic fungus *Curvularia* Sp. isolated from *Rauwolfia macrophylla. PLoS One, 14*(6), e0217627. https://doi.org/10.1371/journal.pone.0217627

Keller, N. P. (2019). Fungal secondary metabolism: Regulation, function and drug discovery. *Nature Reviews. Microbiology, 17*(3), 167–180. https://doi.org/10.1038/s41579-018-0121-1

Keller, N. P., Turner, G., & Bennett, J. W. (2005). Fungal secondary metabolism—From biochemistry to genomics. *Nature Reviews. Microbiology, 3*(12), 937–947. https://doi.org/10.1038/nrmicro1286

Khan, A. A., Bacha, N., Ahmad, B., Lutfullah, G., Farooq, U., & Cox, R. J. (2014). Fungi as chemical industries and genetic engineering for the production of biologically active secondary metabolites. *Asian Pacific Journal of Tropical Biomedicine, 4*(11), 859–870. https://doi.org/10.12980/APJTB.4.2014APJTB-2014-0230

Khare, S., Singh, N. B., Singh, A., Hussain, I., Niharika, K., Yadav, V., Bano, C., Yadav, R. K., & Amist, N. (2020). Plant secondary metabolites synthesis and their regulations under biotic and abiotic constraints. *Journal Of Plant Biology, 63*(3), 203–216. https://doi.org/10.1007/s12374-020-09245-7

Kildgaard, S., Mansson, M., Dosen, I., Klitgaard, A., Frisvad, J., Larsen, T., & Nielsen, K. (2014). Accurate dereplication of bioactive secondary metabolites from marine-derived fungi by UHPLC-DAD-QTOFMS and a MS/HRMS library. *Marine Drugs, 12*(6), 3681–3705. https://doi.org/10.3390/md12063681

Kour, D., Rana, K. L., Yadav, N., Yadav, A. N., Rastegari, A. A., Singh, C., Negi, P., Singh, K., & Saxena, A. K. (2019). Technologies for biofuel production: Current development, challenges, and future prospects. In A. A. Rastegari, A. N. Yadav, & A. Gupta (Eds.), *Prospects of renewable bioprocessing in future energy systems* (Vol. 10, pp. 1–50). Springer. https://doi.org/10.1007/978-3-030-14463-0_1

Książek, E. (2023). Citric acid: Properties, microbial production, and applications in industries. *Molecules, 29*(1), 22. https://doi.org/10.3390/molecules29010022

Kumar, A., & Kumar, A. (2019). Synthesis and regulation of fungal secondary metabolites. In P. K. Arora (Ed.), *Microbial technology for the welfare of society* (Vol. 17, pp. 25–52). Springer. https://doi.org/10.1007/978-981-13-8844-6_2

Lin, L., & Xu, J. (2020). Fungal pigments and their roles associated with human health. *Journal of Fungi, 6*(4), 280. https://doi.org/10.3390/jof6040280

Liu, X., Xu, Y., Zhan, X., Xie, W., Yang, X., Cui, S. W., & Xia, W. (2020). Development and properties of new kojic acid and chitosan composite biodegradable films for active packaging materials. *International Journal of Biological Macromolecules, 144*, 483–490. https://doi.org/10.1016/j.ijbiomac.2019.12.126

Macheleidt, J., Mattern, D. J., Fischer, J., Netzker, T., Weber, J., Schroeckh, V., Valiante, V., & Brakhage, A. A. (2016). Regulation and role of fungal secondary metabolites. *Annual Review of Genetics, 50*(1), 371–392. https://doi.org/10.1146/annurev-genet-120215-035203

Macías-Rubalcava, M. L., & Sánchez-Fernández, R. E. (2017). Secondary metabolites of endophytic *Xylaria* species with potential applications in medicine and agriculture. *World Journal of Microbiology and Biotechnology, 33*(1), 15. https://doi.org/10.1007/s11274-016-2174-5

Mapari, S. A. S., Meyer, A. S., & Thrane, U. (2009). Photostability of natural orange–red and yellow fungal pigments in liquid food model systems. *Journal of Agricultural and Food Chemistry, 57*(14), 6253–6261. https://doi.org/10.1021/jf900113q

Margalith, P. (1993). Enhancement of carotenoid synthesis by fungal metabolites. *Applied Microbiology and Biotechnology, 38*(5). https://doi.org/10.1007/BF00182807

Meyer, M., & Slot, J. (2023). The evolution and ecology of psilocybin in nature. *Fungal Genetics and Biology, 167*, 103812. https://doi.org/10.1016/j.fgb.2023.103812

Murillo-Cruz, M. C., Chova, M., & Bermejo-Román, R. (2021). Effect of adding fungal β-carotene to picual extra virgin olive oils on their physical and chemical properties. *Journal of Food Processing & Preservation, 45*(2). https://doi.org/10.1111/jfpp.15186

Niu, X., Thaochan, N., & Hu, Q. (2020). Diversity of linear non-ribosomal peptide in biocontrol fungi. *Journal of Fungi, 6*(2), 61. https://doi.org/10.3390/jof6020061

Nkadimeng, S. M., Nabatanzi, A., Steinmann, C. M. L., & Eloff, J. N. (2020). Phytochemical, cytotoxicity, antioxidant and anti-inflammatory effects of *Psilocybe natalensis* magic mushroom. *Plants, 9*(9), 1127. https://doi.org/10.3390/plants9091127

Pang, X. J., Zhang, S. B., Xian, P. J., Wu, X., Yang, D. F., Fu, H. Y., & Yang, X. L. (2018). Emericellins A and B: Two sesquiterpenoids with an unprecedented tricyclo [4,4,2,1] hendecane scaffold from the liquid cultures of endophytic fungus *Emericella* sp. XL 029. *Fitoterapia, 131*, 55–58. https://doi.org/10.1016/j.fitote.2018.10.022

Papaspyridi, L. M., Aligiannis, N., Topakas, E., Christakopoulos, P., Skaltsounis, A. L., & Fokialakis, N. (2012). Submerged fermentation of the edible mushroom *Pleurotus ostreatus* in a batch stirred tank bioreactor as a promising alternative for the effective production of bioactive metabolites. *Molecules, 17*(3), 2714–2724. https://doi.org/10.3390/molecules17032714

Park, J. T., & Strominger, J. L. (1957). Mode of action of penicillin: Biochemical basis for the mechanism of action of penicillin and for its selective toxicity. *Science, 125*(3238), 99–101. https://doi.org/10.1126/science.125.3238.99

Patil, R. H., Krishnan, P., & Maheshwari, V. L. (2011). Production of lovastatin by wild strains of *Aspergillus terreus*. *Natural Product Communications, 6*(2), 1934578X1100600207. https://doi.org/10.1177/1934578X1100600207

Patra, B., Schluttenhofer, C., Wu, Y., Pattanaik, S., & Yuan, L. (2013). Transcriptional regulation of secondary metabolite biosynthesis in plants. *Biochimica et Biophysica Acta (BBA) – Gene Regulatory Mechanisms, 1829*(11), 1236–1247. https://doi.org/10.1016/j.bbagrm.2013.09.006

Pina, J. R. S., Silva-Silva, J. V., Carvalho, J. M., Bitencourt, H. R., Watanabe, L. A., Fernandes, J. M. P., Souza, G. E. D., Aguiar, A. C. C., Guido, R. V. C., Almeida-Souza, F., Calabrese, K. D. S., Marinho, P. S. B., & Marinho, A. M. D. R. (2021). Antiprotozoal and antibacterial activity of ravenelin, a xanthone isolated from the endophytic fungus *Exserohilum rostratum*. *Molecules, 26*(11), 3339. https://doi.org/10.3390/molecules26113339

Polturak, G., & Osbourn, A. (2021). The emerging role of biosynthetic gene clusters in plant defense and plant interactions. *PLoS Pathogens, 17*(7), e1009698. https://doi.org/10.1371/journal.ppat.1009698

Prabhasankar, P., Ganesan, P., Bhaskar, N., Hirose, A., Stephen, N., Gowda, L. R., Hosokawa, M., & Miyashita, K. (2009). Edible Japanese seaweed, wakame (*Undaria pinnatifida*) as an ingredient in pasta: Chemical, functional and structural evaluation. *Food Chemistry, 115*(2), 501–508. https://doi.org/10.1016/j.foodchem.2008.12.047

Prasad, P., & Kochhar, A. (2014). Active packaging in food industry: A review. *IOSR Journal of Environmental Science, Toxicology and Food Technology, 8*(5), 01–07. https://doi.org/10.9790/2402-08530107

Qin, J. C., Zhang, Y. M., Hu, L., Ma, Y. T., & Gao, J. M. (2009). Cytotoxic metabolites produced by *Alternaria* no.28, an endophytic fungus isolated from *Ginkgo biloba*. *Natural Product Communications, 4*(11), 1934578X0900401106. https://doi.org/10.1177/1934578X0900401106

Qin, Y. Y., Huang, X. S., Liu, X. B., Mo, T. X., Xu, Z. L., Li, B. C., Qin, X. Y., Li, J., Schäberle, T. F., & Yang, R. Y. (2020). Three new andrastin derivatives from the endophytic fungus *Penicillium vulpinum*. *Natural Product Research*, 1–9. https://doi.org/10.1080/14786419.2020.1853725

Radhi, M., Ashraf, S., Lawrence, S., Tranholm, A. A., Wellham, P. A. D., Hafeez, A., Khamis, A. S., Thomas, R., McWilliams, D., & De Moor, C. H. (2021). A systematic review of the biological effects of cordycepin. *Molecules, 26*(19), 5886. https://doi.org/10.3390/molecules26195886

Rana, K. L., Kour, D., Kaur, T., Sheikh, I., Yadav, A. N., Kumar, V., Suman, A., & Dhaliwal, H. S. (2020). Endophytic microbes from diverse wheat genotypes and their potential biotechnological applications in plant growth promotion and nutrient uptake. *Proceedings of the National Academy of Sciences, India Section B: Biological Sciences, 90*(5), 969–979. https://doi.org/10.1007/s40011-020-01168-0

Rathod, N. B., Phadke, G. G., Tabanelli, G., Mane, A., Ranveer, R. C., Pagarkar, A., & Ozogul, F. (2021). Recent advances in bio-preservatives impacts of lactic acid bacteria and their metabolites on aquatic food products. *Food Bioscience, 44*, 101440. https://doi.org/10.1016/j.fbio.2021.101440

Redondo-Blanco, S., Fernández, J., López-Ibáñez, S., Miguélez, E. M., Villar, C. J., & Lombó, F. (2020). Plant phytochemicals in food preservation: Antifungal bioactivity: A review. *Journal of Food Protection, 83*(1), 163–171. https://doi.org/10.4315/0362-028X.JFP-19-163

Ren, G., Xu, L., Lu, T., & Yin, J. (2018). Structural characterisation and antiviral activity of lentinan from *Lentinus edodes* mycelia against infectious hematopoietic necrosis virus. *International Journal of Biological Macromolecules, 115*, 1202–1210. https://doi.org/10.1016/j.ijbiomac.2018.04.132

Reshi, Z. A., Ahmad, W., Lukatkin, A. S., & Javed, S. B. (2023). From nature to lab: A review of secondary metabolite biosynthetic pathways, environmental influences, and in vitro approaches. *Metabolites, 13*(8), 895. https://doi.org/10.3390/metabo13080895

Rincão, V. P., Yamamoto, K. A., Silva Ricardo, N. M. P., Soares, S. A., Paccola Meirelles, L. D., Nozawa, C., & Carvalho Linhares, R. E. (2012). Polysaccharide and extracts from *Lentinula edodes*: Structural features and antiviral activity. *Virology Journal, 9*(1), 37. https://doi.org/10.1186/1743-422X-9-37

Rousta, N., Ferreira, J. A., & Taherzadeh, M. J. (2021). Production of L-carnitine-enriched edible filamentous fungal biomass through submerged cultivation. *Bioengineered, 12*(1), 358–368. https://doi.org/10.1080/21655979.2020.1863618

Sahasrabudhe, N. A., & Sankpal, N. V. (2001). Production of organic acids and metabolites of fungi for food industry. *Applied Mycology and Biotechnology, 1*, 387–425. https://doi.org/10.1016/S1874-5334(01)80016-2

Saini, S., Chutani, P., Kumar, P., & Sharma, K. K. (2020). Development of an eco-friendly deinking process for the production of bioethanol using diverse hazardous paper wastes. *Renewable Energy, 146*, 2362–2373. https://doi.org/10.1016/j.renene.2019.08.087

Santiago, C., Sun, L., Munro, M. H. G., & Santhanam, J. (2014). Polyketide and benzopyran compounds of an endophytic fungus isolated from Cinnamomum mollissimum: Biological activity and structure. *Asian Pacific Journal of Tropical Biomedicine, 4*(8), 627–632. https://doi.org/10.12980/APJTB.4.2014APJTB-2014-0030

Santos, S. N., Ferraris, F. K., De Souza, A. O., Henriques, M. D. G., & Melo, I. S. (2012). Endophytic fungi from *Combretum leprosum* with potential anti-cancer and antifungal activity. *Symbiosis, 58*(1–3), 109–117. https://doi.org/10.1007/s13199-012-0218-7

Scopel, M., Mothes, B., Lerner, C. B., Henriques, A. T., Macedo, A. J., & Abraham, W. R. (2017). Arvoredol—An unusual chlorinated and biofilm inhibiting polyketide from a marine *Penicillium* sp. of the Brazilian coast. *Phytochemistry Letters, 20*, 73–76. https://doi.org/10.1016/j.phytol.2017.04.010

Shwab, E. K., & Keller, N. P. (2008). Regulation of secondary metabolite production in filamentous ascomycetes. *Mycological Research, 112*(2), 225–230. https://doi.org/10.1016/j.mycres.2007.08.021

Singh, B., & Sharma, R. A. (2015). Plant terpenes: Defense responses, phylogenetic analysis, regulation and clinical applications. *3 Biotech, 5*(2), 129–151. https://doi.org/10.1007/s13205-014-0220-2

Singh, A., Singh, D. K., Kharwar, R. N., White, J. F., & Gond, S. K. (2021). Fungal endophytes as efficient sources of plant-derived bioactive compounds and their prospective applications in natural product drug discovery: Insights, avenues, and challenges. *Microorganisms, 9*(1), 197. https://doi.org/10.3390/microorganisms9010197

Stasiuk, M., & Kozubek, A. (2010). Biological activity of phenolic lipids. *Cellular and Molecular Life Sciences, 67*(6), 841–860. https://doi.org/10.1007/s00018-009-0193-1

Stierle, A. A., & Stierle, D. B. (2000). Bioactive compounds from four Endophytic Penicillium sp. of a northwest pacific yew tree. *Studies in Natural Products Chemistry, 24*, 933–977. https://doi.org/10.1016/S1572-5995(00)80058-7

Stockmann-Juvala, H., & Savolainen, K. (2008). A review of the toxic effects and mechanisms of action of fumonisin B1. *Human & Experimental Toxicology, 27*(11), 799–809. https://doi.org/10.1177/0960327108099525

Szymański, M., Chmielewska, S., Czyżewska, U., Malinowska, M., & Tylicki, A. (2022). Echinocandins – structure, mechanism of action and use in antifungal therapy. *Journal of Enzyme Inhibition and Medicinal Chemistry, 37*(1), 876–894. https://doi.org/10.1080/14756366.2022.2050224

Ueno, Y. (1978). Mode of action of trichothecenes. *Mycotoxins in Foodstuffs, 3*, 1737–1745. https://doi.org/10.1016/B978-0-08-021194-7.50009-1

Vining, L. C. (2007). Roles of secondary metabolites from microbes. In D. J. Chadwick & J. Whelan (Eds.), *Novartis foundation symposia* (pp. 184–198). Wiley. https://doi.org/10.1002/9780470514344.ch11

Wang, T., & Lin, T. (2007). Monascus rice products. In *Advances in food and nutrition research* (Vol. 53, pp. 123–159). Academic Press. https://doi.org/10.1016/S1043-4526(07)53004-4

Wisbeck, E., Facchini, J. M., Alves, E. P., Silveira, M. L. L., Gern, R. M. M., Ninow, J. L., & Furlan, S. A. (2017). A polysaccharide fraction extracted from *Pleurotus ostreatus* mycelial biomass inhibit Sarcoma 180 tumor. *Anais da Academia Brasileira de Ciências, 89*(3), 2013–2020. https://doi.org/10.1590/0001-3765201720150635

Xu, K., Wei, X. L., Xue, L., Zhang, Z. F., & Zhang, P. (2020). Antimicrobial meroterpenoids and erythritol derivatives isolated from the marine-algal-derived endophytic fungus *Penicillium chrysogenum* XNM-12. *Marine Drugs, 18*(11), 578. https://doi.org/10.3390/md18110578

Yadav, A. N., Kour, D., Rana, K. L., Yadav, N., Singh, B., Chauhan, V. S., Rastegari, A. A., Hesham, A. E. L., & Gupta, V. K. (2019). Metabolic engineering to synthetic biology of secondary metabolites production. In *New and future developments in microbial biotechnology and bioengineering* (pp. 279–320). https://doi.org/10.1016/B978-0-444-63504-4.00020-7

Zapaśnik, A., Sokołowska, B., & Bryła, M. (2022). Role of lactic acid bacteria in food preservation and safety. *Food, 11*(9), 1283. https://doi.org/10.3390/foods11091283

Zhang, Y., Han, T., Ming, Q., Wu, L., Rahman, K., & Qin, L. (2012). Alkaloids produced by endophytic fungi: A review. *Natural Product Communications, 7*(7). https://doi.org/10.1177/1934578X1200700742

Zhan, X., Chen, Z., Chen, R., & Shen, C. (2022). Environmental and genetic factors involved in plant protection-associated secondary metabolite biosynthesis pathways. *Frontiers in Plant Science, 13*, 877304. https://doi.org/10.3389/fpls.2022.877304

Zhao, D. L., Wang, D., Tian, X. Y., Cao, F., Li, Y. Q., & Zhang, C. S. (2018). Anti-phytopathogenic and cytotoxic activities of crude extracts and secondary metabolites of marine-derived fungi. *Marine Drugs, 16*(1), 36. https://doi.org/10.3390/md16010036

Zhu, J., Wu, F., Yue, S., Chen, C., Song, S., Wang, H., & Zhao, M. (2019). Functions of reactive oxygen species in apoptosis and ganoderic acid biosynthesis in *Ganoderma lucidum*. *FEMS Microbiology Letters, 366*(23), fnaa015. https://doi.org/10.1093/femsle/fnaa015

Zorrilla, J. G., & Evidente, A. (2022). Structures and biological activities of alkaloids produced by mushrooms, a fungal subgroup. *Biomolecules, 12*(8), 1025. https://doi.org/10.3390/biom12081025

Chapter 4
Natural Colorants and Pigments Derived from Fungal Metabolites

Santhoshkumar Subramaniam, Devi Priya Arumugam, V. H. Vinuthana, S. Monika Sri, B. Sanjeev, and Sivakumar Uthandi

1 Introduction

Color perception in the human visual system results from the interface between light and an object, governed by various influential factors, viz., spectral composition of light, surface gloss, transparency, haze, and turbidity. The application of color by humans has a long and rich history, dating back approximately 150,000–200,000 years. During the Ice Age, early humans practiced ritualistic burial customs using red ochre, symbolizing a belief in the life-affirming properties of the color red. Prehistoric cave paintings in Altamira, Spain estimated to be over 20,000 years old, depicting animals rendered in vivid hues of red earth, black, and yellow ocher. In ancient Egypt, the use of malachite green as a cosmetic pigment was recorded before the dynastic period, around 5000 B.C. Egyptian women commonly employed natural substances for personal adornment, including henna for hair dyeing, carmine for lip coloring, and kohl (an antimony-based compound) for enhancing eyebrows, eyelids, and lashes. In ancient India, as early as one million years ago, it was customary to use saffron to color the face yellow and henna to stain the feet red. Similarly, traditional Chinese practices used plant-based dyes to tint the cheeks, feet, and tongue tips. Roman cultures also used cosmetic practices, applying white lead and chalk to the face, and used blue and gold dyes to color their hair and beards.

Humans are probably the only living beings who can think in color because we have given various names to various colors. Till the middle nineteenth century, all colorants used were derived from nature, and their names were also inspired by their natural occurrence, for example, in the Indo-European language, Sanskrit "Rudhra"

S. Subramaniam · D. P. Arumugam · V. H. Vinuthana · S. Monika Sri · B. Sanjeev · S. Uthandi (✉)
Biocatalysts Laboratory, Department of Agricultural Microbiology, Tamil Nadu Agricultural University (TNAU), Coimbatore, Tamil Nadu, India

B. P. Singh et al. (eds.), *Fungal Additives and Bioactives in Food Processing Industries*, Fungal Biology, https://doi.org/10.1007/978-3-032-04519-5_4

means blood and stands for the color red, the first part of the word appears in Greek and Latin and stands for red, i.e., erthlyros and rutilus. Among modern languages, Rot stands for red in German. Before the mid-nineteenth century, natural colorants were widely used in food and cosmetic products to improve their appearance, restore color lost during processing, and, in certain instances, aid in product adulteration. In pharmaceuticals, these colorants primarily served to distinguish between different medicinal preparations and reduce the likelihood of confusion. Natural colorants were predominantly derived from botanical, zoological, and mineral sources.

The emergence of synthetic organic dyes started in 1856 with Sir William Henry Perkin's discovery of "mauve," the first synthetic dye. This breakthrough sparked the swift advancement and widespread adoption of synthetic colorants across various artistic media (Meier, 1993). Shortly thereafter, their application extended to food, pharmaceuticals, and cosmetics, particularly in Europe. Notably, as early as 1860, French wines were artificially colored using fuchsine, a triphenylmethane dye. In the United States, Congress formally approved the use of synthetic dyes in food, initially allowing their use in butter. By the early twentieth century, many artificially colored foods such as ketchup, jellies, cordials, butter, cheese, ice cream, candies, sausages, noodles, and wine had become a regular part of the American diet. At the same time, synthetic colorants were increasingly incorporated into pharmaceutical and cosmetic products (Meier, 1993). However, the widespread adoption of these color additives soon raised public health concerns. Toxic compounds were often incorporated into consumables, with dyes used to enhance appearance, mask substandard quality, and increase the apparent weight or volume of certain products.

In early twentieth-century London, particularly around 1900, adding yellow dye to diluted and skimmed milk was so widespread that consumers, especially homemakers, often perceived natural, whole milk as abnormal and were reluctant to purchase it. Historically, some initial applications of food colorants were deceptive and hazardous. For example, in 1820, Fredrick Accum documented fatalities among women who regularly consumed pickles artificially colored green with copper sulphate. Additionally, a tea shop in Manchester reportedly possessed toxic substances such as copper arsenate, lead chromate, and indigo, which were used to recolor spent tea leaves for resale (Meier, 1993). Such harmful practices prompted the development of regulatory frameworks for using colorants in food, pharmaceuticals, and cosmetics. The evolution of FDA oversight illustrates the gradual restriction of synthetic dyes over the past century, culminating in a contemporary environment where manufacturers generally avoid synthetic colorants unless absolutely necessary. Dyes hold significant relevance in both microbiology and chemistry due to multiple aspects. As genetic factors govern dye production, their synthesis often serves as an initial indicator of the concurrent biosynthesis of antibiotic compounds.

2 Biosynthesis of Fungal Pigments

Fungi synthesize diverse pigments that fulfill various functions, including shielding against environmental stress and contributing to their pathogenic mechanisms. These pigments also have industrial and pharmaceutical applications due to their antimicrobial, antioxidant, and dyeing properties. These pigments are produced via various metabolic pathways, primarily including polyketide synthases (PKS), non-ribosomal peptide synthetases (NRPS), and the shikimate and mevalonate pathways.

2.1 Polyketide Synthase (PKS) and Non-ribosomal Peptide Synthetase (NRPS) Pathways

The PKS and NRPS pathways drive the production of many fungal pigments. Polyketide synthases (PKSs) are large, multifunctional enzymes constructing polyketide pigments through the sequential condensation of acetyl-CoA and malonyl-CoA units (Mukherjee et al., 2017). These enzymes are iterative in fungi, meaning the same module is used repeatedly to extend the growing polyketide chain. The variations in modifications, such as oxidation and methylation, contribute to different pigment structures (Brakhage, 2013). Some well-known pigments produced by this pathway include melanins, anthraquinones, and azaphilones.

Similarly, NRPSs produce non-ribosomal peptides, which can sometimes combine with polyketides to form hybrid compounds. These hybrid pigments have unique structural and functional properties (Brakhage, 2013). Some of these pigments play essential roles in fungal virulence and survival in harsh environments.

2.2 Shikimate and Mevalonate Pathways

Apart from the PKS and NRPS pathways, the shikimate and mevalonate pathways contribute to fungal pigment biosynthesis. The shikimate pathway is essential for producing aromatic amino acids, which are precursors for several fungal pigments. Enzymes like dehydroquinate synthase help form intermediates like chorismate, which is later converted into quinones and melanins. On the other hand, the mevalonate pathway plays a central role in the biosynthesis of terpenoid pigments, notably carotenoids. This pathway converts acetyl-CoA into isopentenyl pyrophosphate (IPP), the building block for many isoprenoid-derived pigments (Knox & Keller, 2015).

3 Types of Fungal Pigments and Natural Colorants

Fungi synthesize pigments as secondary metabolites in response to nutrient depletion in the culture medium or when environmental conditions are unfavorable for growth (Mukherjee et al., 2017). Various colors like red, yellow, orange, blue, pink, and violet are produced as secondary metabolites by *Aspergillus*, *Fusarium*, *Monascus*, *Penicillium*, *Talaromyces*, *Trichoderma*, etc. (Lin & Xu, 2020). Although several plants and animals produce pigments, fungal pigments are industrially preferred due to their eco-friendly nature, cost-effectiveness, and versatility (Venil et al., 2020). There are several classes of pigments, like carotenoids, melanins, polyketides, azaphilones, flavins, phenazines, quinones, and monastics. Most fungi produce water-soluble pigments that are perfect for industrial production since they are simple to extract without organic solvents and can be readily scaled up in industrial fermenters. Below is a detailed description of each pigment (Afroz Toma et al., 2023).

3.1 Carotenoids

The word "Carotene" is from the Latin word "carota," coined by Wackenroder in 1831, who extracted orange color pigment from the carrot (*Daucus carota*) (Ruiz-Sola & Rodríguez-Concepción, 2012). Over 200 carotenoid-producing fungal species have been documented (Aksu & Eren, 2005). Carotenoids are lipophilic molecules, mainly C_{40} terpenoids that act as scavengers for oxygen and peroxyl radicals, thereby protecting the membrane. Usually, fungal carotenoids are C_{40} or C_{40} derivatives, such as astaxanthin, β-carotene, neurosporaxanthin, torularhodin, and torulene (Rapoport et al., 2021). Carotenoids have chemical structures composed of isoprenoids and C_{40} tetraterpenes, formed from eight isoprene units. They possess an extensive system of conjugated double bonds, creating π-electron resonance systems that extend along the polyene chain. These structural characteristics make carotenoids highly reactive compounds that predominantly absorb visible electromagnetic radiation in the 360–780 nm range. Carotenoids can undergo various chemical modifications, including hydrogenation, dehydrogenation, cyclization, double bond migration, chain extension or shortening, molecular rearrangement, isomerization, substitution reactions, and oxidation. These transformations can significantly change their core structure and impact their physical, chemical, and biological characteristics (Hernández-Almanza et al., 2014; Igreja et al., 2021).

Carotenoids are classified into different categories based on their chemical and nutritional characteristics. Based on chemical structure, they are classified as carotenes and xanthophylls. Carotenes are composed of carbon (C) and hydrogen (H) atoms in their chemical structure. Examples of carotenes include α-carotene, β-carotene, γ-carotene, torulene, and lycopene. In contrast, xanthophylls are carotenoids that contain oxygen, carbon, and hydrogen. Examples of xanthophyll include zeaxanthin, lutein, torularhodin, and canthaxanthin. From a nutritional standpoint, carotenoids are categorized into pro-vitamin A types such as α-carotene, β-carotene, and β-cryptoxanthin; non-provitamin A types including zeaxanthin, lycopene, and lutein; and ketocarotenoids such as canthaxanthin and astaxanthin (Cataldo et al., 2018; Toti et al., 2018). The spectrum of carotenoids synthesized by fungi varies considerably across species and is influenced by environmental factors. These compounds are vital in shielding fungal cells against oxidative damage and non-ionizing radiation, particularly ultraviolet (UV) light.

3.2 *Melanin*

The term "melanin" is from the Greek word "melanos," meaning "black", and was first used by Berzelius, a Swedish chemist, in 1840 (Riley, 1997; Solano, 2016). Melanins are typically high molecular weight, dark-brown, or black pigments that are hydrophobic and negatively charged. They are created by oxidative polymerization of phenols or indols (Chatragadda & Dufossé, 2021). Melanin is well recognized for its unique and noteworthy physicochemical and biological properties. These include its characteristic black coloration, insoluble nature in organic solvents, high resistance to acids, capacity to reduce ammoniacal silver nitrate solutions, and propensity to interact favorably with polyphenolic compounds. Fungal melanin is superior to other forms of melanin in terms of resistance, durability, and multifunctionality, which makes it extremely useful in harsh and hostile conditions. Fungal melanin integrates into the fungal cell wall to produce an obstacle that protects against UV radiation, oxidative stress, high temperatures, and chemical or enzymatic attacks. This contrasts with animal or plant melanin, primarily in UV protection and coloration. Strengthening resistance to antifungal medications and immune responses increases the virulence of pathogenic fungi and improves their survival in various settings, such as deep soils and radioactive zones. Melanin has gained more scientific attention recently due to its biological and functional qualities. Based on the chemical composition of the monomer subunit, melanins are classified into several types (Table 4.1).

3.3 *Azaphilones*

Azaphilones are structurally similar natural fungal compounds with a chiral quaternary center and a highly oxygenated bicyclic pyronequinone structure. "Azaphilone" comes from their easy reaction with amines to produce vinylogous γ-pyridones (Williams et al., 2021). Many ascomycetes and basidiomycetous fungi produce azaphilones, highly oxygenated pyranoquinone bicyclic core fungal polyketide pigments. For ages, azaphilone colors have been made in Asia from rice fermented with Monascus. A promising class of fungal pigments is being studied for industrial

Table 4.1 Types of fungal melanins based on their chemical composition

Pigment	Color	Chemical composition	Precursor	Microorganism	References
Eumelanins	Dark brown or black	6–9% N and 0–1% S	Dopaquinone	*Auricularia auricula, Cryptococcus neoformans, Candida albicans, Paracoccidioides brasiliensis,* and *Sporotrix schenckii*	Ozeki et al. (1997)
Pheomelanins	Yellow-red pigments	8–11% N and 9–12% S and benzothiazine monomer units	Dopaquinone	*Auricularia auricula* and *Tuber melanosporum*	Ozeki et al. (1997)
Allomelanins	Brown to black	di-DHN polymerization results in a heterogeneous collection of nitrogen-free polymers.	Malonyl CoA	*Auricularia auricula, Aspergillus fumigatus, Phoma* sp., *Alternaria alternata, Cladosporium carionii,* and *Fonseceae pedrosoi*	Ozeki et al. (1997)

N - Nitrogen, *S* - Sulphur, *di-DHN* - 1,8- dihydroxynaphthalene

applications mainly due to their distinctive yellow, orange, and red colors. As a result, they attracted more attention from researchers, mainly because of their uses in the printing, cosmetic, and dyeing sectors (Pavesi et al., 2023).

3.4 Quinones

Fungal quinones, particularly anthraquinones and naphthoquinones, have gained significant attention as natural dyes in the textile industry due to their vibrant colors and exceptional light fastness. These compounds, produced by fungi such as *Aspergillus* sp., *Curvularia* sp., *Drechslera* sp., and *Trichoderma* sp., present a sustainable alternative to synthetic dyes. For example, anthraquinones extracted from *Curvularia lunata* have been employed to produce dyes comparable to Acid Green 28 and Disperse Blue 7, highlighting their potential for textile applications (Meruvu & Dos Santos, 2021). Naphthoquinones, present in over 100 varieties from 63 fungal species, exhibit a diverse spectrum of colors and possess antimicrobial properties, rendering them suitable for textile dyeing applications. Using fungal quinones as dyes provides a renewable source of colorants and promotes environmentally sustainable practices within the textile industry (Venil et al., 2020).

3.5 Flavins

Flavins, particularly riboflavin (vitamin B2), are natural pigments produced by various fungi and have been explored as eco-friendly dyes. Several fungal species can synthesize riboflavin in significant quantities, including *Ashbya gossypii*, *Eremothecium gossypii*, and *Candida flareri*. Its bright yellow color positions riboflavin as an ideal candidate for natural dye applications, especially in the food and textile industries. Unlike man-made dyes, which often contain detrimental chemicals and contribute to ecological pollution, flavin-based dyes are biodegradable and nonhazardous. Additionally, riboflavin exhibits fluorescence under UV light, further enhancing its potential applications in bio-imaging and sensor technologies. The production of flavin dyes from fungi presents significant advantages due to their renewable and sustainable characteristics. Furthermore, genetic engineering techniques have increased flavin biosynthesis in fungi, improving yield and efficiency. Recent studies suggest that fungal flavins may also be utilized in antimicrobial coatings and biomedical applications, owing to their ability to generate reactive oxygen species upon exposure to light (Lyu et al., 2021). In light of growing concerns about environmental sustainability, fungal-derived flavin dyes offer a promising natural and eco-friendly alternative for various dyeing applications.

3.6 Phenazines

Fungal phenazines are a class of bioactive pigments characterized by their nitrogen content and recognized for their vibrant colors, which span from yellow to red. While phenazines are predominantly associated with bacteria, certain fungi can produce these compounds. As secondary metabolites, fungal phenazines show diverse biological activities, including antimicrobial, antifungal, and anticancer effects. Specific fungal species typically synthesize these compounds as part of their defense mechanisms or to enhance competition with other microorganisms in their environment (Yan et al., 2021). Environmental factors, including nutrient availability and stress conditions, typically influence the production of phenazines in fungi. Research on fungal phenazines is increasing due to their potential applications in biocontrol as natural antifungal and antibacterial agents and in developing eco-friendly dyes. Furthermore, fungal phenazines have been investigated for their roles in promoting plant growth and protecting plants from pathogens. These compounds are emerging as valuable resources in biotechnology and pharmaceuticals (Yao et al., 2023).

3.7 Monascins

Monascus purpureus, a filamentous fungus, is renowned for its capacity to produce natural pigments, primarily red, orange, and yellow, that are extensively utilized as dyes across various industries, including textiles and food. Monascin is synthesized via complex biosynthetic pathways involving polyketide and fatty acid synthases. Recent studies have concentrated on optimizing fermentation conditions to enhance pigment production, aiming to replace synthetic dyes with natural alternatives. Furthermore, research has explored the stability of Monascin pigments under different processing conditions, which is crucial for their application in food products. Utilizing Monascin pigments in food products encounters challenges concerning stability under varying processing conditions and compliance with safety regulations.

4 Applications of Fungal Pigments

Fungal pigments derived from *Aspergillus*, *Penicillium*, and *Fusarium* play crucial roles in various industries, mainly due to their bioactive properties. These pigments contribute to the aesthetic and sensory attributes of consumer products and exhibit extensive health benefits, including antioxidant, antimicrobial, and anticancer properties. The application of these pigments in pharmaceuticals is particularly noteworthy, as research highlights their potential to mitigate oxidative stress and protect against various diseases. Furthermore, with increasing environmental consciousness, using fungal pigments to produce eco-friendly dyes presents a significant opportunity for sustainable practices across the textile and cosmetic sectors. Such advancements reflect a broader trend towards integrating natural resources into industrial processes, providing an avenue for economic growth and environmental stewardship. Thus, the importance of fungal pigments is multifaceted, making them valuable assets in innovative, health-oriented applications.

4.1 *Antioxidant, Antimicrobial, and Anticancer Properties of Fungal Pigments*

Fungal pigments have attracted considerable interest for their outstanding anticancer, antimicrobial, and antioxidant activities, positioning them as promising candidates for pharmaceutical applications. The bioactivity of these pigments is primarily attributed to their chemical composition, which enables them to neutralize free radicals and reduce oxidative stress, thus offering potential protective effects against various diseases. Additionally, research indicates that fungal pigments demonstrate strong antimicrobial properties, effectively inhibiting the growth of pathogenic bacteria and fungi, thus serving as natural alternatives to synthetic preservatives. Moreover, studies have shown that certain pigments possess anticancer effects, contributing to the suppression of tumor growth and enhancing the efficacy of chemotherapy treatments. As the focus on sustainable and eco-friendly alternative grows, incorporating fungal pigments in pharmaceuticals addresses health benefits and aligns with the increasing global demand for natural and biodegradable products.

4.2 Cosmetics and Textiles

Exploring fungal pigments as eco-friendly dyes represents a significant advancement in the quest for sustainable alternatives to synthetic coloring agents. Fungal pigments, derived from various species, offer vibrant colors and possess intrinsic properties that align with environmental standards. Melanin, a pigment predominantly synthesized by *Aspergillus* sp., has been documented to exhibit antioxidant, antimicrobial, and ultraviolet (UV) protective activities. Due to these functional properties, melanin is incorporated into facial masks, sunglass lenses, and skincare formulations. Similarly, carotenoids such as astaxanthin, canthaxanthin, carotenes, fucoxanthin, lutein, lycopene, and zeaxanthin produced by *Cantharellus cinnabarinus* are utilized as bioactive components in sunscreens, sunblock creams, and antiaging facial treatments. Unilever, L'Oréal, Kanebo Ltd, etc., are a few cosmetic companies that produce lipsticks, face creams, tans, and balms from the pigments produced by *Monascus* sp., *Fusarium*, and *Penicillium*. A recent study demonstrated that the red (rubropunctamine) and yellow (monascinankaflavin) pigments synthesized by *Monascus purpureus* NMCCPF01 can serve as natural and safe enhancers of sun protection factor (SPF) when incorporated into marketable sunscreen formulations and aloe vera extracts. (Liu et al., 2020).

Garments act as carrier material for microorganisms produced due to sweat and other reasons; fungal pigments having antimicrobial activity have more scope as textile dyes. Pigments made from various fungi like *Monascus purpureus, Isaria farinose, Emericella nidulans, Fusarium verticilliodes*, and *Penicillium purpurogenum* used to dye cotton have been demonstrated to have a stronger shade compared to synthetic dye (Molelekoa et al., 2021).

4.3 Food and Beverage Industry

Research utilizing natural pigments derived from fungi has been very popular because of its benefits, including better yield, non-seasonality, and scalability. Red is the most widely used food colorant due to its strong visual appeal to consumers. β-Carotene is gaining prominence as a natural coloring agent in the food industry and is commonly applied at concentrations between 2 and 50 ppm in food products (Orlandi et al., 2022). Natural colorants, when added, not only give color but also antioxidant properties. Similar functional properties have been observed in canthaxanthin, an orange to deep pink pigment produced by *Bradyrhizobium* sp. Additionally, *Aphanizomenon* spp., a genus of cyanobacteria, synthesizes the blue pigment phycocyanin, which is extensively utilized in the food and beverage industry. Fermented rice with *Monascus* species has also been applied to prepare flavored milk products (Bampidis et al., 2024).

5 Production and Extraction Techniques

5.1 Factors Influencing Pigment Production

Many studies have aimed to enhance pigment production in various fungal strains by optimizing fermentation parameters, including orbital speed, light intensity, temperature, pH, and the type and composition of the growth medium (Gmoser et al., 2017). The content and quantity of secondary metabolites from fungal culture can be controlled by analyzing various crucial elements that mark the culture environment. The cultural environment has a substantial effect on the production of biomass and pigment. Different culture configurations, such as agitation, static conditions, aeration, and the use of solid or liquid media, as well as media constituents and culture conditions, such as temperature, pH, salt concentration, light exposure, and inoculum age, can all lead to improved biomass and pigment yields (Kalra et al., 2020). Fungal pigments can be produced using both submerged fermentation (SMF) and solid-state fermentation (SSF) methods. While SSF is well suited to various natural substrates, commercial-scale pigment production is predominantly conducted through submerged fermentation.

A wide range of studies have explored the potential of fungi to produce pigments using various natural substrates, such as rice, corn, wheat, cassava, whole and dehulled sorghum grains, and sorghum bran. Additionally, numerous agro-industrial by-products including feather meal, fish meal, cheese whey, grape pomace, soybean protein and meal, chicken feathers, rice husks, and residues from orange processing have been assessed as economical alternatives for supporting fungal pigment biosynthesis (Lopes et al., 2013). Several studies have also examined how various sugar sources such as glucose, fructose, lactose, sucrose, and maltose influence pigment production. Nitrogen sources play a crucial role in supporting both biomass production and pigment biosynthesis (Stanly Pradeep & Pradeep, 2013). They are essential for gene expression and regulation in synthesizing these compounds (Chatterjee & Chattopadhyay, 2015). Research has shown that incorporating various nitrogen sources can improve pigment yield, modify the fermentation broth color, and improve the light stability of pigments in *Monascus* species (Zhang et al., 2013). Additionally, NaCl has been identified as an effective enhancer that boosts pigment production while inhibiting citrinin synthesis in *M. purpureus*, without negatively impacting fungal growth (Zhen et al., 2019). In addition to chemical factors, physical factors also affect the pigment synthesis since the physiology and metabolism of fungi rely mainly on culture temperature and pH, which play a critical role in pigment formation. Culture temperature and pH regulate growth, fermentation, and the synthesis of primary and secondary metabolites (Méndez et al., 2011). Accordingly, some studies reported that temperature has an impact on how some fungal isolates, including *M. ruber*, *T. purpureogenus* (formerly known as *P. purpurogenum*), *C. aeruginascens*, and others, produce colors (Huang et al., 2017; Stange et al., 2019).

5.2 Extraction and Recovery of Fungal Pigments

According to Rao and co-workers, filamentous fungus is an excellent source of metabolites, including carotenoids, terpenoids, polyphenols, and polyketides (Narsing Rao et al., 2017). One of the most critical steps in effectively recovering these metabolites is the selection of extraction procedures, which mostly depends on the type of metabolite of interest and where it is located in the fungal culture. Researchers have explored various cultivation and extraction techniques to enhance fungal pigment production and recovery. However, developing a cost-effective, efficient, and safe extraction process remains a key challenge in scaling up the production of natural pigments. Various conventional techniques, including organic solvent extraction (such as Soxhlet extraction, homogenization, and shaking), hydrodistillation, centrifugal extraction, and steam distillation, have been explored for extracting pigments from different fungi (Sánchez-Muñoz et al., 2020).

However, scaling up these processes remains a complex challenge. Researchers have investigated and compared several strategies to improve pigment extraction and recovery from submerged fermentation. This involves the application of various surfactants, including Tween 80, Span 20, Triton X-100, and polyethylene glycol 8000, along with a wide range of solvents of varied polarity (Lagashetti et al., 2019). Additionally, advanced extraction techniques, such as pressurized liquid extraction, have been evaluated and validated for their efficiency in enhancing pigment recovery (Lebeau et al., 2017; Morales-Oyervides et al., 2017).

5.3 Influence of Microbial Pigments on Environmental Considerations

In recent years, global temperatures have been rising due to the release of anthropogenic gases from industrial activities, automobiles, and the widespread use of greenhouse gas-emitting systems. As a result, current research has shifted toward sustainable solutions, emphasizing green energy and environmental preservation. The release of harmful gases and wastewater from the production of synthetic colorants and their use in the textile industry has emerged as a significant environmental issue (Hernandez et al., 2016). Consequently, replacing synthetic colorants with microbial pigments presents a sustainable alternative, helping to reduce the release of harmful gases and pollutants associated with their production and use. Some efforts are needed to integrate natural pigments into industries that rely on synthetic colorants to mitigate industrial emissions leading to global warming and climate change. The interaction between green-energy-based industries and pigment-producing microbes offer a promising solution for reducing industrial atmospheric and liquid chemical effluents. It is crucial to prioritize ecosystem preservation over the aesthetic appeal of industrially produced colors, as synthetic colorants contribute to toxic emissions that negatively impact the environment and human health.

Furthermore, replacing synthetic chemical substrates with natural alternatives in fermentation systems can help reduce industrial emissions by lowering the demand for synthetic chemicals and the pollutants released during production. Light-harvesting primary pigments play a crucial role in capturing atmospheric CO_2. Additionally, numerous microbial species in agroecosystems actively contribute to carbon sequestration (Jiao et al., 2014; Tanet et al., 2020; Zhang et al., 2018). However, the role of pigmented microorganisms in CO_2 sequestration remains unexplored, highlighting a significant research gap. Nevertheless, microbial pigments, particularly those from bacteria, may indirectly contribute to CO_2 capture by promoting plant growth and serving as biocontrol agents against pests and diseases of crop plants (Kaur et al., 2019; Sen et al., 2019).

6 Challenges and Future Prospects

Natural pigments are secondary metabolites produced in small quantities as stress responses. Despite its nutritional benefits, the modern food industry has challenges in its production owing to raw material availability, yield, stability, purity, and sensitivity toward external factors. A primary concern is the low yield of fungal pigments. Wild fungal strains usually produce insufficient amounts, necessitating genetic modification, mutagenesis, or fermentation optimization to increase production rates (Dufossé et al., 2014; Keller, 2019). Growth conditions such as pH, light, moisture, temperature, and nutrient availability significantly impact production. It is difficult to optimize these factors, as even slight changes can substantially impact metabolite yield. Methods like the OSMAC (one strain many compounds) approach have been inspected to address this, but their effectiveness varies depending on the type of fungus (Rateb & Ebel, 2011). Additionally, compared to non-pigmented microorganisms, some fungal strains show poor growth after several subcultures, making maintenance more difficult (Ramesh et al., 2022). Another critical limitation is the formation of unwanted by-products, including the synthesis of toxic metabolites like mycotoxin by fungal pigments, which restricts their use directly in the food or as an additive. Additionally, natural pigments are highly sensitive to environmental conditions, resulting in fading and decreased shelf life. It limits their functionality in food matrices and other industrial applications. For instance, carotenoids are unstable under oxygen or light, while chlorophyll deteriorates rapidly under several circumstances (Sen et al., 2019).

Fungal pigment production often presents scalability issues. Switching from laboratory to industrial production requires optimal conditions, such as oxygen supply, pH, temperature, and substrate availability, which can be complex and costly in large bioreactors. Additionally, the design of bioreactors and the need for specific aeration, agitation, and substrate configuration complicate the process. Downstream processing methods for pigment extraction, especially for intracellular pigments, are centrifugation and solvent extraction, but these methods are costly, ineffective, and time-consuming. Effective and scalable extraction methods are constantly being developed. Another difficulty is achieving uniform pigment quality across batches, which remains a persistent hurdle in food and pharmaceutical applications (Baptista et al., 2020; Sen et al., 2019).

Genome mining is a potent technique for exploring the complete metabolic capacity of fungal strains, including dominant gene clusters. For large-scale production, these clusters can be expressed in heterologous hosts or altered to maximize pigment production. Microbial vectors such as *Bacillus subtilis, Corynebacterium glutamicum, Escherichia coli, Pichia pastoris*, and *Pseudomonas putida* can be used to clone the genes involved in pigment production, which is a dependable and economic strategy. For instance, *Rhodotorula mucilaginosa* KC8 has increased its carotenoid (β-carotene and torularhodin) production through metabolic engineering and mutagenesis (Wang et al., 2017). However, enzyme inhibition can hinder efficient biochemical production in engineered microbes. For example, Ma and co-workers found that substrate inhibition of lycopene cyclase significantly limited carotenoid biosynthesis in the yeast *Yarrowia lipolytica* (Ma et al., 2022). After genetic modifications, they successfully enhanced carotenoid production. Despite advancements, identifying the gene is insufficient to obtain high gene expression in the original organism or its host. Overcoming metabolic restrictions remains critical for efficient and commercially sustainable production.

The increasing demand for eco-friendly and natural products has significantly influenced market trends and the acceptance of natural colorants. Industries are looking into sustainable alternatives due to concerns about the detrimental impact of synthetic colorants on health and the environment. The biodegradability, stability, and potential health benefits, including antioxidant and antimicrobial properties of natural pigments, have made them desirable alternatives (Pailliè-Jiménez et al., 2020). Fungal pigments are becoming increasingly prevalent due to their association with healthy aging. As secondary metabolites, these pigments have a variety of bioactivities, including antioxidative, immunosuppressive, cholesterol-lowering, antitumor, anti-obesity, anti-atherosclerotic, and anti-Alzheimer's disease properties. Fungal pigments are appealing for application in the pharmaceutical and nutraceutical sectors because of these health advantages (Afroz Toma et al., 2023).

The market trends indicate a robust and increasing demand for natural colorants derived from fungal metabolites. Consumer acceptance is driven by health and environmental considerations, positioning fungal pigments as viable and sustainable alternatives to synthetic dyes in various industries (Afroz Toma et al., 2023).

7 Conclusion

Fungal metabolites present a rich and untapped resource for producing natural colorants and pigments. As industries seek more sustainable and eco-friendly alternatives to synthetic dyes, fungal pigments offer a viable solution with promising applications across various sectors. While progress has been made in extracting and commercializing these pigments, further research is needed to optimize production methods and overcome pigment stability and cost-effectiveness challenges. By advancing our understanding of fungal pigment biosynthesis and enhancing biotechnological production techniques, we can unlock the full potential of fungal-derived colorants, contributing to developing more sustainable and natural solutions in numerous industries.

References

Afroz Toma, M., Rahman, M. H., Rahman, M. S., Arif, M., Nazir, K. H. M. N. H., & Dufossé, L. (2023). Fungal pigments: Carotenoids, riboflavin, and polyketides with diverse applications. *Journal of Fungi, 9*(4), 454.

Aksu, Z., & Eren, A. T. (2005). Carotenoids production by the yeast *Rhodotorula mucilaginosa*: Use of agricultural wastes as a carbon source. *Process Biochemistry, 40*(9), 2985–2991.

Bampidis, V., Azimonti, G., Bastos, M. D. L., Christensen, H., Dusemund, B., Durjava, M., Kouba, M., López-Alonso, M., Puente, S. L., & Marcon, F. (2024). Safety of a feed additive consisting of hydroxypropyl cellulose for all animal species (Association Management & Regulatory Services Ltd). *EFSA Journal, 22*(2), e8626.

Baptista, S. L., Carvalho, L. C., Romaní, A., & Domingues, L. (2020). Development of a sustainable bioprocess based on green technologies for xylitol production from corn cob. *Industrial Crops and Products, 156*(August), 112867. https://doi.org/10.1016/j.indcrop.2020.112867

Brakhage, A. A. (2013). Regulation of fungal secondary metabolism. *Nature Reviews Microbiology, 11*(1), 21–32. Nature Publishing Group. https://doi.org/10.1038/nrmicro2916

Cataldo, V., Arenas, N., López, J., Camilo, C., & Agosin, E. (2018). Sustainable production of β-Xanthophylls in Saccharomyces cerevisiae. *Microbial Engineering*. https://dc.engconfintl.org/microbial/11

Chatragadda, R., & Dufossé, L. (2021). Ecological and biotechnological aspects of pigmented microbes: A way forward in developing food and pharmaceutical grade pigments. *Microorganisms, 9*(3), 637.

Chatterjee, S., & Chattopadhyay, P. (2015). Characterization of red pigment from *Monascus* in submerged culture red pigment from *Monascus purpureus*. *Journal of Applied Sciences Research, February*, 1–8.

Dufossé, L., Fouillaud, M., Caro, Y., Mapari, S. A. S., & Sutthiwong, N. (2014). Filamentous fungi are large-scale producers of pigments and colorants for the food industry. *Current Opinion in Biotechnology, 26*, 56–61. https://doi.org/10.1016/j.copbio.2013.09.007

Gmoser, R., Ferreira, J. A., Lennartsson, P. R., & Taherzadeh, M. J. (2017). Filamentous ascomycete fungi as a source of natural pigments. *Fungal Biology and Biotechnology, 4*(1), 1–25. BioMed Central Ltd. https://doi.org/10.1186/s40694-017-0033-2

Hernandez, V. A., Galleguillos, F., & Robinson, S. (2016). Fungal pigments from spalting fungi attenuating blue stain in *Pinus* spp. *International Biodeterioration and Biodegradation, 107*, 154–157. https://doi.org/10.1016/j.ibiod.2015.11.020

Hernández-Almanza, A., Montanez, J. C., Aguilar-Gonzalez, M. A., Martínez-Ávila, C., Rodríguez-Herrera, R., & Aguilar, C. N. (2014). *Rhodotorula glutinis* as source of pigments and metabolites for food industry. *Food Bioscience, 5*, 64–72.

Huang, T., Tan, H., Chen, G., Wang, L., & Wu, Z. (2017). Rising temperature stimulates the biosynthesis of water-soluble fluorescent yellow pigments and gene expression in *Monascus ruber* CGMCC10910. *Amb Express, 7*(1), 134. Springer. https://doi.org/10.1186/s13568-017-0441-y

Igreja, W. S., Maia, F. D. A., Lopes, A. S., & Chisté, R. C. (2021). Biotechnological production of carotenoids using low cost-substrates is influenced by cultivation parameters: A review. *International Journal of Molecular Sciences, 22*(16), 8819.

Jiao, N., Robinson, C., Azam, F., Thomas, H., Baltar, F., Dang, H., Hardman-Mountford, N. J., Johnson, M., Kirchman, D. L., Koch, B. P., Legendre, L., Li, C., Liu, J., Luo, T., Luo, Y. W., Mitra, A., Romanou, A., Tang, K., Wang, X., et al. (2014). Mechanisms of microbial carbon sequestration in the ocean-future research directions. *Biogeosciences, 11*(19), 5285–5306. https://doi.org/10.5194/bg-11-5285-2014

Kalra, R., Conlan, X. A., & Goel, M. (2020). Fungi as a potential source of pigments: Harnessing filamentous fungi. *Frontiers in Chemistry, 8*. https://doi.org/10.3389/fchem.2020.00369

Kaur, S., Panesar, P. S., Gurumayum, S., Rasane, P., & Kumar, V. (2019). Optimization of aqueous extraction of orevactaene and flavanoid pigments produced by *Epicoccum nigrum*. *Pigment and Resin Technology, 48*(4), 297–304. https://doi.org/10.1108/PRT-08-2018-0082

Keller, N. P. (2019). Fungal secondary metabolism: Regulation, function and drug discovery. *Nature Reviews Microbiology, 17*(3), 167–180. https://doi.org/10.1038/s41579-018-0121-1

Knox, B. P., & Keller, N. P. (2015). Biosynthesis and molecular genetics of fungal secondary metabolites. In *Biosynthesis and molecular genetics of fungal secondary metabolites* (Vol. 2, pp. 67–79). https://doi.org/10.1007/978-1-4939-2531-5

Lagashetti, A. C., Dufossé, L., Singh, S. K., & Singh, P. N. (2019). Fungal pigments and their prospects in different industries. *Microorganisms, 7*(12). https://doi.org/10.3390/microorganisms7120604

Lebeau, J., Venkatachalam, M., Fouillaud, M., Petit, T., Vinale, F., Dufossé, L., & Caro, Y. (2017). Production and new extraction method of polyketide red pigments produced by ascomycetous fungi from terrestrial and marine habitats. *Journal of Fungi, 3*(3). https://doi.org/10.3390/jof3030034

Lin, L., & Xu, J. (2020). Fungal pigments and their roles associated with human health. *Journal of Fungi, 6*(4), 280.

Liu, Y., Zhang, Y., Yu, Z., Qi, C., Tang, R., Zhao, B., Wang, H., & Han, Y. (2020). Microbial dyes: Dyeing of poplar veneer with melanin secreted by *Lasiodiplodia theobromae* isolated from wood. *Applied Microbiology and Biotechnology, 104*, 3367–3377.

Lopes, F. C., Tichota, D. M., Pereira, J. Q., Segalin, J., de Oliveira Rios, A., & Brandelli, A. (2013). Pigment production by filamentous fungi on agro-industrial byproducts: An eco-friendly alternative. *Applied Biochemistry and Biotechnology, 171*(3), 616–625. https://doi.org/10.1007/s12010-013-0392-y

Lyu, X., Gonzalez, R., Horton, A., & Li, T. (2021). Immobilization of enzymes by polymeric materials. *Catalysts, 11*(10), 1211. https://doi.org/10.3390/catal11101211

Ma, Y., Liu, N., Greisen, P., Li, J., Qiao, K., Huang, S., & Stephanopoulos, G. (2022). Removal of lycopene substrate inhibition enables high carotenoid productivity in *Yarrowia lipolytica*. *Nature Communications, 13*(1). https://doi.org/10.1038/s41467-022-28277-w

Meier, J. (1993). Handbook of U.S. colorants: Foods, drugs, cosmetics and medical devices. *Toxicon, 31*(1), 93. https://doi.org/10.1016/0041-0101(93)90370-x

Méndez, A., Pérez, C., Montañéz, J. C., Martínez, G., & Aguilar, C. N. (2011). Red pigment production by *Penicillium purpurogenum* GH2 is influenced by pH and temperature. *Journal of Zhejiang University: Science B, 12*(12), 961–968. https://doi.org/10.1631/JZUS.B1100039

Meruvu, H., & Dos Santos, J. C. (2021). Colors of life: A review on fungal pigments. *Critical Reviews in Biotechnology, 41*(8), 1153–1177.

Molelekoa, T. B. J., Regnier, T., da Silva, L. S., & Augustyn, W. (2021). Production of pigments by filamentous fungi cultured on agro-industrial by-products using submerged and solid-state fermentation methods. *Fermentation, 7*(4), 295. https://doi.org/10.3390/fermentation7040295

Morales-Oyervides, L., Oliveira, J., Sousa-Gallagher, M., Méndez-Zavala, A., & Montañez, J. C. (2017). Perstraction of intracellular pigments through submerged fermentation of *Talaromyces* spp. in a surfactant rich media: A novel approach for enhanced pigment recovery. *Journal of Fungi, 3*(3). https://doi.org/10.3390/jof3030033

Mukherjee, G., Mishra, T., & Deshmukh, S. K. (2017). Fungal pigments: An overview. In *Developments in fungal biology and applied mycology* (pp. 525–541). Springer.

Narsing Rao, M. P., Xiao, M., & Li, W. J. (2017). Fungal and bacterial pigments: Secondary metabolites with wide applications. *Frontiers in Microbiology, 8*(Jun). https://doi.org/10.3389/FMICB.2017.01113/FULL

Orlandi, V. T., Martegani, E., Giaroni, C., Baj, A., & Bolognese, F. (2022). Bacterial pigments: A colorful palette reservoir for biotechnological applications. *Biotechnology and Applied Biochemistry, 69*(3), 981–1001.

Ozeki, H., Ito, S., Wakamatsu, K., & Ishiguro, I. (1997). Chemical characterization of pheomelanogenesis starting from dihydroxyphenylalanine or tyrosine and cysteine.: Effects of tyrosinase and cysteine concentrations and reaction time. *Biochimica et Biophysica Acta (BBA)-General Subjects, 1336*(3), 539–548.

Pailliè-Jiménez, M. E., Stincone, P., & Brandelli, A. (2020). Natural pigments of microbial origin. *Frontiers in Sustainable Food Systems, 4*. Frontiers Media S.A. https://doi.org/10.3389/fsufs.2020.590439

Pavesi, C., Flon, V., Genta-Jouve, G., Pramil, E., Escargueil, A., Nasir, A., Montier, T., Franck, X., & Prado, S. (2023). Azaphilones pigments from the fungus *Penicillium hirayamae*. *Colorants, 2*(1), 31–41.

Ramesh, C., Prasastha, V. R., Venkatachalam, M., & Dufossé, L. (2022). Natural substrates and culture conditions to produce pigments from potential microbes in submerged fermentation. *Fermentation, 8*(9). https://doi.org/10.3390/fermentation8090460

Rapoport, A., Guzhova, I., Bernetti, L., Buzzini, P., Kieliszek, M., & Kot, A. M. (2021). Carotenoids and some other pigments from fungi and yeasts. *Metabolites, 11*(2), 92.

Rateb, M. E., & Ebel, R. (2011). Secondary metabolites of fungi from marine habitats. *Natural Product Reports, 28*(2), 290–344. https://doi.org/10.1039/c0np00061b

Riley, P. A. (1997). Melanin. *The International Journal of Biochemistry & Cell Biology, 29*(11), 1235–1239.

Ruiz-Sola, M. Á., & Rodríguez-Concepción, M. (2012). Carotenoid biosynthesis in *Arabidopsis*: A colorful pathway. *The Arabidopsis Book/American Society of Plant Biologists, 10*, e0158–e0158.

Sánchez-Muñoz, S., Mariano-Silva, G., Leite, M. O., Mura, F. B., Verma, M. L., da Silva, S. S., & Chandel, A. K. (2020). Production of fungal and bacterial pigments and their applications. In *Biotechnological production of bioactive compounds* (pp. 327–361). Elsevier. https://doi.org/10.1016/B978-0-444-64323-0.00011-4

Sen, T., Barrow, C. J., & Deshmukh, S. K. (2019). Microbial pigments in the food industry—Challenges and the way forward. *Frontiers in Nutrition, 6*. Frontiers Media S.A. https://doi.org/10.3389/fnut.2019.00007

Solano, F. (2016). Photoprotection versus photodamage: Updating an old but still unsolved controversy about melanin. *Polymer International, 65*(11), 1276–1287.

Stange, S., Steudler, S., Delenk, H., Werner, A., Walther, T., & Wagenführ, A. (2019). Influence of the nutrients on the biomass and pigment production of *Chlorociboria aeruginascens*. *Journal of Fungi, 5*(2). https://doi.org/10.3390/jof5020040

Stanly Pradeep, F., & Pradeep, B. V. (2013). Optimization of pigment and biomass production from *Fusarium moniliforme* under submerged fermentation conditions. *International Journal of Pharmacy and Pharmaceutical Sciences, 5*(Suppl 3), 526–535.

Tanet, L., Martini, S., Casalot, L., & Tamburini, C. (2020). Reviews and syntheses: Bacterial bioluminescence-ecology and impact in the biological carbon pump. *Biogeosciences, 17*(14), 3757–3778. https://doi.org/10.5194/bg-17-3757-2020

Toti, E., Chen, C.-Y. O., Palmery, M., Villaño Valencia, D., & Peluso, I. (2018). Non-provitamin A and provitamin A carotenoids as immunomodulators: Recommended dietary allowance, therapeutic index, or personalized nutrition? *Oxidative Medicine and Cellular Longevity, 2018*(1), 4637861.

Venil, C. K., Velmurugan, P., Dufossé, L., Renuka Devi, P., & Veera Ravi, A. (2020). Fungal pigments: Potential coloring compounds for wide ranging applications in textile dyeing. *Journal of Fungi, 6*(2), 68.

Wang, Q., Liu, D., Yang, Q., & Wang, P. (2017). Enhancing carotenoid production in *Rhodotorula mucilaginosa* KC8 by combining mutation and metabolic engineering. *Annals of Microbiology, 67*(6), 425–431. https://doi.org/10.1007/s13213-017-1274-2

Williams, K., Greco, C., Bailey, A. M., & Willis, C. L. (2021). Core steps to the azaphilone family of fungal natural products. *Chembiochem, 22*(21), 3027–3036.

Yan, J., Liu, W., Cai, J., Wang, Y., Li, D., Hua, H., & Cao, H. (2021). Advances in phenazines over the past decade: Review of their pharmacological activities, mechanisms of action, biosynthetic pathways and synthetic strategies. *Marine Drugs, 19*(11), 610.

Yao, X., Guo, H., Zhang, K., Zhao, M., Ruan, J., & Chen, J. (2023). *Trichoderma* and its role in biological control of plant fungal and nematode disease. *Frontiers in Microbiology, 14*, 1160551.

Zhang, X., Wang, J., Chen, M., & Wang, C. (2013). Effect of nitrogen sources on production and photostability of *Monascus* pigments in liquid fermentation. *IERI Procedia, 5*, 344–350. https://doi.org/10.1016/j.ieri.2013.11.114

Zhang, C., Dang, H., Azam, F., Benner, R., Legendre, L., Passow, U., Polimene, L., Robinson, C., Suttle, C. A., & Jiao, N. (2018). Evolving paradigms in biological carbon cycling in the ocean. *National Science Review, 5*(4), 481–499. https://doi.org/10.1093/nsr/nwy074

Zhen, Z., Xiong, X., Liu, Y., Zhang, J., Wang, S., Li, L., & Gao, M. (2019). NaCl inhibits citrinin and stimulates *Monascus* pigments and monacolin K production. *Toxins, 11*(2), 118. https://doi.org/10.3390/toxins11020118

Chapter 5
Role of Fungal Metabolites in the Development of Functional Foods

R. Preethi, J. Arunkumar, N. Unnamalai, and Sayantani Dutta

1 Introduction

Fungi are an extraordinarily diverse and vital part of our planet's ecosystems, with numerous organisms ranging from microscale yeasts to larger fleshy mushroom. These organisms possess unique biological characteristics and a remarkable capacity to influence a broad spectrum of industrial and ecological processes. From the depths of ancient forests to the fermenting vats of breweries, fungi exist everywhere. Their diversity forms the foundation for our knowledge of ecosystems, human nutrition, and the culinary arts in addition to being a testimony to evolution (Copetti, 2019; Marđetko et al., 2025; Ramachandraiah et al., 2021). With more than 1.5 million species identified, fungi play a key role in the cycling of nutrients, form symbiotic interactions with plants, and act as both pathogens and decomposers. Their adaptability and ecological significance are highlighted by its ability to survive in a variety of environments, from deteriorating wood, soil and in the human gut (Marđetko et al., 2025; Martin & van Der Heijden, 2024). Fungi have a big impact on human nutrition and culinary traditions in addition to their ecological significance. Many edible mushrooms are valued for their flavor and nutritional significance, and they are key ingredients in a wide variety of international cuisines. Fungi are also essential to biotechnology, helping to create antibiotics, enzymes, and other products that enhance our quality of life (Ambhore et al., 2024; Navarro-Simarro et al., 2024; Valverde et al., 2015).

R. Preethi (✉) · J. Arunkumar · N. Unnamalai
Shri A.M.M. Murugappa Chettiar Research Centre (MCRC), Chennai, Tamil Nadu, India
e-mail: preethir@mcrc.murugappa.org

S. Dutta
Grupo de Investigación en Polifenoles, Universidad de Salamanca, Campus Miguel de Unamuno, Salamanca, Spain

B. P. Singh et al. (eds.), *Fungal Additives and Bioactives in Food Processing Industries*, Fungal Biology, https://doi.org/10.1007/978-3-032-04519-5_5

In the early twentieth century, the discovery of penicillin transformed healthcare and proved fungi can be used to create drugs that save lives. Likewise, there are several fungal metabolites, which influence our eating habits. Especially, secondary metabolites produced from fungi—also known to as "natural products"—possess an application in food science, health, and ecological interactions. The importance of fungal metabolites in food science is increasing due to their ability to improve flavor. However, food safety and preservation also depend on certain fungal compounds because they contain antibacterial properties that inhibit the development of foodborne pathogens (Bahram & Netherway, 2022; Prescott et al., 2023; Zhang et al., 2023).

Conducting scientific studies on fungal variety is becoming more and more important now-a-days due to requirement for nutritious and healthy food production as worldwide population increases. Fermented foods and edible mushrooms that are high in vitamins, minerals, and bioactive substances throw an insight to address nutritional deficiencies and improve diet quality. Moreover, fungi play an integral part in biotechnological applications where fungal enzymes and bioactives are used to digest food, improve nutritional profiles, and ensure food safety (Li et al., 2023; Losoya-Sifuentes et al., 2025). Other than their benefits, certain fungi can produce harmful mycotoxins, which cause serious risks to human health. The dual nature of fungi, i.e., both beneficial and harmful, is crucial for maximizing their potential in food science while mitigating associated health risks. More research should be focused in the area of fungal genetics, metabolism, and ecology continues so the scope of their applications in food science and nutrition expands, that lead an innovative approach to food production, preservation, and health (Awuchi et al., 2021b; Daou et al., 2021). In light of these factors, a comprehensive overview of fungal diversity and their metabolites illustrates not only their significance in food science but also emphasizes the need for ongoing research and responsible utilization. Exploring the multifaceted relationships between fungi, food, and human health ultimately reveals the potential for fungi to contribute to sustainable solutions in our ever-evolving dietary landscape.

2 Types of Fungal Metabolites

2.1 Primary Metabolites: Nutritional Benefits

Fungi are rich in primary metabolites, which include carbohydrates, proteins, dietary fiber, and lipids. The growth and metabolism of the organism depend on these metabolites. They are essential for numerous applications in the food enterprise, especially in the area of functional foods, besides fungal development. These metabolites may significantly enhance the health benefits as functional foods, which are foods that provide beyond basic nutrition. Among the fungi family, mushrooms are the ones that usually proliferate in large-scale. Mushrooms have been scientifically proven to provide health benefits because of their high protein content and plenty of antioxidants, vitamins, fiber, carbohydrates, and minerals (Pouris et al., 2024). They are lower in fat (especially low in mono- and polyunsaturated fats), cholesterol, and calories. Their nutritional value and functional qualities enable them to operate as functional food; they serve as prebiotic, promoting the development of beneficial gut flora and offering a number of health advantages (Barzee et al., 2021).

One of the key carbohydrates produced by fungi is polysaccharides, such as chitin and glucans. The food industry uses chitin, which is mostly found in fungal cell walls, as a thickening agent and source of dietary fiber. Glucans, particularly β-glucans, have garnered attention for their health-promoting properties, including their ability to modulate the immune system and lower cholesterol levels. These polysaccharides can be incorporated into various food products as gelling agents, binders, thickeners, coagulants, humectants, stabilizers, and texturizers in bakery, low-fat, and dairy products, boosting their nutritional profile and contributing to functional attributes that promote health and well-being (Selvasekaran et al., 2021).

Fungal cell walls are primarily made up of indigestible polysaccharides like chitin and β-glucans, unlike plant cell walls. The β-glucans found in mushrooms such as *Agaricus bisporus*, *Pleurotus spp*, *Auricularia auricula*, *Grifola frondose, Flammulina velutipes*, *Hericium erinaceus, Schizophyllum commune, Pleurotus citrinopileatus, Ganoderma lucidum*, and *Lentinus edodes* have been suggested for use in food applications. However, only around 10% of intestinal epithelial cells are able to efficiently absorb β-glucans, which are difficult to digest and need to be fermented by bacteria; hence it acts as source of prebiotic and have potential health benefits. The benefits include better insulin and glucose responses, as well as a decrease in total and low-density lipoprotein (LDL) cholesterol (Table 5.1). As a result, β-glucans are widely used in food processing industry as emulsifiers and fat substitutes in products like yogurt and low-fat ice cream (Vaclav et al., 2018).

β-Glucans can produce short-chain fatty acids by forming viscous liquids and emulsions that promote healthy colonic microbiota. This reduces the risk of colorectal cancer by improving intestinal peristalsis and improving the colon's elimination of toxic chemicals. Furthermore, β-glucans reduce calorie intake by acting as fat mimetics, which makes products enhanced with them appropriate for both treating and preventing obesity. Zhu et al. (2016) studied the replacement in shortening in shortbread biscuits with oat and yeast β-glucans to improve health benefits. Replacing 5–15% of shortening with β-glucan increased the dietary fiber content and reduced energy and unhealthy fatty acids. *Lentinus edodes* mushrooms contain 514 g kg^{-1} β-glucans, which were used in cake formulation instead of wheat flour as a high fiber and low-calorie substitute. The study showed no significant differences compared to control (wheat-based cake), as the resulting cakes offer 1 g of glucan per serving (Kim et al., 2011). Similarly, in muffin preparation 40% of fat was replaced with yeast β-glucans in order to increase the nutritional profile. Adding 3% or more β-glucans while reducing fat (60% or more) increased batter hardness and degraded muffin crumb quality after baking and during storage (Zbikowska et al., 2022).

Table 5.1 List of fungal β-glucans and their bioactivity

Glucan	Fungus name	Bioactivity	References
Lentinan	*Lentinula edodes*	Antioxidant, anticancer properties, immunostimulatory, prevent obesity-associated cognitive, immunomodulatory	Pan et al. (2021), Wu et al. (2021), and Zhang et al. (2021)
Schizophyllan	*Schizophyllum commune*	Anti-inflammatory	Thongsiri et al. (2021)
Krestin	*Trametes versicolor*	Immune-boosting, antiviral and act as prebiotic	Maehara et al. (2012)
Grifolan	*Grifola frondosa*	Antitumor, supporting the treatment of diabetes	Chen et al. (2015) and Ramle et al. (2021)
Pleuran	*Pleurotus ostreatus*	Antiviral property	Urbancikova et al. (2020)
Scleroglucan	*Sclerotium glucanicumSclerotium rolfsii*	Prebiotics, immunostimulants, immunomodulators, antimicrobial and antiviral agents	Rahman et al. (2021)
β-glucans	*Lentinula edodes*	Prevention and treatment for COVID 19	Mirończuk-Chodakowska et al. (2021)

While the protein and polysaccharides in mushrooms can be used instead of red meat to help people shed weight by making them feel fuller, the fungal cell wall functions as a prebiotic by producing short-chain fatty acids in the large intestine during fermentation, which helps to control the gut microbiota and enhance health benefits (Colosimo et al., 2021). Chitin and chitosan have been extracted from edible mushrooms like *L. edodes*, *P. ostreatus*, *Pleurotus eryngii*, *Hypsizygus tessulatus*, *Flammulina velutipes*, *A. bisporus*, *G. frondosa*, *Pleurotus pulmonarius*, *Trametes versicolor*, and *Morchella esculenta*. Those polysaccharides are particularly attractive because it is vegan friendly alternative and free from the allergens when compared with marine-sourced chitosan (Alimi et al., 2023). Similarly, Osemwegie et al. (2020) showed that galactosaminoglucan from *Aspergillus fumigatus*, along with chitin and chitosan from *Candida albicans* and *Zygosaccharomyces rouxii*, improve food quality by acting as thickeners and stabilizers, enhancing texture and shelf-life in fermented foods like yogurt, cheese, and sour cream. Additionally, they play a major role in moisture retention, viscosity enhancement, and sugar crystallization in confectionery products. Furthermore, these polysaccharides possess bioactive properties like anti-tumor and immunomodulatory effects, which can enhance the health benefits of food products.

The shortage of protein, especially an essential animal amino acids and minerals like iron, zinc, fatty acids, and vitamins in a vegetarian or vegan diet makes it unbalanced and increases the risk of developing a number of illnesses. Hence, it is important to identify the suitable healthy alternative, where microbial or plant-based protein might be a sustainable solution. Since many species of fungi contain all of the essential amino acids, they are a great source of proteins in the same way. *Saccharomyces cerevisiae*, *Schizosaccharomyces pombe*, *Debaryomyces hansenii*, *Candida utilis*, *Yarrowia lipolytica*, *Wickerhamomyces anomalus*, *Hanseniaspora uvarum*, *Candida kefyr*, *Kluyveromyces lactis*, *Kluyveromyces marxianus* and the molds *Fusarium venenatum*, *F. graminearium*, *Aspergillus oryzae*, and *Monascus purpureus* are the various types of primary yeast that produces fungal proteins. These fungal proteins have unique properties: specific enzyme activity for various applications, effective biodegradation for sustainability, and significant nutritional benefits providing essential amino acids vital for human health. They are promising as a sustainable and nutritious food source. Mycoprotein, derived from the fermentation of fungi like *Fusarium venenatum*, is a prime example of a high-protein, low-fat, and environmentally sustainable alternative to meat.

The growing demand for plant-based and protein-rich diets has led to increased interest in these fungal proteins as functional food ingredients. Additionally, mycoprotein contains bioactive compounds that may help reduce the risk of chronic diseases, making it a valuable addition to the functional food market. Only one of the SCP fungal strains—the yeast *Yarrowia lipolytica*—has been approved for use in food by the Novel Food Regulation ((EU) 2017/2470). The fungal protein not only provides the nutrition but also acts as texture modifiers, preservatives, which provides pleasant taste and aroma. The mycoprotein of *Fusarium venenatum* is used in the Quorn brand, a widely accepted meat alternative (Amara & El-Baky, 2023; Derbyshire & Delange, 2021; Pobiega et al., 2024; Wang et al., 2023). A Nature publication evaluated the environmental benefits of replacing animal meat with fungal mycoprotein. The study modeled 20%, 50%, and 80% replacements of animal meat consumption. The results demonstrated that about 20% replacement could significantly reduce annual deforestation and CO_2 emissions and cut methane gas emissions. It also suggested that this shift would stabilize animal meat demand from 2025, countering expected rises in global pasture feed requirements (Derbyshire & Delange, 2021).

Protein digestibility is a key aspect of its bioavailability, which is associated with improved health benefits. A number of variables, including the food's structure, protein solubility, enzyme content, sources, and processing techniques, affect how effectively proteins are metabolized. Due to the presence of chitinous walls in fungi, it is relatively easy to digest and act as good source for dietary nutrients. However, the absorption of proteins from fungi, subsequently increases the essential amino acids in the bloodstream, which occurs at a slower rate than milk proteins. The primary cause of this is the fiber found in fungal cell walls, which increases nutrient absorption in the gastrointestinal tract and delays stomach emptying (Li et al., 2023). Al-Dalain (2018) demonstrated how Shiitake mushrooms can be beneficial by incorporating in beef sausage. Replacing 30% of the beef with Shiitake mushrooms showed the remarkable results by increasing essential amino acids by 1.11 times. The sausage with Shiitake mushrooms also received higher taste ratings in sensory evaluation. Flavor is an important factor in food products along with taste and texture contributing to how food is judged. Among the five basic tastes, most people like umami the most, which is related to meaty, salty, or broth-like flavors.

The thread-like structure of fungi was incorporated in food to make food feel like meat or can be changed to look like meat. *Agaricus bisporus Suksung*, which is a good source of protein (19–38% dry weight basis), is used as meat substitutes as it gives a good texture, firmness, stretchiness, chewiness, and savory flavor (Li et al., 2023; Usman et al., 2021). Additionally, mushroom protein crisps are regarded as a healthier alternative to conventional starchy snacks, owing to their high levels of protein, vitamins, minerals, and beneficial compounds such as beta-glucans, which are known for their immune-boosting properties (Balan et al., 2021).

Stale bread, which would otherwise go to waste, and brewers' spent grain, a byproduct of beer production, were successfully turned into nutrient-rich and protein-packed food products with the help of two types of edible fungi called *Neurospora intermedia* and *Rhizopus oryzae*. These fungi are particularly effective at transforming carbohydrates into proteins, and in this study, they worked to break down stale sourdough bread. Through this process, they converted approximately 65% of the starch in the bread into a protein content of about 21%, while also improving the balance and variety of amino acids in the final product, making it more nutritious. The process also enriched the food with important minerals such as copper, iron, and zinc. It also added valuable vitamins like vitamin E and vitamin D2, both important for overall health. Furthermore, the process introduced α-linolenic acid, an important fatty acid that has various benefits for the human body, including supporting heart health. The texture of the product was greatly improved during the process, resulting in a consistency that was very similar to that of commercially available soybean-based burgers (Assemie & Abaya, 2022).

In addition to carbohydrates and proteins, fungi produce healthy fats, including polyunsaturated fatty acids (PUFAs) such as omega-3 and omega-6 fatty acids, which is gaining recognition in the realm of functional foods (Sande et al., 2019). These healthy fats are known for their numerous health benefits, like balancing cholesterol levels and supporting cardiovascular health, reducing inflammation, preventing excessive blood clotting, and supporting overall body function. The bioactive compounds in fungi complement the fats and enhance the health benefits through synergistic effect. Fungi such as *Ganoderma lucidum* provide healthy fats along with variety of vitamins, minerals, and antioxidants that support metabolic rate and weight management. The oleaginous fungi like *Cunninghamella echinulata* produced 46.6% of total lipids, among that 14% are found to be gamma linolenic acid. Similarly, high levels of polyunsaturated fatty acids (PUFAs) were extracted from *Mortierella* sp., whereas *M. hyalina* and *M. elongata* produce 23% arachidonic acid and along with oleic acid, and 35% of ω-3 and ω-6 PUFAs were found in *Ashbya gossypii*. Shiitake mushrooms are well known to produce hydroxydocosanoic acid, hydroxytricosanoic acid, hydroxytetracosanoic acid and oleic acid. Various research studies demonstrate that incorporating fungal base functional foods into a diet can help in controlling obesity and metabolic disorders. Enhancing the nutritional profiles of food products through mushrooms and other fungi can effectively contribute to improved health outcomes (Cerone & Smith, 2021; Łysakowska et al., 2023; Rizzo et al., 2023).

Certain fungal species, such as *Mortierella alpina*, are efficient at producing these valuable fats. Incorporating these fungal lipids into food products not only enhances the fatty acid profile but also promotes health benefits that align with consumer trends favoring meals rich in essential fats. Patinho et al. (2021) studied *Agaricus bisporus* mushrooms as a fat substitute in beef burgers. The addition of mushrooms improved moisture and yield while reducing cooking loss. The lowest fat content occurred with 15% mushroom replacement, which also enhanced oxidative stability. Fungi like *Aspergillus wentii*, *Penicillium citrinum*, and *Candida lipolytica* produce acids used in food for flavor and preservation. These acids add tartness and control pH, preventing spoilage and extending shelf life, especially in jams and jellies. This natural method is safe, effective, and avoids synthetic preservatives, making it valuable for food manufacturers. Additionally, they reduce the acidity of beverages and reduce the enzymatic browning. Whereas the enzymes like amylases produced from *Aspergillus awamori*, *A. niger,* and *Aspergillus oryzae* have wide application including bread improvement, milk coagulation, and meat tenderization. Invertase is produced from *Saccharomyces spp* used as artificial honeys and soft candies (Ali et al., 2025; Copetti, 2019; Mishra et al., 2021). Many filaments' fungi can produce pigments that are able to replace the artificial colorants used in food industry. For examples carotenoids, melanins, monascins, violacein or indigo, lycopene and β-carotene, monascorubramine, and rubropunctatin, which also possess biological activity like antifungal, anticancer, antioxidant and antiobesity (Afroz Toma et al., 2023; Barzee et al., 2021). Moreover, fungi can transform raw ingredients into nutrient-dense products by enhancing digestibility and bioavailability. For example, fermentation with fungi can increase the levels of vitamins and bioactive compounds in foods, such as B vitamins and antioxidants. Vitamin B12 is an essential vitamin for humans, which can be obtained only from animal sources, but notably, tempeh can have significant amounts of B12 due to the presence of bacteria during fermentation, contributing up to 12.4 mcg of B12 per 100 g of tempeh, whereas the fungi like *Aspergillus oryzae* or *R. oligosporus* inoculated along with the bacteria lead to synthesis of Vitamin B1 and B2 and D2. However, its content depends on the substrate and environmental condition (Rizzo et al., 2023).

In conclusion, the use of fungal primary metabolites in the development of functional foods represents a promising and innovative approach to address the increasing demand for nutrition-focused products that promote health and well-being. These metabolites, which include essential compounds such as organic acids, amino acids, and vitamins, not only enhance the nutritional profile of foods but also provide bioactive properties that support various health benefits. By leveraging fungi as sustainable and efficient biofactories, this approach aligns with both consumer preferences for natural, health-oriented food products and global efforts toward sustainable food production. As research continues to advance, the integration of fungal primary metabolites into functional foods could play a pivotal role in improving public health while contributing to the evolution of the functional food industry.

2.2 Secondary Metabolites: Bioactive Compounds

Secondary metabolites produced by fungi play an essential role in food processing and offer notable health benefits and pharmacological effects immune boosting, anticancer, antioxidant, antiviral, antibacterial, antineoplastic, antiproliferative, antidementia, hepatoprotective, lower the risk of cardiovascular diseases, reduce cholesterol, treating Parkinson's disease and muscular degeneration. A study demonstrated that fermented rice had the highest nutritional and bioactive values, indicating a rich profile of secondary metabolites that enhance the rice's functional properties. Among various fungi, *Ganoderma*, *Phylloporia,* and *Sanghuangporus* have high medicinal values since the fruiting bodies, mycelia and basidiospores contain more than 400 bioactive compounds including triterpenoids, sterol, steroids, styryl, pyrones, and polychlorinated compounds with effective medical properties (Luo et al., 2024). Likewise, study identifies and characterizes 26 compounds, including eight new lanostane triterpenoids from *Ganoderma weberianum.* In traditional Chinese medicine, this particular fungus has a long history of usage. It is believed to have properties that help fight cancer and strengthen the body's immune system. Those isolated compounds act as potent antidiabetic in HepG2 cells and exhibit cytotoxicity against HeLa cells (Yang et al., 2022).

Most of the extensively researched pigments of fungi were extracted from four genera such as *Aspergillus, Penicillium, Paecilomyces,* and *Monascus*. Certain pigments like riboflavin (yellow), β-carotene (orange-yellow), lycopene (yellow to red), and canthaxanthin are globally marketed as food-grade colorants are approved and globally marketed (Afroz Toma et al., 2023; Di Salvo et al., 2023; Lin & Xu, 2020). Besides, pigments like polyketides, anthraquinones, and azaphilones are commonly synthesized by fungi such as *Monascus*, *Aspergillus*, *Penicillium*, and others. Polyketide pigments from *Monascus*, provide a natural and safer alternative to synthetic dyes (Lin & Xu, 2020). *Monascus* pigments create natural red, orange, yellow, green, and blue colors for foods. They are potent antioxidants, possibly reducing chronic disease risk. These pigments are used in foods like ice pops, yogurt, and candy.

Pigments like azaphilones enhances the functional properties in the food through enhancing light stability, water solubility, antioxidant properties, and even anti-atherogenic activities. This pigment also exhibits antibacterial and antifungal properties, which can contribute to food preservation (Afroz Toma et al., 2023). Fungal anthraquinone pigments are bacteriostatic, antiviral, and insecticidal, and can be used safely in a variety of food products such as milk products, ice cream, and confectionaries, up to specific regulatory limits (Qun et al., 2023). Few natural metabolites like bikaverin and immunomodulator endocrocin are produced from *Fusarium* sp., which was applied in nutraceuticals and functional food (Shankar & Sharma, 2022). By harnessing the unique benefits of fungal metabolites, food scientists can engineer products that cater to contemporary dietary needs, promote wellness, and reduce reliance on traditional animal-based food systems. In this way, fungi play a crucial role in the future of functional food development, combining sustainability with health benefits. There are different bioactive compounds that play an important role in food. This will be explained in detail along with their mechanism as follows.

3 Health Benefits and Mechanism of Action

From centuries fungal metabolites are familiar for their potential to improve human nutrition and health. Filamentous fungi are good example in this context; they are widely used in the food industry as well as in the production of functional foods and nutraceuticals, in the form of biomass or their metabolites. For various health benefits, several well-known species of microscopic filamentous fungi, including Penicillium, Aspergillus, and Fusarium, are also used in functional foods (Takahashi et al., 2020).

3.1 Immunomodulatory and Anticancer Effects

With increasing prevalence of tumor and cancer, the trend of research has been shifted toward natural antitumor and anticancer compounds. One of the functional pathways of these compounds is boosting the human immune system. Fungi, also known as medicinal mushrooms, are used to make nutraceutical supplements or capsules and are sold commercially as extracts or powders. Consumption of these products are reportedly capable of improving human immune system, resulting in faster recovery from diseases and enhanced resistance to infectious diseases. In fact, Davis and his team has reported 17 species of macroscopic fungi, which are capable of dual functionalities of anticancer activity through immune-activation and anti-inflammatory activity to stabilize the health of the cancer patients. These species include *Ganoderma applanatum, Cordyceps militaris, A. brazei, Ganoderma oregonense, F. officinalis, F. fomentarius, G. lucidum,* etc. (Zhao et al., 2020). Table 5.2 summarizes the therapeutic potentials of different species of edible mushrooms (macroscopic fungi).

Compounds produced by macroscopic fungi having immunomodulatory potency primarily include fungal immunomodulatory protein (FIP), lectins, terpenes and terpenoids, and polysaccharides (Zhao et al., 2020). Verticillins, a subgroup of fungal alkaloids, are reported to exhibit chemotherapeutic potential by inhibiting histone methyltransferases and affecting apoptosis and immune recognition (Wadhwa et al., 2024). Anticancer activity was reported for other alkaloids chaetomugilides isolated from *Chaetomium globosum TY1*, and vinblastine produced by *Catharanthus roseus, Fusarium oxysporum* (Kousar et al., 2022). Table 5.3 represents the cytotoxic potential of some fungal secondary metabolites against human tumor cell lines, and a comparison of their potency to that of the control used in the research.

Table 5.2 Summary of the therapeutic potentials of edible mushrooms

Edible mushroom	Therapeutic potentials
Agaricus bisporus	Reduction in development of breast cancer Lectin functions as immune stimulant
Agaricus blazei (sun mushroom)	Antimutagenic Anticarcinogenic Immunomodulation
Agaricus subrufescens (almond mushroom)	Anticancer Anti-diabetes Anti-hyperlipidemia, anti-atherosclerotic Chronic hepatitis Antimicrobial Immuno-stimulator Antitumor Antiallergic Antiviral
Pleurotus citrinopileatus	Lectins are immunomodulator, and possess antiproliferative and antitumor potencies Phenolics exhibit antioxidant activities Polysaccharide-peptides act as anticancer agent β-Glucans have antibacterial, antitumor, antiviral, immunomodulatory, and antioxidant properties
Pleurotus djamor	
Pleurotus eryngii	
Pleurotus flabellatus	
Pleurotus florida	
Pleurotus ostreatus	
Pleurotus sajor-caju	
Pleurotus pulmonarius	
Flammulina velutipes	Polysaccharides, dietary fiber, and antioxidants lower blood cholesterol and sugar levels
Auricularia polytricha	Helps in the reduction of bad cholesterol, low-density lipoproteins, and atherosclerosis Controls low blood sugar, tumor growth, and acts as anticoagulant
Panellus serotinus (Mukitake)	Prevention from non-alcoholic fatty liver disease
Antrodia cinnamomea	Treatment of hypertension, initial cancer, diarrhea, abdominal pain, intoxication with drug and food, and skin-itching
Cordyceps sinensis	Enhancement of cellular oxygen utilization capacity, production of ATP Antibacterial Blood sugar control Reduction in asthma Controls blood pressure Protects from kidney, liver, and heart diseases Neuroprotective
Cordyceps militaris	Antitumor, anticancer, insecticidal, antimetastatic, antibacterial, and anti-inflammatory Effective for influenza A and chronic bronchitis
Grifola frondosa	Activity against human gastric carcinoma

(continued)

Table 5.2 (continued)

Edible mushroom	Therapeutic potentials
Ganoderma sp.	Immunomodulatory, antiallergic, antioxidant effects Antiradiation activity Suppress tumor Anti-inflammatory and antiparasitic actions Used to treat cardiovascular, endocrine, and respiratory malfunction
Lentinus edodes	Lentinan and β-glucan suppress lymphocytic leukemia and tumors Protection from microbial infections by lentinan Hypocholesterolemic activity During immune response stimulation of nonspecific inflammatory responses, helper and cytotoxic T cells activities

Adapted from Behar et al. (2024) with modification

Interestingly, the same fungi *L. edodes* can produce different extracts with different but complementary health promoting activities. The fungus's crude aqueous extract has polysaccharides with antiviral and antioxidant properties that can activate macrophages and increase the synthesis of nitric oxide, cytokines, and proteins linked to phagocytosis. Lentinan, produced from the same fungi, has antitumor activity, and heterogalactan has anti-inflammatory properties (Zhao et al., 2020).

Similar to *L. edodes*, "Chaga" or *Inonotus obliquus* possess both antidiabetic and anticancer activities. The pigment produced by Chaga is used in folk medicine for a long time. The pigment contains both triterpenoids (e.g., inotodiol) and a great amount of melanins. The aqueous extract of the fungi exhibited antitumor and anticancer activities in several research; as example, reduction in the viability of human hepatoma (HepG2) cells in vitro, protection of DNA from damages caused by cancer, in vivo inhibition of tumor growth from melanoma B16–F10 cells, and suppression of proliferation of human epithelial (HeLa 229) cells. Apart from these activities, similar water-soluble pigment has also shown hypoglycemic activity and enhanced insulin sensitivity (Lin & Xu, 2020).

3.1.1 Protein and Protein-Conjugate

Production of Tumor Necrosis Factor (TNF)-α, interleukin (IL)-2, IL-3, IL-4, interferon (IFN)-γ are found to be increased by FIP-gsi isolated from *Ganoderma sinensis*. Inhibition of the proliferation of HeLa, MCF7 (breast cancer cell line), and A549 (lung carcinoma epithelial) cells are reported by FIP-Irh produced from *Lignosus rhinocerotis*. Similarly, *Ganoderma lucidum* produces FIP-glu1 (LZ-8) that is able to enhance the transcription of interferon and interleukin and TNF-α. Several other FIPs are found to display immunomodulatory effects by inducing the expression of different cytokines or enhance human peripheral blood lymphocytes. Various FIPs have also exhibited antitumoral activities in tissue culture (Zhao et al., 2020).

Table 5.3 Cytotoxic potency of specific fungal secondary metabolites against human tumor cell lines

Metabolite	Fungal origin	Yield	Cytotoxic activity			
			Human tumor cell	IC_{50}	Control	IC_{50}
7-Desmethyl-6-methylbostrycoidin	*F. solani*	7 mg/ 4.3 g extract	Breast (MDA MB 231)	0.73μM	Doxorubicin	0.07μM
			Pancreatic (MIA PaCa2)	0.64μM		0.04μM
			Cervical (HeLa)	0.71μM		0.05μM
			Non-small-cell lung (NCI HI975)	0.34μM		0.03μM
			Lung fibroblast (WI38)	6.42μM		0.35μM
7-Desmethylscorpinone	*F. solani*	5 mg/ 4.3 g extract	Breast (MDA MB 231)	1.51μM	Doxorubicin	0.07μM
			Pancreatic (MIA PaCa2)	0.98μM		0.04μM
			Cervical (HeLa)	0.96μM		0.05μM
			Non-small-cell lung (NCI HI975)	0.61μM		0.03μM
			Lung fibroblast (WI38)	5.84μM		0.35μM
Ergosterol	*Penicillium chrysogenum*	NI	Breast (MCF-7)	0.10 mM	NI	NI
	V. volvacea	500 mg/ 725 g extract	Prostate (PC-3 M)	27.98 ± 0.97μM	5-Fluorouracil	64.35μM
Ergosterol peroxide	*V. volvacea*	15 mg/ 725 g extract	Prostate (PC-3 M)	23.15 ± 1.54μM	5-Fluorouracil	64.35μM
3β, 5α, 9α, 14 α-Tetrahydroxy-ergosta-7,22-dien-6-one	*V. volvacea*	5.9 mg/ 725 g extract	Liver (HepG2)	20.72 ± 0.76μM	5-Fluorouracil	54.74μM

3β, 5α, 9α-Trihydroxy-ergosta-7,22-dien-6-one	*V. volvacea*	12.5 mg/ 725 g extract	Gastric (SGC-7901)	12.03 ± 0.77μM	5-Fluorouracil	75.05μM
			Liver (HepG2)	5.90 ± 0.44μM		54.74μM
Hypocriol A	*Hypocrea* sp.	50.3 mg/ 63.2 g extract	Colorectal (HCT116)	18.6 ± 0.7μM	Cisplatin	18.8 ± 1.9μM
			Cervical (HeLa)	7.7 ± 0.4μM		14.7 ± 0.8μM
			Lung (A549)	25.3 ± 2.5μM		13.8 ± 1.2μM
			Breast (MCF-7)	19.7 ± 0.4μM		17.6 ± 2.4μM
Hypocriol F	*Hypocrea* sp.	11.2 mg/ 63.2 g extract	Colorectal (HCT116)	2.7 ± 0.6μM	Cisplatin	18.8 ± 1.9μM
			Cervical (HeLa)	4.6 ± 0.1μM		14.7 ± 0.8μM
			Lung (A549)	15.3 ± 1.6μM		13.8 ± 1.2μM
			Breast (MCF-7)	23.6 ± 1.3μM		17.6 ± 2.4μM
Terrein	*A. terreus*	537.26 ± 23.42 g/ kg extract	Colorectal (HCT-116)	12.13μM	Doxorubicin	0.11μM
			Hepatocellular (HepG2)	22.53μM		0.85μM
Trichocarane E	*I. fumosorosea*	30 mg/ 200 g extract	Breast (MDA)	0.13μg/mL	Cisplatin	2.90μg/mL
			Breast (MCF-7)	2.46μg/mL		1.14μg/mL
			Ovary (SKOV-3)	1.01μg/mL		3.80μg/mL
			Cervical (HeLa)	2.32μg/mL		2.24μg/mL
			Lung (A549)	1.40μg/mL		2.13μg/mL
			Liver (HepG2)	1.87μg/mL		0.62μg/mL

(continued)

Table 5.3 (continued)

Metabolite	Fungal origin	Yield	Cytotoxic activity			
			Human tumor cell	IC_{50}	Control	IC_{50}
Trichocarane F	*I. fumosorosea*	41 mg/ 200 g extract	Breast (MDA)	0.89μg/mL	Cisplatin	2.90μg/mL
			Breast (MCF-7)	4.38μg/mL		1.14μg/mL
			Ovary (SKOV-3)	1.46μg/mL		3.80μg/mL
			Cervical (HeLa)	4.57μg/mL		2.24μg/mL
			Lung (A549)	1.66μg/mL		2.13μg/mL
			Liver (HepG2)	3.66μg/mL		0.62μg/mL

Takahashi et al. (2020), under open access Creative Common license
NI Not informed

Several macroscopic fungi are known to produce different lectins. Lectins, which are polypeptides conjugated with specific carbohydrates, are reported to demonstrate immunomodulatory properties by upregulation of the expressions of TNF-α and interleukins, stimulation of nitrite production, stimulation of the production of macrophage-activating factors, activating lymphocytes, etc. Lectin produced from *Hericium erinaceus*, *Hericium erinaceus* agglutinin (HEA), inhibits proliferation of HepG2 and MCF7 cells. Other lectins *Russula lepida* lectin (RLL) and *Inocybe umbrinella* lectin (IUL) also inhibit proliferation of HepG2 and MCF7 cells. *Clitocybe nebularis* lectin (CNL) inhibits proliferation of human leukemic T cells; *Lignosus rhinocerotis* lectin (LRL) inhibits proliferation of A549, MCF7, and HeLa cells. Lectin from *G. frondosa* exhibited in vitro cytotoxicity effect against HeLa cells. Antioxidant, antiviral, and antimicrobial properties are also reported from the lectins (Zhao et al., 2020).

3.1.1.1 Mechanism of Action

The quantity of FIPs produced by each macroscopic fungi is very low which is a major limitation in their research and commercialization, especially when these FIPs exhibit multifunctional activities in our immune system (Zhao et al., 2020). Structurally these protein molecules are similar to the variable region of the immunoglobulin heavy chains. Among its functions, FIPs can activate the phosphorylation of p38/MAPK resulting in enhanced production of nuclear factor (NF)-κB, also it can activate B cells and macrophages, and stimulate the production of Th1 (helper T cells) and Th2 cells from the differentiation and proliferation of Th0 cells (Wang et al., 2004). Other than p38/MAPK and NF-κB, FIPs are also found to induce the human peripheral blood monocyte for the activation of PI3K/Akt signaling pathway and produce IFN-γ (Zhao et al., 2020). Lectins exhibit immunomodulatory properties either by activation of immune system or through cytotoxic effects against cells. Lectins TML-1 and TML-2 produced from *L. mongolica* (syn. *T. mongolicum*) induce the production of TNF-α and nitrite and produce activation factors of macrophage IFN-γ and other cytokines (Zhao et al., 2020).

3.1.2 Polyketides

Polyketides are a group of diverse bioactive compounds having several important health benefits including antibiotic and immunomodulatory properties. Most of the fungal pigments produced by ascomycetous fungi (e.g., genera *Monascus* and *Neurospora*) fall in the polyketide group (Lin & Xu, 2020). Polyketides are synthesized by polyketide synthase (PKS) enzyme complex. These compounds include mycotoxins such as aflatoxin as well as pharmaceuticals such as lovastatin (Wadhwa et al., 2024). Anthraquinones along with hydroxyanthraquinones are familiar groups of polyketide pigments having anticancer potency. Around 700 anthraquinone compounds are identified in plants, lichens and fungi that are responsible for the different colors of the hosts (Lin & Xu, 2020).

Some polyketides produced by different endophytic fungi such as *Guignardia* sp. (IFBE028), *Penicillium chrysogenum* (QEN-24S), *Chaetomium globosum* are guignasulfide, penicitide A, and chaetoglobosin X, respectively. These compounds are known to possess anticancer activities against different cancer cells including HepG2, HCC (hepatocellular liver carcinoma), MFC cells, etc. (Behera, 2024). Bikaverin, a red pigment of naphthoquinone class, is produced by *Fusarium fujikuroi*, and has exhibited potent antineoplastic potency, which can be used against carcinoma, lymphoma, or sarcoma (Lin & Xu, 2020).

3.1.2.1 Azaphilones

Azaphilones are polyketide-based fungal metabolites containing a chiral quaternary center, a pyrone-quinone core and hydroxyl groups as substituents (Pimenta et al., 2021). *Aspergillus, Penicillium, Phomopsis, Talaromyces, Pestalotiopsis, Chaetomium* are among the genera that produce azaphilones (Lin & Xu, 2020). *Chaetomium globosum* is known to be a rich source of azaphilones. A few compounds of azaphilones from the strain MP4-S01–7 of *C. globosum* exhibited high anticancer potencies again gastric cancer cells MGC803 and AGS, which was higher than that of paclitaxel (Pimenta et al., 2021). Chaetoviridin-A, also produced by *C. globosum* has shown antitumor activity by inhibiting progression of tumor via 12-O-tetradecanoylphorbal-13-acetate in two-phase carcinogenesis (Lin & Xu, 2020).

3.1.2.2 Mechanism of Action

Hypothemycin produced by *Hypomyces trichothecoides* and *Coriolus versicolor* are known to selectively inhibit mitogen-activated protein kinase (MEK). Radicicol is another anticancer drug that inhibits the heat shock protein 90 (HSP90), which is a broad-spectrum target of anticancer agents since preventing HSP90 can cause immediate destabilization of hundreds of oncoproteins (Yuan et al., 2020).

3.1.3 Non-ribosomal Peptides (NRP)

Non-ribosomal peptides (NRPs), as the name suggests, are low molecular weight peptides synthesized by a mechanism different than that of protein synthesis in ribosome (Avalos & Limón, 2021). These NRPs are produced from proteinogenic and nonproteinogenic amino acids using the enzyme complex non-ribosomal peptidase synthase (NRPS) (Daley et al., 2017). NRPs are mRNA independent and experience changes in their amino acids during formation or after their release by tailoring enzymes through acylation, hydroxylation, glycosylation, or halogenation, which is different than the proteins that undergo modifications after their synthesis only (Avalos & Limón, 2021; Bills & Gloer, 2016). NRPs exhibit several types of therapeutic effects such as anticancer, antibiotic, and immunosuppressants. Destruxins are found to have anticancer and antiproliferative potentials and are obtained from *Metarhizium anisopliae*. Immunosuppressive NRP compounds include Cyclosporine A obtained from *Tolypocladium inflatum*, and gliotoxins produced by *Aspergillus fumigates* and *Trichoderma virens* (Bhattarai et al., 2021).

3.1.3.1 Hybrid NRP/PK

Several fungal hybrid polyketide synthase-non-ribosomal peptide synthetase (PKS/NRPS) are responsible for the production of different complex NRP/PK molecules (Bhattarai et al., 2021). A number of these hybrid compounds exhibit immunomodulatory properties, e.g., bleomycin and epothilone as anticancer agents (Iacovelli et al., 2021), and Pseurotin A as immunosuppressive agent (Bhattarai et al., 2021).

3.1.3.2 Mechanism of Action

Fungal NRPs vary with the length of the peptide and functions of domains (Bhattarai et al., 2021). Cyclosporine A is a widely used drug that is utilized to treat patients with organ transplantation. This drug especially works to inhibit the rejection of the transplanted organ or tissue graft (Daley et al., 2017). Cyclosporine A binds to cyclophilin A, an intracellular immunophilin, and forms a complex that inhibits the activity of calcineurin, which is a calcium- and calmodulin-dependent phosphatase. Calcineurin is the enzyme that dephosphorylates the nuclear factor of activated T cells, and in turn it activates transcription of interleukin 2. Inhibition of calcineurin thus suppresses the proliferation of IL-2 dependent T cells (Bills & Gloer, 2016; Colombo & Di Pietro, 2012; Das et al., 2024) (Fig. 5.1). Cyclosporine A also hinders the p38 and JNK signaling cascade (Das et al., 2024).

3.1.4 Terpenes and Terpenoids

Reportedly, terpenes and terpenoids stimulate the gene expression for the proteins associated with mitogen-activated protein kinases and NF-κB pathway. Monoterpenoids produced by *Pleurotus cornucopiae* inhibit HeLa and HepG2 cells proliferation, sesquiterpenoids such as anthracophyllone and nambinones C from *Anthracophyllum* sp. *BCC18695* and *Neonothopanus nambi*, respectively, are able to inhibit the proliferation of NCI-H187 cells (Zhao et al., 2020). *Eurotium chevalieri* and *Neosartorya pseudofischeri* produce a terpenoid eurochevalierine, which is reported to display cytostatic property in silico in lung cancer cells A549 and human glioblastoma cells U373 (Behera, 2024).

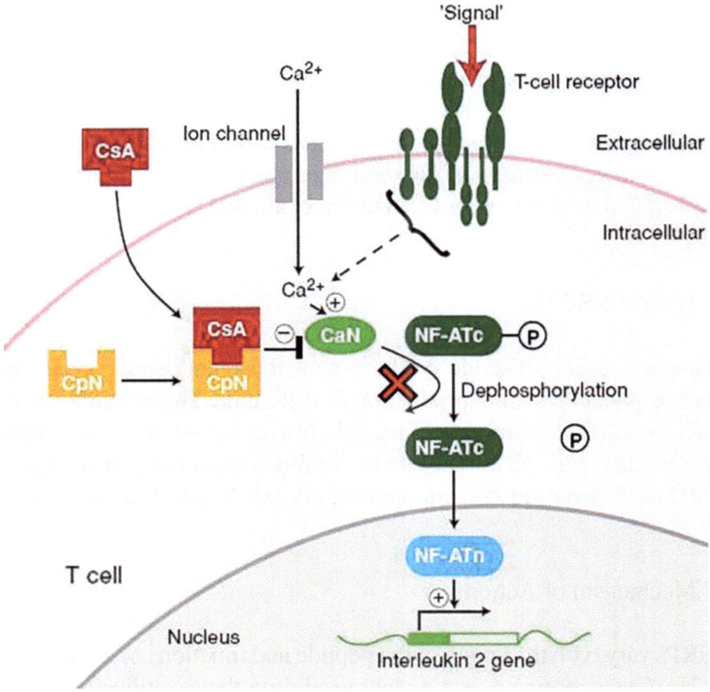

Fig. 5.1 Mechanism of action of cyclosporine A (CsA). CsA attaches to cyclophylin (CpN), and the complex binds and thus blocks enzyme calcineurin (CaN). Consequently, CaN cannot dephosphorylate the cytoplasmic component of the nuclear factor of activated T cells (NF-ATc), and in turn inhibits NF-ATc transport to the nucleus and binding of NF-ATc to the nuclear component of the nuclear factor of activated T cells (NF-ATn). Consequently, T cells do not produce IL-2. (Reproduced from Colombo and Di Pietro (2012) with slight modification; Open Access under Creative Commons Attribution 3.0 License)

Diterpene fusicoccanes produced from phytopathogenic fungi *Drechslera gigantean* and *Fusicoccum amygdali* have showed antitumor activity in vivo without any toxicity to the host (Behera, 2024). Diterpenoids including cyathins H, 11-*O*-acetylcyathatriol, neosarcodonin O obtained from *Cyathus africanus* inhibit the proliferation of K562 and HeLa cells. Different triterpenoids produced from *Ganoderma lucidum*, *Ganoderma orbiforme BCC 22324* are found to inhibit the proliferation of T-47D, LLC, Meth-A, and Sarcoma 180 cells, and NCIH187, MCF7, and KB cells, respectively (Zhao et al., 2020).

Paclitaxel is a potent chemotherapeutic drug from the class taxanes. It is approved by the U.S. Food and Drug Administration (FDA) and is produced by endophytic fungi *Taxomyces andreanae*. Paclitaxel is widely used to treat different cancers including lung, ovarian, and breast cancers by interfering with the microtubules and restricting the proliferation of tumor cells (Wadhwa et al., 2024). Other fungi that produce paclitaxel (Taxol) are *Pestalotiopsis versicolor*, which is found to produce the highest amount of paclitaxel, *P. neglecta*, *P. microspore,* etc., that also displayed anticancer properties (Behera, 2024).

3.1.4.1 Mechanism of Action

The mechanism of immunomodulatory effect of terpenes and terpenoids produced from macroscopic fungi is not yet well understood. The triterpenoids produced by *G. lingzhi* and *G. lucidum* are the most researched upon among this group of compounds, which show potent immunomodulatory properties (Zhao et al., 2020). Ophiobolins are terpenoids of fusicoccanes family that arrest the cell cycle at the G0/G1 phase and reduce viability of cancer cells. Along with p38/MAPK, ophiobolins also modify ERK (extracellular signal-regulated kinase) and JNK (c-Jun NH2-terminal kinase) (Behera, 2024).

Paclitaxel is able to activate several signaling pathways to display immunomodulatory and proapoptotic properties, these pathways are also used by paclitaxel for resistance (Kousar et al., 2022). At low concentration, this drug activates Raf-1 kinase or p53/p21, depending on its dose, that stimulates G0 and G1/S-phase apoptosis, whereas at high concentration, paclitaxel arrests mitosis at G2/M phase. Reportedly, paclitaxel can cause mitotic arrest even at very low dose but with more than 24 h exposure (Kampan et al., 2015). Figure 5.2 displays a summary of mechanism of action of paclitaxel.

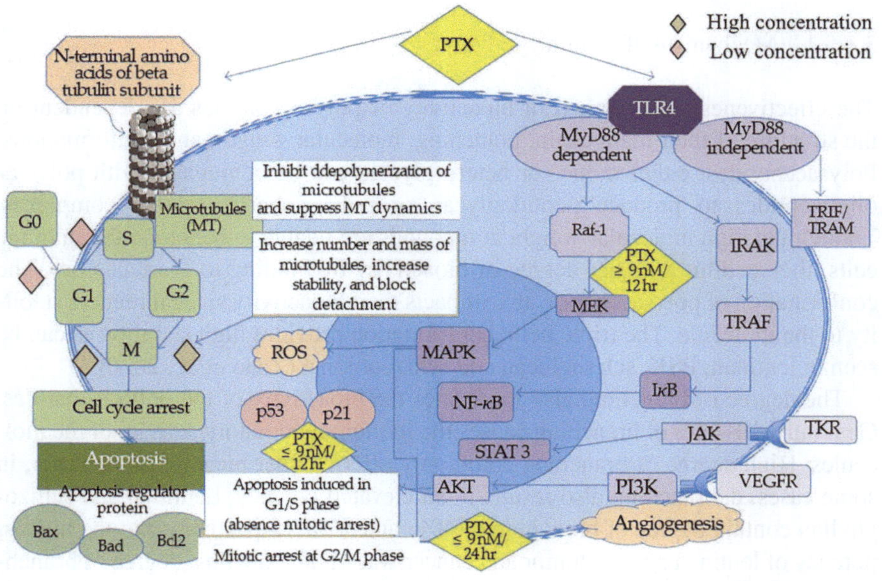

Fig. 5.2 Mechanism of action of paclitaxel. PTX Paclitaxel, G0 resting phase, G1 cells enlarge and make new protein, S phase DNA replication, G2 preparation for division, M phase cell division/mitosis, TLR4 Toll-like receptor 4, MEK/MAPK mitogen activated protein kinase, Raf-1 Raf kinase family, IRAK IL-1 receptor associated kinase, TRAF TNFR associated factor, NF-κB nuclear factor κB, TRIF/TRAM TIR-domain-containing adapter inducing interferon-β, TKR tyrosine kinase receptor, VEGFR vascular endothelial growth receptor, PI3K phosphoinositide 3-kinase, JAK janus kinase, STAT signal transducer and activator of transcription factor. (Reproduced from Kampan et al. (2015); Open Access under the Creative Commons Attribution License)

3.1.5 Polysaccharides

Immunomodulatory properties displayed by macroscopic fungi are mostly obtained from the polysaccharides; lentinan and schizophyllan being the highly known among them. Lentinan is isolated from *L. edodes* and effective for gastric cancer; schizophyllan is obtained from *Schizophyllum commune* and responsive against head and neck cancer. *Trametes versicolor* produces two polysaccharide-conjugates, a Krestin bound β-glucan polysaccharide K (PSK) and a polysaccharide peptide (PSP). PSK is commercially available and being utilized since 1977 as an aid to the cancer therapies. On the other hand, PSP contains monosaccharides arabinose and rhamnose, and is also found in Chinese market since 1987. The modes of operation of PSK are inducing apoptosis of cancer cells, inhibition of metastasis, or improving the gene expression of inflammatory cytokines. PSP is used for pain relief and improving the immune system of cancer patients by stimulating immune cell production, increasing infiltration of tumor, and improving the symptoms of chemotherapy. Several other polysaccharides are also reported that can stimulate NK and dendritic cells, macrophages, induces cancer cell apoptosis, TNF, IFN-γ, and IL-8 production (Zhao et al., 2020).

3.1.5.1 Mechanism of Action

The effectiveness and method of bioactivity of polysaccharides are dependent on the structure of their main chain, branching, molecular weight, and conformations. Polysaccharides, either mono- or hetero-glycan, can be conjugated with poly- or oligopeptides to produce peptidoglycan or polysaccharide-protein complexes. Though the high molecular weight compounds are unable to penetrate the immune cells, they exhibit a higher degree of bioactivity by binding to the receptors. The conformation of polysaccharide also impacts their bioactivity since it renders stability to the molecule. The triple helix conformation provides high stability, as can be seen in lentinan, PSK, scleroglucan and schizophyllan (Zhao et al., 2020).

The degree of branching also modulates the bioactivity of the polysaccharides. Generally, 20–40% of branching causes the immunomodulatory activity of the molecules. High degree of branching commonly offers higher bioactivity; however, in some cases, debranching also results in an elevated potency. Lentinan and schizophyllan contain degree of branching of 40% and 33%, respectively, but the highest potency of lentinan against tumor and cancer was obtained with a degree of branching of 32%. To enhance the bioactivity of these molecules, chemical modification is a general method that can increase their contact with the immune cell receptors. These modifications include formyl-methylation, carboxymethylation, aminoethylation, hydroxylation, and sulfation. Glucan obtained from *L. edodes* cell wall, post-modification by sulfation exhibited better anticancer and immunomodulatory activity compared to the unmodified molecule, which could be owing to the higher solubility of the sulphated compound (Zhao et al., 2020).

3.2 Antimicrobial Potency

3.2.1 Antibiotic

In 1929, Sir Alexander Fleming was the first person who has recognized the role of fungi in manufacturing antibiotic agents (Das et al., 2024). Examples of some common antibiotics derived from fungi are broad-spectrum penicillins from *Penicillium* sp. and cephalosporins from *Cephalosporium* sp. (Daley et al., 2017). Sir Alexander Fleming discovered penicillin from *Penicillium notatum* (Devi et al., 2020). Citrinin is another antibiotic obtained from *Penicillium* sp., which has shown its effect not only against Gram positive bacteria but also against Leishmania. Fusidic acid is another important antibiotic from *Fusidium coccineum* that works against Gram positive bacteria. Streptomycin is a common drug against Gram negative bacteria isolated from *Streptomyces griseus* (Das et al., 2024).

Diterpenoid pleuromutilin isolated from *Clitophilus scyphoides* generates retapamulin which is a topical antibacterial agent (Bills & Gloer, 2016). Acridone alkaloids are known mostly for their antibacterial activity. Several indole alkaloids screened from *Penicillium aurantiogriseum* have also shown potent antibacterial property (Devi et al., 2020). Mycotoxin trichothecenes are also reported to be antibiotics (Das et al., 2024).

Many of the fungal antibiotics work by preventing the cell wall synthesis. Some of the compounds are penicillins (Penicillin G, Penicillin F, Penicillin X, Ticarcillin, Ampicillin, Nafcillin, Amoxicillin), and cephalosporin (Etebu & Arikekpar, 2016). Another mechanism of antibacterial activity followed by fungal metabolites is restricting the bacterial protein synthesis. This mechanism is found to be used by retapamulin, which inhibits peptidyl transferase by binding to the 50S ribosome subunit (Prescott et al., 2023).

3.2.2 Antifungal

Ironically, several antifungal compounds have been isolated from different fungi. Griseofulvin is an early example of the antifungal drugs, first reported in 1938 (Prescott et al., 2023). A few antifungal secondary metabolites researched upon are benzomalvin C isolated from *Penicillium raistrick*, herbarin-A obtained from *Candida albicans*, indole alkaloids extracted from *Aspergillus flavus* and *Fusarium solani*, fusaric acid isolated from *Trichoderma harzianum*, etc. Likewise, flavonoids extracted from *Helminthosporium* sp., β-1,3-glucanolytic from *F. solani*, sesterterpenoids from *Aspergillus insuetus*, and flavonoids and naphthoquinones isolated from *Colletotrichum acutatum* and *Edenia gomezpompae* were reported to possess antifungal activities (Devi et al., 2020). Ibrexafungerp is a synthetic antifungal derivative of the triterpene glycoside enfumafungin, which has been found in *Endoconidioma carpetanum* (Prescott et al., 2023).

Griseofulvin is a topical antifungal agent produced by *Penicillium griseofulvin* that inhibits fungal growth by restricting fungal mitosis (Daley et al., 2017). This compound is utilized against fungal skin diseases (e.g., ringworms, athlete's foot disease). Another mechanism of antifungal effect is found to be the inhibition of glycan biosynthesis by cancidas-R, obtained from *Glarea lozoyensis* (Das et al., 2024).

3.2.3 Antiviral

Viral diseases, in the absence of proper drug/vaccine, can create severe epidemics/pandemics. Several fungal metabolites have shown antiviral properties. In general, antiviral compounds work by inhibiting the function of viral enzymes and nucleic acid formation (Das et al., 2024).

The mycotoxin trichothecenes are reported to be used against herpes simplex virus (HSV). Polysaccharides produced from Basidiomycetes have also shown potential against HSV and human immunodeficiency virus (HIV). *Ganoderma lucidum* produces triterpenoids ganoderic acid B, ganoderiol F, and ganodermanontriol, which showed antiviral effect against HIV-type I. Anti-HIV potency has also been reported for ganodermadiol from *G. pfeifferi*, lignin obtained from Chaga, Stachybosin D isolated from *Stachybotrys chartarum*, and mycelia of *L. edodes* (Das et al., 2024; Esheli et al., 2022). Apart from these compounds, hinnuliquinone, a bisindolyl quinone, has been reported to inhibit wild and clinically resistant HIV-1 protease. Activity against influenza A virus (H1N1) was observed from terpenoid stachyflin; meroterpenoid rhodatin isolated from basidiomycete *Rhodotus palmatus* showed potency against hepatitis-C (Esheli et al., 2022).

Different in silico and in vitro studies have reported a few fungal metabolites and their derivatives to hold potential against SARS-CoV-2. Among them are podophyllotoxin, lovastatin along with its derivative simvastatin, flaviolin, pyranonigrin A, jasmonic acid, cordycepin, etc. (Takahashi et al., 2020).

3.3 Activity Against Mental-Illness and Neuroprotection

There are several drugs available to treat mental illness; however, they are often associated with different side effects, therefore alternative compounds are being researched upon. Several fungal metabolites are found to be able to treat mental illness. Among them the potential components are psilocybin, ergine, ayahuasca, and lysergic acid diethylamide LSD, which act as the psychedelic substances. Treatment of brain function disorder requires the drug to cross the blood brain barrier (BBB) and has structural similarity with neurotransmitters (Wadhwa et al., 2024).

Psilocybin isolated from magic mushroom Psilocybe is a central element of the hallucinogenic prodrug psilocin. Psilocybin possesses good affinity toward dopamine receptors and serotonin transporter protein. Both psilocybin and psilocin are able to cross BBB. Ibotenic acid and muscimol are psychoactive compounds isolated from *Amanita muscaria.* Both these compounds can cross BBB and work by binding with glutamate receptors in the central nervous system (CNS). In spinal cord injuries, muscimol is used to reduce neuropathic pain (Wadhwa et al., 2024).

The ergot alkaloids obtained from fungi are known for their antimigraine property and used to treat Parkinson's disease owing to their effects on neuro-modulatory dopamine receptors. *Penicillium*, *Neotyphodium*, *Aspergillus*, and *Claviceps* are the sources of these compounds (Behar et al., 2024). Bromocriptine, a derivative of peptide alkaloid, is used along with L-dopa treatment to treat Parkinsonism (Das et al., 2024). Hericenones and erinacines are neurotrophic compounds obtained from *Hericium erinaceus.* These compounds are also used for treatment of Parkinson's and Alzheimer's disease since they can cross BBB, enhance nerve growth factor (NGF), and inhibit amyloid beta (Aβ) aggregation (Wadhwa et al., 2024).

Polyphenols obtained from different endophytic fungi exhibited potential as neuroprotective agent. Further, a triterpenoid produced by *Inonotus obliquus*, 2α-hydroxy-inotodiol (2α-HI), activates Nrf2 and BDNF/TrkB/ERK/CREB signaling pathways to reduce oxidative stress and apoptosis in SH-SY5Y cells, and thus, it can be used as dietary supplement to treat neurodegenerative diseases. Interestingly, *Alternaria eureka* biotransforms cyclocephagenol molecules in neuroprotective compounds. Cordycepin is a purine analog used in neurodegenerative diseases. Reportedly, dietary medicinal mushrooms are able to delay neurodegeneration and dementia by enhancing memory and cognitive functions in age-related neurodegenerative disorders. However, detailed research is further required to understand the mechanism of these compounds and further their therapeutic applications (Wadhwa et al., 2024).

4 Applications of Fungal Metabolites in Functional Foods

The integration of fungal metabolites into traditional foods has emerged as a groundbreaking approach in the development of functional foods, allowing the fusion of ancient culinary customs with the advancements of modern science. Fungal metabolites, such as polysaccharides, vitamins, enzymes, phenols, and other bioactive compounds, offer exceptional health-promoting properties. These compounds have been extensively studied for their potential to enhance physiological functions, boost immune health, and reduce the risk of chronic diseases (Yu et al., 2025). By incorporating beneficial fungal metabolites into commonly consumed traditional foods, it is possible to not only preserve cultural heritage but also create nutritionally superior products that cater to contemporary health demands. This innovative practice aligns perfectly with the growing consumer interest in functional foods that contribute to overall well-being.

Traditional foods, which are deeply rooted in various cultures, serve as an ideal platform for blending fungal metabolites to create value-added, nutrient-dense products. These foods provide a familiar foundation for functional food development, ensuring wide acceptance among diverse populations. For instance, staple foods such as bread, fermented beverages, yogurt, soups, and even rice-based dishes can incorporate fungal derivatives like beta-glucans or vitamin D-enriched fungal extracts. The addition of these metabolites not only enhances the nutritional profile of the food but also offers targeted health benefits, such as improved digestion, enhanced immunity, or better cardiovascular health (Corbu et al., 2023; Rousta et al., 2024). This approach represents a sustainable, innovative pathway to repurpose traditional foods while maintaining their flavor, texture, and cultural significance, bridging the gap between tradition and modern nutrition science. The following traditional foods can be blended with the fungal metabolites to develop a significant functional food product.

4.1 Incorporation in Traditional Foods

The incorporation of fungi directly into the fermentation process improves the sensory attributes and improve the nutritional value of the food. These fermented foods mentioned in Table 5.4 are linked to various health advantages like antioxidant, anti-diabetic, anti-inflammatory, anti-lipidemic properties. Jiuqu is one the major raw material in preparation of Huangjiu (Chinese rice wine). The incorporation of 2.5% *Lentinula edodes* (shiitake mushrooms) into traditional Jiuqu significantly enhances microbial diversity, approximately 1593 genera and 5507 species. Among these, key microorganisms such as *Aspergillus*, *Paecilomyces*, *Citrobacter*, *Erwinia*, *Raoultella*, *Pediococcus*, *Serratia*, *Latilactobacillus Lactobacillus*, and *Vibrio* play crucial role in improving nutritional value and enriching the flavor profile. This process promotes the production of specific volatile compounds, including alcohols, esters, and furans like 3-octanone and 1-octen-3-ol, which contribute to an enhanced sensory experience (Geng et al., 2024). Using *Aspergillus oryzae* to

Table 5.4 Incorporation of fungi in fermented foods

Food product	Microorganisms	Reference
Tempeh/Koji	*A. sojae, A. tamari, Rhizopus spp/Aspergillus oryzae, A. luchuensis* mut. *kawachii*	Tamang et al. (2022)
Glutinous rice tape	*Saccharomyces boulardii*	Yulianto et al. (2022)
Tauco	*Aspergillaceae Moromi, Aspergillaceae, Trichosporon, Candida*	Herlina and Setiarto (2025)
Saga tauco	*Rhizopus oligosporus and Rhizopus oryzae*	Amar et al. (2021)
Oncum	*Neurospora sitophila or Neurospora intermedia ascomycete, or Rhizopus oryzae*	Nurmilah et al. (2024)
Pliek	*Aspergillus niger, Aspergillus flavus, Penicillium citrinum, Thielaviopsis paradoxa, and Geotrichum candidum*	Asril et al. (2019)

ferment soybeans in soy sauce production improves the vitamins and amino acids, enhancing the nutritional value, assimilability, and absorbability of the final product (Dimidi et al., 2019).

The oyster mushroom was dried and incorporated into multiple chutney powders, which enhance the traditional taste along with the improved nutrition. The mushroom powder was mixed with traditional herbs (like Brahmi and Moringa leaves) and seeds (e.g., flax, sesame, groundnut), and coconut as ready-to-eat chutney powder with shelf life of 3 months. These chutneys show a wide range of nutritional profiles: protein content varies from 1.31% to 28%, fat content from 1.47% to 18.55%, and the powders contain essential minerals such as phosphorus, potassium, calcium, magnesium, iron, manganese, copper, and zinc. This development is a significant step in combining traditional ingredients with modern convenience, creating both nutritional and economic opportunities (https://www.iihr.res.in/mushroom-value-added-products-blend-tradition-nutrition-and-taste).

Researchers studied the liquid fermentation of *Ganoderma lucidum*, a medicinal mushroom, and extracted two types of natural compounds called polysaccharides: EPS-2-1 (found outside the mushroom cells during fermentation) and IPS-2-1 (found inside the cells). These compounds were tested for their ability to act as antioxidants. The results showed that both EPS-2-1 and IPS-2-1 act as potent antioxidants, but EPS-2-1 was more effective at protecting cells from damage (Luo & Xu, 2021). In biotechnology, bio-capsules, often made using a co-immobilization process, contain microorganisms like yeast and filamentous fungi. These structures allow the organisms to engage in synergistic activity while maintaining their functional capabilities. In the context of the beverages industry, especially in wine making, such co-immobilization of *P. chrysogenum* and *S. cerevisiae* enhances fermentation processes, offering new insights and potential applications for improving wine production techniques (Ogawa et al., 2019).

4.2 As Plant-Based Alternatives

Fermented mycelium became a crucial ingredient in producing plant-based meat alternatives instead of relying on traditional animal protein. This shift is motivated by the growing global population and concerns that there will not be enough animal protein to meet the demand. Mycelium is particularly promising because it is healthy and provides all the essential amino acids that our body requires, low fat, and a taste similar to meat owing to its natural flavor components. It has even been shown to have a better texture and taste compared to some plant-based protein alternatives. As a result, using mycelium to create simulated meats is considered an innovative and sustainable way to produce healthier food that can address future challenges in food supply and environmental sustainability. Mycelia from several mushroom species (*Bisporus*, *Pleurotus*, *Shiitake*, *Flammulina*, and *Volvariella*) serve as protein sources in meatless and dairy-free products like chicken patties, sausages, and cream cheese offering both high flavor and nutritional value (Li et al., 2023; Yan et al., 2023).

Edible mushrooms such as *Morchella esculenta*, *Pleurotus eryngii*, *Oudemansiella radicata*, *Lentinus edodes*, and *Grifola frondosa* contain all essential amino acids, including flavor-enhancing aromatic amino acids. Additionally, it is naturally low in fat, making it a healthy option. One of mycelium's unique features is its natural content of monosodium glutamate (MSG) like amino acid, 5′-nucleotides (5′-guanosine monophosphate (5′-GMP), 5′-inosine monophosphate (5′-IMP), 5′-xanthosine monophosphate (5′-XMP), and 5′-adenosine monophosphate (5′-AMP), the compounds that are responsible for savory, meat-like taste. This is a great choice for creating meat alternatives that are both flavorful and nutritious (Sun et al., 2020; Zhao et al., 2023). *Agaricus bisporus* have been used as a commercial option for the purpose of meat replacement due to their sensory properties, such as umami flavor, which contributes to a meat-like taste. Quantitative descriptive analysis of raw and roasted *Agaricus bisporus* mushrooms revealed that roasting significantly increased ($p < 0.05$) the intensity of dark meat flavor compared to raw mushrooms (Du et al., 2021). Using *Agaricus bisporus* mushrooms in place of animal fat in burgers results in a lighter, redder, juicier, more tender, and flavorful product (Patinho et al., 2021).

4.3 In Novel Food Formulation

Food enrichment and fortification are strategies to increase the nutritional value of foods. Enrichment enhances nutrients lost during food processing, while fortification enhances nutrients not originally present to address or prevent dietary deficiencies. A variety of methods and substances are used for fortification (Ahmad & Ahmed, 2019), in this context fungi were incorporated or replaced in various food products resulting in enormous functional and health benefits (Table 5.5). A study by Ng et al. (2017) found that adding 8% oyster mushroom (*Pleurotus sajor-caju*) powder to biscuits improves the fiber up to 8.62% and lowered the glycemic index to 49. This is attributed to the mushroom fiber reducing starch granule size and digestibility. Fungal metabolites enhance food functionality and exhibit various bioactivities, including potential benefits for diabetes management, cholesterol reduction, nerve regeneration, and improved memory and focus, as previously discussed. Hence, this fungal based food act as a functional food, targeted supplements, and in improving the gut health, especially in the field of personalized nutrition, where the dietary recommendations are based on an individual's specific needs, genetic profile, and microbiome.

As technology improves, researchers are combining the fields of biotechnology (for improving fungal metabolites' quantity and quality) and food technology (creating and enhancing food products) (Ajayeoba & Ijabadeniyi, 2025). This combination has opened new possibilities for enhancing nutrition in more personalized manners. Dutch food designer Chole Rutzerveld's "Edible Growth" uses 3D printing to create a snack where layered food, within 5 days, allows plants and

Table 5.5 Fungi-enriched food products

Food products	Type of fungi	Impact	Reference
Functional cookies and steamed bun	10–30% of *Lentinula edodes*	10% fungal polysaccharides in buns were preferred, while 20% polysaccharides in cookies performed better than the control. The study highlights the potential polysaccharides as a low-cost antioxidant flour for functional foods.	Ahmad et al. (2018)
Instant noodles	2–10% oyster mushroom (*P. ostreatus*) powder	Mushroom-incorporated noodles showed more protein and fiber as compared to control. Protein content and DPPH inhibition (70.11%) improved even at 2% addition	Arora et al. (2018)
Muffins	Shiitake mushroom blend	Higher polyphenol and carotenoid contents were found in enriched muffins with high antioxidant activity	Olawuyi and Lee (2019)
Extruded snacks	Straw mushroom (*Volvariella volvacea*) and phoenix mushroom (*Pleurotus pulmonarius*)	Showed higher hedonic scores	Tepsongkroh et al. (2020)
Biscuits	*Agaricus bisporus* polysaccharide flour/ *Pleurotus sajur-caju* (PSC) mushroom/ oyster mushroom	Mushroom-incorporated biscuit contains high protein, dietary fibers, amino acids, and minerals and resulted in low glycemic index	Bello et al. (2017), Owheruo et al. (2023), and Sulieman et al. (2019)
Pasta	Mushroom powder (white button, shiitake, porcini)	Incorporating mushroom powder into durum wheat pasta enhances its nutritional profile, lowers glycemic response, and boosts antioxidant capacity through increased fiber, protein, and phenolic compounds	Lu et al. (2018)
Mushroom–egg white patty	White and crimini (*Agaricus bisporus*) mushrooms at 0%, 10%, 20%, and 30%	Up to 20% mushroom substitution was well-accepted, pyrazines, furans, and pyrroles considerably increased.	Du et al. (2021)
Ketchup	Button mushroom (*Agaricus bisporus* L.)	Ketchup with a 50% tomato and 50% mushroom mix had the highest sensory scores. Increasing mushroom pulp increased protein, crude fiber.	Kumar and Barman Ray (2016)
Sausage	*Lentinula edodes* (0%, 25%, 50%, 75%, and 100%); straw mushrooms (*Volvariella volvacea*); *Flammulina velutipes*	Increased moisture, dietary fiber, methionine, glutamic, cysteine, total phenolic content, and antioxidant activity. Mushroom act as potential source for partial replacement of lean meat in sausages; enhanced the amount of amino acids and volatile compounds	Wang et al. (2018, 2019a, b)

(continued)

Table 5.5 (continued)

Food products	Type of fungi	Impact	Reference
Nuggets	Enoki *(Flammulina velutipes)* mushroom 2%, 4%, and 6%	Improved emulsion activity, fiber, mineral, and phenolics content. 4% of mushroom powder serve as a functional ingredient to make a healthier, nutrient-enriched meat product.	Banerjee et al. (2020)
Chicken patty	Oyster mushroom	Chicken patties with 10%, 20%, and 30% of mushroom showed significant changes to texture, hardness, and chewiness with high phenolic content.	Wan-Mohtar et al. (2020)
Bread	*Auricularia auricula* (0–10%)	5% mushroom flour improved dough viscosity, water retention, and gluten network structure.	Yuan et al. (2017)
Cookies	*Calocybe indica* mushroom (0, 5, 10%, 15%, and 20%)	Significantly increased protein, fiber, minerals (Fe, Zn, K), β-glucan, phenolic, flavonoid, and antioxidant contents	Rathore et al. (2019)
Frankfurters	Porcini mushroom (*Boletus edulis*)	Adding the mushroom slowed the growth of aerobic mesophilic microorganisms which increased the shelf life and improved the antioxidant property	Novakovic et al. (2020)

fungi to grow and yeast to ferment, transforming the solid interior to liquid (Eswaran et al., 2023). A study focuses on the potential of edible filamentous fungal (FF) pellets as a novel biomaterial for cell immobilization in food technology. Researchers examined the attachment and growth of yeast (*Saccharomyces cerevisiae*) and myoblast (C2C12) cells on FF pellets treated by heat or chemical inactivation. The results showed that yeast have proliferated about sixfolds over 48 h, which can be applicable in fermented beverages and cultivated meat production (Ogawa et al., 2019).

4.4 Fungi as Food Additives

Gas chromatography, electronic nose, and sensory analysis identified 1-octen-3-ol, 3-octanol, and 3-octanone as the major aroma compounds in *Pleurotus ostreatus*. In *Pleurotus eryngii*, benzaldehyde, contributing an almond aroma, was the most prevalent volatile compound. This aromatic compound is responsible for producing meat like flavor (Lavelli et al., 2018). A study examined the potential of mushrooms as a healthy substitute for meat in carne asada and taco blends, it is also an innovative method to reduce sodium without compromising consumer preference (Guinard et al., 2016). Phosphates are used in a lot of processed meats, often up to 65% of the product. They help the meat hold water, mix ingredients smoothly, maintain a good color, and prevent it from going bad quickly. The phosphates help keep the meat juicy and tender and give it a better color and longer shelf life. Without phosphates, the meat products might be drier, less appealing in color, and spoil more quickly. So, they are a key ingredient in making many processed meats that consumers are used to buying. However, alternative solutions are being required to maintain these quality characteristics without relying on synthetic substances (Thangavelu et al., 2019). A study demonstrated the use of winter mushroom powder (WMP) as a phosphate replacement in emulsion-type sausages. WMP also helped in reduction of fat and inhibits the lipid oxidation (Choe et al., 2018). Similarly, *Pleurotus ostreatus* was used as a phosphate in red tilapia fillet sausage and emulsion type sausage (Hleap-Zapata & Rodriguez-de-la-Pava, 2021; Jung et al., 2022). The food industry often uses sodium compounds in meat processing to enhance product quality. A study explored using plasma-treated winter mushroom powder as a replacement for synthetic nitrite and phosphate, commonly used in ground ham. The results indicated that the mushroom powder could achieve similar outcomes to sodium nitrite in improving the ground ham's characteristics. This suggests a potential for a more natural alternative to traditional meat processing additives (Jo et al., 2020). This study analyzed the composition, antioxidant activity, and microbial stability of low-nitrite Cantonese sausages with varying levels of chicken drumstick mushroom *Coprinus comatus* and backfat. *Coprinus* potentially replaces backfat and improves the flavor, microbiological stability, and reduced lipid oxidation, in Cantonese sausages (Qu et al., 2023). To replace the additive sodium tripolyphosphate in meat, shiitake mushroom (*Lentinus edodes* P.) powder was used in the pork patties. Due to the presence of unique flavor compounds such as lanthionine and 5′ribonucleotides in mushroom, it enhances the pork flavor (Chun et al., 2020). Shiitake, portabella, and Champignon mushroom extracts act as monosodium glutamate (MSG) replacers due to the presence of 22 free amino acids, five 5′-nulceotides, and 125 volatiles, which is responsible for the umami taste in the plant-based meat alternatives. Yeast and mushroom extract as MSG replacer improved sensory qualities like overall flavor, meaty taste, and umami in chicken soup (Wang et al., 2019c, d).

Alnoumani et al. (2017) demonstrated that adding *Agaricus bisporus* powder to refrigerated cooked ground beef reduced both lipid and protein oxidation; likewise in beef burger, incorporation of AB acts as a natural antioxidant, fat, and salt substitute (Patinho et al., 2021). Another study by Kurt and Gençcelep (2018) also examined mushroom powder (MP) as a meat emulsion enhancer. Adding 2% MP improved stability, texture, viscosity, heat resistance, and protein adsorption, creating a more structured emulsion. Shiitake mushroom extract was used as a taste enhancer in beef burgers. Mushroom extracts did not enhance saltiness at 0% or 75% NaCl reductions but improved salt perception at 50%, with 20% extract along with improved aroma, texture, and flavor (Mattar et al., 2018).

5 Challenges and Considerations

The role of fungal metabolites in the development of functional foods has numerous challenges and concerns. It exhibits various scientifically proven bioactive properties with numerous health benefits, such as antioxidants, anti-viral, anti-allergic, anti-tumor, anti-diabetes, anti-cancer, anti-obesity, and neuroprotective properties (Yu et al., 2025). However, ensuring the safeness and potency of these metabolites is essential, as some fungi produce several types of mycotoxins, which can be dangerous or life-threatening when consumed. Additionally, the inconsistency in the concentration of metabolites affects when it produces under various factors like temperature, relative humidity, water activity, pH, substrates, different strains, cultivation methods which cause difficulties in uniformity during the development of functional food products (Awuchi et al., 2021a; Chen et al., 2020; Khan, 2024). Existing regulatory issues related to the classification and approval of the fungal-based food products, as well as consumer acceptance and knowledge of fungal functional foods are a major obstacle for the budding entrepreneur or industrialist (Copetti, 2019; Corbu et al., 2023).

5.1 Safety and Regulatory Issues

Several significant variables play a role in the safety and regulatory issues regarding fungal metabolites in the development of functional foods as follows.

(a) *Toxicity Assessments*: Several research extensively studied and detailed the risk or hazards caused by mycotoxins that exist in food like aflatoxins, ochratoxins, and fumonisins, which is naturally occurring toxic compounds produced by molds. Apart from considering the individual toxin, a special focus is required on the issue of combined mycotoxin exposure, which has not received a considerable attention or research. The combined mycotoxin may have synergistic effects, which causes a severe adverse effects like carcinogenic risk, immunotoxicity, and neurotoxicity than the expected limits. Hence, an extensive comprehensive approach is required to evaluate the risk of combined mycotoxin, more intensive research is required on understanding the mechanisms of this multiple mycotoxin exposure on human health for the future risk management (Kumar et al., 2021; Stoev, 2024). As there are various limitations with existing conventional techniques such as TLC, HPLC, GC, ELISA, and lateral flow assays, high-performance biosensors such as immunosensors and aptamer-based biosensors have difficulty in achieving stability and false positives. More advanced biosensor technology and interdisciplinary collaboration are essential to ensure the safety of the food. Moreover, fungal-based companies competing in the market should also be labeled that their mycotoxin levels are within established limits. This requires serious effective monitoring and testing (Khan et al., 2024; Mafe & Büsselberg, 2024; Szelenberger et al., 2024).

(b) *Regulatory compliance:* The safety and regulatory issues associated with fungal metabolites in the advancement of nutrient dense food involve several key aspects. Fungal metabolites categorized in the section of the Novel Food Regulation (EU) 2015/2283. The European Food Safety Authority (EFSA) conducts careful safety evaluations regarding permissible levels of fungal metabolites. This includes toxicological studies, allergenic potential, and exposure assessments. Clinical evaluation is an essential and fundamental criterion for validating the applications and safety of fungal-based functional foods for human consumption. A double-blind randomized controlled trial (RCT) with placebo controls is normally part of the validation process because RCTs reduce bias and make it possible to make clear cause-and-effect decisions on health claims with ethical approval (Hariton & Locascio, 2018). However, studies carried out on animals and in vitro provide valuable data, scientists have been impossible to determine human-specific results. To ensure their validity of the research, results must be published in publications with peer review. Before any claims accepted or approved, scientific evidence supporting them must be authorized by regulatory organizations like the FDA or EFSA. However, in order to overcome practical and logistical challenges, RCTs require a lot of time, resources, and good planning to produce a reliable result.

Products like fungal based or its metabolites consider under novel foods, which require extensive safety evaluations and pre-market approval. Evaluations should also be focussed on the protocol of extraction, solvents subjected for metabolite extraction and production processes. Moreover, companies involved in production of fungal-based products need to provide substantial scientific evidence to support the safety of the metabolites, which satisfied the consumer fears about the products. However, adapting to different national legislation among EU member states is also a major challenge for both new entrepreneurs and established companies, which makes it tougher to bring commodities to market (Aryee et al., 2024; Ashwell et al., 2023; Granato et al., 2020).

(c) *Labeling and Claims:* Manufacturers have a responsibility to ensure their product labels provide clear, comprehensive, and easily understandable information regarding the sourcing and composition of ingredients, particularly those derived from fungi. This includes a detailed explanation of the fungal species used, their cultivation methods, and any processing steps involved. Crucially, the labeling should clearly communicate the specific health benefits and advantages of incorporating these fungal ingredients, substantiated by scientific evidence whenever possible, avoiding misleading or exaggerated claims. Transparency in labeling builds consumer trust and allows informed purchasing decisions. It is important that labels should follow the labeling regulation so the customer can understand what they are purchasing. Being transparent about the ingredients and processing methods makes it easier to gain the trust of consumers. When consumers understand how a product is made and what it contains, they are more likely to feel confident in buying the products. This degree of transparency develops a stronger bond between companies and customers. Food Safety and Standards Authority of India (FSSAI) in India intend to follow that label includes lot, code or batch number to assure the traceability during manufacturing and distribution for the novel foods. Moreover, the storage conditions like temperature, relative humidity, and shelf life must be mentioned (Priya & Alur, 2024; Sharma et al., 2024; Shobana et al., 2024). Foods containing fungal components are regulated by the EU Food Information to Consumers Regulation (EU FIC) in the EU. According to this law, all food labels must include allergens, nutritional information, and any potential health hazards. The European Food Safety Authority (EFSA) must approve any health claims referring to these fungal foods. Additionally, manufacturers must adhere to strict guidelines on the acceptable quantities of mycotoxins in food products (Khan et al., 2024; Kumar et al., 2021). For example, country like Japan follows "Foods with Function Claims" (FFC) refer to the functional ingredients other than nutrients specified, according to FFC, manufactures selling functional foods with FFC labels must state adverse health outcomes. This report is essential for evaluating the product quality and safety. Similarly, South Korea's amended 2014 Functional Health Food Act, where operators should register for tracking functional health foods from production to sale, creates a protocol for active traceability in case of any health issues. Additionally, the Ministry of Food and Drug Safety (MFDS) has set up a Health Functional Food Side Effect Reporting System to collect costumer reports on side effects. Consumers can report adverse effects using the number 1577–2488 found on product labels (Intrasook et al., 2024).

(d) *Long term study:* Companies like Host Defense, Four Sigmatic, OM Mushroom super food, Mushroom wisdom, Mycoformulas, Real mushrooms, Life Cykel marketing fungal-based bioactive products in the global market (Table 5.6). Hence, long-term R&D is recommended to determine the health outcome with respect to long term consumption of products containing fungal metabolites to ensure that they do not cause any chronic effects over a time.

5.2 *Consumer Awareness*

The market for functional foods is rapidly booming with fungal and fungal-based metabolites, first due to the need of alternative meat and second, their health benefits and nutrient profiles. This trend also reflects the growing awareness among consumers in several aspects: (a) Health benefits: Consumers have become more aware of the beneficial effects of fungal metabolites for their health, including anti-oxidant, anti-inflammatory, anti-cancer etc., (b) Sustainability: Growing interest among consumers in renewable and sustainable food sources has been reflected in an urge for items made from plants and fungi. (c) Genuineness and Transparency: Today's consumers have some knowledge about the product, which created an impact on their buying choices by expecting transparency about the ingredient, processing, and health claims made by food products. (d) Overall Wellness: Now-a-days, consumers are being educated to look for foods that help with prevention of diseases, fitness,

Table 5.6 Global companies incorporating mushrooms into innovative products

Company	Location	Description
Beyond Meat	El Segundo, California, USA	Uses mushrooms to create plant-based burgers and sausages
Impossible Foods	Redwood City, California, USA	Utilizes mushrooms to enhance the flavor and aroma of their plant-based burgers
The Mushroom Meat Co.	Los Angeles, California, USA	Produces various products including burgers, sausages, and mince, made primarily from mushrooms
Four Sigmatic	Los Angeles, California, USA	Offers a range of mushroom-infused coffees, hot chocolates, and elixirs that feature medicinal mushrooms like chaga, lion's mane, and reishi
Mondelez International	Deerfield, Illinois, USA	Developed a line of mushroom chips and crackers that cater to consumers seeking healthier snack options
MycoTechnology	Aurora, Colorado, USA	Developed a mycelium-based protein called PureTaste, a sustainable and complete protein source for plant-based foods. The company also produces ClearTaste, which uses mushroom-derived technology to block bitterness in foods and beverages
Meati Foods	Boulder, Colorado, USA	Specializes in whole-cut mushroom meats made from mycelium, such as chicken breasts and steak
Smallhold	Brooklyn, New York, USA	An urban farming startup that uses innovative technology to grow mushrooms in modular, small-scale farms located in urban environments. Their mini mushroom farms installed in grocery stores and restaurants, allow consumers to purchase freshly harvested mushrooms directly from the source
MyForest Foods	Green Island, New York, USA	Specializes in creating innovative plant-based products using mushrooms. Their flagship product is MyBacon, a bacon alternative made from a blend of oyster mushrooms and other plant-based ingredients
NexVeg	Flagstaff, Arizona, USA	Burgers and sausages feature mushrooms as a key ingredient, combined with beans, grains, and vegetables
Rritual Superfoods	Vancouver, Canada	A wellness company specializing in functional mushroom-based products designed to support well-being
Quorn	Stokesley, UK	A pioneering company in the field of meat substitutes, using mycoprotein derived from *Fusarium venenatum*. Quorn's range includes burgers, sausages, nuggets, and mince, all made from mycoprotein
The Vegetarian Butcher	Utrecht, Netherlands	Is renowned for its high-quality meat alternatives from plant-based ingredients, including mushrooms
Shandi Global	Singapore	A biotechnology company that develops plant-based meat alternatives using mushrooms and other plant ingredients

(continued)

Table 5.6 (continued)

Company	Location	Description
Fable Food Co.	Sunshine Coast, Australia	Specializes in creating mushroom-based meat alternatives. The company uses shiitake mushrooms as the primary ingredient for braised beef-style products and pulled pork alternatives
Life Cykel	Byron Bay, Australia	Offers a range of mushroom extracts and supplements. The company embraces a circular economy model by utilizing coffee waste to grow mushrooms

Reproduced from Contato and Conte-Junior (2025); with permission from Elsevier

and general well-being that led a trend toward natural health approaches (Davis et al., 2019; Dufossé et al., 2021; Tetiana Krupodorova et al., 2024a, b; Priya & Alur, 2024; Wadhwa et al., 2024).

5.3 *Market Trend and Conclusion*

Nowadays, market trends are increasingly shaped by consumer demand for natural and holistic dietary approaches that prioritize health and well-being. To align with these preferences, several key aspects need to be addressed. First, product innovation is essential, with growing interest in fungal metabolite-infused products such as beverages, dietary supplements, and snacks; companies like Four Sigmatic, Lifecykel, and RYZE are notable examples leading this space. Second, the popularity of these products is being driven by the effective use of e-commerce platforms to highlight their health benefits and increase consumer awareness. Third, there is a shift toward target-specific product development, with brands focusing on functional foods aimed at gut health, stress reduction, and other wellness areas—creating opportunities for fungal metabolite-based products due to their broad-spectrum health benefits (Abbaspour, 2024; Agrawal & Verma, 2021; Moura et al., 2023; Wadhwa et al., 2024).

To achieve success in this emerging market, four strategic steps are crucial: (i) investing in research and development to create safe, toxin-free fungal-based products supported by strong scientific validation and safety protocols; (ii) maintaining regular and professional communication with regulatory authorities to ensure compliance and address concerns early in the approval process; (iii) enforcing strict quality control measures to guarantee product safety and consistency, thereby enhancing consumer trust; and (iv) raising awareness through well-designed marketing campaigns that educate consumers on the health advantages and safety of fungal metabolites. In conclusion, the vast diversity of fungi and their metabolites underscores their growing importance in food science and nutrition. From enhancing flavor and delivering essential nutrients to promoting health and supporting sustainability, fungi play a vital role in our dietary practices. As ongoing research continues to reveal the potential of fungal biology, the field stands on the threshold of innovative breakthroughs that could significantly impact food systems, human nutrition, and overall well-being.

References

Abbaspour, N. (2024). Fermentation's pivotal role in shaping the future of plant-based foods: An integrative review of fermentation processes and their impact on sensory and health benefits. *Applied Food Research, 4*, 100468. https://doi.org/10.1016/j.afres.2024.100468

Afroz Toma, M., Rahman, M. H., Rahman, M. S., Arif, M., Nazir, K. H. M. N. H., & Dufossé, L. (2023). Fungal pigments: Carotenoids, riboflavin, and polyketides with diverse applications. *Journal of Fungi, 9*, 454.

Agrawal, K., & Verma, P. (2021). Chapter 1 – Fungal metabolites: A recent trend and its potential biotechnological applications. In J. Singh & P. Gehlot (Eds.), *New and future developments in microbial biotechnology and bioengineering* (pp. 1–14). Elsevier. https://doi.org/10.1016/B978-0-12-821005-5.00001-6

Ahmad, A., & Ahmed, Z. (2019). Chapter 3 – Fortification in beverages. In A. M. Grumezescu & A. M. Holban (Eds.), *Production and management of beverages* (pp. 85–122). Woodhead Publishing. https://doi.org/10.1016/B978-0-12-815260-7.00003-1

Ahmad, R., Zain, N. A. M., Hassan, N. A. M., Peryasamy, J., & Halim-Lim, S. A. (2018). Fruiting-body-base flour from an oyster mushroom—A waste source of antioxidative flour for developing potential functional cookies and steamed-bun. *AIMS Agriculture and Food, 3*, 481–492.

Ajayeoba, T. A. & Ijabadeniyi, O. A. (2025). Transforming food for the future: Precision fermentation as a key to sustainability, nutrition, and health. SSRN, 23. https://doi.org/10.2139/ssrn.5119686

Al-Dalain, S. Y. A. (2018). Utilization of mushroom fungi in processing of meat sausage. *Research on Crops, 19*, 294–299.

Ali, S. S., Elsamahy, T., Al-Tohamy, R., & Mahmoud, Y. A.-G. (2025). Chapter 15 – Properties and applications of fungal polysaccharides. In R. S. Singh & R. Bhari (Eds.), *Fungal biotechnology* (pp. 415–439). Academic Press. https://doi.org/10.1016/B978-0-443-13263-6.00015-8

Alimi, B. A., Pathania, S., Wilson, J., Duffy, B., & Frias, J. M. C. (2023). Extraction, quantification, characterization, and application in food packaging of chitin and chitosan from mushrooms: A review. *International Journal of Biological Macromolecules, 237*, 124195. https://doi.org/10.1016/j.ijbiomac.2023.124195

Alnoumani, H., Ataman, Z. A., & Were, L. (2017). Lipid and protein antioxidant capacity of dried Agaricus bisporus in salted cooked ground beef. *Meat Science, 129*, 9–19. https://doi.org/10.1016/j.meatsci.2017.02.010

Amar, A., Makosim, S., Sukotjo, S., Ahadiyanti, N., & Weisman, E. (2021). Growth dynamics of mold-yeast and bacteria during the production process of saga tauco [Adenanthera pavonina]. *IOP Conference Series: Earth and Environmental Science*. IOP Publishing, *741*, 012019.

Amara, A. A., & El-Baky, N. A. (2023). Fungi as a source of edible proteins and animal feed. *Journal of Fungi, 9*, 73.

Ambhore, J. P., Adhao, V. S., Rafique, S. S., Telgote, A. A., Dhoran, R. S., & Shende, B. A. (2024). A concise review: Edible mushroom and their medicinal significance. *Exploration of Foods and Foodomics, 2*, 183–194.

Arora, B., Kamal, S., & Sharma, V. P. (2018). Nutritional and quality characteristics of instant noodles supplemented with oyster mushroom (P. ostreatus). *Journal of Food Processing & Preservation, 42*, e13521.

Aryee, A. N. A., Ofori, K. F., Otchere, E. K., Dare, K. O., & Akanbi, T. O. (2024). Food-derived bioactive components and health claims: The role of regulatory agencies. In *Plant food phytochemicals and bioactive compounds in nutrition and health* (pp. 498–529). CRC Press.

Ashwell, M., Hickson, M., Stanner, S., Prentice, A., & Williams, C. M. (2023). Nature of the evidence base and strengths, challenges and recommendations in the area of nutrition and health claims: A Position Paper from the Academy of Nutrition Sciences. *British Journal of Nutrition, 130*, 221–238.

Asril, M., Perdana, A. T., Mahyarudin, Asmarany, A., & A'yun, Q. (2019). *Isolation of the fungus that plays a role in the process of making pliek u (aceh typical fermented food)*. Majalah Ilmiah Biologi Biosfera, 26-34. https://doi.org/10.20884/1.mib.2019.36.1.807

Assemie, A., & Abaya, G. (2022). The effect of edible mushroom on health and their biochemistry. *International Journal of Microbiology, 2022*, 8744788.

Avalos, J., & Limón, M. C. (2021). Fungal secondary metabolism. *Encyclopedia, 2*, 1–13.

Awuchi, C. G., Ondari, E. N., Eseoghene, I. J., Twinomuhwezi, H., Amagwula, I. O., & Morya, S. (2021a). Fungal growth and mycotoxins production: Types, toxicities, control strategies, and detoxification. In *Fungal reproduction and growth* (p. 100207). IntechOpen.

Awuchi, C. G., Ondari, E. N., Ogbonna, C. U., Upadhyay, A. K., Baran, K., Okpala, C. O. R., Korzeniowska, M., & Guiné, R. P. F. (2021b). Mycotoxins affecting animals, foods, humans, and plants: Types, occurrence, toxicities, action mechanisms, prevention, and detoxification strategies—A revisit. *Food, 10*, 1279.

Bahram, M., & Netherway, T. (2022). Fungi as mediators linking organisms and ecosystems. *FEMS Microbiology Reviews, 46*, fuab058.

Balan, V., Novak, D., Knudson, W., Jones, A. D., Iñiguez-Franco, F. M., Auras, R., Cho, S., Rodgers, A., & Ubanwa, B. (2021). Nutritious mushroom protein crisp–healthy alternative to starchy snack. *Food Production, Processing and Nutrition, 3*, 1–15.

Banerjee, D. K., Das, A. K., Banerjee, R., Pateiro, M., Nanda, P. K., Gadekar, Y. P., Biswas, S., McClements, D. J., & Lorenzo, J. M. (2020). Application of enoki mushroom (Flammulina Velutipes) stem wastes as functional ingredients in goat meat nuggets. *Food, 9*, 432.

Barzee, T. J., Cao, L., Pan, Z., & Zhang, R. (2021). Fungi for future foods. *Journal of Future Foods, 1*, 25–37. https://doi.org/10.1016/j.jfutfo.2021.09.002

Behar, N., Sharma, S., Parihar, R., Dubey, S. K., Mehta, S., & Pandey, V. (2024). Chapter 26 – Role of fungal metabolites in pharmaceuticals, human health, and agriculture. In K. A. Abd-Elsalam & H. I. Mohamed (Eds.), *Fungal secondary metabolites* (pp. 519–535). Elsevier. https://doi.org/10.1016/B978-0-323-95241-5.00005-8

Behera, A. (2024). Chapter 13 – Fungal metabolites as anticancer agent and their agricultural applications. In K. A. Abd-Elsalam & H. I. Mohamed (Eds.), *Fungal secondary metabolites* (pp. 259–282). Elsevier. https://doi.org/10.1016/B978-0-323-95241-5.00019-8

Bello, M., Oluwamukomi, M. O., & Enujiugha, V. N. (2017). Nutrient composition and sensory properties of biscuit from mushroom-wheat composite flours. *Archives of Current Research International, 9*, 1–11.

Bhattarai, K., Kabir, M. E., Bastola, R., & Baral, B. (2021). Fungal natural products galaxy: Biochemistry and molecular genetics toward blockbuster drugs discovery. *Advances in Genetics, 107*, 193–284.

Bills, G. F., & Gloer, J. B. (2016). Biologically active secondary metabolites from the fungi. *Microbiology Spectrum, 4*, 10–1128.

Cerone, M., & Smith, T. K. (2021). A brief journey into the history of and future sources and uses of fatty acids. *Frontiers in Nutrition, 8*, 570401.

Chen, Y.-H., Lee, C.-H., Hsu, T.-H., & Lo, H.-C. (2015). Submerged-culture mycelia and broth of the maitake medicinal mushroom Grifola frondosa (higher basidiomycetes) alleviate type 2 diabetes-induced alterations in immunocytic function. *International Journal of Medicinal Mushrooms, 17*, 541.

Chen, B., Sun, Y., Luo, F., & Wang, C. (2020). Bioactive metabolites and potential mycotoxins produced by Cordyceps fungi: A review of safety. *Toxins (Basel), 12*, 410.

Choe, J., Lee, J., Jo, K., Jo, C., Song, M., & Jung, S. (2018). Application of winter mushroom powder as an alternative to phosphates in emulsion-type sausages. *Meat Science, 143*, 114–118. https://doi.org/10.1016/j.meatsci.2018.04.038

Chun, S., Chambers, E., IV, & Chambers, D. H. (2020). Effects of shiitake (Lentinus edodes P.) mushroom powder and sodium tripolyphosphate on texture and flavor of pork patties. *Food, 9*, 611.

Colombo, D., & Di Pietro, A. (2012). *Systemic cyclosporin in the treatment of psoriasis*. INTECH Open Access Publisher.

Colosimo, R., Warren, F. J., Edwards, C. H., Ryden, P., Dyer, P. S., Finnigan, T. J. A., & Wilde, P. J. (2021). Comparison of the behavior of fungal and plant cell wall during gastrointestinal digestion and resulting health effects: A review. *Trends in Food Science and Technology, 110*, 132–141. https://doi.org/10.1016/j.tifs.2021.02.001

Contato, A. G., Conte-Junior, C. A. (2025). Mushrooms in innovative food products: Challenges and potential opportunities as meat substitutes, snacks and functional beverages. Trends in Food Science & Technology, 104868. https://doi.org/10.1016/j.tifs.2025.104868

Copetti, M. V. (2019). Fungi as industrial producers of food ingredients. *Current Opinion in Food Science, 25*, 52–56. https://doi.org/10.1016/j.cofs.2019.02.006

Corbu, V. M., Gheorghe-Barbu, I., Dumbravă, A. Ş., Vrâncianu, C. O., & Şesan, T. E. (2023). Current insights in fungal importance—A comprehensive review. *Microorganisms, 11*, 1384.

Daley, D. K., Brown, K. J., & Badal, S. (2017). Fungal metabolites. In *Pharmacognosy* (pp. 413–421). Elsevier.

Daou, R., Joubrane, K., Maroun, R. G., Khabbaz, L. R., Ismail, A., & El Khoury, A. (2021). Mycotoxins: Factors influencing production and control strategies. *AIMS Agriculture and Food, 6*, 416–447.

Das, S., Das, M., Nath, R., Nath, D., Patra, J. K., Talukdar, A., & Das. (2024). Fungal metabolites and their importance in pharmaceutical industry. In *Entrepreneurship with microorganisms* (pp. 89–120). Elsevier.

Davis, K. A., Gonzalez, A., Loukine, L., Qiao, C., Sadeghpour, A., Vigneault, M., Wang, K. C., & Ibañez, D. (2019). Early experience analyzing dietary intake data from the Canadian Community Health Survey—Nutrition using the National Cancer Institute (NCI) method. *Nutrients, 11*, 1908.

Derbyshire, E. J., & Delange, J. (2021). Fungal protein–what is it and what is the health evidence? A systematic review focusing on mycoprotein. *Frontiers in Sustainable Food Systems, 5*, 581682.

Devi, R., Kaur, T., Guleria, G., Rana, K. L., Kour, D., Yadav, N., Yadav, A. N., & Saxena, A. K. (2020). Fungal secondary metabolites and their biotechnological applications for human health. In *New and future developments in microbial biotechnology and bioengineering* (pp. 147–161). Elsevier.

Di Salvo, E., Lo Vecchio, G., De Pasquale, R., De Maria, L., Tardugno, R., Vadalà, R., & Cicero, N. (2023). Natural pigments production and their application in food, health and other industries. *Nutrients, 15*, 1923.

Dimidi, E., Cox, S. R., Rossi, M., & Whelan, K. (2019). Fermented foods: Definitions and characteristics, impact on the gut microbiota and effects on gastrointestinal health and disease. *Nutrients, 11*, 1806.

Du, X., Sissons, J., Shanks, M., & Plotto, A. (2021). Aroma and flavor profile of raw and roasted Agaricus bisporus mushrooms using a panel trained with aroma chemicals. *LWT, 138*, 110596. https://doi.org/10.1016/j.lwt.2020.110596

Dufossé, L., Fouillaud, M., & Caro, Y. (2021). Fungi and fungal metabolites for the improvement of human and animal nutrition and health. *Journal of Fungi, 7*, 274.

Esheli, M., Thissera, B., El-Seedi, H. R., & Rateb, M. E. (2022). Fungal metabolites in human health and diseases—An overview. *Encyclopedia, 2*, 1590–1601.

Eswaran, H., Ponnuswamy, R. D., & Kannapan, R. P. (2023). Perspective approaches of 3D printed stuffs for personalized nutrition: A comprehensive review. *Annals of 3D Printed Medicine, 12*, 100125. https://doi.org/10.1016/j.stlm.2023.100125

Etebu, E., & Arikekpar, I. (2016). Antibiotics: Classification and mechanisms of action with emphasis on molecular perspectives. *International Journal of Applied Microbiology and Biotechnology Research, 4*, 90–101.

Geng, J., He, S., Zhang, S., Tian, H., & Jin, W. (2024). Impact of incorporating shiitake mushrooms (Lentinula edodes) on microbial community and flavor volatiles in traditional Jiuqu. *Food, 13*, 1019.

Granato, D., Barba, F. J., Bursać Kovačević, D., Lorenzo, J. M., Cruz, A. G., & Putnik, P. (2020). Functional foods: Product development, technological trends, efficacy testing, and safety. *Annual Review of Food Science and Technology, 11*, 93–118.

Guinard, J.-X., Myrdal Miller, A., Mills, K., Wong, T., Lee, S. M., Sirimuangmoon, C., Schaefer, S. E., & Drescher, G. (2016). Consumer acceptance of dishes in which beef has been partially substituted with mushrooms and sodium has been reduced. *Appetite, 105*, 449–459. https://doi.org/10.1016/j.appet.2016.06.018

Hariton, E., & Locascio, J. J. (2018). Randomised controlled trials—The gold standard for effectiveness research. *BJOG, 125*, 1716.

Herlina, V. T., & Setiarto, R. H. B. (2025). Unveiling the cultural tradition and science of Indonesian fermented ethnic soybean paste: Tauco. *Journal of Ethnic Foods, 12*, 3.

Hleap-Zapata, J. I., & Rodriguez-de-la-Pava, G. C. (2021). Evaluation of oyster mushroom powder (Pleurotus ostreatus) as a partial substitute for phosphates in red tilapia fillet sausage. *British Food Journal, 123*, 2107–2122.

Iacovelli, R., Bovenberg, R. A. L., & Driessen, A. J. M. (2021). Nonribosomal peptide synthetases and their biotechnological potential in Penicillium rubens. *Journal of Industrial Microbiology & Biotechnology, 48*, kuab045.

Intrasook, J., Tsusaka, T. W., & Anal, A. K. (2024). Trends and current food safety regulations and policies for functional foods and beverages containing botanicals. *Journal of Food and Drug Analysis, 32*, 112.

Jo, K., Lee, S., Jo, C., Jeon, H. J., Choe, J. H., Choi, Y.-S., & Jung, S. (2020). Utility of winter mushroom treated by atmospheric non-thermal plasma as an alternative for synthetic nitrite and phosphate in ground ham. *Meat Science, 166*, 108151. https://doi.org/10.1016/j.meatsci.2020.108151

Jung, D. Y., Lee, H. J., Shin, D.-J., Kim, C. H., & Jo, C. (2022). Mechanism of improving emulsion stability of emulsion-type sausage with oyster mushroom (Pleurotus ostreatus) powder as a phosphate replacement. *Meat Science, 194*, 108993.

Kampan, N. C., Madondo, M. T., McNally, O. M., Quinn, M., & Plebanski, M. (2015). Paclitaxel and its evolving role in the management of ovarian cancer. *BioMed Research International, 2015*, 413076.

Khan, R. (2024). Mycotoxins in food: Occurrence, health implications, and control strategies-A comprehensive review. *Toxicon, 248*, 108038.

Khan, R., Anwar, F., & Ghazali, F. M. (2024). A comprehensive review of mycotoxins: Toxicology, detection, and effective mitigation approaches. *Heliyon, 10*, e28361.

Kim, J., Lee, S. M., Bae, I. Y., Park, H., Gyu Lee, H., & Lee, S. (2011). (1–3)(1–6)-β-Glucan-enriched materials from Lentinus edodes mushroom as a high-fibre and low-calorie flour substitute for baked foods. *Journal of the Science of Food and Agriculture, 91*, 1915–1919.

Kousar, R., Naeem, M., Jamaludin, M. I., Arshad, A., Shamsuri, A. N., Ansari, N., Akhtar, S., Hazafa, A., Uddin, J., & Khan, A. (2022). Exploring the anticancer activities of novel bioactive compounds derived from endophytic fungi: Mechanisms of action, current challenges and future perspectives. *American Journal of Cancer Research, 12*, 2897.

Krupodorova, T., Barshteyn, V., Dzhagan, V., Pluzhnyk, A., Zaichenko, T., & Blume, Y. (2024a). Enhancement of antioxidant activity and total phenolic content of Fomitopsis pinicola mycelium extract. *Fungal Biology and Biotechnology, 11*, 18.

Krupodorova, T., Butkevych, T., Barshteyn, V., Sevindik, M., Popovych, V., & Polova, Z. (2024b). Effect of the composition of a biologically active dietary supplement with macrofungi mycelia on its antioxidant activity. *Regulatory Mechanisms in Biosystems, 15*, 932–938.

Kumar, K., & Barman Ray, A. (2016). Development and shelf-life evaluation of tomato-mushroom mixed ketchup. *Journal of Food Science and Technology, 53*, 2236–2243.

Kumar, A., Pathak, H., Bhadauria, S., & Sudan, J. (2021). Aflatoxin contamination in food crops: Causes, detection, and management: A review. *Food Production, Processing and Nutrition, 3*, 1–9.

Kurt, A., & Gençcelep, H. (2018). Enrichment of meat emulsion with mushroom (Agaricus bisporus) powder: Impact on rheological and structural characteristics. *Journal of Food Engineering, 237*, 128–136. https://doi.org/10.1016/j.jfoodeng.2018.05.028

Lavelli, V., Proserpio, C., Gallotti, F., Laureati, M., & Pagliarini, E. (2018). Circular reuse of bio-resources: The role of Pleurotus spp. in the development of functional foods. *Food & Function, 9*, 1353–1372.

Li, K., Qiao, K., Xiong, J., Guo, H., & Zhang, Y. (2023). Nutritional values and bio-functional properties of fungal proteins: Applications in foods as a sustainable source. *Food, 12*, 4388.

Lin, L., & Xu, J. (2020). Fungal pigments and their roles associated with human health. *Journal of Fungi, 6*, 280.

Losoya-Sifuentes, C., Cruz, M., del Refugio Rocha-Pizaña, M., Loredo-Treviño, A., & Belmares, R. (2025). Edible mushrooms: A nutrient-rich ingredient for healthier food products–A review. *Current Nutrition Reports, 14*, 9.

Lu, X., Brennan, M. A., Serventi, L., Liu, J., Guan, W., & Brennan, C. S. (2018). Addition of mushroom powder to pasta enhances the antioxidant content and modulates the predictive glycaemic response of pasta. *Food Chemistry, 264*, 199–209.

Luo, Q., & Xu, J. (2021). *Monosaccharide composition and antioxidant activity of extracellular polysaccharides and intracellular polysaccharides in liquid ferment of Ganoderina lucidum*. Acta Edulis Fungi, 135–142. https://doi.org/10.16488/j.cnki.1005-9873.2021.06.017

Luo, X., Jiang, J.-H., Liu, S.-L., Gao, J.-Y., & Zhou, L.-W. (2024). Metabolomics analysis of rice fermented by medicinal fungi providing insights into the preparation of functional food. *Food Chemistry, 459*, 140372.

Łysakowska, P., Sobota, A., & Wirkijowska, A. (2023). Medicinal mushrooms: Their bioactive components, nutritional value and application in functional food production—A review. *Molecules, 28*, 5393.

Maehara, Y., Tsujitani, S., Saeki, H., Oki, E., Yoshinaga, K., Emi, Y., Morita, M., Kohnoe, S., Kakeji, Y., & Yano, T. (2012). Biological mechanism and clinical effect of protein-bound polysaccharide K (KRESTIN®): Review of development and future perspectives. *Surgery Today, 42*, 8–28.

Mafe, A. N., & Büsselberg, D. (2024). Mycotoxins in food: Cancer risks and strategies for control. *Food, 13*, 3502.

Marđetko, N., Kolakušić, A., Trontel, A., Novak, M., Pavlečić, M., Dobrinčić, A., Tominac, V. P., & Šantek, B. (2025). Usage of the fungus mucor indicus and the bacterium rhodovulum adriaticum in a biorefinery system for biochemical production on grass hydrolysates. *Polymers (Basel), 17*, 369.

Martin, F. M., & van Der Heijden, M. G. A. (2024). The mycorrhizal symbiosis: Research frontiers in genomics, ecology, and agricultural application. *New Phytologist, 242*, 1486–1506.

Mattar, T. V., Gonçalves, C. S., Pereira, R. C., Faria, M. A., de Souza, V. R., & de Deus Souza Carneiro, J. (2018). A shiitake mushroom extract as a viable alternative to NaCl for a reduction in sodium in beef burgers: A sensory perspective. *British Food Journal, 120*, 1366–1380.

Mirończuk-Chodakowska, I., Kujawowicz, K., & Witkowska, A. M. (2021). Beta-glucans from fungi: Biological and health-promoting potential in the COVID-19 pandemic era. *Nutrients, 13*, 3960.

Mishra, B., Mishra, A. K., Kumar, S., Mandal, S. K., Nsv, L., Kumar, V., Baek, K.-H., & Mohanta, Y. K. (2021). Antifungal metabolites as food bio-preservative: Innovation, outlook, and challenges. *Metabolites, 12*, 12.

Moura, M. A. F., Gomes, D. C., & Takahashi, J. A. (2023). Applications of fungi secondary metabolites in the food industry. In M. Carocho, S. A. Heleno, & L. Barros (Eds.), *Natural secondary metabolites: From nature, through science, to industry* (pp. 739–776). Springer International Publishing. https://doi.org/10.1007/978-3-031-18587-8_24

Navarro-Simarro, P., Gómez-Gómez, L., Ahrazem, O., & Rubio-Moraga, Á. (2024). *Food and human health applications of edible mushroom by-products* (Vol. 81, pp. 43–56). New Biotechnology.

Ng, S. H., Robert, S. D., Wan Ahmad, W. A. N., & Wan Ishak, W. R. (2017). Incorporation of dietary fibre-rich oyster mushroom (Pleurotus sajor-caju) powder improves postprandial glycaemic response by interfering with starch granule structure and starch digestibility of biscuit. *Food Chemistry, 227*, 358–368. https://doi.org/10.1016/j.foodchem.2017.01.108

Novakovic, S., Djekic, I., Klaus, A., Vunduk, J., Đorđević, V., Tomovic, V., Koćić-Tanackov, S., Lorenzo, J. M., Barba, F. J., & Tomasevic, I. (2020). Application of porcini mushroom (Boletus edulis) to improve the quality of frankfurters. *Journal of Food Processing & Preservation, 44*, e14556.

Nurmilah, S., Frediansyah, A., Cahyana, Y., & Utama, G. L. (2024). Biotransformation and health potential of isoflavones by microorganisms in Indonesian traditional fermented soy products: A review. *Journal of Agriculture and Food Research, 18*, 101365.

Ogawa, M., Bisson, L. F., García-Martínez, T., Mauricio, J. C., & Moreno-García, J. (2019). New insights on yeast and filamentous fungus adhesion in a natural co-immobilization system: Proposed advances and applications in wine industry. *Applied Microbiology and Biotechnology, 103*, 4723–4731.

Olawuyi, I. F., & Lee, W. Y. (2019). Quality and antioxidant properties of functional rice muffins enriched with shiitake mushroom and carrot pomace. *International Journal of Food Science and Technology, 54*, 2321–2328.

Osemwegie, O. O., Adetunji, C. O., Ayeni, E. A., Adejobi, O. I., Arise, R. O., Nwonuma, C. O., & Oghenekaro, A. O. (2020). Exopolysaccharides from bacteria and fungi: Current status and perspectives in Africa. *Heliyon, 6*, e04205.

Owheruo, J. O., Edo, G. I., Oluwajuyitan, D. T., Faturoti, A. O., Martins, I. E., Akpoghelie, P. O., & Agbo, J. J. (2023). Quality evaluation of value-added nutritious biscuit with high antidiabetic properties from blends of wheat flour and oyster mushroom. *Food Chemistry Advances, 3*, 100375. https://doi.org/10.1016/j.focha.2023.100375

Pan, W., Jiang, P., Zhao, J., Shi, H., Zhang, P., Yang, X., Biazik, J., Hu, M., Hua, H., & Ge, X. (2021). β-Glucan from Lentinula edodes prevents cognitive impairments in high-fat diet-induced obese mice: Involvement of colon-brain axis. *Journal of Translational Medicine, 19*, 1–17.

Patinho, I., Selani, M. M., Saldaña, E., Bortoluzzi, A. C. T., Rios-Mera, J. D., da Silva, C. M., Kushida, M. M., & Contreras-Castillo, C. J. (2021). Agaricus bisporus mushroom as partial fat replacer improves the sensory quality maintaining the instrumental characteristics of beef burger. *Meat Science, 172*, 108307.

Pimenta, L. P. S., Gomes, D. C., Cardoso, P. G., & Takahashi, J. A. (2021). Recent findings in azaphilone pigments. *Journal of Fungi, 7*, 541.

Pobiega, K., Sękul, J., Pakulska, A., Latoszewska, M., Michońska, A., Korzeniowska, Z., Macherzyńska, Z., Pląder, M., Duda, W., & Szafraniuk, J. (2024). Fungal proteins: Sources, production and purification methods, industrial applications, and future perspectives. *Applied Sciences, 14*, 6259.

Pouris, J., Kolyva, F., Bratakou, S., Vogiatzi, C. A., Chaniotis, D., & Beloukas, A. (2024). The role of fungi in food production and processing. *Applied Sciences, 14*, 5046.

Prescott, T. A. K., Hill, R., Mas-Claret, E., Gaya, E., & Burns, E. (2023). Fungal drug discovery for chronic disease: History, new discoveries and new approaches. *Biomolecules, 13*, 986.

Priya, K. M., & Alur, S. (2024). Examining nutrition label knowledge, self-efficacy, and nutrition facts panel usage. *IRJMS, 5*, 251–262.

Qu, H., Dong, Y., Liu, C., Yang, M., & Liu, X. (2023). Effects of Coprinus comatus (chicken drumstick mushroom) on the quality characteristics of low nitrite-backfat Cantonese sausages. *LWT, 187*, 115339. https://doi.org/10.1016/j.lwt.2023.115339

Qun, T., Zhou, T., Hao, J., Wang, C., Zhang, K., Xu, J., Wang, X., & Zhou, W. (2023). Antibacterial activities of anthraquinones: Structure–activity relationships and action mechanisms. *RSC Medicinal Chemistry, 14*, 1446–1471.

Rahman, H. U., Asghar, W., & Khalid, N. (2021). Scleroglucan and schizophyllan: Microbial polysaccharides of functional importance. In *Polysaccharides of microbial origin: Biomedical applications* (pp. 1–28). Springer.

Ramachandraiah, A., Kaushik, J., Gowda, L., Chigadannavar, P. S., Krishnappa, R., Venkatesh, S., Sivadas, S., Judith, S., Ramesh, U. M., Quadri, Z., Dey, P., Chinthala, M., Gola, D., Chauhan, N., & Bharti, R. K. (2021). Chapter 11 – Prospect of biofuel production by fungus. In V. K. Sharma, M. P. Shah, S. Parmar, & A. Kumar (Eds.), *Fungi bio-prospects in sustainable agriculture, environment and nano-technology* (pp. 285–313). Academic Press. https://doi.org/10.1016/B978-0-12-821394-0.00011-1

Ramle, N. A., Zulkunain, M., & Ismail-Fitry, M. R. (2021). Replacing animal fat with edible mushrooms: A strategy to produce high-quality and low-fat buffalo meatballs. *International Food Research Journal, 28*, 905–915.

Rathore, H., Sehwag, S., Prasad, S., & Sharma, S. (2019). Technological, nutritional, functional and sensorial attributes of the cookies fortified with Calocybe indica mushroom. *Journal of Food Measurement and Characterization, 13*, 976–987. https://doi.org/10.1007/s11694-018-0012-1

Rizzo, G., Baroni, L., & Lombardo, M. (2023). Promising sources of plant-derived polyunsaturated fatty acids: A narrative review. *International Journal of Environmental Research and Public Health, 20*, 1683.

Rousta, N., Aslan, M., Yesilcimen Akbas, M., Ozcan, F., Sar, T., & Taherzadeh, M. J. (2024). Effects of fungal based bioactive compounds on human health. *Critical Reviews in Food Science and Nutrition, 64*, 7004–7027.

Sande, D., de Oliveira, G. P., e Moura, M. A. F., de Almeida Martins, B., Lima, M. T. N. S., & Takahashi, J. A. (2019). Edible mushrooms as a ubiquitous source of essential fatty acids. *Food Research International, 125*, 108524. https://doi.org/10.1016/j.foodres.2019.108524

Selvasekaran, P., Mahalakshmi, Roshini, F., Angalene, L. A., Chandini, Sunil, T., & Chidambaram, R. (2021). Fungal exopolysaccharides: Production and biotechnological industrial applications in food and allied sectors. In A. N. Yadav (Ed.), *Recent trends in mycological research: Volume 2: Environmental and industrial perspective* (pp. 311–357). Springer International Publishing. https://doi.org/10.1007/978-3-030-68260-6_12

Shankar, A., & Sharma, K. K. (2022). Fungal secondary metabolites in food and pharmaceuticals in the era of multi-omics. *Applied Microbiology and Biotechnology, 106*, 3465–3488.

Sharma, N., Singh, M., Bahurupi, Y., & Aggarwal, P. (2024). Front-of-package labeling in India: A key strategy for combatting non-communicable diseases and promoting healthy diets. *Preventive Medicine: Research & Reviews, 1*, 273–274.

Shobana, S., Sangavi, G., Wuni, R., Priyanka, B., Leelavady, A., Kayalvizhi, D., Mohan Anjana, R., Krishnaswamy, K., Vimaleswaran, K. S., & Mohan, V. (2024). Assessment of front and back of pack nutrition labels of selected convenience food products and snacks available in the Indian market. *PLoS One, 19*, e0314819.

Stoev, S. D. (2024). Food security and foodborne mycotoxicoses—What should be the adequate risk assessment and regulation? *Microorganisms, 12*, 580.

Sulieman, A. A., Zhu, K.-X., Peng, W., Hassan, H. A., Obadi, M., Siddeeg, A., & Zhou, H.-M. (2019). Rheological and quality characteristics of composite gluten-free dough and biscuits supplemented with fermented and unfermented Agaricus bisporus polysaccharide flour. *Food Chemistry, 271*, 193–203. https://doi.org/10.1016/j.foodchem.2018.07.189

Sun, L., Zhang, Z., Xin, G., Sun, B., Bao, X., Wei, Y., Zhao, X., & Xu, H. (2020). Advances in umami taste and aroma of edible mushrooms. *Trends in Food Science and Technology, 96*, 176–187. https://doi.org/10.1016/j.tifs.2019.12.018

Szelenberger, R., Cichoń, N., Zajaczkowski, W., & Bijak, M. (2024). Application of biosensors for the detection of mycotoxins for the improvement of food safety. *Toxins (Basel), 16*, 249.

Takahashi, J. A., Barbosa, B. V., Martins, B. D. A., Guirlanda, C. P., & Moura, A. F. (2020). Use of the versatility of fungal metabolism to meet modern demands for healthy aging, functional foods, and sustainability. *Journal of Fungi, 6*, 223.

Tamang, J. P., Anupma, A., Jones, N., & Shangpliang, H. (2022). Ethno-microbiology of Tempe, an Indonesian fungal-fermented soybean food and Koji, a Japanese fungal starter culture. *Current Opinion in Food Science, 48*, 100912. https://doi.org/10.1016/j.cofs.2022.100912

Tepsongkroh, B., Jangchud, K., Jangchud, A., Chonpracha, P., Ardoin, R., & Prinyawiwatkul, W. (2020). Consumer perception of extruded snacks containing brown rice and dried mushroom. *International Journal of Food Science and Technology, 55*, 46–54.

Thangavelu, K. P., Kerry, J. P., Tiwari, B. K., & McDonnell, C. K. (2019). Novel processing technologies and ingredient strategies for the reduction of phosphate additives in processed meat. *Trends in Food Science and Technology, 94*, 43–53. https://doi.org/10.1016/j.tifs.2019.10.001

Thongsiri, C., Nagai-Yoshioka, Y., Yamasaki, R., Adachi, Y., Usui, M., Nakashima, K., Nishihara, T., & Ariyoshi, W. (2021). Schizophyllum commune β-glucan: Effect on interleukin-10 expression induced by lipopolysaccharide from periodontopathic bacteria. *Carbohydrate Polymers, 253*, 117285.

Urbancikova, I., Hudackova, D., Majtan, J., Rennerova, Z., Banovcin, P., & Jesenak, M. (2020). Efficacy of pleuran (β-Glucan from Pleurotus ostreatus) in the management of herpes simplex virus type 1 infection. *Evidence-based Complementary and Alternative Medicine, 2020*, 8562309.

Usman, M., Murtaza, G., & Ditta, A. (2021). Nutritional, medicinal, and cosmetic value of bioactive compounds in button mushroom (Agaricus bisporus): A review. *Applied Sciences, 11*, 5943.

Vaclav, V., Petr, S., & Luca, V. (2018). Chapter 8 – Beta Glucan as therapeutic food. In A. M. Holban & A. M. Grumezescu (Eds.), *Therapeutic foods* (pp. 239–256). Academic Press. https://doi.org/10.1016/B978-0-12-811517-6.00008-8

Valverde, M. E., Hernández-Pérez, T., & Paredes-López, O. (2015). Edible mushrooms: Improving human health and promoting quality life. *International Journal of Microbiology, 2015*, 376387.

Wadhwa, K., Kapoor, N., Kaur, H., Abu-Seer, E. A., Tariq, M., Siddiqui, S., Yadav, V. K., Niazi, P., Kumar, P., & Alghamdi, S. (2024). A comprehensive review of the diversity of fungal secondary metabolites and their emerging applications in healthcare and environment. *Mycobiology, 52*, 1–53.

Wang, P.-H., Hsu, C.-I., Tang, S.-C., Huang, Y.-L., Lin, J.-Y., & Ko, J.-L. (2004). Fungal immunomodulatory protein from Flammulina velutipes induces interferon-γ production through p38 mitogen-activated protein kinase signaling pathway. *Journal of Agricultural and Food Chemistry, 52*, 2721–2725.

Wang, X., Zhou, P., Cheng, J., Chen, Z., & Liu, X. (2018). Use of straw mushrooms (Volvariella volvacea) for the enhancement of physicochemical, nutritional and sensory profiles of Cantonese sausages. *Meat Science, 146*, 18–25. https://doi.org/10.1016/j.meatsci.2018.07.033

Wang, L., Guo, H., Liu, X., Jiang, G., Li, C., Li, X., & Li, Y. (2019a). Roles of Lentinula edodes as the pork lean meat replacer in production of the sausage. *Meat Science, 156*, 44–51. https://doi.org/10.1016/j.meatsci.2019.05.016

Wang, X., Xu, M., Cheng, J., Zhang, W., Liu, X., & Zhou, P. (2019b). Effect of Flammulina velutipes on the physicochemical and sensory characteristics of Cantonese sausages. *Meat Science, 154*, 22–28. https://doi.org/10.1016/j.meatsci.2019.04.003

Wang, S., Tonnis, B. D., Wang, M. L., Zhang, S., & Adhikari, K. (2019c). Investigation of monosodium glutamate alternatives for content of umami substances and their enhancement effects in chicken soup compared to monosodium glutamate. *Journal of Food Science, 84*, 3275–3283.

Wang, S., Zhang, S., & Adhikari, K. (2019d). Influence of monosodium glutamate and its substitutes on sensory characteristics and consumer perceptions of chicken soup. *Food, 8*, 71.

Wang, B., Shi, Y., Lu, H., & Chen, Q. (2023). A critical review of fungal proteins: Emerging preparation technology, active efficacy and food application. *Trends in Food Science and Technology, 141*, 104178.

Wan-Mohtar, W. A. A. Q. I., Halim-Lim, S. A., Kamarudin, N. Z., Rukayadi, Y., Abd Rahim, M. H., Jamaludin, A. A., & Ilham, Z. (2020). Fruiting-body-base flour from an Oyster mushroom waste in the development of antioxidative chicken patty. *Journal of Food Science, 85*, 3124–3133.

Wu, X., Zheng, Z., Guo, T., Wang, K., & Zhang, Y. (2021). Molecular dynamics simulation of lentinan and its interaction with the innate receptor dectin-1. *International Journal of Biological Macromolecules, 171*, 527–538.

Yan, M.-Q., Feng, J., Liu, Y.-F., Hu, D.-M., & Zhang, J.-S. (2023). Functional components from the liquid fermentation of edible and medicinal fungi and their food applications in China. *Food, 12*, 2086.

Yang, L., Kong, D.-X., Xiao, N., Ma, Q.-Y., Xie, Q.-Y., Guo, J.-C., Deng, C. Y., Ma, H.-X., Hua, Y., & Dai, H.-F. (2022). Antidiabetic lanostane triterpenoids from the fruiting bodies of Ganoderma weberianum. *Bioorganic Chemistry, 127*, 106025.

Yu, W., Zhang, Y., Lu, Y., Ouyang, Z., Peng, J., Tu, Y., & He, B. (2025). Recent research on the bioactivity of polyphenols derived from edible fungi and their potential in chronic disease prevention. *Journal of Functional Foods, 124*, 106627.

Yuan, B., Zhao, L., Yang, W., McClements, D. J., & Hu, Q. (2017). Enrichment of bread with nutraceutical-rich mushrooms: Impact of Auricularia auricula (Mushroom) flour upon quality attributes of wheat dough and bread. *Journal of Food Science, 82*, 2041–2050.

Yuan, S., Gopal, J. V., Ren, S., Chen, L., Liu, L., & Gao, Z. (2020). Anticancer fungal natural products: Mechanisms of action and biosynthesis. *European Journal of Medicinal Chemistry, 202*, 112502. https://doi.org/10.1016/j.ejmech.2020.112502

Yulianto, W. A., Pujimulyani, D., & Pratami, C. A. (2022). The potential of glutinous rice tape added with lactobacillus plantarum dad-13 and saccharomyces boulardii as a probiotic food. *Journal of Functional Food and Nutraceutical*. https://doi.org/10.33555/jffn.v4i1.96

Zbikowska, A., Kowalska, M., Zbikowska, K., Onacik-Gür, S., Łempicka, U., & Turek, P. (2022). Study on the incorporation of oat and yeast β-glucan into shortbread biscuits as a basis for designing healthier and high quality food products. *Molecules, 27*, 1393.

Zhang, Y., Liu, Y., Zhou, Y., Zheng, Z., Tang, W., Song, M., Wang, J., & Wang, K. (2021). Lentinan inhibited colon cancer growth by inducing endoplasmic reticulum stress-mediated autophagic cell death and apoptosis. *Carbohydrate Polymers, 267*, 118154.

Zhang, K., Zhang, T.-T., Guo, R.-R., Ye, Q., Zhao, H.-L., & Huang, X.-H. (2023). The regulation of key flavor of traditional fermented food by microbial metabolism: A review. *Food Chemistry: X, 19*, 100871.

Zhao, S., Gao, Q., Rong, C., Wang, S., Zhao, Z., Liu, Y., & Xu, J. (2020). Immunomodulatory effects of edible and medicinal mushrooms and their bioactive immunoregulatory products. *Journal of Fungi, 6*, 269.

Zhao, J., Wang, T., Zhang, C., Han, X., Yan, J., & Gan, B. (2023). A comparative analysis of the umami taste of five fresh edible mushrooms by simulating the chemical environment of oral digestion in vitro. *LWT, 176*, 114522. https://doi.org/10.1016/j.lwt.2023.114522

Zhu, F., Du, B., & Xu, B. (2016). A critical review on production and industrial applications of beta-glucans. *Food Hydrocolloids, 52*, 275–288. https://doi.org/10.1016/j.foodhyd.2015.07.003

Further Reading

http://www.foursigmatic.com/
http://www.hostdefense.com/
http://www.lifecykel.com/
http://www.mushroomwisdom.com/
http://www.mycoformulas.com/
http://www.ommyco.com/
http://www.realmushrooms.com/
https://www.iihr.res.in/mushroom-value-added-products-blend-tradition-nutrition-and-taste

Chapter 6
Fungal Metabolites as Probiotic Ingredients in Dairy Products and Supplements

Hagar Fathy

1 Introduction

Specialized proteins known as enzymes act as catalysts in nature, driving essential biochemical reactions in plants, animals, and microorganisms (Akanbi et al., 2020). These remarkable molecules reduce the activation energy required to quicken reactions, making processes faster, more efficient, and critical for sustaining life. Each enzyme is particular and designed to catalyze a specific reaction or a group of closely related ones (Aryee et al., 2022). Beyond their natural roles, enzymes have become indispensable in food production, pharmaceuticals, and environmental cleanup (Abellanas et al., 2023). However, to maximize their potential in these fields, enzymes often need to be stabilized for repeated use, highlighting the importance of enzyme immobilization (Chen et al., 2023). Immobilization of enzymes is a biotechnological method that involves attaching enzymes to a provisional material or carrier (Akanbi et al., 2020). This technique offers several key benefits compared to using enzymes in a liberated form. Enzymes that are immobilized are more reusable, stable, and selective, making them highly efficient tools for industrial applications (Gowda et al., 2018; Silva et al., 2023). Immobilization significantly extends their lifespan and enhances their performance by protecting enzymes against deterioration, high temperatures, and other harsh circumstances. These advantages have made immobilized enzymes essential across various industries (Abellanas et al., 2023; Akanbi et al., 2020). In the manufacturing of dairy products, enzymes are essential, offering distinct advantages over traditional chemical methods (Basso & Serban, 2019; Taheri-Kafrani et al., 2021). They are widely used to modify milk for creating various dairy-based products, including whey protein concentrate (WPC),

H. Fathy (✉)
Animal Production Research Institute (APRI), Agricultural Research Center (ARC), Dokki, Giza, Egypt
e-mail: hagarfathy@pg.cu.edu.eg

B. P. Singh et al. (eds.), *Fungal Additives and Bioactives in Food Processing Industries*, Fungal Biology, https://doi.org/10.1007/978-3-032-04519-5_6

whey protein isolate (WPI), milk fats, and hydrolyzed infant formulas (Akanbi et al., 2013, 2019). For example, whey is commonly produced through rennet-based enzymatic coagulation of cow's milk. Whey protein hydrolysates are also produced by breaking down cow and goat milk proteins with enzymes like *neutralize*, *alcalase*, *trypsin*, *pepsin*, *bromelain*, *proteinase*, and *papain* (Mazorra-Manzano et al., 2020). Enzymes also enable that bioactive peptides are produced from milk proteins and by-products and specialized lipids with functional properties through hydrolysis, esterification, and interesterification (Gunathilake et al., 2022a; Lermen et al., 2023). These applications require food-grade enzymes and suitable immobilization supports that need to be robust, stable, inert, and cost-effective (Forsan, 2023; Gunathilake et al., 2022b; Singh et al., 2019). Enzyme applications include supplementing dairy products with bioactive lipids, and enhancing wellness and sensory attributes (Dong et al., 2024; Silva et al., 2023). Common functional lipids added to dairy products include omega-3, −6, and − 9 polyunsaturated fatty acids (PUFAs) such as EPA (eicosapentaenoic acid), OA (oleic acid),LA (linoleic acid), ALA (alpha-linolenic acid), and DHA (docosahexaenoic acid) (Kleekayai et al., 2021). For instance, studies show that yogurt fortified with PUFAs can provide neuroprotective benefits, counteracting neurotoxicity, and improve brain dopamine and serotonin levels (Forsan & Hasan, 2023; Gowda et al., 2018; Hagar & Forsan, 2024; Xia et al., 2020). Fortified yogurts also demonstrate high bioavailability for EPA and DHA, enhancing their intestinal absorption (Zhang et al., 2021). Additional dairy products enhanced with omega-3 fatty acids and lipids that function were also reported (León-López et al., 2022). However, these PUFAs are typically required in pure forms, and frequently used in industrial techniques to extract or concentrate oils like fractional distillation and urea complexation (Xiaojing Du et al., 2022; Monari et al., 2019). Unfortunately, these processes are environmentally harmful and can degrade the lipids (Gunathilake et al., 2021). Enzymatic methods offer a superior alternative, as they are milder, environmentally friendly, and preserve lipid quality (Wróblewska et al., 2004). Moreover, enzymes are considered natural food processing aids, aligning with regulatory standards for food safety (Forsan, 2024a, b; Gan et al., 2021; Oliveira et al., 2022).

2 Enzyme Immobilization Methods

Binding enzymes to a solid outward is known as the immobilization of enzymes, which improves their activity, stability, reusability, and overall efficiency. This method also helps reduce costs and enzyme loss through reactions, as recovery of immobilized enzymes is simpler compared to their non-immobilized counterparts (Akanbi & Barrow, 2017). There are various methods to immobilize enzymes, such as entrapment, adsorption, cross-linking, and covalent bonding. The selection of the appropriate technique is dependent on many variables including the enzyme's characteristics and its intended application (Maghraby et al., 2023; Mohammadi et al., 2023). Chemical or physical methods can be used to immobilize enzymes, with varying strengths of attachment to the support surface (Imam et al., 2021). In physical methods, enzymes may be adsorbed onto the surface (adsorption) or trapped within the supporting material (entrapment) without creating chemical bonds. On the other hand, chemical techniques entail the creation of covalent bonds between the support and the enzyme or the use of cross-linkers to connect the enzyme to functional components on the support (Guisan et al., 2022; Sicard, 2023).

2.1 Physical Methods of Enzyme Immobilization

2.1.1 Entrapment

Encasing the enzyme in a porous substance that permits free passage of products, and substrates but stops the enzyme from escaping is known as immobilization via entrapment. The support material needs to have the right pore size to securely hold the enzyme. If the material is too porous or has large pores, the enzyme can leak out, resulting in lower immobilization efficiency and reduced activity. To avoid this, the enzyme is mixed with a polymer that creates a complex structure, trapping the enzyme inside. This is prevented by combining the enzyme with a polymer that forms a complex structure and traps the enzyme inside. It is crucial to remember that the support merely serves as a physical barrier to stop leakage; the enzyme is not chemically bonded to it. To produce fructooligosaccharides, for instance, a *β-fructofuranosidase* enzyme from *Aspergillus japonicus* was trapped on Amberlite IRA 900 resin, a food-grade resin, and rendered immobile (Forsan, 2023; Forsan et al., 2022; Hooper et al., 2020; Safwat et al., 2021). The study showed that the immobilized enzyme produced higher fructooligosaccharides yields, maintained better activity, and could be reused for up to 12 batch cycles. Numerous studies have demonstrated that enzymes immobilized through entrapment tend to be highly reusable, which is beneficial for industrial applications. However, some research has shown that entrapment can enable the enzyme to gradually become inactive or stiff.

2.1.2 Adsorption

Using weak interactions including hydrogen bonds, van der Waals forces, and hydrophobic interactions, enzymes are attached to a solid support in the adsorption process. Enzymes stick to the surface of the support during this method without forming strong covalent bonds (Xuran Du et al., 2023). It is a straightforward approach that does not require any chemical reaction of the support material or the enzyme. Studies have shown that enzymes immobilized using this method often maintain high activity because they remain in an open conformation on the support (Binhayeeding et al., 2020). For example, research on *Candida antarctica* lipase adsorbed by physical adsorption on an octadecyl resin found that it had higher immobilization yields and enzyme activity compared to covalent immobilization

(Akanbi & Barrow, 2017). But before there was a noticeable decrease in activity, the enzyme was able to be utilized for roughly five cycles. One of the main drawbacks of the adsorption process is that weak connections between the support and the enzyme may cause the enzyme to leak, reducing both stability and reusability. Despite this, the method remains popular for certain applications where a simple immobilization technique is preferred.

2.2 Chemical Methods of Enzyme Immobilization

2.2.1 Crosslinking

Cross-linking involves establishing a solid covalent connection between a support material and enzymes. This technique is fairly simple and typically consists of two key steps. First, enzymes get precipitated utilizing *tert*-butanol, ethanol, ammonium sulfate, glycol, acetone, isopropanol, and acetonitrile. Once the enzymes are precipitated, they are cross-linked to form either a cross-linked enzyme aggregate or crystal (CLEA-CLEC) (Forsan, 2024c; Forsan & Awad, 2023; Imam et al., 2021; McCleary, 2023). Common cross-linkers used in this process include benzoquinone, glutaraldehyde, dextran–polyaldehyde, carbodiimide, and polyethyleneimine with glutaraldehyde being the most widely used (Rojas et al., 2019). This cross-linker creates a strong binding between the support and enzyme, making the immobilization process more effective. Crosslinking is a well-established method that typically results in stable and reusable enzymes. However, one downside is that enzymes can become denatured during precipitation, especially if they are not resistant to solvents. Therefore, selecting a precipitant that will not damage the enzyme is crucial (Abellanas et al., 2023). In a recent study, lipase from *Pseudomonas stutzeri* was immobilized using crosslinking, resulting in a stronger, more stable enzyme that showed improved activity in deep eutectic solvents (DESs) that the free enzyme could not tolerate. A further study used glutaraldehyde as the crosslinker to immobilize *invertase* on chitosan. According to the results, this enzyme exhibited improved thermal stability, pH stability, and reusability, retaining 70% of its activity after nine reuses. It was also quite successful in producing bioethanol (Abellanas et al., 2023).

2.2.2 Covalent Bond

A common technique for immobilizing enzymes is the establishment of covalent bonds, which provide the enzyme with a solid and long-lasting link with the support material and stop it from leaching (Imam et al., 2021). This method relies on how the reactive groups on the support interact with carboxyl groups or amino or of the enzyme, which can form different types of linkages (Thangaraj & Solomon, 2019). Vinyl sulfone, glyoxyl, epoxide, and glutaraldehyde are examples of common active groups on the support (Rodrigues et al., 2021). Since covalent bonding is irreversible, enzymes immobilized in this way are stable and will not detach during use, making it the perfect technique for situations where the enzyme must be completely removed from the finished product. Additionally, covalent binding enhances enzyme selectivity, reusability, thermos stability, and tolerance to organic solvents, offering versatility for various applications (Wahab et al., 2020).

3 Fungal Enzymes in Dairy Industry

The dairy industry is another key sector where enzymes play a crucial role. These enzymes are widely used in food production to improve and enhance the aroma, color, flavor, and overall yield of dairy products.

3.1 Proteases

Proteolytic enzymes, commonly known as *peptidases*, *proteases*, or *proteinases*, are specialized enzymes capable of breaking down protein molecules by hydrolyzing peptide bonds. *Proteases* are classified into two main types: *endopeptidases* and *exopeptidases* (Singh & Kumar, 2019). These enzymes are sourced from a variety of organisms, including animals, plants, bacteria, and fungi. However, on an industrial scale, bacterial and fungal proteases are the most widely used. Microorganisms produce both intracellular and extracellular proteases, which can be obtained through solid-state or submerged fermentation processes. Among microbial sources, fungi such as *Aspergillus flavus*, *Aspergillus fumigatus*, *Aspergillus niger*, and *Aspergillus oryzae* are particularly recognized for their efficient production of protease enzymes (Singh & Kumar, 2019). Proteases have numerous applications, including speeding up the ripening process of cheese and modifying milk proteins to reduce their allergenic properties in cow's milk products, making them more suitable for infants (Qureshi et al., 2015).

3.2 Lipase

Lipids that encompass oils, fats, and related compounds are a broad category of biomolecules. *Lipase* is one of the main enzymes involved in lipid processing (Mustafa et al., 2023). *Lipases* help lipids be hydrolyzed, esterified, and interesterified. Immobilized *lipases* are essential in modifying lipids for various applications, particularly within the food industry. *Lipases* primarily function through hydrolysis, breaking down free fatty acids from triglycerides. Immobilized *lipases* are usually utilized in the hydrolysis of oils to concentrate specific fatty acids (Zhu et al., 2022).

These enzymes exhibit fatty acid- and positional specificities, allowing them to hydrolyze certain fatty acids from the glycerides of oils selectively (Dovale-Rosabal et al., 2021). For instance, 70% EPA was extracted from microalgal oil using *lipase* from *Alcaligenes* species that were adsorbed on Accurel MP1000. Such pure EPA formulas are now being incorporated into cheese, yogurt, and nutritional drinks. Similarly, *Candida antarctica lipase A* (CAL-A) immobilized on an octadecyl-activated resin was used to concentrate 89% DHA from thraustochytrid oil by hydrolyzing its saturated fatty acids (Khoonin et al., 2022). The high saturated fatty acid content makes this oil a perfect source for the concentration of DHA. This demonstrates that *lipases* are effective in producing premium concentrates of DHA for fortifying food, including in dairy products like butter, cheese, and infant formula (Akanbi et al., 2014). *Lipases* also catalyze esterification, in which esters are created from fatty acids. This activity is crucial in producing flavor esters and various food additives (Ahmad et al., 2019). In some latest studies, the unpleasant rancid and tart flavors of goat milk were eliminated by esterifying the milk fat with immobilized *Thermomyces lanuginosus lipase* (Zanuso et al., 2021). The resulting esters gave the milk sweet, floral, and fruity flavors, reducing the off-flavors by 75%. Interesterification is another process where *lipases* rearrange fatty acids within triglycerides, modifying lipids' chemical and physical characteristics (Bilal & Iqbal, 2019; Forsan et al., 2025; Forsan & Hassan, 2023; Zanuso et al., 2021). This technique is commonly applied to produce structured lipids and customized fats with certain ratios of fatty acids. These modified lipids have superior stability, better nutritional profiles, and increased functional qualities than natural fats. Food companies frequently use interesterification to produce shortenings, margarines, and other specialized fats. Beneficial monounsaturated, saturated, and long-chain PUFAs can be found in human breast milk fat, and are necessary for the development and growth of infants (Charles, 2019). The way these fatty acids are arranged on the glycerol structure of milk fat influences how bioavailable they are to babies. Research has shown that *lipases* are useful for creating alternatives for human milk fats (HMFS), created to replicate the distinct fatty acid positional distribution found in human milk (Sarnoab & Iuliano, 2020). For example, OA and other unsaturated fatty acids are found at the sn-1 and sn-3 positions on the glycerol structure, whereas palmitic acid is usually esterified at the sn-2 position. These alternatives are especially high in long-chain PUFAs, which are important for the health of infants and include EPA, ARA, and DHA (Mehrotra et al., 2021). Immobilized *lipases* enable the production of HMFSs with specific fatty acid profiles that resemble human milk fat. Furthermore, immobilized lipases are essential in esterifying fatty acids with non-lipid compounds such as polyphenols, leading to the formation of phenolipids (Cieh et al., 2024). These substances combine the advantageous qualities of lipids and polyphenols, such as greater bioavailability, better solubility in fat-soluble environments, and enhanced antioxidant activity. Phenolipids are being used in the food, cosmetics, and pharmaceutical industries. These substances combine the advantageous qualities of lipids and polyphenols, such as greater bioavailability, better solubility in fat-soluble environments, and enhanced antioxidant activity. Phenolipids are being used in the food, cosmetics, and pharmaceutical industries (Pourjoula et al., 2020).

3.3 Lactase

Lactase, also known as *β-galactosidase*, is an enzyme that breaks down lactose into simpler sugars: glucose and galactose. This enzyme can be derived from various sources, including animals, plants, bacteria, fungi, yeasts, and molds. Commercially, lactase production typically involves *Aspergillus niger*, *Aspergillus oryzae*, and *Kluyveromyces lactis*. Lactase plays a crucial role as a digestive aid and is used to improve the solubility and sweetness of milk products (Qureshi et al., 2015). A key application is in reducing or removing lactose from milk, making it suitable for lactose-intolerant individuals. This helps prevent symptoms like diarrhea, severe dehydration, and potentially serious complications. In addition to improving milk, lactase is used in the production of ice cream and the preparation of yogurt, enhancing its texture and taste (Singh & Kumar, 2019).

3.4 Catalase

Another key enzyme in the dairy industry is *catalase*, which plays a vital role in breaking down hydrogen peroxide into water and oxygen. This process helps protect cells from damage caused by reactive oxygen species (Singh & Kumar, 2019). Commercially, *catalase* is typically produced using *Aspergillus niger* through solid-state fermentation. *Catalase* is particularly important in cheese production. For example, during the production of certain cheeses, such as Swiss cheese, hydrogen peroxide a strong oxidizing agent that can be toxic to cells is used as part of the pasteurization process. This approach helps preserve the natural enzymes in milk, which are essential for flavor development and the overall quality of the final cheese product (Abada, 2019).

3.5 Rennet

Rennet is a well-known enzyme used in dairy processing, dating back to 6000 BCE (Abada, 2019). It is a blend of *chymosin* and *pepsin* and plays a crucial role in cheese production by coagulating milk into solid curds and liquid whey. Rennet works in two stages: first through an enzymatic reaction and then through a non-enzymatic process, ultimately leading to milk coagulation. Various microorganisms can produce rennet-like enzymes that can substitute traditional calf rennet. Among them, *Aspergillus oryzae, Endothia parasitica, Irpex lactis, Rhizomucor pusillus,* and *Rhizomucor miehei* are widely used for producing rennet in the cheese-making industry (Qureshi et al., 2015).

3.6 Enzymes for Dairy Waste Processing

Using immobilized enzymes to treat lipid-rich waste offers a promising solution for improving industrial waste management and by-products. Unlike free enzymes, immobilized enzymes provide unique advantages, making them highly effective in tackling the challenges of lipid-rich waste streams. Among these, lipase stands out as a key enzyme, playing a crucial role in this process. For instance, immobilized lipases can break down triglycerides in lipid waste into valuable fatty acids and glycerol. This not only transforms waste into useful materials but also significantly reduces the environmental impact of untreated lipid-rich waste (Nimkande & Bafana, 2022; Zieniuk et al., 2022). The fatty acids produced can serve as precursors for biofuels or biochemicals, highlighting the potential of this approach to optimize resource use and minimize environmental harm. However, despite its benefits, there are hurdles to Immobilized enzyme technologies are being widely used for the treatment of lipid-rich waste. The current study is focused on improving immobilization methods, identifying alternative sources of enzymes, and evaluating the consequences of these processes on the environment. With increasing industrial interest in addressing the economic and environmental issues tied to lipid waste, immobilized enzymes could drive more sustainable and efficient waste management practices (Cárdenas-Alcaide et al., 2022; Rather et al., 2022). Several studies underline the effectiveness of lipases in processing lipid waste. For example, immobilized lipases have been used to break down fats and oils to extract dangerous contaminants from trash that is high in lipids. One method involved immobilizing *Acinetobacter haemolyticus lipase* on eggshell membranes successfully hydrolyzed various oils, including olive, canola, and peanut oils, while continuing to function for 19 cycles, from effluent (Chakraborty et al., 2023). Similarly, lipase from *Aspergillus niger* immobilized on a microporous acrylic resin, was used to deacidify high-acid soy sauce residue oil, converting free fatty acids into diacylglycerol-enriched oil suitable for food applications. This enzyme showed excellent activity, thermal stability, and reusability (Işık et al., 2021). Lipid waste from food processing is also being explored as a raw material for biodiesel production. While biodiesel is typically made from edible vegetable oils through lipase-catalyzed transesterification, used cooking oils are increasingly recognized as a viable alternative. Research demonstrates that spent cooking oils or frying can be successfully repurposed into biodiesel, reducing waste and contributing to circular economy practices (Feng et al., 2020).

4 Fungal Enzymes in the Dairy Industry

In dairy processing, enzymes are essential because they help produce a variety of dairy products, including cheese, WPC, WPI, infant formula, and milk fats. This section focuses on how immobilized enzymes are used in the manufacturing of these products.

4.1 Cheese

Lactic acid bacteria are the first of several microorganisms that are essential to the cheese-making procedure curdling the milk and continuing through the maturation stage, where fungi like molds, and yeasts contribute. The biochemical, and microbiological transformations during ripening are key to creating the distinct texture and flavor that define each cheese variety. The two main biochemical processes are proteolysis, which produces amino acids, and short peptides to improve flavor, and lipolysis, which changes lactose into lactate. Numerous factors contribute to these changes, including secondary starters such as molds like *Penicillium roqueforti*, which are frequently seen in blue cheeses (McSweeney & Sousa, 2000). The metabolism of amino acids and fatty acids is a secondary metabolic change. Molds such as *Penicillium roqueforti* in the formation of blue cheese, while *Penicillium camemberti*, and *Geotrichum candidum* are essential for creating the characteristic qualities of soft cheeses like Camembert and Brie (Pouris et al., 2024).

4.1.1 Blue Cheese

Penicillium roqueforti plays a key role in the production of blue cheese, a tradition that dates back to ancient times (Labbe & Serres, 2004, 2009). Historically, this fungus wasn't intentionally added to the cheese; it would appear naturally during the aging process. Blue cheese is easily recognized by its characteristic blue veins, which are created by introducing *Penicillium roqueforti* or *Penicillium glaucum* during the cheesemaking process. These fungi not only give the cheese its distinctive appearance but also contribute to its unique flavor and aroma. This signature taste comes from methyl ketones, which are metabolic by-products of *Penicillium roqueforti* (Pouris et al., 2024).

4.1.2 Camembert Cheese

Camembert cheese production relies on the mold *Penicillium camemberti*, which plays a key role in creating its soft, buttery texture (Michelson, 2010). To achieve this, the surface of the cheese is sprayed with a *Penicillium camemberti* aqueous suspension, and the cheese is left to ripen for several weeks. This ripening process develops the cheese's edible rind, creamy interior, and signature bloomy (Smith, 2005). Beyond Camembert, *Penicillium camemberti* is also used to enhance the flavor of other foods, such as dry, fermented sausages. Other molds, like *Penicillium roqueforti* and *Sporendonema casei*, contribute to the production of various cheeses. For instance, *Sporendonema casei* is involved in making Cantal and Salers cheeses. Uncooked hard cheeses are frequently home to some molds, including *Scopulariopsis flava*, *Scopulariopsis fusca*, and *Scopulariopsis candida*. Furthermore, *Mucor* species including *Mucor lanceolatus*, *Mucor racemosus*, and *Mucor circinelloides* are essential for making raw hard cheeses like Saint Nectaire and Tomme de Savoie, while *Fusarium domesticum* is utilized to make Saint Nectaire and Reblochon (Dupont et al., 2012; Hermet et al., 2012).

4.2 Whey Protein Isolate (WPI)

WPI is a very processed version of whey protein that is valued for both its nutritional quality and its ability to aid in muscle rehabilitation (Kelkar, 2023). When manufacturing cheese, is made from milk and then processed further to eliminate lipids and carbs, leaving a product with over 90% protein content. Enzymatic hydrolysis is used in WPI production to convert whey proteins into amino acids and peptides. Immobilized enzymes offer significant advantages, providing greater reaction control (Srikanth et al., 2023). As a consequence, WPI becomes purer, having a higher protein concentration, fewer contaminants, and a more purified amino acid structure (Sharma et al., 2022). Additionally, the extended operational life of immobilized enzymes reduces the requirement for regular replacements, increasing the process's sustainability and cost-effectiveness. With rising demand from customers for premium supplements with protein, using immobilized enzymes in WPI production offers an efficient and eco-friendly solution (Mora & Toldrá, 2023; Zaworski et al., 2022).

4.3 Whey Protein Concentrate (WPC)

Due to its high protein concentration (66.5%–95.8%) and nutritional advantages, WPC is a well-liked component and supplement. Protein from whey, a by-product of cheese making, is concentrated using filters and separation strategies in traditional WPC production methods. Immobilized enzymes have enabled more effective and sustainable alternatives, potentially lowering production costs over time due to enzyme reusability (Smacchi & Gobbetti, 2000). These methods can also be fine-tuned to fulfill the necessary product standards by achieving particular concentrations of protein (Pellegrino et al., 2022). *β-Galactosidase* and *Lactase* are the most often utilized enzymes in the manufacturing of WPC because they aid in the breakdown of lactose into glucose and galactose while preserving the proteins. Research has shown that immobilizing these enzymes improves stability and operational efficiency. For instance, when *Kluyveromyces lactis β-galactosidase* became immobile on calcium alginate spheres, it showed improved storage and thermal stability and could be reused for over 10 cycles. Another study found that immobilized *β-galactosidase* hydrolyzed more milk lactose on chitosan than the free enzyme, improving efficiency while reducing by-products and maintaining whey proteins (Argenta et al., 2021; Pires et al., 2021).

4.4 *Hydrolyzed Infant Formula*

Hydrolyzed milk is often utilized in products for infants who have inherited susceptibility to allergy to cow's milk. Research has indicated that partially hydrolyzed milk can lessen the possibility of atopic dermatitis in infants. Immobilized enzymes, such as *lactase*, are commonly utilized to create dairy products without lactose for children with lactose intolerance (del Carmen Tocaa et al., 2022; Holvoet et al., 2021). For example, *lactase* immobilized on chitosan has been used to produce low-lactose milk over multiple cycles, demonstrating greater efficiency than free enzymes. Additionally, immobilized *lipase* has been studied for its potential to improve lipid absorption in preterm infants, enhancing their intestinal structure and tissue growth. According to these results, food-grade enzymes immobilized on various supports can be effectively used in the production of specialized infant formulas (Dipasquale et al., 2020; Goncharova et al., 2017; and He et al., 2022).

4.5 *Synthesis of Bioactive Peptide*

In dairy manufacturing, *proteases* are employed for breaking down milk proteins into free amino acids and smaller peptides, improving digestibility and introducing desirable sensory characteristics (Souza, de Andrade et al., 2023). In cheese-making, for example, *proteases* help break down casein, enhancing the flavor and texture of the cheese. Beyond flavor, some milk-derived peptides also possess bioactive characteristics, such as antimicrobial, and antioxidant effects. Immobilizing enzymes increases their reusability, and stability facilitating calm and effective reactions (Smacchi & Gobbetti, 2000). This approach allows for the targeted production of bioactive peptides with practical uses that aid in the creation of milk and other dairy goods with better sensory appeal and health advantages (Punia et al., 2020; Zanutto-Elgui et al., 2019).

5 The Role of Fungal Enzymes in the Food Sector

5.1 *Baking Industry*

Baking refers to the preparation of baked goods, primarily made using wheat flour that serves as the main component and important source of substrates for enzymes (Miguel et al., 2013; Van Oort, 2009). Common baked goods include cakes, bread, cookies, pastries, crackers, tortillas, and pies. Gluten-free items and rye bread also fall under this category (Miguel et al., 2013). The bread development is a major milestone in human history. Following the 1800s, developments in agricultural technology significantly improved bread quality, making white bread widely accessible (Dupaigne & Westbrook, 1999). However, the baking industry saw its most transformative growth when industrial enzymes have been added to the baking procedure (Miguel et al., 2013). Three primary sources of enzymes are needed for baking: the natural enzymes in wheat flour, the enzymes produced by microorganisms' metabolism, and the exogenous enzymes that are added to the dough (Di Cagno et al., 2003). While in baking, different microbial enzymes are utilized, this discussion focuses on fungal enzymes. One notable group of enzymes in baking is *hemicellulases*, which break down hemicellulose, a type of polysaccharide made up of compounds like arabinoxylan, arabinogalactan, xylan, and xylobiose. Among these, *xylanase* (or *endoxylanase*), originally known as *pentosanase*, plays a critical role (Collins & Ng, 1997). *Xylanases* are *glycosidase* enzymes that catalyze the breakdown of 1,4-β-D-xylosidic bonds in xylan and arabinoxylan. Enzymes such as these are present in bacteria, fungi, and archaea.

5.1.1 *Xylanases*

In the baking sector, *xylanases* are essential since they improve the baked goods' quality in multiple ways. These enzymes strengthen the gluten network, leading to better bread quality (Butt et al., 2008; Gray & Bemiller, 2003). They also assist the dough in adjusting to varying flour qualities and changes in processing conditions. By converting water-insoluble hemicelluloses into a soluble form, *xylanases* allow the dough to retain more water, making it softer and easier to handle. This process creates finer, more uniform crumbs and ensures that the dough does not stick to machinery parts, improving its machinability (Rouau, 1993). Additionally, *xylanases* reduce the effort required for sheeting and greatly expand the amount of baked bread (Dervilly et al., 2002; El-Abd et al., 2015; Forsan, 2024d; Forsan et al., 2024; Harbak & Thygesen, 2002; and Sobhy et al., 2021). These enzymes can also increase the bread's arabinoxylan–oligosaccharide content, which offer health benefits. Overall, *xylanases* are instrumental in improving the texture, structure, and overall quality of baked goods like bread, biscuits, and cakes (Harbak & Thygesen, 2002).

5.1.2 *Phytases*

Phytases are a kind of *phosphatase* enzymes known for their capability to break down phytate by releasing at least one phosphate group under laboratory conditions. Among the different types, microbial *phytases* stand out as particularly valuable for biotechnological applications. These enzymes that are capable of degrading phytate outside the cell have been identified in molds and yeasts. Notable sources of *phytases* include species such as *Aspergillus*, *Humicola*, *Candida*, *Myceliophthora*, *Fusarium*, *Pichia*, *Penicillium*, *Rhizopus*, *Rhizomucor*, *Sporotrichum*, *Thermoascus*,

Schizosaccharomyces, *Yarrowia*, *Williopsis*, and *Zygosaccharomyces* (Gupta et al., 2014). *Phytase* has also earned recognition as an exceptional bread-making enhancer. Research has shown that incorporating commercial fungal *phytase*, particularly from *Aspergillus niger*, into fiber-rich dough, formulations accelerates proofing, softens the crumb, improves the bread's form, and raises its specific volume. These enhancements in the quality of bread are linked to the enzyme's indirect effect on boosting *α-amylase* activity (Greiner & Carlsson, 2006).

5.1.3 *Lipases*

Triacylglycerol acyl hydrolases, another name for *lipases*, are versatile enzymes found across various organisms, including animals, plants, bacteria, and fungi. They are crucial to biological processes and have promising industrial applications. One unique property of lipases is their capacity to operate at the intersection of non-water-based, and water-based environments. Under low water activity, they can create esters from long-chain fatty acids and glycerol (Aravindan et al., 2007). While *lipases* are relatively new to the baking industry compared to enzymes like *α-amylases* and *proteases*, their extracellular production is well-documented in fungi, especially in groups like *hyphomycetes* and *zygomycetes* (Gopinath et al., 2013). Lipolytic activity has been observed in several species, such as *Rhizopus, Yarrowia lipolytica, Aspergillus, Candida antarctica*, and *Candida Antarctica* (Wang et al., 2018). In baking, *lipases* contribute to enhancing the flavor characteristics of baked goods (Pruvot et al., 2012). They also improve bread quality by reducing its initial firmness and increasing its specific volume. Additionally, lipase activity can enhance the texture and softness of baked products, offering a more appealing and enjoyable eating experience (Ray, 2012).

5.1.4 *Laccases*

The copper-containing enzyme *laccase* is essential for oxidizing various compounds like diamines, polyphenols, and methoxy-substituted phenols. It can come from plants, fungi, and bacteria, but fungi are the most common source for biotechnological applications. Over sixty strains, including those from the *Basidiomycota*, *Zygomycota*, and *Ascomycota* groups, are known for their high *laccase* activity. In the baking industry, *laccase* is particularly valuable in bread-making. Its use enhances the bread's volume, texture, flavor, and shelf life while also improving the dough's machinability. Additionally, *laccase* contributes to a softer crumb, better crumb structure, increased dough strength and stability, and reduced stickiness, making it a versatile ingredient for improving baked products (Kiiskinen et al., 2004; Mukherjee & Kumar, 2018).

5.1.5 *Amylases*

Among the most important and frequently utilized enzymes in industrial processes are *amylases*, with a history stretching back to thousands of years. These enzymes convert starch molecules into smaller glucose-based polymers and dextrins, among other products. They are essential to contemporary biotechnology with uses in the food and fermentation sectors. Although *amylases* can be obtained from microbes, animals, and microbial sources, microorganism sources are favored for industrial production because of their capacity to satisfy demands on a large scale. Notably, fungal *α-amylases* have been recognized as safe for use in bread production since 1955 in the U.S. and 1963 in the U.K. (Pritchard, 1992). Most fungal *amylases* are sourced from terrestrial fungi like *Aspergillus* and *Penicillium*, with *Aspergillus oryzae* and *Aspergillus niger* being particularly valued for their high enzyme output. Another significant producer is the thermophilic fungus *Thermomyces lanuginosus*, which is highly efficient at amylase production. For ages, *α-amylases* have been used in the baking industry to produce superior goods. These enzymes increase the rate at which the dough ferments, decreases its viscosity, and improves the baked goods' volume and texture. Additionally, they produce more sugar in the dough, improving the crust's color, flavor, and toasting properties. Additionally, *α-amylases* help keep baked goods softer for longer, extending their shelf life and providing an ant staling effect. These properties make *amylases* indispensable for modern bread production and the creation of consistently superior baked products (de Souza et al., 2011; Yadav, 2015).

5.2 *Beverage Industry*

The beverage industry, also referred to as one of the biggest areas of food manufacturing, is the beverage business, producing a variety of alcoholic drinks, and carbonated beverages. It is broadly divided into two categories: alcoholic and nonalcoholic beverages. Alcoholic beverages include distilled spirits, wine, and beer, while the nonalcoholic segment covers soft drink syrups, tea, fruit, coffee, and juices production. Enzymes are essential to manufacture beverages. Similarly, enzymes are critical in winemaking, helping to preserve wine quality during long-term storage, improve clarity, aid filtration, stabilize the product, and shorten fermentation times (Kashyap et al., 2001). With growing health awareness, the demand for fruit juices has steadily increased. Enzymes as *cellulase*, *pectinase*, and **hemicellulase** are widely used to enhance juice extraction and prevent browning. For instance, in grapefruit, and orange processing, *naringinase* enzymes enhance the recovery of essential oil and washing of pulp and reduce bitterness caused by compounds like limonin, neohesperidin, and naringin (Marwaha et al., 1994; West, 1996). Bitterness has been a major obstacle to the widespread acceptance of citrus juices, making *naringinase* essential for improving flavor. Organisms like *Coniothyrium diplodiella* and *Penicillium* species have been identified as efficient producers of *naringinase* (Fukumoto & Okada, 1973; Takiguchi, 1962). Enzymes also streamline fruit and vegetable processing by breaking down pectin, starch, proteins, and cellulose, resulting in higher yields and shorter processing times (Mojsov, 2012). Acidic *pectinases*, predominantly derived from fungi like *Aspergillus niger*, are crucial in fruit juice production and winemaking. They contribute to increased juice yield sometimes up to 100% as reported for *pectinases* and *cellulases* (Alkorta et al., 1998). *Aspergillus aculeatus* is the source of *rhamnogalacturonase*, a kind of *pectinase*, which has since been identified in other *Aspergillus* species (Schols et al., 1990). When making wine, pectic enzymes—that are isolated from fungi like—*Penicillium notatum*, *Botrytis cinerea*, and *Aspergillus niger* are crucial (Fogarty & Kelly, 2012; Robertson, 1977). Beyond beverages, pectinolytic enzymes play a role in processing fibers like jute, flax, hemp, and coconut coir through retting a fermentation process where fungi decompose pectin to release fibers (Brühlmann et al., 1994; H. Sharma & Robinson, 1983).

5.3 Soy Sauce

Soy sauce, originally from the Orient, has become a beloved condiment worldwide. Its production relies on fermentation, using microorganisms like *Aspergillus sojae*, *Aspergillus oryzae*, acetic acid bacteria, and *Saccharomyces rouxii* to create a tasty beverage with superior health benefits. The process starts with soaking and cooking soybeans that are then mashed and fermented with *Aspergillus oryzae* and *Aspergillus sojae* to develop the characteristic taste and aroma (Moore & Chiu, 2001).

5.4 Alcoholic Beverages

Cider, a popular beverage made from apples, is especially well-loved in the UK that boasts the highest per capita cider consumption in the world and is home to some of the largest cider producers. It is also widely enjoyed in countries like Australia, Canada, India, and New Zealand. One of the most critical factors in crafting high-quality cider is yeast selection. The yeast used to make cider typically falls into two categories: wild, native, commercially developed strains. The most common species employed are *Saccharomyces cerevisiae* and *Saccharomyces bayanus*. Interestingly, populations of wild yeast can vary greatly, frequently containing species such as *Hanseniaspora*, *Candida, Pichia*, *Metschnikowia*, and *Saccharomyces* (Valles et al., 2007). Native yeasts, in particular, often become established in a cidery over time, contributing to the unique flavors and characteristics that define the final product (Bedriñana et al., 2010).

6 Conclusions

Fungi are highly valued as a source of enzymes for a range of sectors, including dairy, baking, and food, due to their versatility and efficiency. These microorganisms are a wealth of resources that can be used in industry, agriculture, and the production of products with added value. Enzymes, especially in dairy and lipid processing, are gaining popularity because of their numerous advantages over chemical alternatives. Unlike chemicals, enzymes work under mild circumstances, for example, neutral pH, and low temperatures resulting in natural products that align with regulatory standards. In addition to increasing product quality, this method reduces the production of undesirable by-products, making it an environmentally friendly option. Enzymes are also particular in their actions, allowing for the making of numerous high-purity, functional dairy, and lipid-based products. This book chapter explores the various uses of immobilized enzymes in these sectors, emphasizing the importance of selecting the appropriate immobilization methods and resources to maintain enzyme stability and activity. Despite their potential, enzymes that are immobilized are still rarely used in dairy processing. However, this technology offers significant industrial benefits, including improved environmental sustainability, economic viability, and operational efficiency. Realizing their full potential on an industrial scale will require addressing key challenges. These include optimizing immobilization methods to preserve enzyme performance and identifying food-grade materials to avoid contamination. Future studies ought to concentrate on creating scalable, and cost-effective immobilization techniques that ensure that enzymes remain active, stable, and selective during processing. By overcoming these hurdles, immobilized enzymes could revolutionize industrial dairy processing, delivering enhanced efficiency and sustainability.

References

Abada, E. A. (2019). Application of microbial enzymes in the dairy industry. In *Enzymes in food biotechnology* (pp. 61–72). Elsevier.

Abellanas-Perez, P., Carballares, D., Fernandez-Lafuente, R., & Rocha-Martin, J. (2023). Glutaraldehyde modification of lipases immobilized on octyl agarose beads: Roles of the support enzyme loading and chemical amination of the enzyme on the final enzyme features. *International Journal of Biological Macromolecules, 248*, 125853.

Ahmad, T., Aadil, R. M., Ahmed, H., ur Rahman, U., Soares, B. C., Souza, S. L., et al. (2019). Treatment and utilization of dairy industrial waste: A review. *Trends in Food Science & Technology, 88*, 361–372.

Akanbi, T. O., & Barrow, C. J. (2017). Candida Antarctica lipase A effectively concentrates DHA from fish and thraustochytrid oils. *Food Chemistry, 229*, 509–516.

Akanbi, T. O., Adcock, J. L., & Barrow, C. J. (2013). Selective concentration of EPA and DHA using Thermomyces lanuginosus lipase is due to fatty acid selectivity and not regioselectivity. *Food Chemistry, 138*(1), 615–620.

Akanbi, T. O., Sinclair, A. J., & Barrow, C. J. (2014). Pancreatic lipase selectively hydrolyses DPA over EPA and DHA due to location of double bonds in the fatty acid rather than regioselectivity. *Food Chemistry, 160*, 61–66.

Akanbi, T. O., Marshall, S. N., & Barrow, C. J. (2019). Polydatin-fatty acid conjugates are effective antioxidants for stabilizing omega 3-containing bulk fish oil and fish oil emulsions. *Food Chemistry, 301*, 125297.

Akanbi, T. O., Ji, D., & Agyei, D. (2020). Revisiting the scope and applications of food enzymes from extremophiles. *Journal of Food Biochemistry, 44*(11), e13475.

Alkorta, I., Garbisu, C., Llama, M. J., & Serra, J. L. (1998). Industrial applications of pectic enzymes: A review. *Process Biochemistry, 33*(1), 21–28.

Aravindan, R., Anbumathi, P., & Viruthagiri, T. (2007). Lipase applications in food industry. *Indian Journal of Biotechnology, 6*(2), 141–158.

Argenta, A. B., Nogueira, A., & Scheer, A. d. P. (2021). Hydrolysis of whey lactose: Kluyveromyces lactis β-galactosidase immobilisation and integrated process hydrolysis-ultrafiltration. *International Dairy Journal, 117*, 105007.

Aryee, A. N., Akanbi, T. O., Nwachukwu, I. D., & Gunathilake, T. (2022). Perspectives on preserving lipid quality and strategies for value enhancement. *Current Opinion in Food Science, 44*, 100802.

Basso, A., & Serban, S. (2019). Industrial applications of immobilized enzymes—A review. *Molecular Catalysis, 479*, 110607.

Bedriñana, R. P., Simón, A. Q., & Valles, B. S. (2010). Genetic and phenotypic diversity of autochthonous cider yeasts in a cellar from Asturias. *Food Microbiology, 27*(4), 503–508.

Bilal, M., & Iqbal, H. M. (2019). Sustainable bioconversion of food waste into high-value products by immobilized enzymes to meet bio-economy challenges and opportunities–A review. *Food Research International, 123*, 226–240.

Binhayeeding, N., Yunu, T., Pichid, N., Klomklao, S., & Sangkharak, K. (2020). Immobilisation of Candida rugosa lipase on polyhydroxybutyrate via a combination of adsorption and cross-linking agents to enhance acylglycerol production. *Process Biochemistry, 95*, 174–185.

Brühlmann, F., Kim, K. S., Zimmerman, W., & Fiechter, A. (1994). Pectinolytic enzymes from actinomycetes for the degumming of ramie bast fibers. *Applied and Environmental Microbiology, 60*(6), 2107–2112.

Butt, M. S., Tahir-Nadeem, M., Ahmad, Z., & Sultan, M. T. (2008). Xylanases and their applications in baking industry. *Food Technology and Biotechnology, 46*(1), 22–31.

Cárdenas-Alcaide, M. F., Godínez-Alemán, J. A., González-González, R. B., Iqbal, H. M., & Parra-Saldívar, R. (2022). Environmental impact and mitigation of micro (nano) plastics pollution using green catalytic tools and green analytical methods. *Green Analytical Chemistry, 3*, 100031.

Chakraborty, D., Chatterjee, S., Althuri, A., Palani, S. G., & Mohan, S. V. (2023). Sustainable enzymatic treatment of organic waste in a framework of circular economy. *Bioresource Technology, 370*, 128487.

Charles, M. (2019). *Applications of lactase and immobilized lactase*. CRC Press.

Chen, N., Chang, B., Shi, N., Yan, W., Lu, F., & Liu, F. (2023). Cross-linked enzyme aggregates immobilization: Preparation, characterization, and applications. *Critical Reviews in Biotechnology, 43*(3), 369–383.

Cieh, N. L., Mokhtar, M. N., Baharuddin, A. S., Mohammed, M. A. P., & Wakisaka, M. (2024). Progress on lipase immobilization technology in edible oil and fat modifications. *Food Reviews International, 40*(1), 457–503.

Collins, R. A., & Ng, T. B. (1997). Polysaccharopeptide from Coriolus versicolor has potential for use against human immunodeficiency virus type 1 infection. *Life Sciences, 60*(25), PL383-PL387.

de Souza, T. S. P., de Andrade, C. J., Koblitz, M. G. B., & Fai, A. E. C. (2023). Microbial peptidase in food processing: Current state of the art and future trends. *Catalysis Letters, 153*(1), 114–137.

de Souza, J. J., Vieira, I. J. C., Rodrigues-Filho, E., & Braz-Filho, R. (2011). Terpenoids from endophytic fungi. *Molecules, 16*(12), 10604–10618.
del Carmen Tocaa, M., Fernándezb, A., Orsic, M., Tabaccod, O., & Vinderolae, G. (2022). Lactose intolerance: Myths and facts. An update. *Archivos Argentinos de Pediatria, 120*(1), 59–66.
Dervilly, G., Leclercq, C., Zimmermann, D., Roue, C., Thibault, J.-F., & Saulnier, L. (2002). Isolation and characterization of high molar mass water-soluble arabinoxylans from barley and barley malt. *Carbohydrate Polymers, 47*(2), 143–149.
Di Cagno, R., De Angelis, M., Corsetti, A., Lavermicocca, P., Arnault, P., Tossut, P., et al. (2003). Interactions between sourdough lactic acid bacteria and exogenous enzymes: Effects on the microbial kinetics of acidification and dough textural properties. *Food Microbiology, 20*(1), 67–75.
Dipasquale, V., Serra, G., Corsello, G., & Romano, C. (2020). Standard and specialized infant formulas in Europe: Making, marketing, and health outcomes. *Nutrition in Clinical Practice, 35*(2), 273–281.
Dong, Z., Cui, Z., Jin, J., Cheng, X., Wu, G., Wang, X., & Jin, Q. (2024). Enzymatic synthesis of structured lipids enriched with medium-and long-chain Triacylglycerols via Pickering emulsion-assisted interfacial catalysis: A preliminary exploration. *Molecules, 29*(4), 915.
Dovale-Rosabal, G., Rodríguez, A., Espinosa, A., Barriga, A., & Aubourg, S. P. (2021). Synthesis of EPA-and DHA-enriched structured acylglycerols at the sn-2 position starting from commercial salmon oil by enzymatic lipase catalysis under supercritical conditions. *Molecules, 26*(11), 3094.
Du, X., Jing, H., Wang, L., Huang, X., Wang, X., & Wang, H. (2022). Characterization of structure, physicochemical properties, and hypoglycemic activity of goat milk whey protein hydrolysate processed with different proteases. *Lwt, 159*, 113257.
Du, X., Wang, L., Li, Y., Wu, J., Chen, G., Liang, H., & Gao, D. (2023). Cooperative adsorption-degradation of benzo [a] pyrene by extracellular enzymes of white rot fungi immobilized on macroporous adsorption resin. *International Biodeterioration & Biodegradation, 178*, 105564.
Dupaigne, B., & Westbrook, A. (1999). *The history of bread*. Harry N. Abrams.
Dupont, S., Lemetais, G., Ferreira, T., Cayot, P., Gervais, P., & Beney, L. (2012). Ergosterol biosynthesis: A fungal pathway for life on land? *Evolution, 66*(9), 2961–2968.
El-Abd, M., Elsabie, W. B., Hamad, E., & Saad, H. F. (2015). Yoghurt enriched with conjugated linoleic acid using PROBIOTIC bacteria bio production and addition of avocado fruit. *Journal of Food and Dairy Sciences, 6*(2), 59–69.
Feng, K., Huang, Z., Peng, B., Dai, W., Li, Y., Zhu, X., et al. (2020). Immobilization of aspergillus Niger lipase onto a novel macroporous acrylic resin: Stable and recyclable biocatalysis for deacidification of high-acid soy sauce residue oil. *Bioresource Technology, 298*, 122553.
Fogarty, W. M., & Kelly, C. T. (2012). *Microbial enzymes and biotechnology*. Springer Science & Business Media.
Forsan, H. F. (2023). Applications of nanomaterials in medicine. In *Sustainable nanomaterials for biomedical engineering: Impacts, challenges, and future prospects* (pp. 3–29). Apple Academic Press.
Forsan, H. F. (2024a). Applications of nanosensors in agriculture and food sectors. In G. A. M. Ali, K. F. Chong, & A. S. H. Makhlouf (Eds.), *Handbook of nanosensors*. Springer.
Forsan, H. F. (2024b). Applications of nanosensors in agriculture and food sectors. In *Handbook of nanosensors: Materials and technological applications* (pp. 1331–1360). Springer.
Forsan, H. F. (2024c). Diet and sleep disorders. In *Nutrition and psychiatric disorders: An evidence-based approach to understanding the diet-brain connection* (pp. 421–443). Springer.
Forsan, H. F. (2024d). Zebrafish as a model for a novel neuroprotective natural product. In *Zebrafish as a Model for Parkinson's Disease* (pp. 215–227). CRC Press.
Forsan, H. F., & Awad, S. S. (2023). Cyanidin: Advances on resources, biosynthetic pathway, bioavailability, bioactivity, and pharmacology. In J. Xiao (Ed.), *Handbook of dietary flavonoids* (pp. 1–50). Springer.

Forsan, H. F., & Hasan, R. S. (2023). Biodegradable Polysaccharides Nanocomposites. In *Handbook of biodegradable materials* (pp. 415–449). Springer.
Forsan, H. F., & Hassan, R. S. (2023). Novel nutraceutical Milk compound in Alzheimer's prevention. In E. Mohamed (Ed.), *Handbook of neurodegenerative disorders*. Springer.
Forsan, H. F., Rahman, A. R. A., Safwat, M., Abduljabar, R. S., & Riad, O. K. (2022). COVID-19: A review of immune-enhancing nutrients and supplements. *Journal of Applied Nutritional Sciences, 1*(1), 1–14.
Forsan, H. F., Abd El-Hak, A. E., Ahmad, E., Alswerky, E. M., Elagezy, F. K., Yassin, M., et al. (2024). Toward better science-based advice on nutrition. In *Nutrition and psychiatric disorders: An evidence-based approach to understanding the diet-brain connection* (pp. 183–204). Springer.
Forsan, H. F., Fayed, M. R., Farahat, N. M., Gabr, W. M., Alswerky, E. M., El-Hak, A. E. A., et al. (2025). Polyphenols of mulberry white (Morus alba L.) leaves as a source of functional food: A review. *Al-Kitab Journal for Pure Sciences, 9*(2), 104. https://doi.org/10.32441/kjps
Fukumoto, J., & Okada, S. (1973). Naringinase production by fermentation. *Japanese Patent, 7306554.*
Gan, J., Bagheri, A. R., Aramesh, N., Gul, I., Franco, M., Almulaiky, Y. Q., & Bilal, M. (2021). Covalent organic frameworks as emerging host platforms for enzyme immobilization and robust biocatalysis–A review. *International Journal of Biological Macromolecules, 167*, 502–515.
Goncharova, K., Kirko, S., Grujic, D., Kardas, M., Grochowska-Niedworok, E., Prykhodko, O., et al. (2017). Enhanced absorption of long-chain polyunsaturated fatty acids following consumption of functional milk formula, pre-digested with immobilized lipase ex vivo, in an exocrine pancreatic insufficient (EPI) pig model. *Journal of Functional Foods, 34*, 422–430.
Gopinath, S. C., Anbu, P., Lakshmipriya, T., & Hilda, A. (2013). Strategies to characterize fungal lipases for applications in medicine and dairy industry. *BioMed Research International, 2013*(1), 154549.
Gowda, A., Sharma, V., Goyal, A., Singh, A., & Arora, S. (2018). Process optimization and oxidative stability of omega-3 ice cream fortified with flaxseed oil microcapsules. *Journal of Food Science and Technology, 55*, 1705–1715.
Gray, J., & Bemiller, J. (2003). Bread staling: Molecular basis and control. *Comprehensive Reviews in Food Science and Food Safety, 2*(1), 1–21.
Greiner, R., & Carlsson, N.-G. (2006). Myo-inositol phosphate isomers generated by the action of a phytate-degrading enzyme from Klebsiella terrigena on phytate. *Canadian Journal of Microbiology, 52*(8), 759–768.
Guisan, J. M., Fernandez-Lorente, G., Rocha-Martin, J., & Moreno-Gamero, D. (2022). Enzyme immobilization strategies for the design of robust and efficient biocatalysts. *Current Opinion in Green and Sustainable Chemistry, 35*, 100593.
Gunathilake, T., Akanbi, T. O., & Barrow, C. J. (2021). Lipase-produced omega-3 acylglycerols for the fortification and stabilization of extra virgin olive oil using hydroxytyrosyl palmitate. *Future Foods, 4*, 100045.
Gunathilake, T., Akanbi, T. O., Bucher, T., & Barrow, C. J. (2022a). Enzymes in nutrition, baby foods, and food safety. In *Value-addition in food products and processing through enzyme technology* (pp. 153–161). Elsevier.
Gunathilake, T., Akanbi, T. O., Van Vuong, Q., Scarlett, C. J., & Barrow, C. J. (2022b). Enzyme technology in the production of flavors and food additives. In *Value-addition in food products and processing through enzyme technology* (pp. 45–55). Elsevier.
Gupta, R., Gangoliya, S., & Singh, N. (2014). Isolation of thermotolerant phytase producing fungi and optimisation of phytase production by aspergillus Niger NRF9 in solid state fermentation using response surface methodology. *Biotechnology and Bioprocess Engineering, 19*, 996–1004.
Hagar, F., Forsan, M. M. E. A., Elsabie, W. B., & Sobhy, H. M. (2024). Apricot addition for enrichment yogurt with amygdalin. *Al-Kitab Journal for Pure Sciences, 8*(1), 63–70.

Harbak, L., & Thygesen, H. (2002). Safety evaluation of a xylanase expressed in Bacillus subtilis. *Food and Chemical Toxicology, 40*(1), 1–8.

He, W. S., Sun, Y., Li, Z., Yang, H., Li, J., Wang, Q., et al. (2022). Enhanced antioxidant capacity of lipoic acid in different food systems through lipase-mediated esterification with phytosterols. *Journal of the Science of Food and Agriculture, 102*(15), 7115–7125.

Hermet, A., Méheust, D., Mounier, J., Barbier, G., & Jany, J.-L. (2012). Molecular systematics in the genus Mucor with special regards to species encountered in cheese. *Fungal Biology, 116*(6), 692–705.

Holvoet, S., Nutten, S., Dupuis, L., Donnicola, D., Bourdeau, T., Hughes-Formella, B., et al. (2021). Partially hydrolysed whey-based infant formula improves skin barrier function. *Nutrients, 13*(9), 3113.

Hooper, L., Martin, N., Jimoh, O. F., Kirk, C., Foster, E., & Abdelhamid, A. S. (2020). Reduction in saturated fat intake for cardiovascular disease. *Cochrane Database of Systematic Reviews, 5*(8), CD011737.

Imam, H. T., Marr, P. C., & Marr, A. C. (2021). Enzyme entrapment, biocatalyst immobilization without covalent attachment. *Green Chemistry, 23*(14), 4980–5005.

Işık, C., Saraç, N., Teke, M., & Uğur, A. (2021). A new bioremediation method for removal of wastewater containing oils with high oleic acid composition: Acinetobacter haemolyticus lipase immobilized on eggshell membrane with improved stabilities. *New Journal of Chemistry, 45*(4), 1984–1992.

Kashyap, D., Vohra, P., Chopra, S., & Tewari, R. (2001). Applications of pectinases in the commercial sector: A review. *Bioresource Technology, 77*(3), 215–227.

Kelkar, N. (2023). Clinical applications of whey protein. In *Nutrition science, marketing nutrition, health claims, and public policy* (pp. 13–22). Elsevier.

Khoonin, W., Shantavasinkul, P. C., Santivarangkna, C., & Trachootham, D. (2022). Loss of Eicosapentaenoic acid (EPA) after retort sterilization of the EPA-BCAA fortified complete nutrition drink. *Food, 11*(14), 2023.

Kiiskinen, L.-L., Kruus, K., Bailey, M., Ylosmaki, E., Siika-Aho, M., & Saloheimo, M. (2004). Expression of Melanocarpus albomyces laccase in Trichoderma reesei and characterization of the purified enzyme. *Microbiology, 150*(9), 3065–3074.

Kleekayai, T., Cermeño, M., & FitzGerald, R. J. (2021). The production of bioactive peptides from milk proteins. In *Agents of change: Enzymes in milk and dairy products* (pp. 447–497).

Labbe, M., & Serres, J. (2004). *Chroniques du Roquefort—De la préhistoire à l'aube industrielle*. Graphi Imprimeur.

Labbe, M., & Serres, J. (2009). *Chroniques du Roquefort-Des hommes, des entreprises, des marques, période moderne*. Graphi Imprimeur.

León-López, A., Pérez-Marroquín, X. A., Estrada-Fernández, A. G., Campos-Lozada, G., Morales-Peñaloza, A., Campos-Montiel, R. G., & Aguirre-Álvarez, G. (2022). Milk whey hydrolysates as high value-added natural polymers: Functional properties and applications. *Polymers, 14*(6), 1258.

Lermen, A. M., Clerici, N. J., Borchartt Maciel, D., & Daroit, D. J. (2023). Characterization and application of a crude bacterial protease to produce antioxidant hydrolysates from whey protein. *Preparative Biochemistry & Biotechnology, 53*(1), 12–21.

Maghraby, Y. R., El-Shabasy, R. M., Ibrahim, A. H., & Azzazy, H. M. E.-S. (2023). Enzyme immobilization technologies and industrial applications. *ACS Omega, 8*(6), 5184–5196.

Marwaha, S., Puri, M., Bhullar, M., & Kothari, R. (1994). Optimization of parameters for hydrolysis of limonin for debittering of kinnow mandarin juice by Rhodococcus fascians. *Enzyme and Microbial Technology, 16*(8), 723–725.

Mazorra-Manzano, M. A., Mora-Cortes, W. G., Leandro-Roldan, M. M., González-Velázquez, D. A., Torres-Llanez, M. J., Ramírez-Suarez, J. C., et al. (2020). Production of whey protein hydrolysates with angiotensin-converting enzyme-inhibitory activity using three new sources of plant proteases. *Biocatalysis and Agricultural Biotechnology, 28*, 101724.

McCleary, B. V. (2023). Measurement of dietary fiber: Which AOAC official method of AnalysisSM to use. *Journal of AOAC International, 106*(4), 917–930. https://doi.org/10.1093/jaoacint/qsad051
McSweeney, P. L., & Sousa, M. J. (2000). Biochemical pathways for the production of flavour compounds in cheeses during ripening: A review. *Le Lait, 80*(3), 293–324.
Mehrotra, T., Dev, S., Banerjee, A., Chatterjee, A., Singh, R., & Aggarwal, S. (2021). Use of immobilized bacteria for environmental bioremediation: A review. *Journal of Environmental Chemical Engineering, 9*(5), 105920.
Michelson, P. (2010). *Cheese: Exploring taste and tradition*. Gibbs Smith.
Miguel, A. M., Martins-Meyer, T. S., Figueiredo, E., Lobo, B. W. P., & Dellamora-Ortiz, G. M. (2013). Enzymes in bakery: Current and future trends. *Food Industry*, 278–321.
Mohammadi, Z. B., Zhang, F., Kharazmi, M. S., & Jafari, S. M. (2023). Nano-biocatalysts for food applications; immobilized enzymes within different nanostructures. *Critical Reviews in Food Science and Nutrition, 63*(32), 11351–11369.
Mojsov, K. (2012). Microbial α-amylases and their industrial applications: A review. *International Journal of Managment, IT and Engineering, 2*(10), 583–609.
Monari, S., Ferri, M., Russo, C., Prandi, B., Tedeschi, T., Bellucci, P., et al. (2019). Enzymatic production of bioactive peptides from scotta, an exhausted by-product of ricotta cheese processing. *PLoS One, 14*(12), e0226834.
Moore, D., & Chiu, S. W. (2001). *Fungal products as food. Bio-exploitation of filamentous fungi* (pp. 223–251). Fungal Diversity Press.
Mora, L., & Toldrá, F. (2023). Advanced enzymatic hydrolysis of food proteins for the production of bioactive peptides. *Current Opinion in Food Science, 49*, 100973.
Mukherjee, A., & Kumar, D. (2018). A review on applications of lignolytic enzymes of fungi. *The World Journal of Pharmaceutical Sciences, 7*, 484.
Mustafa, A., Ramadan, R., Niikura, F., Inayat, A., & Hafez, H. (2023). Highly selective synthesis of glyceryl monostearate via lipase catalyzed esterification of triple pressed stearic acid and glycerin. *Sustainable Energy Technologies and Assessments, 57*, 103200.
Nimkande, V. D., & Bafana, A. (2022). A review on the utility of microbial lipases in wastewater treatment. *Journal of Water Process Engineering, 46*, 102591.
Oliveira, J. P., Gonçalves, L. R., Amorim, K. P., Pinheiro, B. B., Ramos, M. V., Souza, P. F., et al. (2022). Immobilization and characterization of latex cysteine peptidases on different supports and application for cow's milk protein hydrolysis. *Process Biochemistry, 117*, 180–190.
Pellegrino, L., Hogenboom, J. A., Rosi, V., Sindaco, M., Gerna, S., & D'Incecco, P. (2022). Focus on the protein fraction of sports nutrition supplements. *Molecules, 27*(11), 3487.
Pires, A. F., Marnotes, N. G., Rubio, O. D., Garcia, A. C., & Pereira, C. D. (2021). Dairy by-products: A review on the valorization of whey and second cheese whey. *Food, 10*(5), 1067.
Pouris, J., Kolyva, F., Bratakou, S., Vogiatzi, C. A., Chaniotis, D., & Beloukas, A. (2024). The role of fungi in food production and processing. *Applied Sciences, 14*(12), 5046.
Pourjoula, M., Picariello, G., Garro, G., D'Auria, G., Nitride, C., Ghaisari, A. R., & Ferranti, P. (2020). The protein and peptide fractions of kashk, a traditional Middle East fermented dairy product. *Food Research International, 132*, 109107.
Pritchard, P. E. (1992). Studies on the bread-improving mechanism of fungal alpha-amylase. *Journal of Biological Education, 26*(1), 12–18.
Pruvot, B., Quiroz, Y., Voncken, A., Jeanray, N., Piot, A., Martial, J. A., & Muller, M. (2012). A panel of biological tests reveals developmental effects of pharmaceutical pollutants on late stage zebrafish embryos. *Reproductive Toxicology, 34*(4), 568–583.
Punia, H., Tokas, J., Malik, A., Sangwan, S., Baloda, S., Singh, N., et al. (2020). Identification and detection of bioactive peptides in milk and dairy products: Remarks about agro-foods. *Molecules, 25*(15), 3328.
Qureshi, M., Khare, A., Pervez, A., & Uprit, S. (2015). Enzymes used in dairy industries. *International Journal of Applied Research, 1*(10), 523–527.

Rather, A. H., Khan, R. S., Wani, T. U., Beigh, M. A., & Sheikh, F. A. (2022). Overview on immobilization of enzymes on synthetic polymeric nanofibers fabricated by electrospinning. *Biotechnology and Bioengineering, 119*(1), 9–33.
Ray, A. (2012). Application of lipase in industry. *Asian Journal of Pharmacy and Technology, 2*(2), 33–37.
Robertson, G. (1977). *Pectic enzymes and winemaking*. Food Technol NZ.
Rodrigues, R. C., Berenguer-Murcia, Á., Carballares, D., Morellon-Sterling, R., & Fernandez-Lafuente, R. (2021). Stabilization of enzymes via immobilization: Multipoint covalent attachment and other stabilization strategies. *Biotechnology Advances, 52*, 107821.
Rojas, M. J., Amaral-Fonseca, M., Zanin, G. M., Fernandez-Lafuente, R., de Lima Camargo Giordano, R., & Tardioli, P. W. (2019). Preparation of crosslinked enzyme aggregates of a thermostable cyclodextrin glucosyltransferase from Thermoanaerobacter sp. critical effect of the crosslinking agent. *Catalysts, 9*(2), 120.
Rouau, X. (1993). Investigations into the effects of an enzyme preparation for baking on wheat flour dough pentosans. *Journal of Cereal Science, 18*(2), 145–157.
Safwat, M.-A., Abdul-Rahman, A. R., Forsan, H. F., Abduljabar, R., & Riad, O. (2021). COVID-19; immunology, pathology, severity and immunosuppressants. *Azhar International Journal of Pharmaceutical and Medical Sciences, 1*(1), 1–12.
Sarnoab, M., & Iuliano, M. (2020). Biodiesel production from dairy waste scum by using an efficient nano-biocatalyst. *Chemical Engineering Transactions, 79*, 181–186.
Schols, H. A., Geraeds, C. C., Searle-van Leeuwen, M. F., Kormelink, F. J., & Voragen, A. G. (1990). Rhamnogalacturonase: A novel enzyme that degrades the hairy regions of pectins. *Carbohydrate Research, 206*(1), 105–115.
Sharma, H., & Robinson, E. (1983). *Fungal colonization during glyphosate induced desiccation and dew-retting of flax cultivars*. Technical Report, Lambeg Industrial Research Association, No., 1219, 11.
Sharma, K., Saini, R., & Patil, S. (2022). A systematic review on the emerging role of protein powder. *International Journal of Research in Engineering and Science (IJRES), 10*(5), 88–90.
Sicard, C. (2023). In situ enzyme immobilization by covalent organic frameworks. *Angewandte Chemie, 135*(1), e202213405.
Silva, J. M. F., Dos Santos, K. P., Dos Santos, E. S., Rios, N. S., & Gonçalves, L. R. B. (2023). Immobilization of Thermomyces lanuginosus lipase on a new hydrophobic support (streamline phenyl™): Strategies to improve stability and reusability. *Enzyme and Microbial Technology, 163*, 110166.
Singh, P., & Kumar, S. (2019). Microbial enzyme in food biotechnology. In *Enzymes in food biotechnology* (pp. 19–28). Elsevier.
Singh, R., Singh, A., & Sachan, S. (2019). Enzymes used in the food industry: Friends or foes? In *Enzymes in food biotechnology* (pp. 827–843). Elsevier.
Smacchi, E., & Gobbetti, M. (2000). Bioactive peptides in dairy products: Synthesis and interaction with proteolytic enzymes. *Food Microbiology, 17*(2), 129–141.
Smith, R. (2005). *Biodegradable polymers for industrial applications*. CRC Press.
Sobhy, H. M., Abd, M. E., Elsabie, W., & FathyForsan, H. (2021). Study of high nutritive value of almond milk beverage. *Plant Archives, 21*. https://doi.org/10.51470/PLANTARCHIVES.2021.v21.S1.405
Srikanth, D., Gopi, D., Sunil, C., Michael, K., & Rawson, A. (2023). Proteins as fat replacers in the food industry. In *Fat Mimetics for Food Applications* (pp. 133–154).
Taheri-Kafrani, A., Kharazmi, S., Nasrollahzadeh, M., Soozanipour, A., Ejeian, F., Etedali, P., et al. (2021). Recent developments in enzyme immobilization technology for high-throughput processing in food industries. *Critical Reviews in Food Science and Nutrition, 61*(19), 3160–3196.
Takiguchi, Y. (1962). Annual reports of Takamine Institute, Tokyo. *Agricultural Society of Japan, 14*, 101.
Thangaraj, B., & Solomon, P. R. (2019). Immobilization of lipases–A review. Part I: Enzyme immobilization. *ChemBioEng Reviews, 6*(5), 157–166.

Valles, B. S., Bedriñana, R. P., Tascón, N. F., Simón, A. Q., & Madrera, R. R. (2007). Yeast species associated with the spontaneous fermentation of cider. *Food Microbiology, 24*(1), 25–31.
Van Oort, M. (2009). Enzymes in bread making. In *Enzymes in food technology* (pp. 103–143).
Wahab, R. A., Elias, N., Abdullah, F., & Ghoshal, S. K. (2020). On the taught new tricks of enzymes immobilization: An all-inclusive overview. *Reactive and Functional Polymers, 152*, 104613.
Wang, Y., Ma, R., Li, S., Gong, M., Yao, B., Bai, Y., & Gu, J. (2018). An alkaline and surfactant-tolerant lipase from Trichoderma lentiforme ACCC30425 with high application potential in the detergent industry. *AMB Express, 8*, 1–11.
West, S. (1996). *Industrial enzymology*. Stockton Press.
Wróblewska, B., Karamać, M., Amarowicz, R., Szymkiewicz, A., Troszyńska, A., & Kubicka, E. (2004). Immunoreactive properties of peptide fractions of cow whey milk proteins after enzymatic hydrolysis. *International Journal of Food Science & Technology, 39*(8), 839–850.
Xia, Q., Akanbi, T. O., Wang, B., Li, R., Liu, S., & Barrow, C. J. (2020). Investigation of enhanced oxidation stability of microencapsulated enzymatically produced tuna oil concentrates using complex coacervation. *Food & Function, 11*(12), 10748–10757.
Yadav, A. (2015). Bacterial diversity of cold deserts and mining of genes for low temperature tolerance. IARI, New Delhi/BIT, p 234.
Zanuso, E., Gomes, D. G., Ruiz, H. A., Teixeira, J. A., & Domingues, L. (2021). Enzyme immobilization as a strategy towards efficient and sustainable lignocellulosic biomass conversion into chemicals and biofuels: Current status and perspectives. *Sustainable Energy & Fuels, 5*(17), 4233–4247.
Zanutto-Elgui, M. R., Vieira, J. C. S., do Prado, D. Z., Buzalaf, M. A. R., de Magalhães Padilha, P., de Oliveira, D. E., & Fleuri, L. F. (2019). Production of milk peptides with antimicrobial and antioxidant properties through fungal proteases. *Food Chemistry, 278*, 823–831.
Zaworski, K., Woliński, J., Słupecka-Ziemilska, M., Pierzynowski, S., & Pierzynowska, K. (2022). Pre-digestion of the lipids in infant formula affects gut maturation of the preterm pig. *PLoS One, 17*(3), e0265144.
Zhang, R., Yan, X., & Fan, K. (2021). Nanozymes inspired by natural enzymes. *Accounts of Materials Research, 2*(7), 534–547.
Zhu, H., Wang, X., Zhang, W., Zhang, Y., Zhang, S., Pang, X., et al. (2022). Dietary Schizochytrium microalgae affect the fatty acid profile of goat milk: Quantification of docosahexaenoic Acid (DHA) and its distribution at Sn-2 position. *Food, 11*(14), 2087.
Zieniuk, B., Białecka-Florjańczyk, E., Wierzchowska, K., & Fabiszewska, A. (2022). Recent advances in the enzymatic synthesis of lipophilic antioxidant and antimicrobial compounds. *World Journal of Microbiology and Biotechnology, 38*, 1–16.

Chapter 7
Fungal Metabolites as Natural Flavor Enhancers

Ritu Vasisht, Jaishree Yadav, and Shekhar Agnihotri

1 Introduction

Flavor compounds are essential in the food and beverage industries for enhancing sensory appeal and supporting product differentiation (Ghandehari-Alavijeh et al., 2024). Synthetic flavor compounds accounted for a substantial market share, generating a revenue of USD 305.2 million (Grand View Research, 2022). Conversely, increasing consumer preference for clean-label and health-oriented products has fueled substantial expansion in the worldwide flavor compounds industry. Valued at roughly USD 424.61 million in 2022, the market is anticipated to grow at a 5.7% CAGR, potentially attaining USD 660.4 million by 2030 (Grand View Research, 2022). Despite this growth, concerns regarding sustainability, cost, and environmental impact have not yet been addressed. This has led to the exploration of microbial bioprocessing, particularly fungal species, owing to their diverse biosynthetic capabilities and environmentally sustainable production processes (Berger & Zorn, 2004; Frey, 2005).

Molds and yeasts represent diverse fungal species that naturally synthesize numerous flavor and aroma compounds (Yin et al., 2019). These flavors or aromas are nothing but secondary metabolites, which the fungal community tends to produce during its growth and survival under diverse conditions (Liu et al., 2024a; Tian

R. Vasisht · J. Yadav
Department of Agriculture and Environmental Sciences, National Institute of Food Technology Entrepreneurship and Management, Kundli, Sonepat, Haryana, India

S. Agnihotri (✉)
Department of Agriculture and Environmental Sciences, National Institute of Food Technology Entrepreneurship and Management (NIFTEM), Kundli, Sonepat, Haryana, India
e-mail: agnish@niftem.ac.in

B. P. Singh et al. (eds.), *Fungal Additives and Bioactives in Food Processing Industries*, Fungal Biology, https://doi.org/10.1007/978-3-032-04519-5_7

et al., 2024a). Recent advances in fungal genomics have uncovered that merely 20% of potential secondary metabolites have been characterized (Sharma et al., 2023a, b). This suggests that an extensive reservoir of undiscovered compounds remains to be explored. Finding new secondary metabolites with potential benefits for humans is thus an exciting development to harness the full potential of fungal-derived compounds for commercial relevance. Currently, filamentous molds such as *Aspergillus*, *Penicillium*, and *Rhizopus* are well known to produce plethora of volatile compounds including ketones, esters, alkenes, acids, benzenoids, and terpenes (Sun et al., 2025). These bioactive metabolites impart a vast range of flavors and aroma viz. fruity, nutty, and earthy notes, which are used in various food purposes (Inamdar et al., 2025). Whereas, yeasts like *Saccharomyces cerevisiae (S. cerevisae)*, *Pichia pastoris*, and *Candida species*, mainly synthesize alcohols, aldehydes, and sulfur-containing compounds, which provide floral and savory characteristics (Fu, 2024). Furthermore, basidiomycete fungi like *Trametes versicolor* and *Lentinula edodes* (shiitake mushroom) contribute umami-rich compounds, including amino acids and nucleotides, enhancing the depth of flavors in various culinary applications (Yan et al., 2019).

The choice of substrate has a major impact on the quality and yield of flavors produced by a particular fungus (Krings et al., 2009; Rajendran et al., 2023). Moreover, different fungal strains have been shown to produce distinct secondary metabolites using the same substrate. Notable examples of substrates include glucose, starch, lignocellulosic biomass, and agro-food byproducts and waste (e.g., molasses, fruit peels), which serve as carbon sources and influence fungal metabolism for flavor biosynthesis (Xie et al., 2016). Additionally, the use of lipid-rich substrates can enhance the production of fruity esters, whereas proteinaceous substrates may promote the generation of savory and umami notes via amino acid metabolism (Reyes-Reyes et al., 2022; Siebenhaller et al., 2018). It is therefore imperative that substrate optimization during fermentation is not only critical for modulating flavor profiles, but also a decisive factor in achieving the desired yield and productivity of fungal-derived flavors. This, in turn, contributes to the commercial viability and sustainability of the bioprocess, making it a profitable business.

Beyond substrate selection, bioprocessing parameters critically influence the yield, profile, and quality of fungal-derived flavors (Shin et al., 2025). Parameters including pH, temperature, dissolved oxygen (DO) levels, agitation rate, and fermentation mode (e.g., submerged vs solid-state) directly regulate fungal metabolism by modulating enzymatic activity, cellular stress responses, and metabolic flux. For instance, oxygen availability serves as a key determinant, particularly for aerobic fungi, where optimal DO levels (20–40%) support oxidative pathways responsible for producing desirable volatile compounds like alcohols, aldehydes, and ketones (Zhang et al., 2024b). Insufficient oxygen levels, by contrast, drive metabolic shifts toward anaerobic byproducts including esters and organic acids, fundamentally altering the flavor profile. Similarly, pH exerts pronounced effects on enzyme activities of proteases and esterases involved in flavor biosynthesis (Tian et al., 2024a). With lower pHs (2–4) conditions, organic acids are preferentially produced, while sub-neutral pH (5–6.5) ranges promote the formation of esters (isoamyl acetate, fruity-flavor) and alcohols (Alves et al., 2025). Temperature regulation maintains equal importance, as it directly affects the fungal growth kinetics and enzyme activity, with most flavor-producing species exhibiting optimal activity between 25 °C and 30 °C. Modern bioreactor systems now allow precise control of these interdependent parameters, enabling consistent, large-scale production of specific flavor profiles. This controlled optimization has become increasingly vital for industrial applications, where reproducibility and sensory quality standards demand strict parameter management throughout the fermentation process, as demonstrated in recent studies of fungal flavor production systems (Duan et al., 2023).

This section aims to holistically review the potential of fungal biochemical products as sustainable flavoring alternatives in food manufacturing processes. It categorizes the key metabolites involved, discusses bioprocessing parameters that influence flavor variation, and evaluates their quality and suitability for food applications. Furthermore, it highlights existing studies demonstrating the application of fungal-derived flavors in sectors such as beverages, condiments, bakery, and dairy, thereby underscoring their commercial significance. By integrating current insights into bioprocess optimization and fungal metabolism tailored for flavor enhancement, this chapter moves a step closer to redefining modern food innovation, positioning fungal-derived flavor enhancers as a promising solution for clean-label food products.

2 Fungal Metabolites and Associated Fungal Communities

The fungal kingdom serves as a biological aroma factory ascribed to their great biochemical potential to produce them as secondary metabolites. Since fungi propagates usually through the decomposition of organic material, aroma production may either be produced as secondary metabolites via microbial fermentation or de novo synthesis routes (Bills & Gloer, 2016). In addition to these, the metabolic pathways encountered during biotransformation and bioconversion processes may also lead to generation of the aroma constituents. The biotransformation processes consist of single-step reactions, in contrast to bioconversion processes involving multistep reactions (Liu et al., 2024a). Whole cells are required for targeting complex molecules or substrates mixture, in contrast to isolated enzymes that perform well in single-step processes. Among fungal sources, a variety of original aroma volatiles originate majorly from basidiomycetes and a few cases from ascomycetes (Bills & Gloer, 2016; Ranjitha & Oberoi, 2016).

2.1 De novo Synthesis

The ability of whole cells to catabolize proteins, lipids, and carbohydrates and then further transform the breakdown products into more complex flavor molecules has long been utilized in the preparation of fermented foods via de novo synthesis (Rodrigues & de Carvalho, 2022). Primary metabolites are produced in significant quantities by common starting cultures, although more complex flavor compounds are produced only in trace amounts. A co-fermentation strategy was used for enhancing the yield of strawberry wine. First, a combination of four non-Saccharomyces yeast strains was allowed to ferment strawberries for 1–3 days, then inoculated with *S. Cerevisiae*. This sequential fermentation led to decreased organic acid contents by 16–65% with the exception of lactic acid, which has almost doubled, contributing to the soft and sound sour taste of wine. More complexity to the flavors has been brought by enhanced production of amino acids by 0.28–3.4 times (Xu et al., 2025). Yang et al. (2024) produced xanthohumol from glucose by balancing the three parallel biosynthetic pathways. While in de novo synthesis, three pathways have shown simultaneous balancing among them. These included an increase in precursor (de methyl xanthohumol (DMX)) supply by 83-fold, helping fusion of enzymes and engineering of peroxisomes along with prenyl transferase. Prenylation is the rate determining step in DMX biosynthesis. Therefore, metabolic regulations can be done in this step for enhancing DMAPP availability and prenylation efficiency.

2.2 Biotransformation/Bioconversion

Valuable pre-existing flavor compounds, e.g., essential oils and fatty acids, are used as substrates in biotransformation and bioconversion processes for generating high-value flavors (Roy et al., 2022). In biotransformation, compounds are modified structurally by organisms or enzyme systems while retaining their original carbon framework. In contrast, bioconversion entails converting organic materials into energy or other products, with a focus on turning biomass into usable forms (Smitha et al., 2017). The primary biochemical pathways responsible for flavor compound production through biotransformation of essential oils and similar substrates include their β-oxidation and processing via Ehlrich pathway. Although some other biosynthetic pathways, such as dehydration, hydrolysis, oxidation, and reduction, along with generation of new carbon-carbon bonds, have also been investigated as reactions yielding flavors within fungal metabolic pathways. The production of γ-decalactone (GDL) and vanillin are examples of such pathways. GDL is synthesized as a metabolized product of ricinoleic fatty acid when acted upon by yeasts (Singh et al., 2024), while vanillin is produced from ferulic acid in various cereal brands by fungi (Mehmood et al., 2024). Fermentation of oysters (shellfish) is another example in which five bacterial and four fungal genera generated volatiles with *Lactobacillus* and *Kazchstania exigua* being the dominant bacterial and fungal species, respectively. Some of the volatiles are produced by ammonia acid synthesis and degradation pathways. 2-Octenal and 2-undecanone were among the major volatiles produced after fermentation of oysters followed by 33 more volatiles. *Kazchstania exigua* and *S. Cerevisiae* were particularly responsible for volatiles like methyl esters, higher alcohols, ketones, and acids. *S. Cerevisiae* particularly led to lipid oxidation for production of furans, aldehydes, and ketones (Liu et al., 2024b).

Major fungal flavors generated as secondary metabolites during fermentation encompass compounds such as alcohols, aliphatic acids, ketones, sulfur compounds, aldehydes, esters, and lactones (Ranjitha & Oberoi, 2016). *Pichia kudriavzevii* is a non-conventional yeast species found in diverse environments, commonly involved in the natural fermentation process of traditional food and drink products. In this process, *P. kudriavzevii* breaks down organic acids, producing several hydrolases that contribute to the formation of flavor compounds, including esters and fatty acids, as reported by Chu et al. (2023). A variety of these metabolites are produced in large amounts and exhibit a diverse range of acyclic, heterocyclic, cyclic, and aromatic structures with different degrees of oxidation. Edible mushrooms can be broadly classified into three categories based on their distinct flavor profiles. The first one, comprising C_8 derivative compounds, such as octan-1-ol, octan-3-ol, oct-1-en-3-ol, oct-2-en-1-ol, oct-1-en-3-one, and octan-3-one, primarily contributes to distinct mushroom flavor and covers majority of volatile compounds. While the second and third consists of fungal-derived terpenoids and sulfur-containing volatile compounds, respectively (Fraatz & Zorn, 2010). A number of fungal volatiles have been identified and show structural similarities to the key flavor components of higher plants (Table 7.1). The widely researched and commercialized flavor metabolites from fungi have been explained in the following section.

3 Carbon-Eight (C_8) Derivatives

C_8 flavor compounds, identified by their eight-carbon structure, play a key role in shaping the distinct aromas associated with various fungi, especially mushrooms. For example, oct-1-en-3-ol, in conjunction with oct-1-en-3-one, is considered as C_8 metabolite responsible for characteristic fungal flavor. It was initially isolated as the predominant volatile compound of *Tricholoma matsutake*, consequently got its name matsutake alcohol, and was later structurally identified as oct-1-en-3-ol. This compound is prevalent in *Lentinula edodes* (shiitake) (Guo et al., 2024) and *Agaricus bisporus* (button mushroom) species (Su et al., 2022). Another notable C_8 compound, 1-octen-3-one, complements the profile with its metallic, green, and mushroom-like aroma, creating a characteristic sensory experience (Murray et al., 2020). The production of these compounds generally occurs through oxidation of polyunsaturated fatty acids (e.g., arachidonic acid, linoleic acid) by LOX (lipoxygenase) pathway to produce hyperperoxides as intermediates. These intermediates are then processed

Table 7.1 Flavor metabolites and their associated aromas produced by fungal strains

Fungi	Metabolite flavor	Produced metabolites	References
Ischnoderma spp.	Bitter almond flavor	Benzaldehyde	Melini and Melini (2024)
Kluyveromyces spp	Rose-like odor	2-phenylethanol	Adame-Soto et al. (2023)
Penicillium roqueforti	Blue cheese flavor	2-alkanones, alkanols, volatile fatty acids, and aldehydes	Caron et al. (2025)
Pichia ohmeri	Fruity, nutty, and fatty odor	4-alkanolide, lactones with 8–12 carbons, and 5-alkanolide	Zhou et al. (2021)
Saccharomyces spp.	Rose like odor	2-phenylethanol	Villar et al. (2025)
Penicillium caseicolum	Earthy, ketonic, mushroom like	3-octanone	Nwankwo et al. (2023)
Rhodotorula minuta	Rose like odor	Citronellal	Jiang et al. (2021)
Clostridium butyricum	Apple and pineapple	Butyric acid	Pahalagedara et al. (2024)
Saccharomyces cerevisiae	Flavor of many essential oil compounds	(+)- Curcumene	Choi et al. (2024)
Kluyveromyces species	Chocolate flavors	Thaumatin and monellin	Pérez et al. (2022)
Beauveria bassiana	Raspberry	Raspberry ketone	Cancellieri et al. (2024)
Pycnoporus cinnabarinus	Vanilla	Vanillin	Valera et al. (2020)
Yarrowia lipolytica	Peach	γ-Decalactone	Yu et al. (2023)

by hydroperoxide lyase to yield C_8 alcohols and ketones (Karrer et al., 2021; Kinderlerer, 1989). Variations in fungal species, substrate composition, and environmental factors influence the efficiency and outcome of this metabolic pathway. A unique volatile composition of wheat bran was observed by Ghamry et al. (2023) in which *S. Cerevisae*-led fermentation increased GABA up to three times approximately. Additionally, the production of these compounds has been shown to differ across the tissues of *Lentinula edodes*, suggesting that specific biosynthetic pathways operate in various parts of the mushroom (Zhang et al., 2023a).

Processing methods, such as drying, have been demonstrated to significantly affect the concentration and types of C_8 compounds present in mushrooms. Different drying techniques can alter the levels of 1-octen-3-ol and related volatiles. Natural air dried and variable temperature drying fermentation were proved more impactful than constant temperature drying in increasing growth of major fungi contributing to the characteristic flavors of Yunan style sausages, thereby influencing flavor composition of final aroma metabolites (Shi et al., 2024). While the production of C_8 flavors is promising, challenges remain in optimizing fermentation processes and ensuring consistent flavor profiles across different strains and conditions. The study of these compounds is not only important for understanding fungal metabolism but also for their practical applications in flavor production and enhancement. Ongoing research continues to reveal the intricate dynamics of C_8 flavor compounds, emphasizing their potential significance in both natural and industrial contexts.

4 Terpenoids

Terpenoids represent the most abundant and structurally heterogeneous category of natural flavoring compounds. These encompass hemi-, mono-, di-, tri-, tetra-, and sesqui-terpenes. There exist over 30,000 isoprenoid compounds within the plant kingdom. These fulfill a myriad of critical functions in fundamental plant processes, such as photosynthesis, respiration, growth, reproduction, development, defense mechanisms, and resiliency against varying environmental factors. Terpenoids are formed biosynthetically through condensation of IPP (isopentenyl diphosphate) and its isomer DMAPP (dimethylallyl diphosphate), bonded either from head-to-head or head-to-tail fashion as shown in Fig. 7.1 (Ashour et al., 2010; Ranjitha & Oberoi, 2016). The mevalonate (MVA) pathway has recently been supplanted by a two-step isopentenol utilization (IU) pathway that exclusively employs ATP as a cofactor, thereby enhancing the production of primary terpenoid precursors. In *S. cerevisiae*, efficacy of the IU pathway was diminished by isopentenol due to impeding energy metabolism. To circumvent this limitation, an IUPD (IU pathway-dependent) strain was developed. This strain utilized a high-throughput screening methodology for enzyme evolution. It is compelled to augment ATP availability, which is crucial for synthesizing complex terpenoids (Li et al., 2024a).

[*Metabolite abbreviations:* G3P (glyceraldehyde-3-phosphate), PYR (pyruvate), DXP (1-Deoxy-D-xylulose 5-phosphate), MEP (2-C-Methyl-D-erythritol 4-phosphate), CDP-ME (4-(Cytidine 5′-pyrophospho)-2-C-methyl-D-erythritol), CDP-MEP (2-Phospho-4-(cytidine 5′-pyrophospho)-2-C-methyl-D-erythritol), MEcPP (2-C-Methyl-D-erythritol 2,4-cyclopyrophosphate), HMBPP (1-Hydroxy-2-methyl-2-butenyl 4-pyrophosphate), AcAc-CoA (acetoacetyl-CoA), Mev-5-P/P (mevalonate-5-phosphate/pyrophosphate), MVA-3-P/PP (mevalonate-3-phosphate/biphosphate), IP (isopentenyl phosphate), 3-M-G-CoA (3-methylglutacoyl-CoA), 3-M-2-B-CoA (3-methyl-2-butenoyl-CoA), 3-M-2-butenal, DMAP (dimethylallyl phosphate), IPP (isopentenyl pyrophosphate), DMAPP (dimethylallyl pyrophosphate), GPP (geranyl pyrophosphate), FPP (farnesyl pyrophosphate), GGPP (geranylgeranyl pyrophosphate), HEPP (heptaprenyl pyrophosphate), UDPP (undecaprenyl pyrophosphate)] (Adapted from Song et al. (2024), available under Creative Commons Attribution License).

Current understanding of terpenoid biosynthesis in indigenous yeasts has been largely restricted to minimal production levels observed in *S. cerevisiae* and select non-Saccharomyces species such as *Streptomyces* (Kuzuyama, 2017), *Trichoderma*

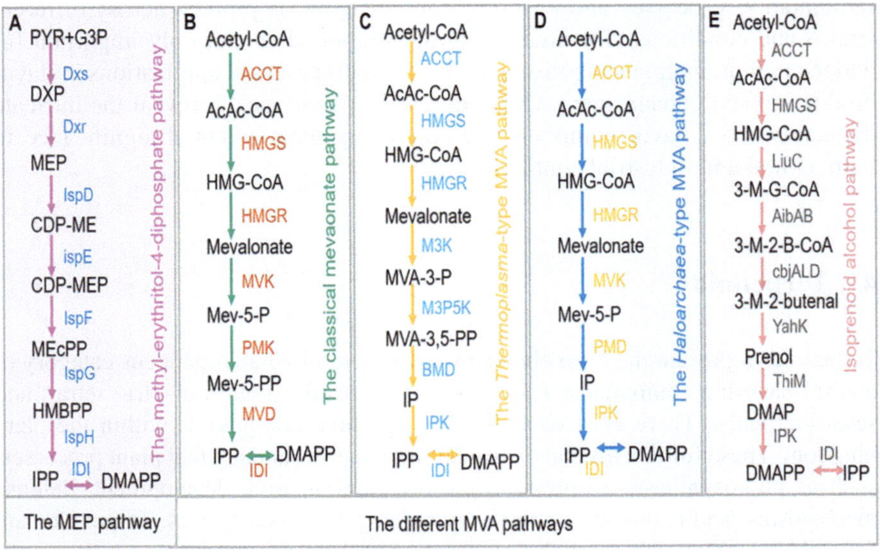

Fig. 7.1 Comparative schematic of five isoprenoid biosynthesis pathways: (**a**) 2-C-Methyl-D-erythritol 4-phosphate (MEP) pathway; (**b**) Canonical mevalonate (MVA) pathway; (**c**) Thermoplasma-type MVA variant; (**d**) Haloarchaea-type MVA variant; (**e**) Isoprenoid alcohol pathway. [*Enzyme abbreviations:* Dxs (1-Deoxy-D-xylulose-5-phosphate synthase), IspC (reductoisomerase), IspD (synthase), IspE (kinase), IspF (cyclopyrophosphate synthase), IspG (hydroxy-2-methyl-2-(E)-butenyl 4-pyrophosphate synthase), IspH (reductase), IDI (isopentenyl pyrophosphate isomerase), IspA (farnesyl pyrophosphate synthase), ACCT (acetoacetyl-CoA thiolase), HMGS/HMGR (3-hydroxy-3-methyl-glutaryl-CoA synthase/reductase), MVK/PMK/MVD (mevalonate/phosphomevalonate kinases, decarboxylase), PMD (phosphomevalonate decarboxylase), M3K/M3P5K (mevalonate 3-kinase/3-phosphate 5-kinase), BMD (bisphosphomevalonate decarboxylase), LiuC (enoyl-CoA hydratase), AibAB (glutaconyl-CoA decarboxylase), cbjALD (acyl-CoA reductase), YahK (alcohol dehydrogenase), ThiM (hydroxyethylthiazole kinase), IPK (phosphomevalonate decarboxylase)]

Trichoderma (Bai et al., 2023), *Torulaspora delbrueckii, Metschnikowia pulcherrima, Kluyveromyces lactis, Ambrosiozyma monospora,* and *Candida stellata* (Kampranis & Makris, 2012). The principal challenges encountered during conventional monoterpene synthesis pertain to the chemical instability of precursor (monoterpene) and its biosynthetic product (terpenoid), low aqueous solubility of monoterpene precursors, suboptimal conversion rates, elevated levels of volatile compounds associated with both precursor and product, alongside consequential cytotoxicity (Krings & Berger, 1998).

The transformation of free terpenes by various yeast species, particularly non-Saccharomyces varieties, has been reported. For example, Yang et al. (2022) utilized combination of Saccharomyces (*S. cerevisiae* ES488) and non-Saccharomyces species (*M. pulcherrima 346*) for enhancing total concentrations of terpenoids along with alcohols and esters. The concentration of the metabolites was optimized by combination of single factor analysis and Box-Behken design. The fermentation conditions leading to more floral aromas comprised of an initial pH of 3.7, sequential inoculation time of 45 h, temperature 18 °C, and an inoculation ratio of *M. pulcherrima* 346: *S. cerevisiae* ES488 (1.5,1). Citral, d-limonene, α-pinene, and β-pinene are some other extensively used flavor compounds utilized for bioconversion. For example, geraniol and nerol can be synthesized by biotransforming citral utilizing *S. cerevisiae*. Consequently, d-limonene originated from citrus peel oils and are among the most extensively researched monoterpenes. It is employed for producing biotransformed products such as α-terpineol, menthol, carveol, perillyl alcohol, and carvone by action of *Penicillium* spp. and *Pseudomonas* spp. (Gamero et al., 2011). Nevertheless, given that fungi typically do not produce monoterpenes, scholarly attention regarding the synthetic pathways has predominantly concentrated on polyterpenoids, such as carotenoids, triterpenoids, and ubiquinone (Fig. 7.2). Despite recent developments aimed at producing sesquiterpenes through engineered yeasts, the food flavoring compounds based on sesquiterpenes are still rare. Song et al. (2024) have recently reviewed challenges while obtaining valencene from natural sources. The interventions required for its microbial transformation and industrial upscaling is still in its infancy.

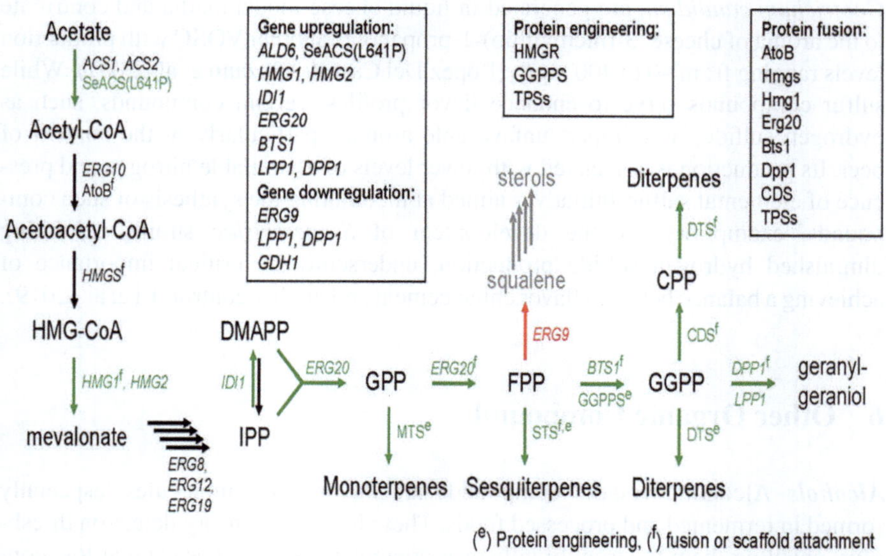

Fig. 7.2 Engineered terpene production pathway in yeast depicting: (i) transcriptional regulation (green/red = up/down-regulated genes), (ii) protein fusion constructs (f), and (iii) enzyme engineering modifications (e). Pathway components include CPP (copalyl diphosphate), CDS (copalyl diphosphate synthase), DTS/MTS (di/monoterpene synthases), SeACS(L641P) (mutant acetyl-CoA synthase), and AtoB (acetoacetyl-CoA thiolase). (Adapted from Kampranis and Makris (2012) with copyright permission)

5 Sulfur-Containing Compounds

Yeasts and fungi produce a wide array of sulfur-containing flavor compounds that significantly contribute to the sensory profiles of various fermented foods and beverages. These compounds are majorly routed through the metabolic pathways involving methionine and cysteine, the sulfur-containing amino acids. Specific yeasts and fungi, including *S. cerevisiae*, *Geotrichum candidum*, and basidiomycetous yeasts, are recognized for their capacity to generate a variety of volatile sulfur compounds (VSCs) that hold considerable value within the food and flavor sectors. In a study, Pérez et al. (2022) demonstrated that the cryotolerant strains of *S. cerevisiae* performed much better in producing flavors like erythritol by faster hexose sugars consumption. *S. uvarum SC2* and *S. kudriavzevii* had improved floral flavor profiles of wines by releasing polyfunctional mercaptans. The strain *S. uvarum* showcased enhanced succinic acid production to 7.3 g/L due to GABA shunt, depending upon the species and isolation origin. Other compounds synthesized by basidiomycetous yeasts, dairy yeasts, and *Geotrichum candidum* include dimethyl disulfide, S-methyl thioacetate, and dimethyl trisulfide, respectively, contributing significantly to the distinctive aroma profile of various cheeses (Zhang, Wang, & Cao, 2023b). S-Methyl thiopropionate and S-methyl thiobutanoate, unique to *Geotrichum candidum*, are generated in liquid cheese model media and contribute to the aroma of cheese. 3-(methylthio)-1-propanol is another VOSC with production levels ranging from 40 to 400 mg/L (López Del Castillo-Lozano et al., 2007). While sulfur compounds serve to enhance flavor profiles, certain compounds, such as hydrogen sulfide, may impart unfavorable aromas, particularly in the context of beer. Its production is associated with lower levels of assimilable nitrogen and presence of elemental sulfur. Initiatives aimed at modulating the synthesis of such compounds, exemplified by the development of *S. cerevisiae* strains exhibiting diminished hydrogen sulfide production, underscore the critical importance of achieving a balance between flavor enhancement and quality control (Li et al., 2019).

6 Other Organic Compounds

Alcohols Alcoholic volatile compounds are flavor-active molecules, especially formed in fermented and processed foods. These have low sensory detection thresholds, enabling them to strongly influence organoleptic properties even at low concentrations. Their biosynthesis primarily results from the oxidative degradation of polyunsaturated fatty acids during food processing and fermentation (Carrau et al., 2016). For example, 1-octen-3-ol is a major aroma volatile known for its mushroom-like and metallic sensory attributes. This compound is produced through the autoxidation of linoleic acid and other unsaturated fatty acids, catalyzed by microbial species such as *S. cerevisiae* and *Aspergillus* spp. (Singh et al., 2020). Lee et al. (2020) has demonstrated first time use of recombinant *S. cerevisiae* for production of (R)-(-)-1-octen-3-ol. Linoleic acid was used as a substrate and reacted with different combinations of gene-encoding proteins for obtaining maximum yield of biosynthesized (R)-(-)-1-octen-3-ol. Lipoxygenase 1 and hydroperoxide lyase proved to be best among all combinations for maximizing yield with minimum fermentation time. The Ehrlich pathway is another critical metabolic route for alcohol production in yeast fermentation. This pathway involves enzymatic reactions such as oxidation, reduction, decarboxylation, and transamination of aromatic amino acids (e.g., tryptophan, phenylalanine), sulfur-containing amino acids (e.g., cysteine, methionine), and branched-chain amino acids (e.g., isoleucine, leucine). In addition, acetyl-CoA/alcohol acetyltransferase reactions contribute to the formation of esters and alcohols, enhancing flavor complexity in fermented products. In alcoholic beverages, advanced alcohols such as n-propanol, isobutanol, and isoamyl alcohol are key contributors to flavor. These higher alcohols enhance sweetness and viscosity with simultaneous production of important metabolites in yeast-driven fermentation. For example, Kazakh-originated yeast facilitates ethanol production by fermenting sugars, acetic acid, and esters, reflecting the complex metabolic networks that influence the flavor profiles of alcoholic beverages. In addition, heptanal has been found as the prominent flavor metabolite having production up to 10 μL/L (Bilal et al., 2024). The sensory qualities of non-beverage products, such as fermented fish, had also been considerably affected by these volatiles. While fermenting hairtail surimi in range 0–5 h, 20 alcohols were produced in major by Monascus along with 118 other aroma metabolites (Li et al., 2023). Finally, alcohols with the molecular structure R–CH(OH)–CH_4CH_2, where R represents alkyl groups ranging from methyl to pentyl, have distinct odor profiles that influence food sensory characteristics. These processes collectively highlight the critical role of volatile alcohols in shaping the flavor profiles of fermented and natural foods.

However, the complete biosynthetic pathways of volatile benzenoids in plants have not been completely determined. Still, some studies about formation of benzyl alcohol confirmed its synthesis during phenylpropanoid synthesis by phenylalanine ammonia-lyase (PAL) enzyme. To our knowledge, this enzyme is rarely found in yeast. Although it has been reported for the *Basidiomycota* and *Ascomycota* fungi such as *Aspergillus*, *Neurospora*, and *Botrytis*; *Rhodotorula graminis*, it has not been found in the subphylum *Saccharomycotina*. For example, in *Bjerkandera adusta*, PAL catalyzes the deamination of L-phenylalanine, yielding trans-cinnamic acid, which is then reduced and oxidized into benzaldehyde and benzyl alcohol. This pathway has been shown to result in a yield of up to 70% benzyl alcohol under optimized conditions, demonstrating its industrial potential (Lapadatescu et al., 2000). Benzyl alcohol has also been produced by some non-Saccharomyces species such as *Hanseniaspora uvarum*, *Candida stellata*, *Schizosaccharomyces*, *Kloeckera apiculata* and *Zygosaccharomyces*. For instance, de novo synthesis of p-hydroxybenzaldehyde, benzyl alcohol, benzaldehyde, and p-hydroxybenzyl alcohol by wine yeast *Hanseniaspora vineae* can be proceeded without any need for precursors derived from grapes. The results of genome-sequenced strains of *H. vineae* showed that neither the phenylalanine ammonia lyase (PAL) nor the tyrosine ammonia lyase (TAL) biosynthesis pathways are present. As a result, a different biosynthesis mandelate pathway has been proposed that starts with chorismate and involves the intermediates of mandelate, phenylpyruvate, and p-hydroxyphenylpyruvate (Valera et al., 2020). In another study by Kirschning (2022), p-hydroxybenzoate was found to function as an intermediary in the synthesis of ubiquinone Q6 in Saccharomyces. In a study, conversion of L-phenylalanine and benzoic acid into benzyl alcohol by *Polyporus tuberaster* K2606 resulted in conversion efficiencies of 60–65%, making it a promising candidate for large-scale production (Kawabe & Morita, 1994). Innovative production strategies have also been explored. For example, *Annulohypoxylon stygium* utilizes the Ehrlich pathway to convert L-phenylalanine into aromatic alcohols, achieving 25–30 mg/L benzyl alcohol production under controlled conditions. Additionally, whole-cell biocatalysis in bioreactor systems, such as stirred-tank and packed-bed reactors, has been shown to improve scalability and catalytic efficiency, with studies reporting up to 90% conversion of aromatic substrates to benzyl alcohol. These advancements further highlight the industrial feasibility of fungal biocatalysis for natural flavor production (Rodrigues & de Carvalho, 2022).

Aldehydes Vanillin is an aldehydic flavor metabolite that is used extensively as a spice in food industry for wide applications. It can be synthesized biosynthetically through carbon sources produced by *Aspergillus Niger*, yeast, and Bacillus (Sharma et al., 2023a, b). One of the well-known biotransformation routes comprises a process including two steps. In this process, first a strain of *Aspergillus niger* is used for conversion of ferulic acid to vanillic acid. While in the next step, produced vanillic acid is reduced by a laccase-deficient strain of *Pycnoporus cinnabarinus* to vanillin. Furthermore, *Ischnoderma resinosum* has demonstrated the ability to synthesize benzaldehyde and benzyl alcohol during submerged fermentation, with yields reaching 35–40 mg/L when utilizing a range of aromatic precursors such as L-phenylalanine and benzoic acid (Wickramasinghe & Munafo, 2020). Among eight aldehydes, hexanal and heptanal were found among major flavor compounds of yeast proteins having concentrations in range 59–195 μg/kg and 76–1176 μg/kg, respectively. These yeast proteins have found their application in flavoring meat analogues for mimicking conventional meat products (Chen et al., 2023).

Ketones The primary methods used for production of ketones include breakdown of amino acids and oxidation of unsaturated fatty acids. High ketone concentrations frequently result into fruity, spicy, and creamy aromas. For instance, 2,3-octanedione, primary ketone posing a distinctive fruity aroma in dried salted ducks, was found to be produced by *Pseudomonas* and *Mucor*. In another study, a distinct off-flavor alike “fresh mushroom” produced during wine fermentation was ascribed to the generation of volatile. Though this volatile seems undesirable to the conventional wine industry, the 1-hydroxyoctan-3-one study underscores the wider implications of fungal-derived volatiles in food and beverages sectors (Tian et al., 2024b).

Esters These compounds are known to produce fruity flavors that can be produced either by non-enzymatic esterification of alcohols and organic acids, or enzymatic catalysis of microorganisms. During enzymatic catalysis, yeasts first synthesize fatty acids by hydrolyzing acyl-CoA derivatives, and then they convert activated fatty acids and alcohols to esters. Fatty acids can undergo oxidation via α-, β-, and w-oxidation pathways. Among these, β- of fatty acids is particularly used by yeast for synthesizing lactones (Romero-Guido et al., 2011). β-oxidation comprises a cyclic oxidation process of fatty acids. It encompasses four principal reactions, viz., dehydrogenation, yielding trans 2,3-enoyl-CoA; hydration of unsaturation, yielding 3-hydroxyacyl-CoA; oxidation to 3-ketoacyl-CoA; and ultimately, the liberation of acetyl-CoA alongside acyl-CoA, which is shortened by two carbons. In instances involving saturated fatty acids, this quartet of reactions is reiterated multiple times until the complete catabolism of the fatty acid into acetyl-CoA molecules is achieved. The conversion of intermediates within the β-oxidation cycle into different lactones is catalyzed by β-oxidation enzymes, such as enoyl-CoA hydratase, OH-acyl-CoA dehydrogenase, acyl-CoA oxidase, and ketoacyl-CoA thiolase, throughout this metabolic process. The response of KU1 group strains, which demonstrate an elevated production of these compounds under conditions of low nitrogen availability, contradicts the prevailing notion articulated in numerous studies that assert a direct relationship between increased nitrogen levels in the must and enhanced ester production. This paradoxical behavior observed in KU1 group strains may also elucidate why several investigations have failed to identify a consistent correlation between yeast assimilable nitrogen (YAN) in grape musts and the concentrations of esters and fatty acids. More intriguingly, the fatty acid and ester profiles generated by the KU1 group bear a notable resemblance to the profiles of higher alcohols and isoacids, contrasting sharply with the profiles exhibited by strains with a high nitrogen demand. Some yeasts, such as *Sporobolomyces odorus, Candida tropicalis,* and *Yarrowia lipolytica*, have the ability to biotransform ricinoleic acid into GDL. In this biotransformation, ricinoleic acid is first metabolized to 4-hydroxy decanoic acid, and then chemically lactonized to GDL and other lactones. The existing literature provides limited insights into the biosynthesis of γ-butyrolactone and ethyl 4-hydroxybutanoate by yeast. Furthermore, the production profile of γ-butyrolactone is similar to that of higher alcohols and isoacids (Carrau et al., 2016; Ranjitha & Oberoi, 2016). A correlation between the production of γ-butyrolactone and ethyl 4-hydroxybutanoate has been previously proposed. Limited information about the production of ethyl 4-hydroxybutanoate by yeast is found in the literature.

6.1 Correlation Between Fungal Communities and Flavor Production

Flavor compounds obtained as "secondary metabolites" are prone to varying fermentation parameters and mixed cultures. Recent investigations have elucidated that increase in the diversity of yeast strain and species in a fermentation system led to increased sensory complexity and diversity in aroma metabolites in the final fermented product. These metabolites perform key roles for yeast, including survival and dispersal strategies, pheromonal signaling and defensive mechanisms, along with "quorum sensing" pathways facilitating cellular communication (Carrau et al., 2016). Humanity has leveraged numerous aspects of these functionalities to augment taste and enhance sensory experiences for a discerning consumer base, a phenomenon termed "yeast domestication." The interaction of microorganisms is mainly divided into synergistic effects and antagonistic effects. Microorganisms possess the capability to both stimulate the biosynthesis of a broader array of flavor compounds and amplify flavor expression via synergistic interactions or to inhibit the formation of certain harmful substances and undesirable flavors through antagonistic interactions (Zhang et al., 2023a). Concerning meat products, molds and yeasts have also demonstrated a synergistic effect on flavor enhancement. Beyond the enhancement of traditional food flavors via synergistic interactions, undesirable odors may also be mitigated through antagonistic effects (Wang et al., 2021). The use of *P. kudriavzevii*, a non-conventional yeast, as antagonist for malic acid, targeted malic acid degradation leading to unfavorable flavor in wine production. The instances related to the effect of the below-explained factors on flavor metabolites have been mentioned in Table 7.2.

1. **Fungal structure and composition:** The interplay among fungal species impacts the metabolite spectrum and resultant flavors (Guo et al., 2024). In high-temperature Daqu used for Chinese Baijiu production, specific fungal genera such as *Thermomyces*, *Thermoascus*, and *Aspergillus* produce distinct flavors linked to sourness, bitterness, or aromatic richness. These interactions highlight how microbial community structure and composition dictate taste and aroma (Ji et al., 2022).
2. **Flavor phenotype:** Functional traits are of paramount importance within a specific ecosystem when the overarching biological mechanisms are taken into account. Notably, since flavor characteristics are not invariably critical for cellular viability, the genetic pathways that underlie these traits exhibit diminished susceptibility to positive selection pressures, which in turn enhances genome-wide variability during the process of microbial evolution, potentially augmenting flavor diversity among yeast species (He et al., 2025; Jin et al., 2019). Furthermore, the wine compounds that influence flavor are significantly more susceptible to polygenic than traditional enological traits, such as minimal acetic acid concentration, ethanol tolerance, or effective fermentation rates at reduced temperatures (Liu et al., 2023; Tian et al., 2024b). Thus, flavor phenotype should

Table 7.2 Impact of fungal strains and process conditions on flavor metabolites

Substrate	Yeast/fungal strain	Fermentation process conditions	Flavor metabolites	Applications	References
Wheat straw	*Yarrowia lipolytica M53-S*	Solid state fermentation (pH 5.0 and 45 °C for 24 h)	Erythritol	Erythritol production up to 190.5 mg/gds was attained	Liu et al. (2021)
Wheat dough	*Saccharomyces cerevisiae*	Fermentation medium in a controlled environment (34 °C, RH 78% for 90 min)	Thymol, carvacrol, and p-cymene	Shelf life was extended up to 7 days in bread due to higher production of EOs	Jurkaninová et al. (2024)
Grape skins	*S. cerevisiae*	Alcoholic fermentation (24 °C)	Tannins	Monomers showed highest percentage of release (87.9%–90.3%) from seed tannins	Wang et al. (2022)
Avocado seed	*A. niger* GH_1	Solid state fermentation (60% RH, 120 h, and particle size ≤2.5 mm)	Gallic acid, catechin, epi catechin, and procyanidin	*A. Niger* GH_1 increased their antioxidant activity for 2,2-diphenyl-1-picrylhydrazyl (DPPH) radical uptake by 77 and 42%	Yepes-Betancur et al. (2021)
Grape pomace	*Actinomucor elegans and Umbelopsis isabellina*	Solid state fermentation	γ-linolenic acid, carotenoids, phenolic compounds, and flavonoids	Fermented grape pomaces possessed 50 mgβ carotene/100 Gdw while lipids with more than 94% nutritionally-valuable PUFA	Dulf et al. (2020)
Brewer's spent grain (cereals and legumes)	*A. awamori*, ***Aspergillus oryzae***, *Aspergillus terreus, A. niger, and R. oryzae*	Solid state fermentation (potato dextrose agar medium at 30 °C/7 days)	Flavonoids (57%), phenolic acids (25%), other polyphenols (6.5%), lignans (6.5%)	Promote cinnamate carboxy-lyase activity that can convert hydroxycinnamic acids into their derivatives by releasing CO_2	da Costa Maia et al. (2020)

(continued)

Table 7.2 (continued)

Substrate	Yeast/fungal strain	Fermentation process conditions	Flavor metabolites	Applications	References
Agro-industrial wastes Enriched with maguey sap	*Aspergillus oryzae DIA–MF*	Solid-state fermentation (30 °C, pH:4.5, 70% initial moisture content)	Fructooligosaccharides (sweetener)	Sugarcane bagasse, coffee husk, pineapple peel, prickle pear peel, and banana peel were valorized for production	de la Rosa et al. (2020)
Red chili	*Nectriaceae and Colletotrichum*	Fermentation (room temperature for 10–14 days)	Ethyl alcohol	The fermentation of Flavour compounds in red chilli (FCRC) can be separated based on flavor: Day 0 is the first stage, days 3–10 are the second stage, and days 15–20 are the third stage.	Ji et al. (2022)
Biochar-PS + waste cooking oil	*Y. lipolytica M53-S*	Solid state fermentation (pH 4.0, 70% moisture content, and NaCl addition 0.02 g/gds)	Erythritol (185.4 mg/gds)	40% sesame meal and 10% waste cooking oil.	Liu et al. (2019)
Apple juice	*S. cerevisiae*	Sequential fermentation inoculate with L. plantarum for 16 h at 39 °C in a static environment after dealcoholizing fermented juice with S. Cerevisiae for 5–6 days at 25 °C.	Propanol, citronellol, ethyl lactate, and hexyl butyrate	*S. cerevisiae* eliminates ethanol and certain fragrance compounds.	Li et al. (2021)

Soybean	*Mucoromycota, Aspergillus*	Solid-state	Glutamic acid and aspartic acid	Strains belonging to *A. oryzae* showed high amino acid and are crucial components for soybean fermentation.	Kim et al. (2024)
Castor oil (3% or 6% (v/v) in ALS medium)	*Saccharomyces cerevisiae, Zygosaccharomycees rouxii*	Submerged	4-hydroxy-2-pentenoic acid γ-lactone, 2,3,4-trimethyl-3-pentanol, 5-(hydroxymethyl) furfural, and γ-butyrolactone	Cell permeabilization increased the formation of ethanol and organic acids through modification in their synthesis pathways	Álvarez et al. (2022)
p-coumaric acid	Oleaginous fungus *Mucor circinelloides*	Submerged fermentation (24–48 h and pH: 4.5)	Naringenin	*Mucor circinelloides* produced 2.2 mg/L naringenin from p-coumaric acid, demonstrating potential as a microbial cell factory for flavonoids.	Zan et al. (2024)

be regarded as an essential characteristic when formulating methodologies for yeast selection. A recent study investigated the genetic modification of *Aspergillus oryzae (A. oryzae)* to enhance both its nutritional and sensory properties. Researchers applied a synthetic biology approach to engineer *A. oryzae* for the overproduction of ergothioneine, a potent antioxidant, and heme, a compound responsible for flavor and color in red meat. The modified strains showed significantly increased ergothioneine levels compared to natural fungal sources, while heme concentrations were comparable to those found in plant-based meat alternatives. This work highlights the potential of genetically engineered fungi to improve the nutritional content and flavor profiles of food products (Maini Rekdal et al., 2024).

3. **Fungal enzymes as flavor precursors:** Accumulation of aroma volatiles is enhanced by microbial metabolism of protein and lipid, particularly through microbial enzymes. Proteases break down proteins to shorter peptides and free amino acids, further biotransformed via well-known pathways to various aromatic flavor precursors (Fig. 7.3) (Ahmed et al., 2024). For example, in fermented products like cheese (López Del Castillo-Lozano et al., 2007) and soy sauce (Mizuno et al., 2024), *Penicillium* and *Aspergillus* species release proteases and amylases that generate free amino acids, sugars, and volatile compounds, contributing to umami, sweetness, and complex aromas.
4. **Microbial Consortia Dynamics**: The diversity within fungal communities also determines the stability and robustness of the fermentation process. Mixed cultures provide better understanding about interplay among different species for their synergistic and antagonistic behavior. In addition, it also helps in comprehending the impact of such mixtures on prevalence and evolution of certain flavors with simultaneous activation of metabolic pathways during fermentation (Fang et al., 2023). However, there are limited studies concerning the impact of mixed culture on flavor quality and safety aspects.

6.2 *Effect of Fungal Bioprocessing on Flavor Production*

6.2.1 Substrates for Flavor Production

Attempts to harness the potential of fungi for flavor production and for the transformation of readily available substrates into valuable flavor compounds started in the early 1950s and have been pursued constantly since then. Simple and inexpensive sources of carbon and nitrogen are prerequisites for fungal growth. Conventionally, substrates such as wheat or rice straw were used for cultivating mushrooms. Although fungi have also been produced in synthetic media enriched with specific amino acids, vitamins, minerals, or other micronutrients (Wei et al., 2023). However, the utilization of agrifood wastes as culture media is being promoted for the biotechnological synthesis of flavors and fragrances, owing to the cost-effectiveness

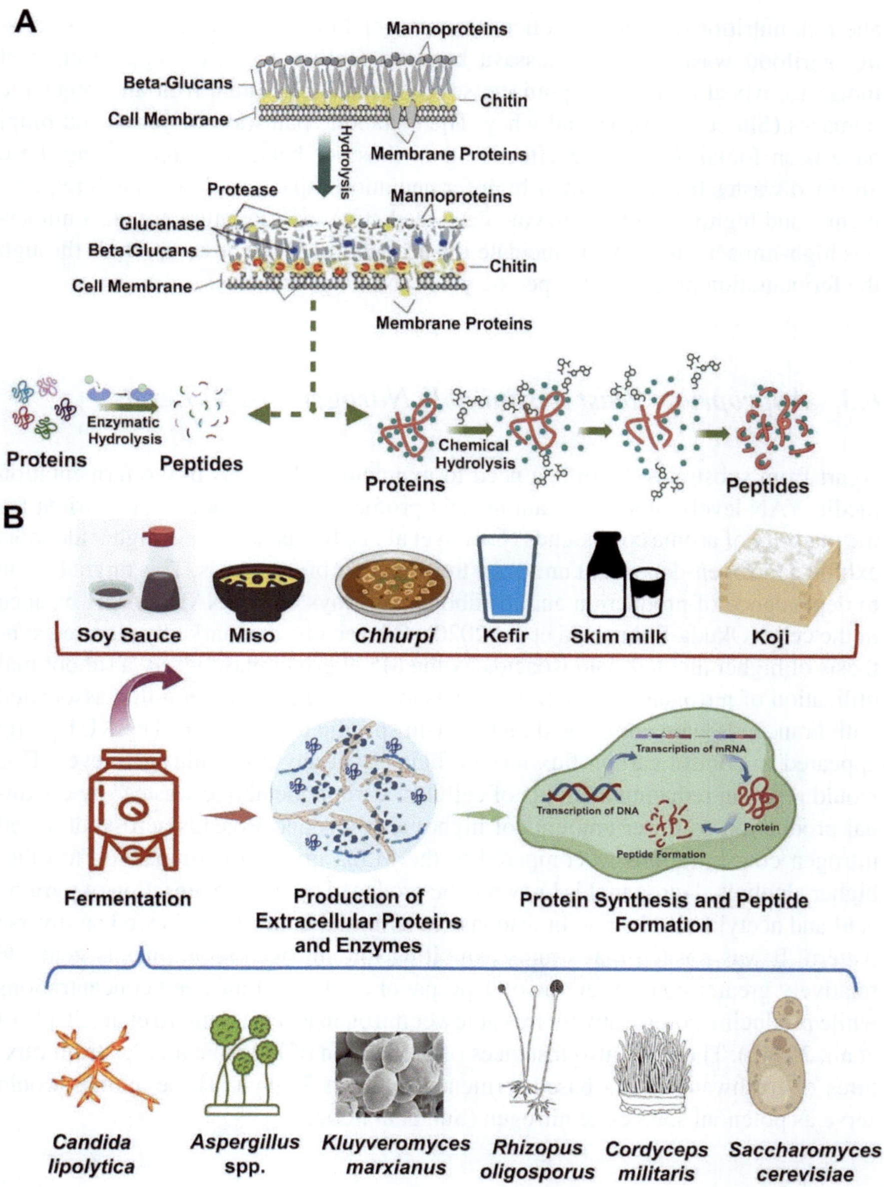

Fig. 7.3 Graphic representation of enzymatic hydrolysis, chemical hydrolysis (**a**), and fermentation of different protein substrates using fungal strains (**b**), resulting in the release of bioactive peptides. (Ahmed et al., 2024, Open Access under Common Creative License 3.0 Agreements)

and rich nutritional profile conducive to microbial proliferation. The most promising agrifood wastes include cassava bagasse, tapioca bagasse, sugar beet, beet molasses, wheat bran, apple pomace, soybean meal, rice bran, fruit and vegetable pomaces (Shi et al., 2023), and whey. The metabolic pathways of yeasts and fungi have been found to be more effective than those of bacteria at fermenting these agrifood wastes because of their high fermentation capacities, low growth requirements, and highly effective enzyme-catalyzed systems. Literature contains numerous high-impact studies that elucidate the production of flavor compounds through the fermentation processes of specific yeast and fungal species.

6.3 Appropriate Yeast Assimilable Nitrogen (YAN) Levels

Apart from substrates, there is a need to maintain YAN levels in the fermentation media. YAN levels, along with amino acid profiles, constitute a critical nutrient for the majority of aroma compounds (Carrau et al., 2016). Isoacids and higher alcohols exhibit a nitrogen-dependent universal trend in their biosynthesis. This might be due to dependence of production and metabolic pathways on the NAD/NADH balance in the cells (Okuda-Shimazaki et al., 2020). Conversely, the markedly elevated synthesis of higher alcohols and isoacids by the M522 group may signify a suboptimal utilization of nitrogen, which in turn leads to an augmented carbon flux associated with branched-chain amino acid metabolism specific to this strain. The KU1 group appeared to control carbon flux more efficiently at any given nitrogen level. This would result in reducing amounts of cellular "carbon metabolic wastes" by continual production of lower amounts of higher alcohols and isoacids across all tested nitrogen concentrations as compared to the M522 group. In comparison to other higher alcohols, 1-propanol is known to be produced by the condensation of pyruvic acid and acetyl CoA. Its rise in response to assimilable nitrogen showed an inverse pattern. It was posited that strains exhibiting low nitrogen requirements generate relatively greater concentrations of 1-propanol at elevated nitrogen concentrations while producing comparatively less at lower nitrogen levels (Gamero et al., 2011; Li et al., 2024b). There are also instances of production of biogenic amines from mixtures of freshwater fishes based fermented product Suanyu. These amines would serve as potential sources of nitrogen (Sun et al., 2022).

6.3.1 Fermentation Process Conditions

1. Fermentation Time

The duration of fungal fermentation significantly influences the production of flavor metabolites, impacting the sensory qualities of fermented foods and beverages. Extended fermentation periods can alter the composition and concentration of these metabolites, leading to diverse flavor profiles (Mizuno et al., 2024). For instance, in kimchi, fungi contribute to the production of metabolites such as alanine, thymine, galacturonic acid, and malonic acid between 30 and 50 days of fermentation. These compounds are associated with flavor development during the fermentation process. Additionally, metabolites like malic acid, oxaloacetic acid, galactitol, glucose, and mannitol are consumed by *Pichia* and *Candida* influencing the flavor profile of kimchi (Seo et al., 2020). Prolonged fermentation times can lead to significant changes in microbial communities and the physicochemical properties of the substrate. For instance, ultra-long fermentation periods in grain-based fermentations have been observed to alter the microbial dynamics in baijiu. *Eurotiales* was present approx. 96–98% initially, while *Saccharomycetales* took over with increasing fermentation time and was present approx. 82–82.5% at the end of fermentation. However, varying times showcased change in fungal genus from *Thermoascus* to *Thermomyces* at 24th hour to *Issatchenika* at 72th hour, responsible for key flavor components, resulting in distinct flavor profiles (Duan et al., 2023).

2. Fermentation Temperature

Temperature significantly impacts fungal metabolism, leading to shifts in flavor profiles. The specific effects vary with the fungal species, fermentation conditions, and the compounds of interest. Adjusting temperatures within optimal ranges can enhance the production of desirable flavors while minimizing unwanted notes (Feng et al., 2024). Impact of initial temperature on flavor metabolites produced during sauce flavored *Baiju* fermentation by *Saccharomyces* and Lactobacillus. Two temperatures groups were chosen for the study viz., low temp (28 °C ± 2 °C) and high temp (37 °C ± 2 °C). Higher initial temperature led to increased total acid including acetic acid with lower amounts of esters and alcohols. The temperature has also increased *Lactobacillus* growth which in turn was responsible for major acidic profile in *Baiju* (Zhang et al., 2021). Unpasteurized beer (Craft beer) also evaluated for its technological variation impact on its color, pH, and storage stability. It has been found that microfiltration at 0.45 μm reduced head retention and flavor stability, leading to oxidation and stale notes. Cold storage (4 ± 0 °C) was found to be the best option for maintaining beer flavor stability.

3. Level of Aeration in Substrates

Aeration during fermentation significantly affects fungal growth and the production of flavor metabolites. Oxygen is crucial for aerobic metabolic pathways in fungi, which produce a variety of aroma compounds, such as alcohols, esters, and aldehydes. Aeration differs in submerged fermentation (SmF) and solid-state fermentation (SSF) (Duan et al., 2023). For instance, *A. oryzae* used in soy sauce production generates higher concentrations of fruity esters and umami-related compounds under optimized aeration during SmF. While in SSF, excessive aeration had shown oxidative stress impacting the flavor profile by favoring production of aldehydes and ketones (Gao et al., 2024).

4. Stirring Rate

Stirring (or mixing) affects the uniformity of nutrient distribution and oxygen transfer, which can influence fungal flavor metabolite production. Regular stirring ensures even oxygen distribution, particularly in SmF, leading to higher yields of volatiles like isoamyl acetate (banana-like) and phenolic compounds. Over stirring, however, may cause cell lysis or shear stress, reducing flavor diversity. In SSF, periodic mixing can improve heat and mass transfer, preventing localized overheating and promoting a consistent flavor profile (Mizuno et al., 2024).

5. Geographical Flavor Diversity

Fungal communities contribute to unique regional and product-specific flavors (Rodríguez-Berríos et al., 2024). Soil fungal diversity is a major predictor of regional flavor variations, influencing both direct and indirect effects on fermentation substrates (Tan et al., 2022). For instance, in wine production, soil and grape-associated fungal microbiomes, including *S. cerevisiae*, interact with environmental factors to create distinct aromatic profiles. *S. cerevisiae* is known to produce different amounts of aroma compounds depending on the must's micronutrients, vitamins, and nitrogen content, temperature, grape variety, inoculum size, redox state, and other factors (Pérez et al., 2022). Nevertheless, there exists a paucity of information concerning non-Saccharomyces species.

7 Applications of Fungal Metabolites in Producing Flavors

In the food industry, the application of fungal flavors is extensively enhancing sensory profiles of food products. The ability of fungi to generate a diverse array of flavor compounds renders them indispensable in culinary applications and food biotechnology (Gong et al., 2024). These flavor compounds are produced through natural metabolic processes of fungi called fermentation. Fermentation represents a widely adopted technique for preparing and preserving food products (Lv et al., 2024). Besides, fermentation of foods has also proven its significance in production of aromas and flavors. Over the past few decades, there has been a surge in biotechnological production of volatile flavor compounds (VFCs) through various methodologies, because this approach has been recognized as a balanced means of supplying natural additives to the food industry. Mushrooms are particularly valued for their umami and earthy flavors while yeast fermentation is crucial in the production of alcoholic beverages and baked goods. Therefore, flavors produced by microbes constitute most of the crucial attributes of food and are finding an array of applications in the food processing industry intrinsically linked to consumer acceptance of products (Shi et al., 2024).

7.1 Beverages

Tea

Tea fermentation, more accurately described as enzymatic oxidation, is a biochemical process in the production of various tea types, influencing flavor, color, and aroma profiles. During fermentation endogenous polyphenol oxidase enzymes catalyze the oxidation of catechins (major polyphenolic compounds in tea leaves) into theaflavins and thearubigins (Abudureheman et al., 2022). According to Rigling et al. (2021), a new starting culture *Wolfiporia cocos* was used to adjust the flavor of green tea infusion in response to Western consumer's growing interest. As consumers have less tolerance to its green color, *Wolfiporia cocos* transformed the distinguishing green smell to a pleasing flowery, jasmine-like, and faintly citrus-like flavor after submerged fermentation. Using stir bar sorptive extraction (SBSE) coupled with gas chromatography-mass spectroscopy-olfactory (GC-MS-O) in conjunction with sensory testing, the resulting pleasant scent was mostly ascribed to geraniol (OAV 118), 2-phenylethanol (OAV 165), linalool (OAV 190), and methyl anthranilate (OAV 802). At the same time, the Ultra-High Performance Liquid Chromatography-Mass Spectrometry (UHPLC-MS)-derived catechin profile revealed varying post-fermentation decrease rates (10–50%) for each particular catechin. However, in the 2,2-diphenyl-1-picrylhydrazyl (DPPH) assay, up to 80% of the in vitro antioxidant activity was maintained and no additional substances such as sweets, spices, or fruits were added in the preparation of this tea.

The production of flowery fungal scents has been evidenced in dark tea. The characteristic flavor notes responsible are produced by the golden-flower fungus *Eurotiwm Cristatum*. This fungus is also known to produce a well-balanced flavor in Golden Flower White Tea (GFWT) by reforming its roughness and coarseness through EC-mediated fermentation. A stripping method was adopted for analyzing the mechanism of production of complex aroma scents of GFWT. The white tea (also known as Shoumei) along with GFWT and their decomposed samples were thoroughly examined using SPME-GC–MS, QDA, and microstructure analysis. The findings demonstrated that EC itself absorbs the fragrance of GFWT's fungal flowers. Additionally, WT's aging was expedited by the EC-mediated flowering process, giving it a stale, jujube, and sweet scent. Leaf was primarily responsible for the stale scent. Together, stem and leaf gave GFWT a sweet and stale scent. Additionally, correlation studies revealed that 1-hexanol, nonanal, 1-heptanol, (E)-2-heptenal, hexanal, ethyl nonanoate, 1-dodecanol, ethyl butyrate, and styrene were the main components that presented the aroma of fungal flowers. Cedrol, 6-methyl-5-hepten-2-one, 2-heptanone, methyl salicylate, and 1,1-diethoxyethane were the main chemicals that presented stale fragrance. Endo-borneol and 2-heptanone were the main chemicals that presented the aroma of jujube. Hotrienol and 1,2,4,5-tetramethylbenzene were the primary compounds that presented a sweet scent. Additionally, unsaturated fatty acids (oleic, palmitoleic, and α-linolenic acids) were the primary source of the fragrance components of the fungal floral scent GFWT (Guo et al., 2025).

Kuding tea (KDT), renowned for its culinary and therapeutic qualities, is currently used rarely because of its unpleasant flavor. Therefore, researcher's goals were to use fungal fermentation to enhance KDT flavor and clarify the process of flavor change by examining the interaction between ingredient and taste. *Aspergillus neoniger* (AN) and *Aspergillus cristatus* (AC) evaluated the flavor characteristics of unfermented and fermented teas using an electronic tongue. The taste-related differential metabolites were analyzed using a constituent-taste relationship approach that combined untargeted metabolomics with taste sensors. The possible processes that change taste were further clarified by the examination of metabolic pathways. The umami and sweetness of KDT changed as a result of the fermentation of both AN and AC. Furthermore, sourness and bitterness were affected by AN fermentation, but saltiness was affected by AC fermentation. Based on the principle of constituent-taste relationships, a total of 31 and 30 metabolites primarily comprising lipids, phenolic acids, flavonoids, amino acids and phenolic glycosides were identified as being significantly correlated with the flavor characteristics of KDT. During the fermentation process, AN and AC were involved in five and four metabolic pathways, respectively, indicating distinct regulatory mechanisms that influence the flavor of KDT. AN was responsible for a marked deglycosylation of glycosides, whereas AC influenced degradation of phenolic acids and lipids. Notably, KDT fermented by two different fungi exhibited significant variations in flavor characteristics and metabolic profiles (Zeng et al., 2024).

The summer green tea (SGT) is inexpensive and produces a lot of tea each year, but its low-quality limits its use. Using *A. niger* SSF for low-value SGT successfully decreased the ester catechin concentration while greatly raising the non-ester catechin content and boosting the amount of free amino acids in fermented tea. A higher concentration of linalool, (E)-linalool oxide (furan type), linalool oxide (pyran type), theapyrrole, and (Z)-α, α,5-trimethyl-5-vinyltetrahydrofuran-2-methanol resulted in a more flowery flavor in the fermented tea. The resulting fermented tea has mild acidity and a sweet, mellow, and fragrant aftertaste. It also successfully improved the flavor of SGT, which set the stage for the creation of functional tea drinks. The purpose of this work is to use fermentation with *A. niger* RAF106 to improve the flavor of SGT while investigating alterations in its metabolites. The findings showed that after fermentation, tea leaves had higher levels of alcohol, alkanes, and nitroxides. Linalool, (Z)-α, α, 5-trimethyl-5-vinyltetrahydrofuran-2-methanol, (E)-linalool oxide (furan type), linalool oxide (pyran type), and theapyrrole were the main volatile chemicals found in tea leaves following a 6-day fermentation period. These chemicals showed significant increases over the non-fermented control group (CK) of 31.48%, 230.43%, 225.12%, 70.71%, and 521.62%, respectively. The fourth day of fermentation was when the levels of non-ester catechins, soluble sugars, and total flavonoids peaked, showing notable increases of 114.8%, 95.59%, and 54.70%, respectively (Cai et al., 2023).

Coffee

Coffee is an essential resource that is used to make high-quality drinks that are consumed by people all around the world. Coffee's flavor profile can be enhanced by the microbiome of coffee beans, which naturally fermented coffee beans to produce fragrant flavor chemicals. In the study given by Murthy et al. (2019) *A. oryzae* (KX522630) was isolated from coffee pulp for the first time, the carboxypeptidase enzyme was recovered. A carboxypeptidase activity of 3.14×10^6 nanokatal per milliliter was obtained using wheat bran as a substrate, despite the fact that coffee pulp showed higher activity due to supply constraints. The inoculum concentration (108), fermentation period (100 h), temperature (30 °C), and moisture (30%) were the optimal complementary bioprocess parameters. When carboxypeptidase was originally added to coffee, it greatly improved the taste characteristics of the pyrazines that contribute to the sensory qualities' sweet, fruity flavors. It has been revealed for the first time that carboxypeptidase added unique flavor notes to the coffee bean.

The ultimate goal for microbial communities was to investigate the interactions and contribute to the development of flavor in coffee in a humid subtropical region. During the fermentation process, predominant species are of two groups, such as the prokaryotic group including *Enterbacter*, *Erwinia, Kluyvera,* and *Pantoea* as well as the eukaryotic group includes *Fusarium, Cladosporium, Kurtzmaniella, Leptoshaerulina,* and *Penicillin.* The main metabolites found in the process were ethanol and the acids such as lactic, acetic, and citric acids. Interestingly, plant-derived metabolites like caffeine and linalool stay stable during fermentation, whereas the bacterial community uses metabolic flexibility to create esters and alcohols, playing a critical role in development of flavors (de Melo Pereira et al., 2024).

The volatile compounds present in coffee like 2,4,6 trichloroanisole, geosmin, and terpenes are responsible for development of off-flavors. Three species were isolated for the experimental study: *Penicillium brevicompactum, Aspergillus luchuensis* (belonging to section Nigri), and *Pencillium* spp.*nov.* (related to *Penicillium crustosum*). VCs developed by fungal inoculated in raw coffee beans were isolated and identified by PME-GC-MS: Solid-Phase Microextraction-Gas Chromatography-Mass Spectrometry. Oct-1-en-3-ol was detected in the raw coffee inoculated with *A. luchuensis.* These compounds can be responsible for the positive characteristics of the beverage, observed in the sensory analysis. After the inoculation, coffee beans showed 100% infection by the species *A. luchuensis, P. brevicompactum,* and *Penicillium* spp. *nov.* The former component is employed in the perfume industry and is present in plants. Several aldehydes, including 2-octenol and decanal, have citrusy, fresh and herbal smells along with ketones. Other compounds like nonan-2-one, non-3-en-2-one, and undecane-2-one presenting fruity aromas were created by *Penicillium brevicopactum* in the growth medium (Iamanaka et al., 2014).

Wine

The grape rot causes deviations in organoleptic deviations in wines, which includes the aroma of fresh mushroom. One molecule that is responsible for this deviation is 1-octen-3-one. The bunches affected by this defect are often contaminated by *Botrytis cinerea (B. cinerea).* However, they also contain a whitish mycelium that cannot be assigned to *B. cinerea.* The focus of this study is the extra fungal proliferation. Utilizing grape bunches from various vineyards in the Alsace and Champagne area of France, *B. cinerea* was isolated multiple times using a culture medium that contained the anti-botrytis fungicide boscalid. According to the Internal Transcribed Spacer (ITS) region sequencing, the majority of isolations were associated with the endophyte basidiomycete *Crustomyces subabruptus*. This fungus is responsible for the emergence of fresh mushroom defects in wines. It also produces considerable amounts of 1-octen-3-one in a warm, humid environment. The results of the research imply that the fresh mushroom scents in wines are caused by this basidiomycete (Shi et al., 2023).

The fresh mushroom off-flavor (FMOff) in wine, primarily caused by C8 compounds such as 1-octen-3-one, 1-octen-3-ol, and 1-hydroxyoctan-3-one. These compounds can be derived from glycosidic precursors found in grape musts, particularly in musts contaminated with *Crustomyces subabruptus*. During alcoholic fermentation, the levels of FM off-flavors glycosidic precursors decrease, while the volatile compounds evolve in a compound- and matrix-dependent manner, with 1-hydroxyoctan-3-one reaching a consistent threshold concentration in wine (Shi et al., 2023).

Duan et al. (2023) provides new insight into intracellular redox management in the yeast *S. cerevisiae*. Since fewer cofactors are available to support the enzymatic breakdown of sugars after the mid-exponential phase, it specifically shows that the concentration of NAD(H) drops with time, which is to be correlated with the decrease in alcoholic fermentation rate. This investigation demonstrated distinct strain specificity, despite the intricate link between redox cofactors and main metabolite synthesis being well understood. The metabolic snowball effect is caused by its inherent high glycerol synthesis. It is reasonable to predict that circumstances responsible for more glycerol production will further decrease ethanol and boost the generation of desired byproducts like succinic acid.

Beer

Beer is one of the most widely consumed and produced beverages in the world, yet the impact of pesticide residues on its quality has not been thoroughly examined. In particular, the fungicide triadimefon (TF), which is used to protect barley crops from infections, can transfer into the wort during brewing and persist into the final product (Pereira de Paula et al., 2021). This compound belongs to the sterol biosynthesis-inhibiting fungicides, which disrupt fungal cell wall formation (Kong et al., 2016a, b). Studies have shown that residual triadimefon in beer can inhibit the growth and metabolic activity of *S. cerevisiae*, the yeast species responsible for fermentation in brewing, winemaking, and baking (Pérez-Lucas et al., 2023). This inhibition leads to undesirable changes in beer color and flavor. To investigate these effects, researchers have employed metabolic profiling techniques, including phenotype microarray technology and ultra-performance liquid chromatography, to identify the pathways affected by TF. At concentrations of 5 mg L^{-1}, TF was found to reduce yeast growth by 45% and alter amino acid metabolism, decreasing the synthesis of flavor-active compounds such as isobutyl alcohol and isoamyl alcohol while increasing ethyl acetate production. Metabolic profiling influenced steroid biosynthesis and metabolism of glutathione, phenylalanine and sphingolipid as they are closely related to the activity of yeast fermentation. Sensory analysis using electronic tongue and nose technology confirmed that TF negatively influences the beer's profile of sugars, amino acids, and volatile compounds, ultimately compromising fermentation quality (Kong et al., 2016a, b).

It is well-known fact that flavor and aroma of beer are largely determined by volatile organic compounds (VOCs), which are produced during fermentation. Research using UV-C induced mutagenesis has identified yeast strains that modify VOC production, leading to different sensory characteristics (Diesler et al., 2019).

For instance, one fungal strain was found to increase terpenoid concentrations, while another elevated levels of 2-ethylhexanol. Some of these VOCs, such as caryophyllene, exceed their odor thresholds and significantly impact flavor, whereas nerolidol remain below detectable levels. The values of acetate esters were tenfolds higher than those in beers and wines produced with other yeast strains. GC-MS was used to identify 31 intracellular metabolites involved in flavor development, including 15 amino acids, four coenzymes, and four intermediates from various metabolic pathways (Jan Jabłoński et al., 2024).

Different yeast strains also contribute distinct sensory profiles to beer. *S. cerevisiae* var. diastaticus, had produced beers with higher alcohol content, reaching up to 8.14% ABV, while *Brettanomyces claussenii* tends to intensify beer color. Chromatographic analyses of pale and dark beers fermented with various strains have detected 58 volatile compounds, with higher alcohols and esters constituting the majority of the aroma profile (Paszkot et al., 2023). Among these, 3-methylbutanol and 2-methylpropanol are key contributors to the sensory characteristics of the final product. *S. cerevisiae var. diastaticus, S. cerevisiae var. bayanus, Brettanomyces claussenii* are some of the distinct varieties of yeasts, showing variation in beer quality (Romero-Rodríguez et al., 2022). Beer which was fermented with *Saccharomyces cerevisiae var. diastaticus* achieved the highest alcoholic grade, i.e., 8.14% (Matukas et al., 2022), whereas *Brettanomyces claussenii fermented* beer showed a high color intensity (Lawton et al., 2021). Overall, the choice of yeast strain and bioprocessing conditions play critical roles in determining beer quality by influencing fermentation efficiency and flavor development. Further research is necessary to optimize brewing conditions and minimize contaminants to ensure consistent and high-quality beer production.

7.2 Condiments

Condiments arise from a diverse class of flavor compounds. These help in seasoning food products for providing them fermented, acidic, sweet, and aromatic notes by their complex flavor profiles. Diverse and deep flavors are known to produce using enzymatically hydrolyzed vegetable proteins, yeast extracts, and nucleotide-based enhancers such as disodium inosinate and disodium guanylate, which contribute to umami perception through synergistic interactions with naturally occurring amino acids (Que et al., 2023). However, food producers have started looking for other sources in recent years other than monosodium glutamate (MSG). Some processing techniques could make yeast extracts naturally having nutritional qualities and flavor exactly like MSG. This study assessed the chemical, functional, and flavor characteristics of yeast extract powders made from *Kluyveromyces marxianus* TGM66, *S. cerevisiae* TGM10, and *S. boulardii* S11. The most protein-rich sample, according to the results, was *S. cerevisiae* TGM10 extract (69.17%), which was followed by *S. boulardii* S11 (66.16%) and *K. marxianus* TGM66 (62.42%) extracts, respectively. Furthermore, *S. cerevisiae* TGM10 extract has a high concentration of flavor-enhancing amino acids such as glycine, glutamic acid, aspartic acid, and alanine (47.41 g/100 g protein) making it likewise the most abundant yeast extract for essential amino acids. The results of the sensory evaluation of yeast extracts showed that they scored lower than MSG for meaty flavor, umami taste, and salty taste, whereas they scored higher for fruity flavor. These results indicate that three yeast strains have the capacity to make yeast extracts, which could improve the flavor and nutritional content of food (Demirgül et al., 2022).

A remarkable fresh impact is produced by the substantial amounts of amino acids and nucleotides found in mushrooms, which enable taste coordination, balancing, flavor reinforcement, odor masking, and enhanced umami perception. The effect of concentrated fungal extract on the flavor peptides of five chicken flavor seasonings was investigated. A clear relationship between the amount of sodium glutamate and disodium inosinate-guanylate (I + G) and the taste intensity of fungus extract was evidenced (Sun et al., 2022). For example, the main component of beef flavor is produced by altering the conditions of glutathione's reaction and its way of interaction during Maillard reaction. In order to generate umami/meat taste items, fungi are heated and their processes are optimized. Furan and pyrazine have been determined to be the main aromatic components by results shown by GC-MS-O (Alim et al., 2019).

7.3 Bakery Products

Fungi play a fundamental role in the development of flavor in bakery products through their biochemical activities. They drive fermentation processes, where carbohydrates are metabolized into various compounds that enhance the sensory attributes of baked goods. The aim of the investigation was to elucidate the sources of contamination while identifying the microorganisms responsible for the bakery products degradation, thereby proposing strategies to mitigate these issues without the application of chemical preservatives in food products. The predominant contaminating microorganisms identified included molds (*Asperigillus species* and *Penicillium species*) and yeast (*Hyphopichia* sp., *Saccharomyces* sp., and *candida* spp.). An assessment of the microbiological quality like disinfection protocol was implemented as a corrective intervention. The influence of ethanol utilization and gas composition in modified atmosphere packaging was critically examined on the prevalence of mold and yeasts. The effectiveness of utilizing sourdough was investigated. Optimal results were achieved with sourdough that was prepared using specific starter cultures and a gas mixture for packaging comprising 30% nitrogen and 70% carbon dioxide, devoid of ethanol. These conditions effectively maintained the quality of the baked products while ensuring their expected shelf life (Diowksz et al., 2024).

Bakery products made by cereal and edible fungi powder have a unique flavor and health benefits, enhancing nutritional value of the products. The antifatigue effect of a novel *trichloma matsutake* cookie (TMC) was investigated by the exhaustive swimming test. Male mice were fed with saline, ordinary cookies, and TMC as they were divided into groups of 10. After the administration, it was found that mice who consumed TMC showed high speed in swimming. TMC could enforce exhaustive swimming tolerance, accelerate the decomposition of sports related metabolites such as lactic acid and urea, and increase the activity of the antioxidant enzyme thereby improving sports related energy storage and relieving fatique (Ma et al., 2020).

7.4 Dairy Products

Flavors incorporated into dairy products play a critical role in modulating sensory perception and consumer acceptance. The selection of flavor compounds is influenced by the physicochemical properties of the dairy matrix, such as fat content, pH, and water activity, which collectively affect flavor release and stability. In fermented dairy products like yogurt and kefir, the presence of lactic acid bacteria not only contributes intrinsic tanginess but also interacts with added flavorings, potentially enhancing aromatic notes (Zhang et al., 2024a). Lipophilic flavor compounds, such as vanilla or fruit esters, exhibit high affinity for high-fat dairy bases like ice cream or cheese, leading to more pronounced and sustained flavor perception (Wang et al., 2024). Moreover, the Maillard reaction and lipid oxidation during processing can generate or modify flavor compounds, contributing to the complexity of the final sensory profile (Martin et al., 2021).

Liu and Crow (2010) examined the production of sulfur flavor concentrate in dairy products as media (milk and cream) through yeast fermentation supplemented with L-methionine. The yeasts analyzed, referred to as dairy yeasts, included *Candida kefyr*, *Kluyveromyces marxianus*, *Debaryomyces hansenii*, *Geotrichum candidum*, and *Yarrowia lipolytica*, alongside wine yeasts *S. cerevisiae* and *S. bayanus*. Methional was identified as the principal volatile flavor compound produced. Other volatile sulfur compounds generated included dimethyl disulfide, S-methyl thioacetate, 3-methylthio-1-propene, and ethyl 3-methylthio-1-propanoate. Additionally, a variety of other volatile flavor compounds were detected, encompassing ethanol, branched-chain esters, branched-chain fatty acids, and branched-chain fatty aldehydes. The sensory characteristics of methionol sulfur flavor concentrate in cream exhibited concentration-dependent attributes, ranging from a subtle savory potato note to a more pronounced potato-savory note, culminating in a cooked cheese-potato note. Due to resource scarcity, there is growing demand for cheese-like aromas derived from plant-based sources as alternatives to traditional cheese. Accordingly, research has explored the production of natural cheese-like aroma through the fermentation of four plant-based proteins and coconut oil with basidiomycetes. Mixtures of soy and sunflower protein with coconut oil have demonstrated the formation of cheese-like aromas after fermentation with *Cyclocybe aegerita* and *Trametes versicolor*, respectively (Wagner et al., 2024). Additionally, identification of oct-1-en-3-ol as the compound responsible for the mushroom flavor in oxidized dairy products underscores its significance in their flavor profile. The alcohol's low flavor threshold values in water, skim milk, and butterfat suggest its potential impact on the sensory qualities of dairy products, particularly when oxidation occurs. Additionally, research indicates that oct-1-en-3-ol may also contribute to the natural flavor of other foods, such as cooked meats—especially pork, which contains higher levels of linoleic acid (Chi et al., 2025).

8 Conclusions and Future Directions

Fungal-derived secondary metabolites present a natural, healthy, and sustainable alternative for imparting complex flavors to diverse food products, offering significant advantages over synthetic counterparts. This chapter has emphasized the immense metabolic diversity of filamentous fungi and yeasts, leveraging their biochemical pathways to synthesize structurally diverse aroma compounds, primarily classified as C8 derivatives, terpenoids, and sulfur-containing compounds. The successful application of fungal bioprocessing hinges on the meticulous optimization of biochemical and bioprocess parameters. Key factors influencing flavor yield and profile include substrate composition (e.g., agro-industrial wastes rich in carbon and nitrogen, lipid-rich substrates for esters, proteinaceous substrates for umami flavors), yeast assimilable nitrogen (YAN) levels, strain physiology, and critical fermentation conditions such as time, temperature, pH, dissolved oxygen (DO), aeration, and stirring rate. The modern bioreactors have the ability to tune these parameters with precision controls, allowing production of the specific flavor for targeted industrial application with desired reproducibility and sensory profiles.

The versatility of fungal metabolites is evident in their successful integration across major food sectors. In beverages, fungi transform tea (e.g., *Wolfiporia cocos* imparting floral notes, *Aspergillus niger* improving summer green tea), coffee (enhancing pyrazines via carboxypeptidase, contributing regional microbial terroir), wine (producing complex aromas, though sometimes introducing off-flavors like "fresh mushroom" from *Crustomyces subabruptus*), and beer (where yeast strain selection significantly impacts VOC profiles and sensory characteristics). Within condiments, fungal processing and enzymatic action generate umami-rich compounds and savory profiles, with yeast extracts emerging as promising MSG alternatives. Bakery products benefit from fungal fermentation for flavor development and shelf-life extension, while also offering functional enhancements like the anti-fatigue properties of *Tricholoma matsutake* cookies. Dairy applications leverage fungi for sulfur flavor concentrates (e.g., methional from dairy yeasts) and the creation of plant-based cheese-like aromas using basidiomycetes, demonstrating the potential to meet evolving consumer preferences for alternative products. The characteristic mushroom flavor, often attributed to oct-1-en-3-ol, highlights the significant contribution of specific fungal volatiles to overall sensory perception.

Despite the promising advancements, several significant challenges impede the full realization of fungal flavors on an industrial scale. A primary hurdle is to achieve a consistent flavor compound production. This inconsistency stems from the inherent complexity of fungal secondary metabolism, governed by intricate biosynthetic gene clusters tightly regulated by environmental cues, epigenetic factors, and the dynamic interplay between primary and secondary metabolic fluxes. Fluctuations in yield and quality are common, hindering reliable scale-up from laboratory settings to large-scale industrial processes. Fine-tuning and strict monitoring of fermentation parameters (oxygen, pH, nutrient supply, shear stress) remain critical but complex tasks, particularly at large scale of several thousand liters. Furthermore, while fungal diversity offers a vast reservoir of undiscovered compounds, unlocking this potential requires overcoming limitations in screening and characterizing novel strains and their metabolites efficiently. Controlling off-flavor production, such as hydrogen sulfide in beer or undesirable aldehydes/ketones under suboptimal aeration, necessitates precise strain selection and process control. The economic viability of large-scale production using complex waste streams also demands further optimization to ensure cost-effectiveness.

Nonetheless, the future of fungal-derived flavor enhancers is bright, backed by converging technological advancements and shifting consumer trends toward clean-label, natural ingredients. Key opportunities lie in:

1. **Advanced Genetic and Metabolic Engineering:** Utilizing synthetic biology tools (CRISPR-Cas9, pathway engineering) to reprogram fungal metabolism. This includes enhancing precursor supply (e.g., acetyl-CoA, malonyl-CoA), overexpressing key biosynthetic enzymes, silencing competing/undesired pathways, and introducing heterologous genes to enable the production of novel or high-value compounds (e.g., valencene, specific terpenoids) not naturally synthesized efficiently by fungi. Engineered strains for improved robustness, substrate utilization efficiency (especially lignocellulosic biomass), and reduced off-flavor formation is equally crucial.
2. **Omics-Guided Systems Biology:** Integrating genomics, transcriptomics, proteomics, and metabolomics provides unprecedented insights into fungal metabolism under different conditions. This systems-level understanding allows for the identification of metabolic bottlenecks, regulatory nodes, and the genetic basis of flavor phenotypes, enabling rational design of strains and bioprocesses for enhanced yield, stability, and targeted flavor profiles.
3. **Innovative Bioprocess Engineering:** Optimizing fermentation strategies like fed-batch, continuous, or mixed-culture systems tailored for a desired flavor production. Implementing real-time monitoring using biosensors for critical parameters (nutrients, metabolites, DO) coupled with adaptive process control algorithms will enhance consistency and efficiency during scale-up. Exploring novel reactor designs for solid-state fermentation (SSF) and submerged fermentation (SmF) to improve mass/heat transfer and oxygen distribution would be vital.

Acknowledgements RV is grateful to National Institute of Food Technology and Entrepreneurship Management, Kundli for providing the Institute doctoral fellowship.

Conflicts of Interest There exists no conflicts of interests among the authors.

References

Abudureheman, B., Yu, X., Fang, D., & Zhang, H. (2022). Enzymatic oxidation of tea catechins and its mechanism. *Molecules, 27*(3), 942. https://doi.org/10.3390/MOLECULES27030942

Alim, A., Song, H., Liu, Y., Zou, T., Zhang, Y., Zhang, S., & Raza, A. (2019). Research of beef-meaty aroma compounds from yeast extract using carbon module labeling (CAMOLA) technique. LWT, 112, 108239. https://doi.org/10.1016/j.lwt.2019.06.006

Adame-Soto, P. J., Aréchiga-Carvajal, E. T., González-Herrera, S. M., Moreno-Jiménez, M. R., & Rutiaga-Quiñones, O. M. (2023). Characterization of mating type on aroma production and metabolic properties wild *Kluyveromyces marxianus* yeasts. *World Journal of Microbiology and Biotechnology, 39*(8), 216. https://doi.org/10.1007/s11274-023-03659-4

Ahmed, T., Juhász, A., Bose, U., Shiferaw Terefe, N., & Colgrave, M. L. (2024). Research trends in production, separation, and identification of bioactive peptides from fungi–A critical review. *Journal of Functional Foods, 119*, 106343. https://doi.org/10.1016/j.jff.2024.106343

Álvarez, A., Gutiérrez, A., Ramírez, C., Cuenca, F., & Bolívar, G. (2022). Aroma compounds produced by liquid fermentation with *Saccharomyces cerevisiae* and *Zygosaccharomyces rouxii* from castor oil through cell permeabilization. *Biocatalysis and Agricultural Biotechnology, 39*, 102243. https://doi.org/10.1016/j.bcab.2021.102243

Alves, R. O., de Oliveira, R. L., de Moraes, M. M., Santos, W. W. V., Gomes da Câmara, C. A., da Silva, S. P., Porto, C. S., & Porto, T. S. (2025). Evaluation of the impact of fermentation conditions, scale up and stirring on physicochemical parameters, antioxidant capacity and volatile compounds of green tea kombucha. *Fermentation, 11*(4), 201. https://doi.org/10.3390/fermentation11040201

Ashour, M., Wink, M., & Gershenzon, J. (2010). Biochemistry of terpenoids: Monoterpenes, sesquiterpenes and diterpenes. *Annual Plant Reviews: Biochemistry of Plant Secondary Metabolism, 40*, 258–303. https://doi.org/10.1002/9781444320503.ch5

Bai, B., Liu, C., Zhang, C., He, X., Wang, H., Peng, W., & Zheng, C. (2023). Trichoderma species from plant and soil: An excellent resource for biosynthesis of terpenoids with versatile bioactivities. *Journal of Advanced Research, 49*, 81–102. https://doi.org/10.1016/J.JARE.2022.09.010

Berger, R. G., & Zorn, H. (2004). *Flavors and fragrances. Advances in fungal biotechnology for industry, agriculture, and medicine* (pp. 341–358). Springer. https://doi.org/10.1007/978-1-4419-8859-1_13

Bilal, Z., Kondybayev, A., Ospanova, A., Tormo, H., Akhmetsadykova, S., Amutova, F., Faye, B., & Konuspayeva, G. (2024). Volatile organic compounds of camel milk and shubat across Kazakhstan's regions, seasons, and breeds. *Heliyon, 10*(15), e35365. https://doi.org/10.1016/j.heliyon.2024.e35365

Bills, G. F., & Gloer, J. B. (2016). Biologically active secondary metabolites from the fungi. *Microbiology Spectrum, 4*(6), 10–1128. https://doi.org/10.1128/microbiolspec.funk-0009-2016

Cai, M., Huang, L., Dong, S., Diao, N., Ye, W., Peng, Z., & Fang, X. (2023). Enhancing the flavor profile of summer green tea via fermentation with *Aspergillus Niger* RAF106. *Food, 12*(18), 3420. https://doi.org/10.3390/foods12183420

Cancellieri, M. C., Nobbio, C., Gatti, F. G., Brenna, E., & Parmeggiani, F. (2024). Applications of biocatalytic C double bond C bond reductions in the synthesis of flavours and fragrances. *Journal of Biotechnology, 390*, 13–27. Elsevier B.V. https://doi.org/10.1016/j.jbiotec.2024.05.006

Caron, T., Crequer, E., Le Piver, M., Le Prieur, S., Brunel, S., Snirc, A., Cueff, G., Roueyre, D., Place, M., & Chassard, C. (2025). Generation of diversity in the blue cheese mold Penicillium roqueforti and identification of pleiotropic QTL for key cheese-making phenotypes. https://doi.org/10.1101/2024.02.22.581506

Carrau, F., Boido, E., & Dellacassa, E. (2016). Yeast diversity and flavor compounds. In *Fungal Metabolites* (pp. 569–597). Springer. https://doi.org/10.1007/978-3-319-19456-1_32-2

Chen, J., Pu, D., Shi, Y., Sun, B., Guo, H., Li, K., & Zhang, Y. (2023). Characterization of the key aroma compounds in different yeast proteins by GC-MS/O, sensory evaluation, and E-nose. *Food, 12*(16), 3136. https://doi.org/10.3390/foods12163136

Chi, X., Zhang, Y., Yang, Q., Zhang, J., Sun, B., & Ai, N. (2025). An insight into specific flavor sensation in fermented milk: Linalool and mushroom alcohol. *Journal of Dairy Science., 108*(6), 5741–5753. https://doi.org/10.3168/JDS.2025-26449

Choi, K. T., Lee, S. H., Kim, Y. J., Choi, J. S., & Lee, S. B. (2024). Improvement of volatile aromatic compound levels and sensory quality of distilled soju derived from Saccharomyces cerevisiae and Wickerhamomyces anomalus co-fermentation. *Food Chemistry: X, 22*, 101368. https://doi.org/10.1016/J.FOCHX.2024.101368

Chu, Y., Li, M., Jin, J., Dong, X., Xu, K., Jin, L., Qiao, Y., & Ji, H. (2023). Advances in the application of the non-conventional yeast *Pichia kudriavzevii* in food and biotechnology industries. *Journal of Fungi, 9*(2), 170. https://doi.org/10.3390/jof9020170

da Costa Maia, I., dos Santos, T., D'Almeida, C., Guimarães Freire, D. M., d'Avila Costa Cavalcanti, E., Cameron, L. C., Furtado Dias, J., & Simões Larraz Ferreira, M. (2020). Effect of solid-state fermentation over the release of phenolic compounds from brewer's spent grain revealed by UPLC-MSE. *LWT, 133*, 110136. https://doi.org/10.1016/j.lwt.2020.110136

de la Rosa, O., Múñiz-Marquez, D. B., Contreras-Esquivel, J. C., Wong-Paz, J. E., Rodríguez-Herrera, R., & Aguilar, C. N. (2020). Improving the fructooligosaccharides production by solid-state fermentation. *Biocatalysis and Agricultural Biotechnology, 27*, 101704. https://doi.org/10.1016/j.bcab.2020.101704

de Melo Pereira, G. V., de Mello Sampaio, V., Wiele, N., da Silva Vale, A., de Carvalho Neto, D. P., de Freitas Diniz de Souza, A., dos Santos, D. V. N., Ruiz, I. R., Rogez, H., & Soccol, C. R. (2024). How yeast has transformed the coffee market by creating new flavors and aromas through modern post-harvest fermentation systems. *Trends in Food Science and Technology, 151*, 104641. https://doi.org/10.1016/j.tifs.2024.104641

Demirgül, F., Şimşek, Ö., Bozkurt, F., Dertli, E., & Sağdıç, O. (2022). Production and characterization of yeast extracts produced by *Saccharomyces cerevisiae*, *Saccharomyces boulardii* and *Kluyveromyces marxianus*. *Preparative Biochemistry and Biotechnology, 52*(6), 657–667. https://doi.org/10.1080/10826068.2021.1983833

Diesler, K., Golombek, P., Kromm, L., Scharfenberger-Schmeer, M., Durner, D., Schmarr, H. G., Stahl, M. R., Briviba, K., & Fischer, U. (2019). UV-C treatment of grape must: Microbial inactivation, toxicological considerations and influence on chemical and sensory properties of white wine. *Innovative Food Science & Emerging Technologies, 52*, 291–304. https://doi.org/10.1016/J.IFSET.2019.01.005

Duan, Z., Wu, Y., Zhang, C., Niu, J., Zhao, J., Li, W., & Li, X. (2023). Comparison of fungal communities and flavour substances in surface and inner layers of fermented grains during stacking fermentation of sauce-flavour baijiu. *Journal of Bioscience and Bioengineering, 136*(4), 295–303. https://doi.org/10.1016/j.jbiosc.2023.06.010

Dulf, F. V., Vodnar, D. C., Toşa, M. I., & Dulf, E.-H. (2020). Simultaneous enrichment of grape pomace with γ-linolenic acid and carotenoids by solid-state fermentation with *Zygomycetes* fungi and antioxidant potential of the bioprocessed substrates. *Food Chemistry, 310*, 125927. https://doi.org/10.1016/j.foodchem.2019.125927

Fang, G.-Y., Mu, X.-J., Huang, B.-W., Wu, G.-Z., & Jiang, Y.-J. (2023). Fungal biodiversity and interaction complexity were the important drivers of multifunctionality for flavor production in a spontaneously fermented vinegar. *Innovative Food Science & Emerging Technologies, 83*, 103259. https://doi.org/10.1016/j.ifset.2022.103259

Feng, Y., Xie, Z., Huang, M., Tong, X., Hou, S., Tin, H., & Zhao, M. (2024). Decoding temperature-driven microbial community changes and flavor regulation mechanism during winter fermentation of soy sauce. *Food Research International, 177*, 113756. https://doi.org/10.1016/J.FOODRES.2023.113756

Fraatz, M. A., & Zorn, H. (2010). Fungal flavours. In *Industrial applications: 2nd edition, the Mycota* (pp. 249–268). Springer. https://doi.org/10.1007/978-3-642-11458-8_12

Frey, C. (2005). Natural flavors and fragrances: Chemistry, analysis, and production. *Natural Flavors And Fragrances, 908*, 3–19, ACS Symposium Series. https://doi.org/10.1021/bk-2005-0908.ch001

Fu, Y. (2024). Yeast extract as a more sustainable food ingredient: Insights into flavor and bioactivity. *Advances in Food and Nutrition Research, 114*, 97–147. https://doi.org/10.1016/bs.afnr.2024.09.002

Gamero, A., Manzanares, P., Querol, A., & Belloch, C. (2011). Monoterpene alcohols release and bioconversion by saccharomyces species and hybrids. *International Journal of Food Microbiology, 145*(1), 92–97. https://doi.org/10.1016/J.IJFOODMICRO.2010.11.034

Gao, X., Han, X., Tang, Y., Song, H., Meng, Q., Chen, R., Yu, J., Yang, P., & Yu, Z. (2024). Exploring changes in key aroma-active compounds of soy sauce due to different fermentation processes. *Journal of Food Composition and Analysis, 133*, 106393. https://doi.org/10.1016/j.jfca.2024.106393

Ghamry, M., Zhao, W., & Li, L. (2023). Impact of *Lactobacillus apis* on the antioxidant activity, phytic acid degradation, nutraceutical value and flavor properties of fermented wheat bran, compared to *Saccharomyces cerevisiae* and *Lactobacillus plantarum*. *Food Research International, 163*, 112142. https://doi.org/10.1016/J.FOODRES.2022.112142

Ghandehari-Alavijeh, S., Can Karaca, A., Akbari-Alavijeh, S., Assadpour, E., Farzaneh, P., Saidi, V., & Jafari, S. M. (2024). Application of encapsulated flavors in food products; opportunities and challenges. *Food Chemistry, 436*, 137743. https://doi.org/10.1016/J.FOODCHEM.2023.137743

Gong, D., Prusky, D., Long, D., Bi, Y., & Zhang, Y. (2024). Moldy odors in food – A review. *Food Chemistry, 458*, 140210. https://doi.org/10.1016/J.FOODCHEM.2024.140210

Grand View Research. (2022). The United States Flavor Compounds Market Size & Outlook, 2030.

Guo, Y., Zhao, J., Wei, H., Gao, Q., Song, S., Fan, Y., Yan, D., Liu, Y., & Wang, S. (2024). Disentangling the tissue-specific variations of volatile flavor profiles of the *Lentinula edodes* fruiting body. *Food, 13*(1), 86. https://doi.org/10.3390/foods13010086

Guo, S., Zhao, Y., Ran, L., Li, Y., Liu, Z., Wang, K., Chen, T., Huang, J., & Zhu, M. (2025). A special "Golden Flower" fungus with yellow cleistothecium: Its impact on the flavor of golden flower loose tea produced from summer-autumn tea. *Journal of Food Composition and Analysis, 137*, 106903. https://doi.org/10.1016/j.jfca.2024.106903

He, K., Peng, X., Li, Y., Zhao, M., & Feng, Y. (2025). Revealing metabolite profiles in soy sauce and exploring their correlation with umami taste using UPLC-Orbitrap-MS/MS and GC-Tof-MS derivatization. *Food Chemistry, 463*, 141303. https://doi.org/10.1016/j.foodchem.2024.141303

Iamanaka, B. T., Teixeira, A. A., Teixeira, A. R. R., Vicente, E., Frisvad, J. C., Taniwaki, M. H., & Bragagnolo, N. (2014). Potential of volatile compounds produced by fungi to influence sensory quality of coffee beverage. *Food Research International, 64*, 166–170. https://doi.org/10.1016/j.foodres.2014.06.017

Inamdar, A. A., Morath, S., & Bennett, J. W. (2025). Fungal volatile organic compounds: More than just a funky smell? *Annual Review of Microbiology, 74*(1), 101–116. https://doi.org/10.1146/annurev-micro-012420

Jan Jabłoński, S., Anna Mielko-Niziałek, K., Leszczyński, P., Gasiński, A., Kawa-Rygielska, J., Młynarz, P., & Łukaszewicz, M. (2024). Examination of internal metabolome and VOCs profile of brewery yeast and their mutants producing beer with improved aroma. *Scientific Reports, 14*, 14582. https://doi.org/10.1038/s41598-024-64899-4

Ji, X., Yu, X., Wu, Q., & Xu, Y. (2022). Initial fungal diversity impacts flavor compounds formation in the spontaneous fermentation of Chinese liquor. *Food Research International, 155*, 110995. https://doi.org/10.1016/j.foodres.2022.110995

Jiang, G., Yao, M., Wang, Y., Xiao, W., & Yuan, Y. (2021). A "push-pull-restrain" strategy to improve citronellol production in Saccharomyces cerevisiae. *Metabolic Engineering, 66*, 51–59. https://doi.org/10.1016/J.YMBEN.2021.03.019

Jin, Y., Li, D., Ai, M., Tang, Q., Huang, J., Ding, X., Wu, C., & Zhou, R. (2019). Correlation between volatile profiles and microbial communities: A metabonomic approach to study Jiang-flavor liquor Daqu. *Food Research International, 121*, 422–432. https://doi.org/10.1016/J.FOODRES.2019.03.021

Jurkaninová, L., Švec, I., Kučerová, I., Havrlentová, M., Božik, M., Klouček, P., & Leuner, O. (2024). The use of thyme and lemongrass essential oils in cereal technology—Effect on wheat dough behavior and bread properties. *Applied Sciences, 14*(11), 4831. https://doi.org/10.3390/app14114831

Kampranis, S. C., & Makris, A. M. (2012). Developing a yeast cell factory for the production of terpenoids. *Computational and Structural Biotechnology Journal, 3*(4), e201210006. https://doi.org/10.5936/CSBJ.201210006

Karrer, D., Weigel, V., Hoberg, N., Atamasov, A., & Rühl, M. (2021). Biotransformation of [U-13C]linoleic acid suggests two independent ketonic- and aldehydic cycles within C8-oxylipin biosynthesis in *Cyclocybe aegerita* (V. Brig.) Vizzini. *Mycological Progress, 20*(8), 929–940. https://doi.org/10.1007/s11557-021-01719-3

Kawabe, T., & Morita, H. (1994). Production of benzaldehyde and benzyl alcohol by the mushroom *Polyporus tuberaster* K2606. *Journal of Agricultural and Food Chemistry, 42*(11), 2556–2560. https://doi.org/10.1021/jf00047a034

Kim, D. H., Chun, B. H., Lee, J.-J., Kim, O. C., Hyun, J., Han, D. M., Jeon, C. O., Lee, S. H., Lee, S.-H., Choi, Y.-H., & Hong, S.-B. (2024). Enzymatic activity and amino acids production of predominant fungi from traditional Meju during soybean fermentation. *Journal of Microbiology and Biotechnology, 34*(3), 654–662. https://doi.org/10.4014/jmb.2309.09008

Kinderlerer, J. L. (1989). Volatile metabolites of filamentous fungi and their role in food flavour. *Journal of Applied Bacteriology Symposium Supplement, 67*(18), 133–144. https://doi.org/10.1111/j.1365-2672.1989.tb03777.x

Kirschning, A. (2022). On the evolution of coenzyme biosynthesis. *Natural Product Reports, 39*(11), 2175–2199. https://doi.org/10.1039/d2np00037g

Kong, Z., Li, M., An, J., Chen, J., Bao, Y., Francis, F., & Dai, X. (2016a). The fungicide triadimefon affects beer flavor and composition by influencing *Saccharomyces cerevisiae* metabolism. *Scientific Reports, 6*(1), 33552. https://doi.org/10.1038/srep33552

Kong, Z., Li, M., Chen, J., Gui, Y., Bao, Y., Fan, B., Jian, Q., Francis, F., & Dai, X. (2016b). Behavior of field-applied triadimefon, malathion, dichlorvos, and their main metabolites during barley storage and beer processing. *Food Chemistry, 211*, 679–686. https://doi.org/10.1016/j.foodchem.2016.05.058

Krings, U., & Berger, R. G. (1998). Biotechnological production of flavours and fragrances. *Applied Microbiology Biotechnology, 49*, 1–8. https://doi.org/10.1007/s002530051129

Krings, U., Lehnert, N., Fraatz, M. A., Hardebusch, B., Zorn, H., & Berger, R. G. (2009). Autoxidation versus biotransformation of a-pinene to flavors with pleurotus sapidus: Regioselective hydroperoxidation of a-pinene and stereoselective dehydrogenation of verbenol. *Journal of Agricultural and Food Chemistry, 57*(21), 9944–9950. https://doi.org/10.1021/jf901442q

Kuzuyama, T. (2017). Biosynthetic studies on terpenoids produced by *Streptomyces*. *Journal of Antibiotics, 70*(7), 811–818. https://doi.org/10.1038/ja.2017.12

Lapadatescu, C., Ginie's, C., Ginie's, G., Le, J.-L., Ré, Q., & Bonnarme, P. (2000). Novel scheme for biosynthesis of aryl metabolites from L-Phenylalanine in the fungus *Bjerkandera adusta*. *Applied and Environmental Microbiology, 66*(4), 1517–1522. https://doi.org/10.1128/aem.66.4.1517-1522.2000

Lawton, M. R., deRiancho, D. L., & Alcaine, S. D. (2021). Lactose utilization by *Brettanomyces claussenii* expands potential for valorization of dairy by-products to functional beverages through fermentation. *Current Opinion in Food Science, 42*, 93–101. https://doi.org/10.1016/J.COFS.2021.05.006

Lee, N. Y., Choi, D. H., Kim, M. G., Jeong, M. J., Kwon, H. J., Kim, D. H., Kim, Y. G., di Luccio, E., Arioka, M., Yoon, H. J., & Kim, J. G. (2020). Biosynthesis of (R)-(-)-1-Octen-3-ol in recombinant *Saccharomyces cerevisiae* with lipoxygenase-1 and hydroperoxide lyase genes from *Tricholoma matsutake*. *Journal of Microbiology and Biotechnology, 30*(2), 296–305. https://doi.org/10.4014/jmb.2001.01049

Li, Y., Zhang, Y., Liu, M., Qin, Y., & Liu, Y. (2019). *Saccharomyces cerevisiae* isolates with extreme hydrogen sulfide production showed different oxidative stress resistances responses during wine fermentation by RNA sequencing analysis. *Food Microbiology, 79*, 147–155. https://doi.org/10.1016/J.FM.2018.10.021

Li, H., Huang, J., Wang, Y., Wang, X., Ren, Y., Yue, T., Wang, Z., & Gao, Z. (2021). Study on the nutritional characteristics and antioxidant activity of dealcoholized sequentially fermented apple juice with *Saccharomyces cerevisiae* and *Lactobacillus plantarum* fermentation. *Food Chemistry, 363*, 130351. https://doi.org/10.1016/j.foodchem.2021.130351

Li, Y., You, S., Cheng, L., Zeng, H., Zheng, B., & Zhang, Y. (2023). Physiochemical quality, microbial diversity, and volatile components of Monascus-fermented hairtail surimi. *Food, 12*(15), 2891. https://doi.org/10.3390/foods12152891

Li, G., Liang, H., Gao, R., Qin, L., Xu, P., Huang, M., Zong, M. H., Cao, Y., & Lou, W. Y. (2024a). Yeast metabolism adaptation for efficient terpenoids synthesis via isopentenol utilization. *Nature Communications, 15*(1), 9844. https://doi.org/10.1038/s41467-024-54298-8

Li, J., Yuan, M., Meng, N., Li, H., Sun, J., & Sun, B. (2024b). Influence of nitrogen status on fermentation performances of non-Saccharomyces yeasts: A review. *Food Science and Human Wellness, 13*(2), 556–567. https://doi.org/10.26599/FSHW.2022.9250050

Liu, S. Q., & Crow, V. L. (2010). Production of dairy-based, natural Sulphur flavor concentrate by yeast fermentation. *Food Biotechnology, 24*(1), 62–77. https://doi.org/10.1080/08905430903562724

Liu, X., Yan, Y., Zhao, P., Song, J., Yu, X., Wang, Z., Xia, J., & Wang, X. (2019). Oil crop wastes as substrate candidates for enhancing erythritol production by modified *Yarrowia lipolytica* via one-step solid state fermentation. *Bioresource Technology, 294*, 122194. https://doi.org/10.1016/j.biortech.2019.122194

Liu, X., Dong, X., Chen, S., Yan, Y., He, J., Xu, J., & Xu, J. (2021). Enhancing erythritol production by wheat straw biochar-incorporated solid-state fermentation of agricultural wastes using defatted *Schizochytrium* sp. biomass as supplementary feedstock. *Industrial Crops and Products, 170*, 113703. https://doi.org/10.1016/j.indcrop.2021.113703

Liu, S., Lou, Y., Li, Y., Zhao, Y., Laaksonen, O., Li, P., Zhang, J., Battino, M., Yang, B., & Gu, Q. (2023). Aroma characteristics of volatile compounds brought by variations in microbes in winemaking. *Food Chemistry, 420*, 136075. https://doi.org/10.1016/J.FOODCHEM.2023.136075

Liu, L., Liu, T., Wang, H., Zhao, Y., Xu, X., & Zeng, M. (2024a). Identification and validation of core microbes for the formation of the characteristic flavor of fermented oysters (*Crassostrea gigas*). *Food Chemistry, 449*, 138970. https://doi.org/10.1016/J.FOODCHEM.2024.138970

Liu, S., Li, D., Zhao, X., Qin, Z., Zeng, W., & Zhou, J. (2024b). Production of food flavor and color by synthetic biology. *Current Opinion in Food Science, 57*, 101168. https://doi.org/10.1016/j.cofs.2024.101168

López Del Castillo-Lozano, M., Delile, A., Spinnler, H. E., Bonnarme, P., & Landaud, S. (2007). Comparison of volatile Sulphur compound production by cheese-ripening yeasts from methionine and methionine-cysteine mixtures. *Applied Microbiology and Biotechnology, 75*(6), 1447–1454. https://doi.org/10.1007/s00253-007-0971-3

Lv, Y., Bai, X., Zhao, H., Xu, Y., Li, J., & Li, X. (2024). Flavor characteristics of large yellow croaker soup served with different dried edible fungi. *Food Chemistry: X, 21*, 101059. https://doi.org/10.1016/J.FOCHX.2023.101059

Ma, N., Tao, H., Du, H., Zhao, L., Hu, Q., & Xiao, H. (2020). Antifatigue effect of functional cookies fortified with mushroom powder (*Tricholoma Matsutake*) in mice. *Journal of Food Science, 85*(12), 4389–4395. https://doi.org/10.1111/1750-3841.15510

Maini Rekdal, V., van der Luijt, C. R. B., Chen, Y., Kakumanu, R., Baidoo, E. E. K., Petzold, C. J., Cruz-Morales, P., & Keasling, J. D. (2024). Edible mycelium bioengineered for enhanced nutritional value and sensory appeal using a modular synthetic biology toolkit. *Nature Communications, 15*(1), 2099. https://doi.org/10.1038/s41467-024-46314-8

Martin, N. H., Torres-Frenzel, P., & Wiedmann, M. (2021). Invited review: Controlling dairy product spoilage to reduce food loss and waste. *Journal of Dairy Science, 104*(2), 1251–1261. https://doi.org/10.3168/JDS.2020-19130

Matukas, M., Starkute, V., Zokaityte, E., Zokaityte, G., Klupsaite, D., Mockus, E., Rocha, J. M., Ruibys, R., & Bartkiene, E. (2022). Effect of different yeast strains on biogenic amines, volatile compounds and sensory profile of beer. *Food, 11*(15), 2317. https://doi.org/10.3390/FOODS11152317/S1

Mehmood, T., Saeed, S., Hussain, N., & Waseem, R. (2024). Biotransformation of wheat straw into biovanillin by solid-state fermentation and optimization of conditions parameters through response surface methodology. *Biomass Conversion and Biorefinery, 14*(6), 7569–7578. https://doi.org/10.1007/s13399-022-02680-1

Melini, F., & Melini, V. (2024). Role of microbial fermentation in the bio-production of food aroma compounds from vegetable waste. *Fermentation, 2024*(10), 132. https://doi.org/10.3390/FERMENTATION10030132

Mizuno, Y., Yoshimura, T., Sawada, K., Tsuge, K., Nagano, Y., Yoshizaki, Y., Goto, M., & Kobayashi, G. (2024). Dynamics of the microbiome and volatile organic compounds during fermentation and aging of soy sauce. *Journal of Bioscience and Bioengineering, 138*(6), 522–532. https://doi.org/10.1016/j.jbiosc.2024.08.009

Murray, A. F., Moore, A. J., & Munafo, J. P. (2020). Key odorants from the American matsutake, tricholoma magnivelare. *Journal of Agricultural and Food Chemistry, 68*(36), 9768–9775. https://doi.org/10.1021/acs.jafc.0c03372

Murthy, P. S., Basavaraj, K., & Kusumoto, K. I. (2019). Modulation of coffee flavor precursors by Aspergillus oryzae serine carboxypeptidases. *LWT, 113*, 108312. https://doi.org/10.1016/j.lwt.2019.108312

Nwankwo, C. S., Ulu, F. O., Okpomor, E. O., Mbachiantim, J. T., Okoyeuzu, C. F., Carew, I. E., Belay, D., Ike, D. C., & Teshome, A. (2023). Natural spices and flavour substitution in zobo tea and drink production: A review. *International Food Research Journal, 30*(4), 832–847. https://doi.org/10.47836/ifrj.30.4.02

Okuda-Shimazaki, J., Yoshida, H., & Sode, K. (2020). FAD dependent glucose dehydrogenases–Discovery and engineering of representative glucose sensing enzymes. *Bioelectrochemistry, 132*, 107414. https://doi.org/10.1016/j.bioelechem.2019.107414

Pahalagedara, A. S. N. W., Gkogka, E., & Hammershøj, M. (2024). A review on spore-forming bacteria and moulds implicated in the quality and safety of thermally processed acid foods: Focusing on their heat resistance. *Food Control, 166*, 110716. https://doi.org/10.1016/J.FOODCONT.2024.110716

Paszkot, J., Gasiński, A., & Kawa-Rygielska, J. (2023). Evaluation of volatile compound profiles and sensory properties of dark and pale beers fermented by different strains of brewing yeast. *Scientific Reports, 13*(1), 6725. https://doi.org/10.1038/s41598-023-33246-4

Pereira de Paula, B., de Souza Lago, H., Firmino, L., Fernandes Lemos Júnior, W. J., Ferreira Dutra Corrêa, M., Fioravante Guerra, A., Signori Pereira, K., & Zarur Coelho, M. A. (2021). Technological features of *Saccharomyces cerevisiae var. boulardii* for potential probiotic wheat beer development. *LWT, 135*, 110233. https://doi.org/10.1016/J.LWT.2020.110233

Pérez, D., Denat, M., Minebois, R., Heras, J. M., Guillamón, J. M., Ferreira, V., & Querol, A. (2022). Modulation of aroma and chemical composition of albariño semi-synthetic wines by non-wine *Saccharomyces* yeasts and bottle aging. *Food Microbiology, 104*, 103981. https://doi.org/10.1016/j.fm.2022.103981

Pérez-Lucas, G., Navarro, G., & Navarro, S. (2023). Comprehensive review on monitoring, behavior, and impact of pesticide residues during beer-making. *Journal of Agricultural Food Chemistry, 71*, 1820–1836. https://doi.org/10.1021/acs.jafc.2c07830

Que, Z., Jin, Y., Huang, J., Zhou, R., & Wu, C. (2023). Flavor compounds of traditional fermented bean condiments: Classes, synthesis, and factors involved in flavor formation. *Trends in Food Science & Technology, 133*, 160–175. https://doi.org/10.1016/J.TIFS.2023.01.010

Rajendran, S., Silcock, P., & Bremer, P. (2023). Flavor volatiles of fermented vegetable and fruit substrates: A review. *Molecules, 28*(7), 3236. https://doi.org/10.3390/molecules28073236

Ranjitha, K., & Oberoi, H. S. (2016). Food flavours from yeasts: Improved productivity through biotechnological interventions and process optimization. *Current Biotechnology, 7*(3), 199–213. https://doi.org/10.2174/2211550105666160218214040

Reyes-Reyes, A. L., Valero Barranco, F., & Sandoval, G. (2022). Recent advances in lipases and their applications in the food and nutraceutical industry. *Catalysts, 12*(9), 960. https://doi.org/10.3390/catal12090960

Rigling, M., Liu, Z., Hofele, M., Prozmann, J., Zhang, C., Ni, L., Fan, R., & Zhang, Y. (2021). Aroma and catechin profile and in vitro antioxidant activity of green tea infusion as affected by submerged fermentation with Wolfiporia cocos (Fu Ling). *Food Chemistry, 361*, 130065. https://doi.org/10.1016/j.foodchem.2021.130065

Rodrigues, C. J. C., & de Carvalho, C. C. C. R. (2022). Process development for benzyl alcohol production by whole-cell biocatalysis in stirred and packed bed reactors. *Microorganisms, 10*(5), 966. https://doi.org/10.3390/microorganisms10050966

Rodríguez-Berríos, R. R., Rodríguez-Vargas, J. A., Colón-Cartagena, F., Maldonado-Silva, U., Ortiz-Colón, Y. W., Escalante-Castaneda, A., Conde-González, A., & Álamo-Diverse, K. Y. (2024). Polyketides from Plakortis sponges around caribbean coastal regions: Collection, isolation, characterization, and bioactivity. *Coasts, 4*(3), 568–593. https://doi.org/10.3390/coasts4030029

Romero-Guido, C., Belo, I., Ta, T. M. N., Cao-Hoang, L., Alchihab, M., Gomes, N., Thonart, P., Teixeira, J. A., Destain, J., & Waché, Y. (2011). Biochemistry of lactone formation in yeast and fungi and its utilisation for the production of flavour and fragrance compounds. *Applied Microbiology and Biotechnology, 89*(3), 535–547. https://doi.org/10.1007/s00253-010-2945-0

Romero-Rodríguez, R., Durán-Guerrero, E., Castro, R., Díaz, A. B., & Lasanta, C. (2022). Evaluation of the influence of the microorganisms involved in the production of beers on their sensory characteristics. *Food and Bioproducts Processing, 135*, 33–47. https://doi.org/10.1016/J.FBP.2022.06.004

Roy, P. S. D., Singh, B., Sharma, V., & Thappa, C. (2022). Biotransformation: A novel approach of modulating and synthesizing compounds. *Journal for Research in Applied Sciences and Biotechnology, 1*(2), 68–82. https://doi.org/10.55544/jrasb.1.2.8

Seo, S. H., Park, S. E., Kim, E. J., Cho, K. M., Kwon, S. J., & Son, H. S. (2020). Effect of fungi on metabolite changes in kimchi during fermentation. *Molecules, 25*(21), 5040. https://doi.org/10.3390/molecules25215040

Sharma, A., Sahu, S., Sharma, S., Singh, G., & Arya, S. K. (2023a). Valorization of agro-industrial wastes into vanillin: A sustainable and bio-economical step towards the indigenous production of flavors. *Biocatalysis and Agricultural Biotechnology, 51*, 102754. Elsevier Ltd. https://doi.org/10.1016/j.bcab.2023.102904

Sharma, N., Takkar, S., Bhatia, K., Vardhan, H., Tripathi, M., Iqbal, K., Shukla, S., Shukla, K., Tyagi, J., Varma, A., & Mishra, A. (2023b). Recent advances in fungal secondary metabolites and their applications. In I. Singh, V. R. Rajpal, & S. S. Navi (Eds.), *Fungal resources for sustainable economy: current status and future perspectives* (pp. 411–432). Springer. https://doi.org/10.1007/978-981-19-9103-5_15

Shi, N., Li, H. Q., Lu, H. C., Tian, M. B., Han, X., He, F., & Wang, J. (2023). Adjusting the pomace ratio during red wine fermentation: Effects of adding white grape pomace and juice runoff on wine flavoromics and sensory qualities. *Food Chemistry: X, 20*, 100939. https://doi.org/10.1016/J.FOCHX.2023.100939

Shi, L., Wang, Q., Xie, Z., Wu, C., Peng, T., Nian, X., Li, L., Li, H., & Chen, T. (2024). The changes of fungal community and flavor substances in Yunnan-style sausages: A comparative analysis of different drying methods. *Food Chemistry, 460*, 140750. https://doi.org/10.1016/j.foodchem.2024.140750

Shin, H. J., Ro, H. S., Kawauchi, M., & Honda, Y. (2025). Review on mushroom mycelium-based products and their production process: From upstream to downstream. *Bioresources and Bioprocessing, 12*(1), 3. https://doi.org/10.1186/s40643-024-00836-7

Siebenhaller, S., Kirchhoff, J., Kirschhöfer, F., Brenner-Weiß, G., Muhle-Goll, C., Luy, B., Haitz, F., Hahn, T., Zibek, S., Syldatk, C., & Ochsenreither, K. (2018). Integrated process for the enzymatic production of fatty acid sugar esters completely based on lignocellulosic substrates. *Frontiers in Chemistry, 6*, 421. https://doi.org/10.3389/fchem.2018.00421

Singh, D., Son, S. Y., & Lee, C. H. (2020). Critical thresholds of 1-octen-3-ol shape inter-species *Aspergillus* interactions modulating the growth and secondary metabolism. *Scientific Reports, 10*(1), 11116. https://doi.org/10.1038/s41598-020-68096-x

Singh, M., Raj, M., Kumar, V., Devi, T., Upadhyay, S. K., Yadav, M., Sehrawat, N., Mishra, P., Kumar, S., & Kumari, M. (2024). 7 Gluconic acid: Strategies for microbial production using organic waste and applications. *Physical Sciences Reviews, 9*(7), 2371–2383. https://doi.org/10.1515/9783110792584-007

Smitha, M. S., Singh, S., & Singh, R. (2017). Microbial biotransformation: A process for chemical alterations. *Journal of Bacteriology & Mycology, 4*(2), 85. https://doi.org/10.15406/jbmoa.2017.04.00085

Song, Y., Liu, H., Quax, W. J., Zhang, Z., Chen, Y., Yang, P., Cui, Y., Shi, Q., & Xie, X. (2024). Application of valencene and prospects for its production in engineered microorganisms. *Frontiers in Microbiology, 15*, 1444099. https://doi.org/10.3389/fmicb.2024.1444099

Su, T., Chen, Y., Liu, H., Gao, Y., Guo, J., Li, Y., Qi, Y., & Qiu, L. (2022). The biosynthesis of 1-octene-3-ol by a multifunctional fatty acid dioxygenase and Hydroperoxide Lyase in *Agaricus bisporus*. *Journal of Fungi, 8*(8), 827. https://doi.org/10.3390/jof8080827

Sun, H., Liu, X., Wang, L., Sang, Y., & Sun, J. (2022). Exploring the fungal community and its correlation with the physicochemical properties of chinese traditional fermented fish (Suanyu). *Food, 11*(12), 1721. https://doi.org/10.3390/foods11121721

Sun, X., Yu, L., Xiao, M., Zhang, C., Zhao, J., Narbad, A., Chen, W., Zhai, Q., & Tian, F. (2025). Exploring Core fermentation microorganisms, flavor compounds, and metabolic pathways in fermented Rice and wheat foods. *Food Chemistry, 463*, 141019. https://doi.org/10.1016/j.foodchem.2024.141019

Tan, Y., Du, H., Zhang, H., Fang, C., Jin, G., Chen, S., Wu, Q., Zhang, Y., Zhang, M., & Xu, Y. (2022). Geographically associated fungus-bacterium interactions contribute to the formation of geography-dependent flavor during high-complexity spontaneous fermentation. *Microbiology Spectrum, 10*(5), e0184422. https://doi.org/10.1128/spectrum.01844-22

Tian, L., Xu, P., Chen, J., Chen, H., Qin, J., Wu, X., Liu, C., He, Z., Liu, Y., & Guan, T. (2024a). Comprehensive analysis of spatial heterogeneity reveals the important role of the upper-layer fermented grains in the fermentation and flavor formation of Qingxiangxing baijiu. *Food Chemistry: X, 22*, 101508. https://doi.org/10.1016/J.FOCHX.2024.101508

Tian, M. B., Hu, R. Q., Liu, Z. L., Shi, N., Lu, H. C., Duan, C. Q., Wang, J., Sun, Y. F., Kong, Q. S., & He, F. (2024b). The pH adjustment of vitis amurensis dry red wine revealed the evolution of organic acids, volatomics, and sensory quality during winemaking. *Food Chemistry, 436*, 137730. https://doi.org/10.1016/J.FOODCHEM.2023.137730

Valera, M. J., Boido, E., Ramos, J. C., Manta, E., Radi, R., Dellacassa, E., & Carraua, F. (2020). The mandelate pathway, an alternative to the phenylalanine ammonia lyase pathway for the synthesis of benzenoids in ascomycete yeasts. *Applied and Environmental Microbiology, 86*(17), 701–720. https://doi.org/10.1128/AEM.00701-20

Villar, N., Pérez-Nevado, F., Andrés, A. I., García-Parra, J., Ramírez, M., Valdés, M. E., & Moreno, D. (2025). Influence of yeast inoculum (*Saccharomyces cerevisiae* and *Torulaspora delbrueckii*) on the production of rosé wines from high hydrostatic pressure-treated musts. *European Food Research and Technology, 251*, 467–482. https://doi.org/10.1007/s00217-024-04646-1

Wagner, T., Pfeifle, H., Hildebrand, G., & Zhang, Y. (2024). Production of a cheese-like aroma via fermentation of plant proteins and coconut oil with the basidiomycetes cyclocybe aegerita and trametes versicolor. *Journal of Agricultural and Food Chemistry, 72*(12), 6544–6553. https://doi.org/10.1021/acs.jafc.4c00219

Wang, H., Su, W., Mu, Y., & Zhao, C. (2021). Correlation between microbial diversity and volatile flavor compounds of suan zuo rou, a fermented meat product from Guizhou, China. *Frontiers in Microbiology, 12*, 736525. https://doi.org/10.3389/fmicb.2021.736525

Wang, Z., Yang, N., Zhang, J., & Yuan, C. (2022). Skin cell wall ripeness alters wine tannin profiles via modulating interaction with seed tannin during alcoholic fermentation. *Food Research International, 162*, 111974. https://doi.org/10.1016/j.foodres.2022.111974

Wang, W., Sun, B., Deng, J., & Ai, N. (2024). Addressing flavor challenges in reduced-fat dairy products: A review from the perspective of flavor compounds and their improvement strategies. *Food Research International, 188*, 114478. https://doi.org/10.1016/j.foodres.2024.114478

Wei, J., Lu, J., Nie, Y., Li, C., Du, H., & Xu, Y. (2023). Amino acids drive the deterministic assembly process of fungal community and affect the flavor metabolites in baijiu fermentation. *Microbiology Spectrum, 11*(2), e0264022. https://doi.org/10.1128/spectrum.02640-22

Wickramasinghe, P. C. K., & Munafo, J. P. (2020). Biosynthesis of benzylic derivatives in the fermentation broth of the edible mushroom, *Ischnoderma resinosum. Journal of Agricultural and Food Chemistry, 68*(8), 2485–2492. https://doi.org/10.1021/acs.jafc.9b07218

Xie, C., Yan, L., Gong, W., Zhu, Z., Tan, S., Chen, D., Hu, Z., & Peng, Y. (2016). Effects of different substrates on lignocellulosic enzyme expression, enzyme activity, substrate utilization and biological efficiency of *Pleurotus Eryngii. Cellular Physiology and Biochemistry, 39*(4), 1479–1494. https://doi.org/10.1159/000447851

Xu, X.-y., Wei, J.-b., Hou, R.-t., Li, M.-q., Wang, Y., Gao, X.-L., Liang, J., & Zhang, H.-w. (2025). Sequential inoculation of non-*Saccharomyces* yeasts and *Saccharomyces cerevisiae* modulates strawberry wine quality: Physicochemical, volatile and sensory analyses. *Applied Food Research, 5*(1), 100890. https://doi.org/10.1016/J.AFRES.2025.100890

Yan, L., Xu, R., Bian, Y., Li, H., & Zhou, Y. (2019). Expression profile of laccase gene family in white- rot basidiomycete *Lentinula edodes* under different environmental stresses. *Genes, 10*(12), 1045. https://doi.org/10.3390/genes10121045

Yang, X., Zhao, F., Yang, L., Li, J., & Zhu, X. (2022). Enhancement of the aroma in low-alcohol apple-blended pear wine mixed fermented with *Saccharomyces cerevisiae* and non-*Saccharomyces* yeasts. *LWT, 155*, 112994. https://doi.org/10.1016/J.LWT.2021.112994

Yang, S., Chen, R., Cao, X., Wang, G., & Zhou, Y. J. (2024). De novo biosynthesis of the hops bioactive flavonoid xanthohumol in yeast. *Nature Communications, 15*(1), 253. https://doi.org/10.1038/s41467-023-44654-5

Yepes-Betancur, D. P., Márquez-Cardozo, C. J., Cadena-Chamorro, E. M., Martinez-Saldarriaga, J., Torres-León, C., Ascacio-Valdes, A., & Aguilar, C. N. (2021). Solid-state fermentation—Assisted extraction of bioactive compounds from hass avocado seeds. *Food and Bioproducts Processing, 126*, 155–163. https://doi.org/10.1016/j.fbp.2020.10.012

Yin, C., Fan, X., Fan, Z., Shi, D., Yao, F., & Gao, H. (2019). Comparison of non-volatile and volatile flavor compounds in six pleurotus mushrooms. *Journal of the Science of Food and Agriculture, 99*(4), 1691–1699. https://doi.org/10.1002/jsfa.9358

Yu, Y., Zhou, Y., Wang, K., Sun, T., Lin, L., Ledesma-Amaro, R., & Ji, X.-J. (2023). Metabolic and process engineering for producing the peach-like aroma compound γ-decalactone in Yarrowia lipolytica. *Journal of Agricultural and Food Chemistry., 71*, 110–120. https://doi.org/10.1021/acs.jafc.2c07356

Zan, X., Xu, Y., Chen, G., Wang, L., Yang, J., Wen, Y., Sun, L., Cui, F., Sun, W., Song, Y., & Koffas, M. A. G. (2024). Biosynthesis of naringenin from *p* -coumaric acid in engineering oleaginous filamentous fungus *Mucor circinelloides*. *ACS Food Science & Technology, 4*(3), 747–756. https://doi.org/10.1021/acsfoodscitech.3c00645

Zeng, Z., Song, C., Hu, X., Zhu, X., Li, Y., Ren, J., Wang, Y., Yang, H., Huang, X., Zhao, M., Huang, R., & Jin, S. (2024). Constituent-taste relationship of kuding tea fermented by *Aspergillus neoniger* and *Aspergillus cristatus:* Unveiling taste characteristics through untargeted metabolomics. *Food Bioscience, 62*, 105027. https://doi.org/10.1016/j.fbio.2024.105027

Zhang, H., Wang, L., Wang, H., Yang, F., Chen, L., Hao, F., Lv, X., Du, H., & Xu, Y. (2021). Effects of initial temperature on microbial community succession rate and volatile flavors during Baijiu fermentation process. *Food Research International, 141*, 109887. https://doi.org/10.1016/J.FOODRES.2020.109887

Zhang, K., Zhang, T. T., Guo, R. R., Ye, Q., Zhao, H. L., & Huang, X. H. (2023a). The regulation of key flavor of traditional fermented food by microbial metabolism: A review. *Food Chemistry: X, 19*, 100871. https://doi.org/10.1016/J.FOCHX.2023.100871

Zhang, Z., Wang, B., & Cao, Y. (2023b). Factors influencing the formation of dimethyl disulfide and dimethyl trisulfide in model systems. *Food Research International, 172*, 113200. https://doi.org/10.1016/J.FOODRES.2023.113200

Zhang, R., Jia, W., Shu, J., Zou, L., & Shi, L. (2024a). Flavor experiences augmentation strategy for fermented dairy products: Perspective of multimodal perception, starter and enhancer, and processing. *Food Reviews International, 40*(8), 2227–2255. https://doi.org/10.1080/87559129.2023.2262031

Zhang, T., Liao, Z., Li, Z., Liu, Y., Liu, Y., Song, Y., & Qin, Y. (2024b). Dynamic changes in dissolved oxygen concentration, microbial communities, and volatile compounds during industrial oak-barrel fermentation of Sauvignon Blanc wine. *Food Research International, 197*, 115250. https://doi.org/10.1016/J.FOODRES.2024.115250

Zhou, M., Li, Y., Kudinha, T., Xu, Y., & Liu, Z. (2021). Kodamaea ohmeri as an emerging human pathogen: A review. *Frontiers in Microbiology, 12*, 736582. Frontiers Media S.A. https://doi.org/10.3389/fmicb.2021.736582

Chapter 8
Fungal Secondary Metabolites in Postharvest Disease Management: A Natural Way to Fight Against Food Loss and Waste

Aarzoo, Nidhi Mandloi, Samadrita Mandal, Abhimanyu Gaur, Vijay Kumar Gahlawat, and Bhim Pratap Singh

1 Introduction

As per the Food and Agricultural Organization (FAO), approximately 33% of food produced globally is lost or wasted. This loss is lesser in developed countries as compared to developing countries due to the availability of technological interventions and adequate infrastructure. Globally, this has become a major challenge in the agriculture sector (Sapna et al., 2024). Agricultural produce is rich in vitamins, minerals, and bioactive substances, with tremendous human health advantages which make them an important part of everyone's plate (Jiang et al., 2020; Cömert et al., 2020). Among all the food loss, fruit and vegetable loss accounts for at least 40% of total food loss produced globally each year (FAO, 2011; Singh & Sharma, 2018). FAO estimated around 14% of food produced in 2016 was lost already reaching the retail stage (FAO, 2020), with an economic value of roughly $400 billion USD (Surucu-Balci & Tuna, 2021). The market and shelf life of fruits and vegetables is influenced by temperature, humidity, atmospheric composition, and infection by microbes or insects (Droby, 2006). Food loss occurs throughout the supply chain, including agricultural production, storage, processing, transport, management, and consumption (Iordăchescu et al., 2019). Improper postharvest handling and storage can lead to significant losses in fruit and vegetable production, accounting for about 25% in industrialized countries and an additional 50% in developing countries (Nunes, 2012). During storage, shelf life decreases due to moisture loss, microbial

Aarzoo · N. Mandloi · S. Mandal · A. Gaur · B. P. Singh (✉)
Department of Agriculture and Environmental Sciences, National Institute of Food Technology Entrepreneurship and Management (NIFTEM), Kundli, Sonepat, Haryana, India
e-mail: bpsingh@niftem.ac.in

V. K. Gahlawat
Department of Interdisciplinary Sciences, National Institute of Food Technology Entrepreneurship and Management-Kundli, Sonepat, Haryana, India

B. P. Singh et al. (eds.), *Fungal Additives and Bioactives in Food Processing Industries*, Fungal Biology, https://doi.org/10.1007/978-3-032-04519-5_8

and rodent decomposition, energy loss, and mineral and vitamin depletion (Singh & Sharma, 2018). Pathogens can cause postharvest decay of fruit and vegetables during flowering or harvesting, as well as handling and storage (Droby, 2006). Fungi such as *Alternaria*, *Botrytis*, *Aspergillus*, *Colletotrichum*, *Penicillium*, *Diplodia*, *Monilinia*, *Phomopsis*, *Rhizopus*, *Mucor*, as well as bacteria such as *Erwinia* and *Pseudomonas*, are responsible for the mass postharvest losses (Sharma et al., 2009). Some of the major diseases in fruits and vegetables are caused by pathogens, as shown in (Table 8.1) which are responsible for postharvest losses.

Pathogens such as bacteria and fungi are liable for various diseases in fruits and vegetables (Sharma & Pongener, 2010). Most postharvest infections require a wound to penetrate the fruit cuticle, unlike field pathogens that can penetrate immediately. These wounds are typically generated during harvest, transportation, packinghouse, and storage activities (Barkai-Golan, 2001). Despite decades of efforts, the fight against the postharvest deterioration of fruits and vegetables remains ongoing. Consumers who buy fresh produce understand the issue of postharvest

Table 8.1 List of major pathogens causing diseases in fruits and vegetables

Name of the diseases	Pathogen	Affected fruits and vegetables	References
Gray mold	*B. cinerea* *P. infestans* *A. solani*	Tomato and onion	Wei et al. (2024)
Black spot	*Alternaria alternata*	Tomato, potato, and brinjal	Balodi et al. (2024)
Purple blotch Black mold	*Alternaria porri* *Aspergillus niger*	Onion	Ali et al. (2024) El-Dawy et al. (2024)
White and basal rot Bulb rots and leaf spots	*Sclerotium cepivorum* *Fusarium oxysporum* *Fusarium solani* *Fusarium oxysporum*, *Penicillium hirsutum dierckx*, and *Alternaria embellisia*	Garlic	Mohy et al. (2024) Anum et al. (2024)
Pink mold rot Ring rot	*Trichothecium roseum* *Botryosphaeria dothidea*	Apple	Dai et al. (2024) Huang et al. (2024)
Blue molds, green molds, and sour rot	*Penicillium* spp., *Geotrichum candidum*	Oranges	Wang et al. (2018) Jing et al. (2020)
Downy blight	*Peronophythora litchii*	Litchi	Xu et al. (2024)
Green mold	*Penicillium digitatum*	Tangerine	Boukaew et al. (2024)

degradation, as they frequently abandon spoilt items (Janisiewicz & Korsten, 2002). According to NABCONS, the latest postharvest losses between 2020 to 2022 for cereal losses varied from 3.89% to 5.92%, fruits from 6.02% to 15.05%, and vegetables from 4.87% to 11.61% (NABCONS, 2023).

Proper postharvest disease management is essential for producing a consistent and profitable food supply. Combining different methods for postharvest disease control can increase efficacy, but some of these methods are not compatible with the environment and health (Eshel et al., 2009). Physical treatments such as heat treatment and irradiation are found to be effective in managing postharvest diseases but can destroy the treated plant tissue and also affect the quality attributes (Hansen et al., 2004). Chemical treatments such as fungicides and pesticides have extended the shelf life of fruits and vegetables by lowering the loss, but concerns about pathogen-resistant populations, public health, and the environment have increased (Wills & Golding, 2016). These approaches are restrained due to their negative influence on human health and environment (Keswani et al., 2019). Biological control holds considerable promise for future postharvest disease management because it is safe for humans and the environment (Droby et al., 2009). Fungal secondary metabolites (SM) from species such as *Trichoderma* play a significant role in postharvest disease management through biological control (Yao et al., 2023). Fungal secondary metabolites have been useful in various areas because of their discrete structural properties (Shu et al., 2024). Fungal secondary metabolites such as alkaloids, polyketides, and non-ribosomal peptides act as a natural pesticide and fungicides, protecting the crop from pathogens and pests (Alam et al., 2021). Fungal secondary metabolites have biological activity that has led to the development of life-saving drugs and agrochemicals (Bills & Gloer, 2016). These metabolites promote sustainable agricultural practices by reducing dependency on fungicides and pesticides (Fan & Shi, 2024).

1.1 Secondary Metabolites and Their Function

According to classical definitions, secondary metabolites are compounds that are not important for the growth or survival of the producing organisms (Bills & Gloer, 2016). Nevertheless, secondary metabolites play an important function in plant growth regulation, antimicrobial, and health by acting as bioactive compounds (Alam et al., 2021). Secondary metabolite's primary purpose is a defence against herbivores and bacteria, but certain secondary metabolites serve as signal molecules to attract pollinating and seed-dispersing animals, or they work in symbiotic relationships with plants and microbes. Specific secondary metabolites are frequently limited to a single systematic unit, but isolated instances can arise in substantially dissimilar taxonomic groups (Wink, 2008). Fungi have the ability to produce a wide number of so-called secondary metabolites with numerous functions (Macheleidt et al., 2016).

2 Fungal Secondary Metabolites: An Overview

Microorganisms, mainly fungi, produce a wide range of bioactive molecules. Biomolecules that are produced from fungal taxa are termed metabolites, helping to promote growth and other physiological activities by organisms (Keller, 2019). Primary metabolites have a wide-ranging characteristic of secondary metabolites, but the line between them is blurry. Occasionally, primary metabolites behave like secondary metabolites, for example, oxalic acid (Bills & Gloer, 2016). Primary metabolites are the compounds that have important roles accompanying photosynthesis, respiration, and growth and development, such as amino acids, nucleotides, and organic acids (Crozier et al., 2006). Primary metabolites are physiological producers as organisms utilize them for their growth and reproduction. On the other side, secondary metabolites, also known as natural products, specialized metabolites, and secondary metabolites, are derived from metabolic pathways and show biological activity that contributes to the survival of organisms in ecological niches by the defence mechanism (Dashora et al., 2021). The mechanism of action has demonstrated biological control over postharvest pathogens by stashing secondary metabolites such as antibiotics. Both types of metabolites have the same intermediates and are produced via similar pathways. Secondary metabolites share the same structural complexity as primary metabolites (Dashora et al., 2021). Secondary metabolites are a diverse set of low-molecular-weight compounds synthesized by polyketide synthetases (PKSs), non-ribosomal peptide synthetases (NRPSs), prenyltransferases, and dimethylallyl tryptophan synthases. The great structural variety of these compounds explains their wide range of actions and functions (Macheleidt et al., 2016).

Most of the fungal biomass genome is keen to produce a comprehensive range of metabolites when they are triggered for the stress condition. This production leads to adaption in their niche and competition for resources or to form beneficial relationships with other organisms (Bills & Gloer, 2016). Based on this, the secondary metabolites are further divided into different classes.

2.1 *Polyketides*

Fungal secondary metabolites produced by endophytic fungi can increase resistance to stress and stimulate growth in plants. Polyketides constitute the amplest and heterogeneous secondary metabolites family and are biosynthesized by Claisen decarboxylation condensation reactions using short-chain Acyl-CoA, Propionyl-CoA, Malonyl-CoA, and Methylmalonyl-CoA. The natural compound of polyketide provides antifungal properties against invasive infections. Polyketides are structurally diverse and consist of four classes: aromatics, macrolides, polyenes, and polyether (Han et al., 2018). Amphotericin B has polyketide compounds and has antifungal activity against *Candida*, *Aspergillus*, and *Cryptococcus* species (Wang et al., 2023).

Recent studies and advancements in genome sequencing and industrial biotechnology have increased the discovery of polyketide pesticides as a novel compound. Almost 70% of the pesticides in the market are based on natural products such as polyketides. Polyketides have a diversity of 7000 chemical structures, which leads to the production of 20 commercially applicable drugs (Li et al., 2021).

2.2 *Terpenes*

Different types of terpenes are produced by insects, plants, bacteria, and fungi. Fungal secondary metabolites have large bioactive groups which make them antifungal and help in protecting plants from pathogenic microorganisms. Trichothecenes, carotenoids, gibberellins, indole-diterpenes, and aristolochene are some of the beneficial types of terpenes that produce antifungal metabolites (Dashora et al., 2021). These natural radical scavenging compounds are the largest class of metabolites in plants. Around 40,000 chemical structures in this group act against environmental damages and stresses; the double bond between chemical structures absorbs high UV light by scavenging them and functions as a defence mechanism. Secretory gland trichomes release these bioactive compounds, but not all the terpene bioactives have scavenging mechanisms, but then some of them act as pollinators. Terpenes such as Limonone, thymol terpineol, and linalool are some of the major groups that act as antifungal agents against *Aspergillus* spp. because of their lipophilic nature, which allows them to interact with cellular and mitochondrial membranes and cause alteration of cell permeability (Loi et al., 2020).

2.3 *Indole Alkaloids*

Indole alkaloids are produced by endophytic fungi and marine fungi. Endophytic fungi produce various categories of alkaloids, including indoles, which shows excellent properties of antimicrobial activity, and this interaction between secondary metabolites of endophytic fungi and plants offers resistance to microbial and environmental stresses (Rassbach et al., 2023). On the other hand, marine endophytic fungi with diverse biological activities and unique metabolic pathways produce many antifungal metabolites *C. globosum* is one of the species to make significant metabolites (Xu et al., 2021a). These secondary metabolites originate from the biosynthesis of the amino acid tryptophan and have a six-membered benzene ring with five-membered nitrogen-containing pyrrole rings; this unique bicyclic structure makes it unique and the largest in alkaloid (Hamid et al., 2017). In terms of postharvest loss management, indole alkaloids indicate their ability to inhibit pathogenic fungal growth and work as fungicides by acting on cell membrane integrity and inhibiting spore germination, causing the loss of function and cell death, and certain indole alkaloids act as fungicides on *Candida albicans* and *Penicillium notatum* and make it a valuable metabolite for postharvest disease management (Xu et al., 2021a).

2.4 Non-Ribosomal Peptides

Non-ribosomal peptide (NRP) are secondary metabolites produced via biosynthesis pathways by NRPSs, that contain properties to reduce postharvest losses during storage and transportation, as it has properties to combat plant pathogens and have medicinal properties (Elhamouly et al., 2022). Non-ribosomal polyketide-like biosurfactants have wide commercial applications as they are less toxic, environmentally friendly, and have the efficacy of biodegradation. Due to this property, these biosurfactants have applications in food and beverages. Biosurfactants such as rhamnolipid act as fungicides and are applied on vegetables, legumes, and fruit crops approved by the Food Drug Administration (FDA). Similarly, Cyclosporine A also has applications in agriculture by acting as an antifungal and immunosuppressive agent to control plant diseases (Martínez-Núñez et al., 2016).

3 Mechanism of Action of Secondary Metabolites in Postharvest Disease Management

The mechanism of action of this secondary metabolite indicates the biochemical interaction in which biomolecules interact with the pathogens and produce their effect against diseases caused in the crops, and these molecules specifically bind to the targeted receptor present on the surface (Enyiukwu et al., 2014). The formation of the microbial secondary metabolite from the primary metabolism branch occurs after trophophase during a stationary phase or idiophase. For the production of interesting metabolites, it is compulsory to provide essential alteration conditions during metabolic pathway regulation to get the maximum yield. The primary growth initiators help the production of secondary metabolites during substrate-limited conditions. This metabolite is not utilized by microorganisms for their own growth or metabolic activity but due to its chemical structure, it shows the activity as an antibacterial, antifungal, and antiviral agent in postharvest (Fig. 8.1). The complex chemical structure of metabolites makes them specific for nutrient availability such as nitrogen, carbon, and phosphorus (Nigam et al., 2014). *Gloeophyllum abietinum*, *Heterobasidion annosum*, *Stereum hirsutum*, *Coprinus micaceus/Coprinus disseminates*, *Aspergillus clavatonanicus*, *Pytium ultimum*, *Eutypa lata*, and *Botryosphaeria obtuse* are some of the fungi that produce polyketides as secondary metabolites and they act as antibacterial and antifungal agents (Xu et al., 2021b).

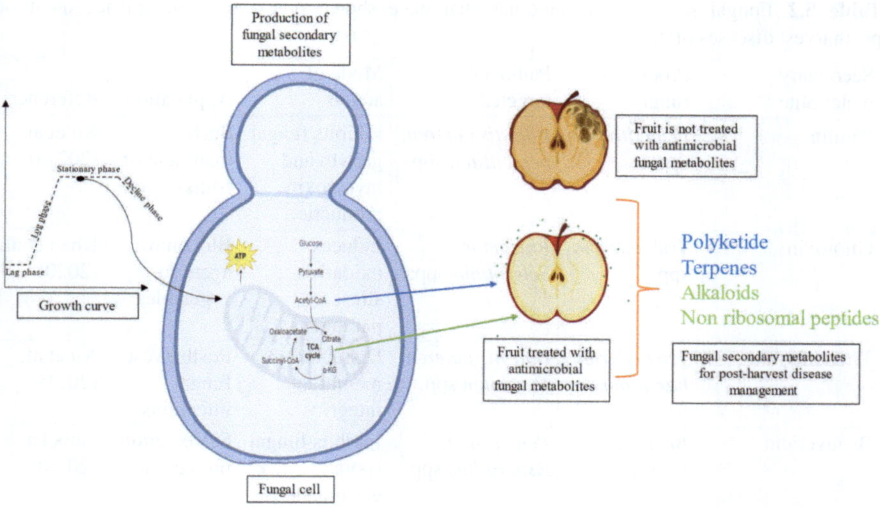

Fig. 8.1 Synthesis of fungal secondary metabolites during the stationary phase of microbial growth curve by the metabolic pathway and their action on apple post application. The potential biological properties of polyketides have a wide range of applications in postharvest disease management. This metabolite is synthesized from Acyl-CoA by a protein of fatty acid synthase that is a type I polyketide synthase. Fungal polyketide metabolites have the property to slow down physiological processes and reduce the ageing of products, antimicrobial activity, and antifungal agents. The precursor substrate is formed by short-chain Acetyl-CoA and Malonyl-CoA. The chain is further elongated by key components of PKs that are (KS), Acyl Transferase (AT), and Acyl Carrier Protein (ACP), where KS perform condensation of Acetyl-CoA and Malonyl-CoA and ACP anchor polyketide chain after getting Malonyl-CoA from AT. Polyketides of fungus repeat this reaction for condensation, then, after the completion of the required length, they get activated. This reaction pathway is concluded as the acetate pathway (Keller et al., 2005)

Terpenoids are odoriferous metabolites synthesized by fungi as well as by plants. Aristolochenes, carotenoids, gibberellins, indole-diterpenes, and trichothecenes are the most common terpenes from fungal groups containing linear or cyclic units of isoprene. The synthesis of this metabolite can be induced by the mevalonic acid (MVA) pathway or by methiylerythritol phosphate (MEP) pathway, where fungi use the MVA pathway for the synthesis of terpenes. The three groups of Acetyl-CoA are condensed and then one enzyme is attached by Claisen condensation. This condensed group gets attached to the aldol group to form 3-hydroxy-3-methylglutaryl-CoA and conversion of MVA into five carbon units (Dashora et al., 2021). The production of secondary metabolites by fungi-targeting pathogens in postharvest disease management with different applications is shown in Table 8.2.

Table 8.2 Fungal secondary metabolites that have shown potential in the management of postharvest diseases of crops

Secondary metabolite	Producing fungi	Pathogen targeted	Mode of action	Application	Reference
Patulin	*Penicillium* spp.	*Botrytis cinerea*, *penicillium* spp.	Inhibits fungal growth and mycotoxin production	Surface treatment of fruits	Xu et al. (2021a)
Gliotoxin	*Trichoderma* spp.	*Rhizoctonia*, *Sclerotinia* spp.	Induces oxidative stress in pathogens	Biocontrol agent for vegetables	Khan et al. (2020)
Trichodermin	*Trichoderma harzianum*	*Botrytis cinerea*, *Fusarium* spp.	Disrupts membrane integrity	Postharvest fungal infections	Xu et al. (2021b)
Beauvericin	*Beauveria bassiana*	*Penicillium*, *Aspergillus* spp.	Inhibits fungal spore germination	Stored grain protection	Patocka (2016)
Iturin A	*Bacillus subtilis*	*Aspergillus niger*, *Fusarium* spp.	Disrupts membrane integrity	Spray treatment for fruits	Fira et al. (2018)
Fusaric acid	*Fusarium* spp.	*Verticillium*, *Phytophthora* spp.	Inhibits cell wall synthesis	Control of root and stem rot	Ghule et al. (2022)
Citrinin	*Penicillium citrinum*	*Aspergillus niger*, *Fusarium* spp.	Inhibits DNA replication	Postharvest fruit treatment	Godana et al. (2023)
Chaetoglobosin A	*Chaetomium globosum*	*Botrytis*, *Rhizoctonia* spp.	Disrupts fungal cytoskeleton	Biocontrol of fruit pathogens	Xu et al. (2021b)
Lovastatin	*Aspergillus terreus*	*Fusarium*, *Rhizoctonia* spp.	Inhibits sterol cytoskeleton	Stored grain protection	Sawant et al. (2022)
Cyclopentanone, cinnamic acid, simvastatin	*Penicillium italicum*	*Colletotrichum*, *Botrytis* spp.	Antifungal via metabolic disruption	Fruit coating and sprays	Li et al. (2024)
Zearalenone	*Fusarium* spp.	*Fusarium Alternaria* spp.	Disrupts hormone metabolism in fungi	Cereal grain storage	Yazar and Omurtag (2008)

4 Importance of Fungal Secondary Metabolites in Disease Management

Crop diseases due to pathogens cause major global food loss, which is more than 20% of the total crop yield (Kim et al., 2004). So, chemical control for these pathogens is widely adopted in agriculture. The exclusive application of the chemicals on crop targeted the wrong organisms and their undesirable toxic effect on the environment causing pollution are subject to risk for use. Based on this issue, antifungal bioactive compounds from endophytic fungi have been analysed. The endophytic fungus has complex characteristics like mutualism protects the host from insects and pathogens. A large number of promising antifungal, antibacterial, insecticidal, and herbicidal secondary metabolites such as terpenoids, alkaloids, polyketides, benzopyranones, quinones, non-ribosomal peptides, applicable for postharvest come from the endophytic fungus group (Xu et al., 2021b). A wide range of endophytic fungi show symbiotic relationships while providing positive benefits to host cells by producing secondary metabolites and gaining advantages simultaneously (Alam et al., 2021). These metabolites act as biocontrol agents to manage postharvest diseases. For example, the Trichoderma species of fungus has shown a suppression effect on phytopathogens and their toxigenic metabolites and has shown an effect on crop productivity and health. Fungus-like *Alternaria alternata*, *Alternaria brassicicola*, and *Aspergillus fumigate* have antifungal and antibacterial properties, making them applicable for postharvest losses caused by pathogens (Elhamouly et al., 2022). The secondary metabolite of *Bacillus* spp. is classified as non-VOCs and VOCs (Volatile Organic Compound) based on the chemical and physical structure and properties. VOCs are produced by 75 foodborne *Bacillus* spp. and significantly help in postharvest disease management by acting as microbial antagonists on pathogen mycelium and increasing resistance in plants against pathogens. VOCs of *Bacillus* spp. make it a good biocontrol agent due to its broad spectrum as an antifungal agent, non-contact, and fast degradation for postharvest disease management of fruits. These VOCs show antimicrobial properties against pathogenic fungi that cause disease in fruits (Ling et al., 2023).

5 Secondary Metabolites Versus Chemical Methods

Secondary metabolites, produced by plants or microorganisms, offer a promising alternative to chemical methods in managing postharvest diseases caused by fungal pathogens. Unlike synthetic chemicals, secondary metabolites are biodegradable and environmentally friendly, reducing the risk of chemical residues on produce and environmental contamination (Barkai-Golan, 2001). This aligns with the increasing consumer demand for organic and sustainable agricultural practices. Secondary metabolites such as alkaloids, phenolics, and terpenoids often exhibit strong antifungal properties, effectively inhibiting the growth of pathogens such as *Botrytis cinerea*, *Penicillium* spp., and *Aspergillus* spp. (Tripathi & Dubey, 2004). For instance, phenolic compounds from plant extracts have been shown to disrupt fungal cell walls and membranes, leading to pathogen death (Mandal et al., 2009). These natural compounds can also work in synergy with existing storage techniques to enhance protection against decay. Another advantage of secondary metabolites is their specificity and reduced likelihood of resistance development in fungal pathogens, which is a common problem with repeated use of synthetic fungicides (Palou et al., 2008). Besides, they can enhance the plant's innate immune responses, contributing to systemic acquired resistance (Tzortzakis & Economakis, 2007). In addition, secondary metabolites often increase the shelf life and quality of fruits and vegetables. For example, essential oils containing terpenes have been shown to reduce fungal infections while maintaining the sensory qualities of produce, such as aroma and flavour (Tripathi & Dubey, 2004). This dual functionality provides an edge over chemical fungicides that may negatively impact the taste and appearance of treated produce. Both plant-derived secondary metabolites and chemical fungicides are utilized in postharvest disease management to control fungal pathogens that degrade the quality and shelf life of fruits, vegetables, and other crops. However, these two approaches differ significantly in terms of their mechanisms of action, environmental impact, resistance potential, and sustainability. The comparison of the two methods emphasizes their roles, advantages, and limitations, as shown in Table 8.3. Postharvest losses can be minimized by using modern storage facilities but losses are also influenced by temperature, genotypes, fruit age, and many other factors (Solanki et al., 2024). To overcome the postharvest losses, several farmers and exporting corporations have chosen to apply chemical fungicides in packhouses or postharvest facilities within 48 hours of fruit picking (Smilanick et al., 2006). The application of chemical fungicides aimed to reduce postharvest losses and disease-causing agents, but it aggravated the issue during pre or postharvest storage (Tayel et al., 2016). Postharvest spoiling management is getting difficult as pesticides or chemicals become scarce and consumers become more concerned about food safety (Tripathi et al., 2022).

6 Current Approaches to Control Postharvest Diseases

Till 2014, almost 170,000 secondary metabolites have been identified from which around 22,500 secondary metabolites are created by microorganisms and from which 38% come from fungus due to its varied range of biological activities, counting antimicrobial, antioxidant and antiphrastic properties, which make them function as nematicides, pesticides, molluscicides, and plant growth promoters (Abdel-Wareth, 2021). Among all the microorganisms, fungi play a vital role as biocontrol agents against postharvest pathogens. The biocontrol property of 50 species of fungi kills nematodes by acting against female cysts by developing a sticky network from mycelium, then fungi start digesting the tissues of trapped nematodes (Qadri et al., 2020). The diverse *Trichoderma* fungus has a wide variety of secondary metabolites such as polyketides, terpenes, and diketopiperazine, demonstrating its antibiotic abilities in biocontrol applications at low concentrations (Hermosa et al., 2014). *Trichoderma* fungi have potential effects as biocontrol agents on postharvest pathogens and play a crucial role in the agriculture sector as a pest management component. The antagonistic effect of *Trichoderma* fungi on soil-borne pathogens, as well as on nematodes and the environment-friendly application of this species, make it effective for postharvest management in agriculture (Sahebani & Hadavi, 2008). Production of biomass is done by solid- and liquid-state

Table 8.3 Comparison of secondary metabolites with chemical fungicides in different aspects

Aspect	Secondary metabolites	Chemical fungicides	References
Mode of action	Induce systemic acquired resistance (SAR), antifungal properties through VOCs, and direct inhibition of pathogen growth	Directly inhibit fungal growth by disrupting cell wall synthesis, metabolic processes, or spore germination	Isah (2019) and Wadhwa et al. (2024)
Applications	Applied through vapor-phase application, sprays, or packaging materials. Enhance plant health and prevent pathogen colonization	Applied via sprays, dips, or coatings to prevent fungal spoilage during storage and transportation	Aelenei et al. (2016)
Environmental impact	Biodegradable, non-toxic to non-target organisms, environmentally friendly, and sustainable	Toxic to non-target organisms, persistent in the environment, and can contaminate soil, air, and water	Wadhwa et al. (2024)
Efficacy	Effective in inducing resistance and reducing pathogen growth, but dependent on environmental conditions and pathogen type	Quick and effective control of fungal diseases with reliable results	Kumar et al. (2021)
Resistance issues	Pathogens cannot directly evolve resistance to secondary metabolites, making them less prone to resistance development	Repeated use can lead to resistance in pathogens, requiring new formulations to manage resistant strains	Zaynab et al. (2018) and Gressel (2011)
Sustainability	Sustainable, eco-friendly alternative to chemical fungicides, ideal for integrated pest management (IPM) systems	Less sustainable due to environmental concerns and toxicity; dependent on continuous use and new formulations	Wu et al. (2023) and Fenibo et al. (2021)
Use in postharvest management	Used to enhance shelf life, reduce pathogen colonization, and improve fruit and vegetable freshness	Widely used in commercial operations for quick disease management and long-term storage control	Ayilara et al. (2023) and Vischetti et al. (2024)
Broad-spectrum activity	Effective against a wide range of pathogens and pests, including fungi, bacteria, and viruses	Primarily effective against fungal pathogens, with limited activity against bacteria and viruses	Bennett and Wallsgrove (1994) and Zhao et al. (2022)
Toxicity to humans and animals	Generally non-toxic to humans and animals due to their natural origin	Can be toxic to humans, animals, and beneficial organisms if not used correctly	Daraban et al. (2023) and Ahmad et al. (2024)
Cost and accessibility	Can be cost-effective for small-scale operations; however, production may require specialized processes	Widely available, relatively low cost, and commonly used in large-scale commercial agriculture	Ayilara et al. (2023) and Malinga and Laing (2023)

fermentation, and further, this formulation is applied as a bio-priming agent against soil-borne pathogens such as *Pythium*, *Phytophthora*, *Rhizoctonia*, *Fusarium*, and *Sclerotium* spp. in several crops. The life and activity of biocontrol agents depend on physical conditions like temperature; Trichoderma virens show their viability at 5 °C temperature in the refrigerator for 6 months and 3 months at room temperature. Mostly, the interaction between host and pathogen creates a competition for nutrition requirements and space to grow, leading to antibiosis, competition, mycoparasitism, cell wall degradation by enzymes, and induced resistance. Plant pathogens such as *Sclerotium rolfsii*, *Rhizoctonia solani*, and *Fusarium solani* cause postharvest losses in apple beans and groundnuts, which are controlled by *Trichoderma* spp. and *Gliocladium* spp. by seed coating. Nano-particle application of *Trichoderma asperellum* has also shown antifungal properties against *Rhizoctonia solani* in comparison to chemical pesticides (Bahadur & Dutta, 2022). Peptaibiotic antimicrobial polypeptides show the activity as antibacterial, antifungal and antiviral and shows resistance towards phytopathogen and become a good sustainable alternative in postharvest. It is derived from fungi and shows a beneficial effect in agriculture by promoting the growth and inhibiting the growth of pathogens such as fungi (Pereira-Dias et al., 2023). The collective pathogens that are assumed to be controlled are *E. coli*, *Staphylococcus aureus*, *Campylobacter*, *Salmonella*, *Clostridium perfringens*, and *Bacillus cereus* by affecting their growth kinetics. Fungal secondary metabolites also play a major role in food packaging due to their wide properties. The inclusion of fungal metabolites and essential oils is done by physical application by blending or chemical immobilization to obtain the antimicrobial action on the surface of packaged harvest (Moeini et al., 2021).

7 Environmental and Health Impact of Postharvest Diseases

Food loss and waste have a substantial influence on society, the economy, and the environment, and lowering it has the capacity to increase food security and promote food system sustainability in conjunction with the food supply chain (Ali et al., 2021). The postharvest diseases result in qualitative and quantitative losses of produce, rendering them unsuitable for human utilization due to potential health risks (Klich, 2007). Quantitative losses in mass (weight) and volume diminish the number of products accessible for consumption. Qualitative losses include a drop in nutritional or energy content (calories), and deterioration of freshness (wilting, yellowing, colour change, and meagre visual appearance) (Chaboud & Daviron, 2017). Agriculture accounts for more than 26% of global Green House Gas (GHG) emissions and 70% of global freshwater use, occupying half of the world's geographical area. Food loss and waste have multiple downstream and upstream environmental consequences, and preventative initiatives may have better environmental paybacks than recovery (Kummu et al., 2012). Reducing agriculture's environmental impact is a key step towards creating a circular economy (Vilariño et al., 2017). A major contribution to improving food security and proper utilization of resources can be made by decreasing food losses (Teferra, 2022). *Aspergillus flavus* and *Aspergillus ochraceus* are saprophytic soil fungi that infect postharvest vegetables and produce the carcinogenic metabolite aflatoxin and ochratoxin A (Ling et al., 2024). They also cause disorders such as aflatoxicosis or liver cancer in animals and human beings by intake of contaminated food and feed, as well as aggressive growth (aspergillosis), which is typically deadly to weak immune human beings (Hedayati et al., 2007; Tripathi et al., 2022). Food losses are most likely a reflection of food supply chain weaknesses, which require a venture to expand production practices and postharvest processing to attain the intended sustainable food arrangement (Poore & Nemecek, 2018).

8 Conclusions and Future Perspectives

Postharvest diseases considerably contribute to food loss and waste worldwide, threatening food security and economic stability. Traditional methods, such as using chemical fungicides and pesticides, are effective but, in the long run, result in disease resistance, environmental degradation, and potential health hazards. To deal with this constant problem, fungal secondary metabolites have been developed as an environmentally and eco-friendly option for postharvest disease management. These bioactive compounds exhibit diverse antifungal, antibacterial, and antioxidant properties produced by various endophytic fungi such as *Trichoderma, Penicillium, Aspergillus*, and *Fusarium* species. Patulin, gliotoxin, and trichodermin are key secondary metabolites that damage cell membranes, inhibit spore germination, and interfere with fungal metabolism. The mode of action is disrupting cell walls and membranes to initiate oxidative stress in microorganisms. Furthermore, these metabolites increase plant resistance, making them appropriate for IPM strategies. Future studies should concentrate on optimizing the production, formulation, and application of fungal secondary metabolites to improve their efficacy, stability, and scalability in real-world agricultural contexts. These natural compounds offer a viable alternative for minimizing food loss and waste while minimizing the environmental effects of chemical-based postharvest disease-control methods. In conclusion, fungal secondary metabolites offer a potential, environmentally friendly approach for postharvest disease management. Integrating these natural compounds into existing agricultural practices can significantly reduce food waste, enhance food security, and promote sustainable farming systems. To achieve their potential and contribute to global efforts towards sustainable agriculture and food preservation, they must be implemented with ongoing research, funding, and collaboration across scientific fields.

Acknowledgements The authors are grateful to the National Institute of Food Technology Entrepreneurship and Management, Kundli, India, for providing all necessary facilities.

Conflict of Interest The authors confirm that they have no conflict of interest to declare for this publication.

References

Abdel-Wareth, M. T. A. (2021). Fungal secondary metabolites: Current research, commercial aspects, and applications. In A. M. Abdel-Azeem, A. N. Yadav, N. Yadav, & M. Sharma (Eds.), *Industrially important fungi for sustainable development: Volume 2: Bioprospecting for biomolecules* (pp. 299–346). Springer International Publishing. https://doi.org/10.1007/978-3-030-85603-8_8

Aelenei, P., Miron, A., Trifan, A., Bujor, A., Gille, E., & Aprotosoaie, A. C. (2016). Essential oils and their components as modulators of antibiotic activity against gram-negative bacteria. *Medicine, 3*(3), Article 3. https://doi.org/10.3390/medicines3030019

Ahmad, M. F., Ahmad, F. A., Alsayegh, A. A., Zeyaullah, M., AlShahrani, A. M., Muzammil, K., Saati, A. A., Wahab, S., Elbendary, E. Y., Kambal, N., Abdelrahman, M. H., & Hussain, S. (2024). Pesticides impacts on human health and the environment with their mechanisms of action and possible countermeasures. *Heliyon, 10*(7). https://doi.org/10.1016/j.heliyon.2024.e29128

Alam, B., Lǐ, J., Gě, Q., Khan, M. A., Gōng, J., Mehmood, S., Yuán, Y., & Gǒng, W. (2021). Endophytic fungi: From Symbiosis to secondary metabolite communications or vice versa? *Frontiers in Plant Science, 12*. https://doi.org/10.3389/fpls.2021.791033

Ali, A., Xia, C., Ouattara, N., Mahmood, I., & Faisal, M. (2021). Economic and environmental consequences' of postharvest loss across food supply chain in the developing countries. *Journal of Cleaner Production, 323*, 129146. https://doi.org/10.1016/j.jclepro.2021.129146

Ali, A., Ölmez, F., Zeshan, M. A., Mubeen, M., Iftikhar, Y., Sajid, A., Abid, M., Kumar, A., Divvela, P. K., & Solanki, M. K. (2024). Yeast-based solutions in controlling plant pathogens. *Biocatalysis and Agricultural Biotechnology, 58*, 103199. https://doi.org/10.1016/j.bcab.2024.103199

Anum, H., Tong, Y., & Cheng, R. (2024). Different Preharvest diseases in garlic and their eco-friendly management strategies. *Plants, 13*(2), Article 2. https://doi.org/10.3390/plants13020267

Ayilara, M. S., Adeleke, B. S., Akinola, S. A., Fayose, C. A., Adeyemi, U. T., Gbadegesin, L. A., Omole, R. K., Johnson, R. M., Uthman, Q. O., & Babalola, O. O. (2023). Biopesticides as a promising alternative to synthetic pesticides: A case for microbial pesticides, phytopesticides, and nanobiopesticides. *Frontiers in Microbiology, 14*. https://doi.org/10.3389/fmicb.2023.1040901

Bahadur, A., & Dutta, P. (2022). Trchoderma spp.: Their impact in crops diseases management. In *Trichoderma-technology and uses*. IntechOpen. https://doi.org/10.5772/intechopen.101846

Balodi, R., Raghavendra, K. V., Singh, P. K., Hussain, Z., Suroshe, S. S., Kumar, P., & Chander, S. (2024). Alternaria alternata species complex impairing solanaceous vegetables in Northern parts of India: An emerging problem in Solanum lycopersicum L. *3 Biotech, 14*(10). Scopus. https://doi.org/10.1007/s13205-024-04059-7.

Barkai-Golan, R. (2001). *Postharvest diseases of fruits and vegetables: Development and control*. Elsevier.

Bennett, R. N., & Wallsgrove, R. M. (1994). Secondary metabolites in plant defence mechanisms. *New Phytologist, 127*(4), 617–633. https://doi.org/10.1111/j.1469-8137.1994.tb02968.x

Bills, G. F., & Gloer, J. B. (2016). Biologically active secondary metabolites from the fungi. *Microbiology Spectrum, 4*(6), 4.6.01. https://doi.org/10.1128/microbiolspec.FUNK-0009-2016

Boukaew, S., Yossan, S., Petlamul, W., Nooprom, K., Sakphet, N., Chumkaew, K., Srinuanpan, S., & Zhang, Z. (2024). Postharvest biocontrol ability and involved mechanism of volatile organic compounds from *Serratia nematodiphila* BC-SKRU-1 against *Penicillium digitatum* tangerine fruit. *Food Control, 166*, 110735. https://doi.org/10.1016/j.foodcont.2024.110735

Chaboud, G., & Daviron, B. (2017). Food losses and waste: Navigating the inconsistencies. *Global Food Security, 12*, 1–7. https://doi.org/10.1016/j.gfs.2016.11.004

Cömert, E. D., Mogol, B. A., & Gökmen, V. (2020). Relationship between color and antioxidant capacity of fruits and vegetables. *Current Research in Food Science, 2*, 1–10. https://doi.org/10.1016/j.crfs.2019.11.001

Crozier, A., Clifford, M. N., & Ashihara, H. (2006). Plant secondary metabolites. In *Occurrence, structure and role in the human diet*. Blackwell-Publishers. https://doi.org/10.1002/9780470988558

Dai, P., Li, N., Li, B., Wang, S., Wang, Y., Meng, X., Li, B., Cao, K., & Hu, T. (2024). An endophytic strain *bacillus velezensis* JZ51 controlled pink mold rot of postharvest apple fruit *via* antagonistic action and disease resistance induction. *Postharvest Biology and Technology, 210*, 112793. https://doi.org/10.1016/j.postharvbio.2024.112793

Daraban, G. M., Hlihor, R.-M., & Suteu, D. (2023). Pesticides vs. biopesticides: From Pest management to toxicity and impacts on the environment and human health. *Toxics, 11*(12), 983. https://doi.org/10.3390/toxics11120983

Dashora, K., Javed, Z., Mishra, M., & Tripathi, G. D. (2021). Role of fungal secondary metabolites in plant protection. In *New and future developments in microbial biotechnology and bioengineering* (pp. 39–46). Elsevier. https://doi.org/10.1016/B978-0-12-821005-5.00003-X

Droby, S. (2006). Biological control of postharvest diseases of fruits and vegetables: Difficulties and challenges. *Phytopathologia Polonica, 39*, 105–117.

Droby, S., Wisniewski, M., Macarisin, D., & Wilson, C. (2009). Twenty years of postharvest biocontrol research: Is it time for a new paradigm? *Postharvest Biology and Technology, 52*(2), 137–145. https://doi.org/10.1016/j.postharvbio.2008.11.009

El-Dawy, E. G. A. M., Hussein, M. A., & El-Nahas, S. (2024). Description and management of aspergillus section Nigri causing post-harvest bulbs rot of onion. *Scientific Reports, 14*(1), 6076. https://doi.org/10.1038/s41598-024-53849-9

Elhamouly, N. A., Hewedy, O. A., Zaitoon, A., Miraples, A., Elshorbagy, O. T., Hussien, S., El-Tahan, A., & Peng, D. (2022). The hidden power of secondary metabolites in plant-fungi interactions and sustainable phytoremediation. *Frontiers in Plant Science, 13*. https://doi.org/10.3389/fpls.2022.1044896

Enyiukwu, D. N., Awurum, A. N., Ononuju, C. C., & Nwaneri, J. A. (2014). Significance of characterization of secondary metabolites from extracts of higher plants in plant disease management. *International Journal of Advance Agricultural Research, 2*, 8–28.

Eshel, D., Regev, R., Orenstein, J., Droby, S., & Gan-Mor, S. (2009). Combining physical, chemical and biological methods for synergistic control of postharvest diseases: A case study of black root rot of carrot. *Postharvest Biology and Technology, 54*(1), 48–52. https://doi.org/10.1016/j.postharvbio.2009.04.011

Fan, Y., & Shi, B. (2024). Endophytic fungi from the four staple crops and their secondary metabolites. *International Journal of Molecular Sciences, 25*(11), Article 11. https://doi.org/10.3390/ijms25116057

FAO, G. (2011). Global food losses and food waste–extent, causes and prevention. *Save food: An initiative on food loss and waste reduction, 9*, 2011. https://www.fao.org/sustainable-food-value-chains/library/details/en/c/266053/. Last accessed: 1 Jan 2025.

FAO. (2020). *Agriculture at a crossroads. Finding and recommendations for future farming*. https://shorturl.at/b5dSy. Last accessed: 15 Oct 2024.

Fenibo, E. O., Ijoma, G. N., & Matambo, T. (2021). Biopesticides in sustainable agriculture: A critical sustainable development driver governed by green chemistry principles. *Frontiers in Sustainable Food Systems, 5*. https://doi.org/10.3389/fsufs.2021.619058

Fira, D., Dimkić, I., Berić, T., Lozo, J., & Stanković, S. (2018). Biological control of plant pathogens by *bacillus* species. *Journal of Biotechnology, 285*, 44–55. https://doi.org/10.1016/j.jbiotec.2018.07.044

Ghule, M. R., Sawant, I. S., Oulkar, D., Hingmire, S., Shabeer, A., & Holkar, S. (2022). Identification of secondary metabolites in mycoparasites Fusarium strains and antifungal activity of fusaric acid against Plasmopara viticola. *Archives of Phytopathology and Plant Protection, 55*(11), 1283–1297. https://doi.org/10.1080/03235408.2021.1973848

Godana, E. A., Yang, Q., Zhang, X., Zhao, L., Wang, K., Dhanasekaran, S., Mehari, T. G., & Zhang, H. (2023). Biotechnological and biocontrol approaches for mitigating postharvest diseases caused by fungal pathogens and their mycotoxins in fruits: A review. *Journal of Agricultural and Food Chemistry, 71*(46), 17584–17596. https://doi.org/10.1021/acs.jafc.3c06448

Gressel, J. (2011). Low pesticide rates may hasten the evolution of resistance by increasing mutation frequencies. *Pest Management Science, 67*(3), 253–257. https://doi.org/10.1002/ps.2071

Hamid, H. A., Ramli, A. N. M., & Yusoff, M. M. (2017). Indole alkaloids from plants as potential leads for antidepressant drugs: A mini review. *Frontiers in Pharmacology, 8*. https://doi.org/10.3389/fphar.2017.00096

Han, J. W., Choi, G. J., & Kim, B. S. (2018). Antimicrobial aromatic polyketides: A review of their antimicrobial properties and potential use in plant disease control. *World Journal of Microbiology and Biotechnology, 34*(11), 163. https://doi.org/10.1007/s11274-018-2546-0

Hansen, J. D., Wang, S., & Tang, J. (2004). A cumulated lethal time model to evaluate efficacy of heat treatments for codling moth *Cydia pomonella* (L.) (Lepidoptera: Tortricidae) in cherries. *Postharvest Biology and Technology, 33*(3), 309–317. https://doi.org/10.1016/j.postharvbio.2004.03.004

Hedayati, M. T., Pasqualotto, A. C., Warn, P. A., Bowyer, P., & Denning, D. W. (2007). Aspergillus flavus: Human pathogen, allergen and mycotoxin producer. *Microbiology, 153*(6), 1677–1692. https://doi.org/10.1099/mic.0.2007/007641-0

Hermosa, R., Cardoza, R. E., Rubio, M. B., Gutiérrez, S., & Monte, E. (2014). Secondary metabolism and antimicrobial metabolites of *Trichoderma*. In V. K. Gupta, M. Schmoll, A. Herrera-Estrella, R. S. Upadhyay, I. Druzhinina, & M. G. Tuohy (Eds.), *Biotechnology and biology of Trichoderma* (pp. 125–137). Elsevier. https://doi.org/10.1016/B978-0-444-59576-8.00010-2

Huang, Y., Li, J., Shan, X., Wang, H., & Duan, Y. (2024). Bioactivities evaluation of an endophytic bacterial strain *bacillus tequilensis* QNF2 inhibiting apple ring rot caused by *Botryosphaeria dothidea* on postharvest apple fruits. *Food Microbiology, 123*, 104590. https://doi.org/10.1016/j.fm.2024.104590

Iordăchescu, G., Ploscutanu, G., Pricop, E. M., Baston, O., & Barna, O. (2019). Postharvest losses in transportation and storage for fresh fruits and vegetables sector. *Journal of International Scientific Publications, 7*, 244–251.

Isah, T. (2019). Stress and defense responses in plant secondary metabolites production. *Biological Research, 52*. https://doi.org/10.1186/s40659-019-0246-3

Janisiewicz, W. J., & Korsten, L. (2002). Biological control of postharvest diseases of fruits. *Annual Review of Phytopathology, 40*(1), 411–441. https://doi.org/10.1146/annurev.phyto.40.120401.130158

Jiang, Q., Zhang, M., & Xu, B. (2020). Application of ultrasonic technology in postharvested fruits and vegetables storage: A review. *Ultrasonics Sonochemistry, 69*, 105261. https://doi.org/10.1016/j.ultsonch.2020.105261

Jing, J., Zhang, H., Xue, Y., & Zeng, K. (2020). Effects of INA on postharvest blue and green molds and anthracnose decay in citrus fruit. *Journal of Integrative Agriculture, 19*(5), 1396–1406. https://doi.org/10.1016/S2095-3119(20)63169-0

Keller, N. P. (2019). Fungal secondary metabolism: Regulation, function and drug discovery. *Nature Reviews Microbiology, 17*(3), 167–180. https://doi.org/10.1038/s41579-018-0121-1

Keller, N. P., Turner, G., & Bennett, J. W. (2005). Fungal secondary metabolism—From biochemistry to genomics. *Nature Reviews Microbiology, 3*(12), 937–947. https://doi.org/10.1038/nrmicro1286

Keswani, C., Singh, H. B., Hermosa, R., García-Estrada, C., Caradus, J., He, Y.-W., Mezaache-Aichour, S., Glare, T. R., Borriss, R., Vinale, F., & Sansinenea, E. (2019). Antimicrobial secondary metabolites from agriculturally important fungi as next biocontrol agents. *Applied Microbiology and Biotechnology, 103*(23), 9287–9303. https://doi.org/10.1007/s00253-019-10209-2

Khan, R. A. A., Najeeb, S., Hussain, S., Xie, B., & Li, Y. (2020). Bioactive secondary metabolites from Trichoderma spp. against phytopathogenic fungi. *Microorganisms, 8*(6), Article 6. https://doi.org/10.3390/microorganisms8060817

Kim, Y.-M., Lee, C.-H., Kim, H.-G., & Lee, H.-S. (2004). Anthraquinones isolated from *Cassia tora* (Leguminosae) seed show an antifungal property against Phytopathogenic fungi. *Journal of Agricultural and Food Chemistry, 52*(20), 6096–6100. https://doi.org/10.1021/jf049379p

Klich, M. A. (2007). Aspergillus flavus: The major producer of aflatoxin. *Molecular Plant Pathology, 8*(6), 713–722. https://doi.org/10.1111/j.1364-3703.2007.00436.x

Kumar, J., Ramlal, A., Mallick, D., & Mishra, V. (2021). An overview of some biopesticides and their importance in plant protection for commercial acceptance. *Plants, 10*(6), Article 6. https://doi.org/10.3390/plants10061185

Kummu, M., de Moel, H., Porkka, M., Siebert, S., Varis, O., & Ward, P. J. (2012). Lost food, wasted resources: Global food supply chain losses and their impacts on freshwater, cropland, and fertiliser use. *Science of the Total Environment, 438*, 477–489. https://doi.org/10.1016/j.scitotenv.2012.08.092

Li, S., Yang, B., Tan, G.-Y., Ouyang, L.-M., Qiu, S., Wang, W., Xiang, W., & Zhang, L. (2021). Polyketide pesticides from actinomycetes. *Current Opinion in Biotechnology, 69*, 299–307. https://doi.org/10.1016/j.copbio.2021.05.006

Li, S., Hu, J., Ning, S., Li, W., Jiang, R., Huang, J., & Li, Y. (2024). *Bacillus velezensis* HY19 as a sustainable preservative in post-harvest citrus (*Citrus reticulata Blanco* L.) fruit management. *Food Control, 155*, 110068. https://doi.org/10.1016/j.foodcont.2023.110068

Ling, L., Li, Y., Jiang, K., Wang, Y., Luo, H., Cheng, W., Pang, M., Feng, L., Yue, R., & Zhou, Y. (2023). Volatile organic compounds of *bacillus* spp. as an emerging antifungal resource play a significant role in fruit postharvest disease control. *Food Bioscience, 56*, 103201. https://doi.org/10.1016/j.fbio.2023.103201

Ling, L., Feng, L., Li, Y., Yue, R., Wang, Y., & Zhou, Y. (2024). Endophytic fungi volatile organic compounds as crucial biocontrol agents used for controlling fruit and vegetable postharvest diseases. *Journal of Fungi, 10*(5), Article 5. https://doi.org/10.3390/jof10050332

Loi, M., Paciolla, C., Logrieco, A. F., & Mulè, G. (2020). Plant bioactive compounds in pre- and postharvest Management for Aflatoxins Reduction. *Frontiers in Microbiology, 11*, 243. https://doi.org/10.3389/fmicb.2020.00243

Macheleidt, J., Mattern, D. J., Fischer, J., Netzker, T., Weber, J., Schroeckh, V., Valiante, V., & Brakhage, A. A. (2016). Regulation and role of fungal secondary metabolites. *Annual Review of Genetics, 50*(1), 371–392. https://doi.org/10.1146/annurev-genet-120215-035203

Malinga, L., & Laing, M. (2023). Cost analysis of biopesticides and chemical insecticides: Implications for cotton farmers in South Africa. *Entomology and Applied Science Letters, 10*(4–2023), 44, Article 4–2023. https://doi.org/10.51847/YA2wqC4r3i

Mandal, S., Mallick, N., & Mitra, A. (2009). Salicylic acid-induced resistance to *Fusarium oxysporum* f. Sp. *Lycopersici* in tomato. *Plant Physiology and Biochemistry, 47*(7), 642–649. https://doi.org/10.1016/j.plaphy.2009.03.001

Martínez-Núñez, M. A., López, V. E. L., & y. (2016). Nonribosomal peptides synthetases and their applications in industry. *Sustainable Chemical Processes, 4*(1), 13. https://doi.org/10.1186/s40508-016-0057-6

Moeini, A., Germann, N., Malinconico, M., & Santagata, G. (2021). Formulation of secondary compounds as additives of biopolymer-based food packaging: A review. *Trends in Food Science & Technology, 114*, 342–354. https://doi.org/10.1016/j.tifs.2021.05.040

Mohy, S., Elameen, T. M., Khalaphallah, R., & Hassan, N. M. (2024). Biological control of garlic white and basal rot pathogens in vitro and under field conditions. *SVU-International Journal of Agricultural Sciences, 6*, 12. https://doi.org/10.21608/svuijas.2024.319563.1393

NABCONS. (2023). *Study to determine postharvest losses of agri-produces*. https://www.nabcons.com/. Last accessed: 16 Dec 2024.

Nigam, P. S., Singh, A., & Lyngby, D. (2014). Production of secondary metabolites–fungi. *FEBS Letters*, 1319–1328. Elsevier. https://doi.org/10.1016/B978-0-12-384730-0.00202-0

Nunes, C. A. (2012). Biological control of postharvest diseases of fruit. *European Journal of Plant Pathology, 133*(1), 181–196. https://doi.org/10.1007/s10658-011-9919-7

Palou, L., Smilanick, J. L., & Droby, S. (2008). Alternatives to conventional fungicides for the control of citrus postharvest green and blue moulds. *Stewart Postharvest Review, 2*(2), 1–16. https://doi.org/10.2212/spr.2008.2.2

Patocka, J. (2016). Bioactive metabolites of entomopathogenic fungi Beauveria bassiana. *Military Medical Science Letters, 85*(2), 80.

Pereira-Dias, L., Oliveira-Pinto, P. R., Fernandes, J. O., Regalado, L., Mendes, R., Teixeira, C., Mariz-Ponte, N., Gomes, P., & Santos, C. (2023). Peptaibiotics: Harnessing the potential of microbial secondary metabolites for mitigation of plant pathogens. *Biotechnology Advances, 68*, 108223. https://doi.org/10.1016/j.biotechadv.2023.108223

Poore, J., & Nemecek, T. (2018). Reducing food's environmental impacts through producers and consumers. *Science, 360*(6392), 987–992. https://doi.org/10.1126/science.aaq0216

Qadri, R., Azam, M., Khan, I., Yang, Y., Ejaz, S., Akram, M. T., & Khan, M. A. (2020). Conventional and modern technologies for the management of post-harvest diseases. In *Plant disease management strategies for sustainable agriculture through traditional and modern approaches* (pp. 137–172). Springer International Publishing. https://doi.org/10.1007/978-3-030-35955-3_7

Rassbach, J., Hilsberg, N., Haensch, V. G., Dörner, S., Gressler, J., Sonnabend, R., Semm, C., Voigt, K., Hertweck, C., & Gressler, M. (2023). Non-canonical two-step biosynthesis of anti-oomycete indole alkaloids in Kickxellales. *Fungal Biology and Biotechnology, 10*(1), 19. https://doi.org/10.1186/s40694-023-00166-x

Sahebani, N., & Hadavi, N. (2008). Biological control of the root-knot nematode *Meloidogyne javanica* by *Trichoderma harzianum*. *Soil Biology and Biochemistry, 40*(8), 2016–2020. https://doi.org/10.1016/j.soilbio.2008.03.011

Sanzani, S. M., Reverberi, M., & Geisen, R. (2016). Mycotoxins in harvested fruits and vegetables: Insights in producing fungi, biological role, conducive conditions, and tools to manage postharvest contamination. *Postharvest Biology and Technology, 122*, 95–105. https://doi.org/10.1016/j.postharvbio.2016.07.003

Sapna, Sharma, C., Pathak, P., Yadav, S. P., & Gautam, S. (2024). Potential of emerging "all-natural" edible coatings to prevent post-harvest losses of vegetables and fruits for sustainable agriculture. *Progress in Organic Coatings, 193*, 108537. https://doi.org/10.1016/j.porgcoat.2024.108537

Sawant, A., Sajwan, H., & Vamkudoth, K. R. (2022). Fungal metabolites: Industrial applications and challenges. In *Sustainable utilization of fungi in agriculture and industry* (pp. 159–201). Bentham Science Publishers. https://www.benthamdirect.com/content/books/9789815040340.chap10

Sharma, R., & Pongener, A. (2010). Natural products for postharvest decay control in horticultural produce: A review. *Stewart Postharvest Review, 6*. https://doi.org/10.2212/spr.2010.4.1

Sharma, R. R., Singh, D., & Singh, R. (2009). Biological control of postharvest diseases of fruits and vegetables by microbial antagonists: A review. *Biological Control, 50*(3), 205–221. https://doi.org/10.1016/j.biocontrol.2009.05.001

Shu, H., Chen, X., Jiang, Q., Wang, Y., Wan, Z., Xu, J., & Wang, P. (2024). Optimization of fungal secondary metabolites production via response surface methodology coupled with multiparameter optimized artificial neural network model. *Bioresource Technology, 413*, 131495. https://doi.org/10.1016/j.biortech.2024.131495

Singh, D., & Sharma, R. R. (2018). Postharvest diseases of fruits and vegetables and their management. In *Postharvest disinfection of fruits and vegetables* (pp. 1–52). Elsevier. https://doi.org/10.1016/B978-0-12-812698-1.00001-7

Smilanick, J. L., Mansour, M. F., & Sorenson, D. (2006). Pre- and postharvest treatments to control green mold of citrus fruit during ethylene degreening. *Plant Disease, 90*(1), 89–96. Scopus. https://doi.org/10.1094/PD-90-0089

Solanki, M. K., Wang, Z., Kaushik, A., Singh, V. K., Roychowdhury, R., Kumar, M., Kumar, D., Singh, J., Singh, S. K., Dixit, B., & Kumar, A. (2024). From orchard to table: Significance of fruit microbiota in postharvest diseases management of citrus fruits. *Food Control, 165*, 110698. https://doi.org/10.1016/j.foodcont.2024.110698

Surucu-Balci, E., & Tuna, O. (2021). Investigating logistics-related food loss drivers: A study on fresh fruit and vegetable supply chain. *Journal of Cleaner Production, 318*, 128561. https://doi.org/10.1016/j.jclepro.2021.128561

Tayel, A. A., Moussa, S. H., Salem, M. F., Mazrou, K. E., & El-Tras, W. F. (2016). Control of citrus molds using bioactive coatings incorporated with fungal chitosan/plant extracts composite. *Journal of the Science of Food and Agriculture, 96*(4), 1306–1312. Scopus. https://doi.org/10.1002/jsfa.7223

Teferra, T. F. (2022). The cost of postharvest losses in Ethiopia: Economic and food security implications. *Heliyon, 8*(3). https://doi.org/10.1016/j.heliyon.2022.e09077

Tripathi, P., & Dubey, N. K. (2004). Exploitation of natural products as an alternative strategy to control postharvest fungal rotting of fruit and vegetables. *Postharvest Biology and Technology, 32*(3), 235–245. https://doi.org/10.1016/j.postharvbio.2003.11.005

Tripathi, A. N., Tiwari, S. K., Behera, T. K., Tripathi, A. N., Tiwari, S. K., & Behera, T. K. (2022). Postharvest diseases of vegetable crops and their management. In *Postharvest technology—Recent advances, new perspectives and applications*. IntechOpen. https://doi.org/10.5772/intechopen.101852

Tzortzakis, N. G., & Economakis, C. D. (2007). Antifungal activity of lemongrass (*Cympopogon citratus* L.) essential oil against key postharvest pathogens. *Innovative Food Science & Emerging Technologies, 8*(2), 253–258. https://doi.org/10.1016/j.ifset.2007.01.002

Vilariño, M. V., Franco, C., & Quarrington, C. (2017). Food loss and waste reduction as an integral part of a circular economy. *Frontiers in Environmental Science, 5*. https://doi.org/10.3389/fenvs.2017.00021

Vischetti, C., Feliziani, E., Landi, L., De Bernardi, A., Marini, E., & Romanazzi, G. (2024). Effectiveness of four synthetic fungicides in the control of post-harvest gray mold of strawberry and analyses of residues on fruit. *Agronomy, 14*(1), Article 1. https://doi.org/10.3390/agronomy14010065

Wadhwa, K., Kapoor, N., Kaur, H., Abu-Seer, E. A., Tariq, M., Siddiqui, S., et al. (2024). A comprehensive review of the diversity of fungal secondary metabolites and their emerging applications in healthcare and environment. *Mycobiology*, 1–53. https://doi.org/10.1080/12298093.2024.2416736

Wang, W., Liu, S., Deng, L., Ming, J., Yao, S., & Zeng, K. (2018). Control of citrus post-harvest green molds, blue molds, and sour rot by the Cecropin A-Melittin hybrid peptide BP21. *Frontiers in Microbiology, 9*, 2455. Scopus. https://doi.org/10.3389/fmicb.2018.02455

Wang, L., Lu, H., & Jiang, Y. (2023). Natural polyketides act as promising antifungal agents. *Biomolecules, 13*(11), 1572. https://doi.org/10.3390/biom13111572

Wei, X., Nie, Q., Medison, R. G., Zheng, T., Meng, X., Sun, Z., & Zhou, Y. (2024). Evaluation of *Pseudomonas koreensis* B17-12 as a potential biological control agent against postharvest diseases of tomato. *Physiological and Molecular Plant Pathology, 132*, 102311. https://doi.org/10.1016/j.pmpp.2024.102311

Wills, R. B. H., & Golding, J. (2016). *Advances in postharvest fruit and vegetable technology*. CRC Press.

Wink, M. (2008). Plant secondary metabolism: Diversity, function and its evolution. *Natural Product Communications, 3*(8), 1934578X0800300801. https://doi.org/10.1177/1934578X0800300801

Wu, P.-H., Chang, H.-X., & Shen, Y.-M. (2023). Effects of synthetic and environmentally friendly fungicides on powdery mildew management and the phyllosphere microbiome of cucumber. *PLoS One, 18*(3), e0282809. https://doi.org/10.1371/journal.pone.0282809

Xu, K., Li, X.-Q., Zhao, D.-L., & Zhang, P. (2021a). Antifungal secondary metabolites produced by the fungal endophytes: Chemical diversity and potential use in the development of biopesticides. *Frontiers in Microbiology, 12*, 689527. https://doi.org/10.3389/fmicb.2021.689527

Xu, X., Qu, R., Wu, W., Jiang, C., Shao, D., & Shi, J. (2021b). Applications of microbial co-cultures in polyketides production. *Journal of Applied Microbiology, 130*(4), 1023–1034. https://doi.org/10.1111/jam.14845

Xu, D., Niu, Q., Gao, L., Wang, Q., Lan, C., Jiang, Z., & Qiao, F. (2024). Biocontrol potential of *Burkholderia gladioli* STPT16 on postharvest decay of litchi fruit caused by *Peronophythora litchii* and *Colletotrichum fructicola*. *Food Control*, 110956. https://doi.org/10.1016/j.foodcont.2024.110956

Yao, X., Guo, H., Zhang, K., Zhao, M., Ruan, J., & Chen, J. (2023). Trichoderma and its role in biological control of plant fungal and nematode disease. *Frontiers in Microbiology, 14*. https://doi.org/10.3389/fmicb.2023.1160551

Yazar, S., & Omurtag, G. Z. (2008). Fumonisins, trichothecenes and zearalenone in cereals. *International Journal of Molecular Sciences, 9*(11), Article 11. https://doi.org/10.3390/ijms9112062

Zaynab, M., Fatima, M., Abbas, S., Sharif, Y., Umair, M., Zafar, M. H., & Bahadar, K. (2018). Role of secondary metabolites in plant defense against pathogens. *Microbial Pathogenesis, 124*, 198–202. https://doi.org/10.1016/j.micpath.2018.08.034

Zhao, X., Zhou, J., Tian, R., & Liu, Y. (2022). Microbial volatile organic compounds: Antifungal mechanisms, applications, and challenges. *Frontiers in Microbiology, 13*. https://doi.org/10.3389/fmicb.2022.922450

Chapter 9
Fungal Metabolites and Their Application in the Food Industry: Challenges and Solutions

Abhimanyu Gaur, Poojal Chhirang, Nikita Singh, Vijay Kumar Gahlawat, Bhim Pratap Singh , and G. V. Prasanna Kumar

1 Introduction

Fungal metabolites (FMs) are a diverse group of natural compounds that fungi produce. They encompass various bioactive molecules that play crucial roles in various biological processes (Dufossé et al., 2021). These metabolites can be broadly categorized into primary and secondary metabolites (Machiani et al., 2022). Primary metabolites are essential for the growth and metabolism of fungi, including compounds such as amino acids, nucleotides, and organic acids. These are typically involved in basic cellular functions and are produced during the active growth phase of the fungi. On the other hand, secondary metabolites are not directly involved in the growth and reproduction of fungi but play crucial ecological roles. These compounds often have diverse and potent bioactivities, including antimicrobial, antifungal, and anti-inflammatory properties. Secondary metabolites include well-known compounds such as antibiotics (e.g., penicillin), enzymes (e.g., cellulases), mycotoxins (e.g., aflatoxins), and pigments (e.g., melanin) (Keller, 2019; Wadhwa et al., 2024). Additionally, certain FMs act as natural antioxidants, enhancing the shelf life and safety of food items by inhibiting the growth of spoilage microorganisms and pathogens (Krupodorova et al., 2024).

The food industry, in particular, has harnessed the potential of FMs for various applications, including enhancing flavour, preservation, and nutritional value of food products (Rao et al., 2017; Wadhwa et al., 2024). Compounds such as citric acid, produced by *Aspergillus niger*, and kojic acid, produced by *Aspergillus*

A. Gaur · N. Singh · B. P. Singh · G. V. Prasanna Kumar (✉)
Department of Agriculture and Environmental Sciences, National Institute of Food Technology Entrepreneurship and Management (NIFTEM), Kundli, Sonepat, Haryana, India

P. Chhirang · V. K. Gahlawat
Department of Interdisciplinary Sciences, National Institute of Food Technology Entrepreneurship and Management – Kundli, Sonepat, Haryana, India

B. P. Singh et al. (eds.), *Fungal Additives and Bioactives in Food Processing Industries*, Fungal Biology, https://doi.org/10.1007/978-3-032-04519-5_9

oryzae, are widely used to enhance the flavour of various food products (Behera, 2020; Li et al., 2022). Additionally, FMs like vanillin, derived from the fermentation of lignin by certain fungi, are used as natural flavouring agents in foods and beverages, providing a desirable aroma and taste (D'Arrigo et al., 2024). Another important application of FMs in the food industry is as preservatives. Citric acid is commonly used as a preservative to extend the shelf life, aroma, and flavour of ice creams, marmalades, and juices (Behera, 2020). These acids inhibit the growth of spoilage microorganisms and pathogens by lowering the pH of the food environment, thereby creating unfavourable conditions for microbial growth (Coban, 2019). Furthermore, natamycin, a polyene antifungal compound, is used to prevent the growth of mould and yeast on cheese and other dairy products (Meena et al., 2021). Using these natural preservatives ensures the safety and quality of food products. It meets the increasing demand for natural alternatives to synthetic additives, clean-labelled and minimally processed foods (Mishra et al., 2022). In addition, FMs also play a significant role in improving the nutritional profile of food products. Enzymes produced by fungi, such as amylases, proteases, and lipases, are widely used in the food industry to modify the texture, flavour, and nutrients of various food items (El-Gendi et al., 2022). For instance, fungal amylases produce high-fructose corn syrup (HFCS), a widely used sweetener in the beverage industry (Amaral-Fonseca et al., 2021). On the other hand, fungal proteases are used to break down proteins into peptides and amino acids, improving the digestibility and nutritional value (Naeem et al., 2022). Moreover, FMs like ergosterol, a precursor of vitamin D2, are used to fortify foods with essential vitamins (Hu et al., 2021).

Despite the numerous benefits of FMs, their use in the food industry presents several challenges. This includes potential toxicity, safety concerns, regulatory hurdles, and consumer acceptance (Stoev, 2015; Mafe & Büsselberg, 2024). One of the primary concerns is the safety and toxicity of certain FMs, particularly mycotoxins. Mycotoxins are toxic secondary metabolites produced by fungi that can contaminate food and agricultural products (Stoev, 2024). These compounds can pose serious health risks to humans and animals, leading to severe health issues such as liver damage, immune suppression, and carcinogenic effects (Awuchi et al., 2021; Khan, 2024). Common mycotoxins include aflatoxins produced by *Aspergillus* species, fumonisins by *Fusarium* species, and ochratoxin-A by *Penicillium* and *Aspergillus* species (El-Sayed et al., 2022). The presence of mycotoxins in food and feed is a major concern, necessitating stringent regulatory measures and quality control practices to ensure safety and limit the presence of mycotoxins in food products. Additionally, advancements in biotechnological tools and techniques have enabled the development of genetically modified fungi that produce desirable metabolites while minimizing the production of harmful mycotoxins (Liu et al., 2020). Another challenge associated with the use of FMs is consumer acceptance (Li et al., 2023). There is often a perception that FMs, particularly those produced through fermentation or biotechnological processes, are unnatural or unsafe. To overcome this challenge, promoting transparency in the labelling and marketing of food products is essential. Furthermore, the development of innovative and sustainable production methods can help address consumer concerns and build trust in the use of FMs.

This chapter aims to explore the complex world of FMs and provide an overview of their applications, challenges, and solutions within the food industry. It highlights the significance of FMs in developing safe, nutritious, and sustainable food products.

2 Production of FMs

The production of FMs involves the cultivation of fungi in controlled environments, where specific growth conditions such as temperature, pH, and nutrient availability are optimized (Singh & Kumar, 2023). These are also produced through fermentation processes, where fungi are grown in bioreactors under controlled conditions (Svobodová & Novotný, 2018). Factors such as aeration, agitation, and substrate concentration are meticulously regulated to maximize the desired metabolite yield (Ganeshan et al., 2021). These conditions facilitate the synthesis of both primary and secondary metabolites, which are then extracted and purified for various industrial applications, including food and pharmaceuticals.

2.1 Natural Production of FMs

Natural production of FMs occurs in the fungi's native environments, such as soil, decaying organic matter, or within plant tissues. These environments provide the natural substrates and conditions necessary for the fungi to produce metabolites (Tiwari & Bae, 2022; Wadhwa et al., 2024). The production rates can be variable and are influenced by nutrient availability, temperature, pH, and microbial interactions (VanderMolen et al., 2013). Examples of naturally occurring FMs are:

- *Penicillin* is produced by *Penicillium chrysogenum and P. notatum* in a decaying organic matter, which has been historically significant as a natural antibiotic (Kardos & Demain, 2011).
- *Aflatoxins* are produced by *Aspergillus flavus* on crops like peanuts and corn under favourable environmental conditions, posing health risks (Wu et al., 2008).

Natural production methods are often limited by the inconsistency and unpredictability of environmental conditions, making it challenging to consistently produce large quantities of metabolites.

2.2 *Industrial Production of FMs*

Industrial production of FMs involves using controlled biotechnological processes to cultivate fungi and produce metabolites on a large scale (Nielsen & Nielsen, 2017). Techniques such as submerged and solid-state fermentation are commonly employed (Kumar et al., 2021). Industrial production offers several advantages.

- *Consistency and control*: Precise control over growth conditions such as temperature, pH, aeration, and nutrient supply ensures consistent and high-yield production.
- *Scale and efficiency*: Industrial setups can produce large quantities of metabolites efficiently, meeting the demands of various industries.
- *Purification and quality*: Advanced extraction and purification techniques ensure high purity and quality of the final products.

Examples of industrially produced FMs are:

- *Citric acid*: It is produced by *Aspergillus niger* through submerged fermentation. It is widely used as a flavouring agent and preservative in the food industry (Tong et al., 2019).
- *Enzymes*: These include amylases and proteases produced by fungi. Enzymes are used in food processing, detergents, and pharmaceutical applications (Rana et al., 2013; da Silva, 2017).

Industrial production allows for the genetic modification of fungi to enhance the yield and quality of desired metabolites, reduce the production of harmful by-products, and tailor the production process to specific industrial needs (Mohamed et al., 2021). While the natural production of FMs relies on the native environment of fungi and can be inconsistent, industrial production offers controlled, efficient, and scalable methods to produce high-quality FMs for various applications.

FMs are utilized in various sectors, both at industrial and laboratory scales. The main sectors where FMs are used include agriculture, biomedical, food processing, and biotechnology (Table 9.1).

3 Applications of FMs in the Food Industry

FMs have gained massive attention in the food industry due to their bioactivities, such as antimicrobial, antiviral, and antioxidant. They are utilized as bioactive compounds, food colorants, flavour enhancers, nutritional supplements, fermentation processes, preservatives, and biocontrol agents.

Table 9.1 Fungal metabolites and their applications

Fungi	Metabolites	Sector	Application role	Reference
Aspergillus oryzae	Amylase	Food processing	Enzymes are used in brewing, baking, and producing HFCS by converting starches into sugars	Sun et al. (2024)
	Glucose oxidase	Fermentation	Improved yield by conversion of glucose into gluconic acid	
Aspergillus niger	Tannase	Beverage industry	Reduction in bitter taste and improvement in flavour, especially in coffee, tea, and wine	Papadaki and Mantzouridou (2023)
	Citric acid	Food and beverage industry	Acidulant, flavour enhancer, and preservative	Behera (2020)
Aspergillus terreus	Lovastatin, Statins	Nutraceuticals	A cholesterol-lowering drug that inhibits HMG-CoA reductase	Alberts (1990) and Lins et al. (2024)
Penicillium sp.	Ethyl acetate extract	Post-harvest preservation	Insecticidal for *S. litura* and *C. quinquefasciatus*	Arunthirumeni et al. (2023)
Penicillium spp.	Patulin	Food industry	Contribute to complex flavours in fermented foods like cheese	Yin et al. (2021)
Monascus purpureus	Rubropunctatin, Monacolin K	Food industry	Colorants	Chaudhary et al. (2022) and Mohankumari et al. (2021)
Saccharomyces cerevisiae	Ethanol, carbon dioxide	Food industry	Used in baking goods as a leavening agent. Production of CO_2 and ethanol during fermentation	Chávez et al. (2023) and Chugh et al. (2023)
Mucor circinelloides	Proteases	Food industry	Used in cheeses, breaking of proteins for flavour enhancement	Bensmail et al. (2023)
Fusarium spp.	Mycoprotein	Food industry	Used as a meat analogue in vegetarian food products	Khan et al. (2024)
Trichoderma spp.	Chitinase	Food waste management	Degrades chitin from crustacean shells	Sharma et al. (2023)
	Cellulases, hemicellulases	Food and beverage industry	Enhance the yield of fruit juices and alcoholic beverages	Bhardwaj et al. (2021)

3.1 Bioactive Compounds

Kingdom fungi are a centre of traction due to their anti-cancerous, immunosuppressing, and antibiotic secondary metabolites. Fungi are assumed to have 130 various medicinal activities, including antioxidant, antitumor, antiviral, cardioprotective, immunomodulation, radical scavenging, and so many (Lull et al., 2005; Sugiharto et al., 2016; Hameed et al., 2017; Deshmukh et al., 2018; Uzma et al., 2018; Bains et al., 2021; Chopra et al., 2021; Chugh et al., 2022). The polysaccharides present in mushrooms, like β-glucans, micronutrients and antioxidants like ergosterols, flavonoids, glycoproteins, and triterpenoids, can help to resist infections and toxins. β-lactam antibiotics, including penicillins, cephalosporins, and carbapenems, inhibit the synthesis of the peptidoglycan layer of the bacterial cell wall. Omega-3 fatty acids, produced by oleaginous fungi, are important for heart health and are incorporated into various food products to provide these beneficial fats (Moura et al., 2023).

3.2 Food Colorants

FMs are increasingly used as natural colorants in the food industry, offering a sustainable and safe alternative to synthetic dyes. Natural pigments enhance the visual appeal of food products and provide additional health benefits due to their antioxidant and antimicrobial properties (Ezati & Rhim, 2021; Martínez-Rodríguez et al., 2022). Pigments extracted from the *Monascus* anka, *M. ruber*, and *M. purpureus* are broadly classified into three categories: (1) orange pigments (rubropunctatin and monascorubrin), (2) red pigments (rubropunctamin and monascurubramin), and (3) yellow pigments (monascin and ankaflavin) (Shi et al., 2017; Darwesh et al., 2020; Mohankumari et al., 2021). Pigments from *Monascus purpureus* have a potent cholesterol-lowering property, and are used to colour-fermented yeast rice that imparts a vibrant red hue (Kalaivani et al., 2010; Burke, 2015; Bruno et al., 2018). The red pigment of *Penicillium purpurogenum* has been reported to extend the shelf life of minced meat due to its antimicrobial property (Salama et al., 2021). *Blakeslea trispora* produces lycopene that can be used to impart a red colour to food products (Sevgili & Erkmen, 2019). Bikaverin is a red pigment produced by *Fusarium* spp. that possesses potential antibacterial properties (Son et al., 2008; Deshmukh et al., 2014). These natural colourants are biodegradable, environmentally friendly, and free from associated health risks. As research continues to uncover new FMs with colourant properties, their application in the food industry is expected to expand, promoting healthier and more sustainable food production practices.

3.3 Flavour Enhancers

FMs play a significant role as flavour enhancers in the food industry, offering natural alternatives to synthetic additives (He, 2025). These metabolites, produced by various fungi, impart a wide range of flavours, including fruity, floral, malty, and umami, enhancing the sensory appeal of food products. One of the most widely used FMs is citric acid, produced by *Aspergillus niger* (Lotfy et al., 2007; Tong et al., 2019; Behera, 2020). Citric acid is added to beverages, candies, and processed foods for a tangy and refreshing taste. Its ability to balance sweetness and enhance other flavours makes it indispensable in the food industry. Another important flavour enhancer is kojic acid, derived from *Aspergillus oryzae* (Yamada et al., 2014; Sun et al., 2024). It is a precursor for flavour enhancers, produced during fermented products like soy sauce and miso (Nohynek et al., 2004; El-Kady et al., 2014).

Enzymes produced by fungi also play a significant role in flavour enhancement. For example, amylases and proteases are used in the baking industry to break down starches and proteins, releasing sugars and amino acids that contribute to the Maillard reaction during baking (Liu et al., 2024). This reaction is responsible for developing rich, complex flavours and the golden-brown colour of baked goods (El-Gendi et al., 2022). In the brewing industry, fungal enzymes like glucoamylases are used to break down complex carbohydrates into fermentable sugars, enhancing the flavour and alcohol content of beer (Wayllace et al., 2023). FMs such as vanillin, derived from the fermentation of lignin by certain fungi, are used as natural flavouring agents in foods and beverages, providing a desirable aroma and taste (Martău et al., 2021). As consumer preferences continue to shift towards natural and sustainable ingredients, the importance of FMs in flavour enhancement is likely to grow, driving innovation and development in the food industry.

3.4 Nutritional Supplements

FMs are increasingly being recognized for their potential as nutritional supplements in the food industry, offering a natural and sustainable source of essential nutrients (Takahashi et al., 2020). These include vitamins, essential fatty acids, and bioactive compounds that contribute to human health. Omega-3 fatty acids, produced by oleaginous fungi, are important for heart health and are incorporated into various food products to provide these beneficial fats (Shanmuga, 2021).

Additionally, FMs such as beta-glucans and chitin have been shown to have immune-boosting and cholesterol-lowering properties, respectively (Murphy et al., 2020). These compounds are added to functional foods and dietary supplements to promote overall health and well-being. FMs as nutritional supplements align with the increasing consumer demand for natural and clean-label ingredients, making them a potent alternative in the food industry.

3.5 Fermentation Processes

FMs play a crucial role in fermentation processes within the food industry, enhancing the production of various food products and improving their quality (Chai et al., 2022). These metabolites, produced by fungi such as *Saccharomyces*, *Penicillium*, and *Aspergillus* species, are integral to the fermentation of beverages, bakery products, dairy products, and soy-based foods (Pouris et al., 2024). In alcoholic fermentation, *Saccharomyces cerevisiae* converts sugars into ethanol and carbon dioxide, producing beverages such as beer and wine (Walker & Stewart, 2016). This process not only generates alcohol but also contributes to the development of complex flavours and aromas (Parapouli et al., 2020). Similarly, yeast fermentation in bread production causes dough to rise, resulting in a light and airy texture (Heitmann et al., 2018). FMs are also essential in the fermentation of dairy products like cheese and yogurt. *Penicillium* species are used to produce blue cheese, imparting distinct flavours and textures (Dumas et al., 2020). In dairy products' fermentation, lactic acid bacteria and yeast give their characteristic tangy taste and creamy texture (Narvhus & Gadaga, 2003; Turner et al., 2017). Soy-based fermented foods such as tempeh and miso rely on fungal fermentation to break down complex proteins and carbohydrates, enhancing their nutritional value and digestibility (do Prado et al., 2022). Overall, FMs are indispensable in fermentation processes, improving the sensory qualities, nutritional value, and shelf life of various food products. Their natural and sustainable characteristics align with consumer preferences for clean-label ingredients, making them a valuable asset in the food industry.

3.6 Preservatives and Biocontrol Agents

FMs are increasingly used as natural preservatives and biocontrol agents in the food industry, offering a safer and more sustainable alternative to synthetic additives (Mishra et al., 2022). These metabolites possess antimicrobial, antifungal, and insecticidal properties that help to protect crops from pests and diseases. This reduces the need for synthetic chemicals and extends the storage period of edibles by preventing pathogen contamination (Redondo-Blanco et al., 2020). For instance, citric acid is produced by fungi like *Aspergillus niger*, commonly used to prevent colour and flavour deterioration in beverages and frozen goods (Behera, 2020). Antifungal compounds like kojic acid are used to protect fruits, vegetables, and other perishable foods from fungal infections and spoilage (Kim et al., 2012; Zhu et al., 2022). Bikaverin, a polyketide produced by certain fungi, has shown potential as a natural fungicide (Deshmukh et al., 2014). Antibacterial compounds, such as penicillin and streptomycin, derived from *Penicillium* and *Streptomyces* species (Miller et al., 2022). It inhibits the pathogenic activities in fruits and vegetables, extending their shelf life and ensuring food safety. Additionally, volatile organic compounds (VOCs) produced by fungi can act as natural insect repellents, protecting stored grains and other food products from insect infestations (Alam et al., 2022; Lahlali et al., 2022).

Synthetic pesticides pose a significant threat to the environment, while FMs represent a natural and effective alternative against diseases, infections, insects, and nematodes, making them useful in integrated pest management. Wadhwa et al. (2024) explained that FMs such as beauvericin, destruxins, and oxalic acid can penetrate insect exoskeletons. Metabolites of *Beauveria bassiana*, *Cladosporium* spp., *Isaria fumosorosea*, *Metarhizium anisopliae*, *Nomuraea rileyi*, and *Purpureocillium* spp. have been reported to overcome the insect immunity, leading to the killing of locusts, aphids, and other crop pests (Elbanhawy et al., 2019). FMs of *Clodosporium cladosporioides* and *Verticillium lecanii* extracts demonstrated effective toxicity for armyworms with LC_{50} values of 229 and 341 ppm, respectively, suggesting their potential as natural pesticides (Abdullah et al., 2024). FMs of *Alternaria alternata*, *A. tenuissima*, *Chaetomium dentatus*, and *Nigrospora oryzae* have bioherbicide features (Mohammed & Badawy, 2020). A total of 183 phytotoxic FMs were categorized into five categories: polyketides, terpenoids, nitrogenous metabolites, phenols and phenolic acids, and miscellaneous chemicals (Bendejacq-Seychelles et al., 2023). These FMs can be formulated into effective and eco-friendly bioherbicides.

4 Challenges in the Use of FMs

The use of FMs in the food industry offers numerous benefits, but it also presents several challenges that must be addressed. One of the primary challenges is ensuring consistency and quality control. Variations in growth conditions, substrate composition, and fungal strains can lead to inconsistencies in the quality and yield of FMs (Singh & Kumar, 2023). Standardization of production processes is crucial for meeting industrial standards. Bioreactor design, nutrient supply, and aeration must be optimized to maintain productivity and efficiency (Moreira et al., 2003). FMs used as food additives or preservatives must undergo rigorous regulatory approval to ensure their safety and efficacy (Mishra et al., 2022). Consumer acceptance is another hurdle. Despite the benefits associated with FMs, there is a controversy surrounding the products derived from metabolites biosynthesized by toxin-producing fungi (Takahashi et al., 2020). For example, Agmatine and Terrein, both metabolites are obtained from *Fusarium graminearum* and *Aspergillus terreus* (Gardiner et al., 2010; Raghuvanshi & Katiyar, 2024). These fungi are reported to be associated with the biosynthesis of mycotoxins. To address these challenges, ongoing research, technological advancements, collaboration among scientists and industry stakeholders, and regulatory bodies are essential. By overcoming these challenges, we can unlock the tremendous potential of FMs, paving the way for the creation of food products that are not only healthier but also more sustainable.

4.1 *Mycotoxin Risks*

Mycotoxin risks are a significant challenge in using FMs in the food industry. These are the toxic secondary metabolites produced by *Aspergillus*, *Fusarium*, and *Penicillium* species, which can contaminate various food crops and pose severe health risks. These toxins can cause various health issues, including carcinogenic, neurotoxic, and immunosuppressive effects (Mafe & Büsselberg, 2024). Plant pathogens, including *Alternaria*, *Trichoderma*, and *Trichothecium*, are reported as potential food contaminants (Table 9.2) (Richard, 2007; Ashiq, 2015). Contamination can occur at various stages, from the field to storage, especially under warm and humid conditions that favour fungal growth. Aflatoxin, citrinin, fumonisins, ochratoxins, patulin, sterigmatocystin, trichothecenes, and zearalenone are well-known potential mycotoxins (Awuchi et al., 2021; Pouris et al., 2024). The crop foods that are most sensitive to mycotoxins include cereals, millets, and pulses such as corn, peanuts, soybeans, sorghum, wheat, and their by-products (Awuchi et al., 2021).

4.1.1 Aflatoxins

Aflatoxins (AFs) are primarily FMs of *Aspergillus* spp. (Becker-Algeri et al., 2016). These consist of 20 similar polycyclic structures that belong to the furanocoumarins class of compounds. Major sources of AFs include bovine milk, oilseeds, nuts, peanuts, and cereals such as maize and wheat (Klich, 2007; Yu, 2012). Pathogenic fungi primarily produced AFs during preharvest heat or drought stress phases of plants (Bennett & Klich, 2003; Kebede et al., 2012). AFs develop in conditions of excessive moisture during storage at temperatures between 28 and 35 °C, which includes the presence of insect activity (Amaike & Keller, 2011; Yu, 2012). AFs have four major groups: AFB1, AFB2, AFG1, and AFG2 (de Ruyck et al., 2015). AFs are extremely thermostable carcinogenic metabolites with teratogenic and mutagenic properties (Yu, 2012; Kew, 2013).

Aflatoxin M1, a hydroxylated metabolite of AFB1 and AFB2, has been detected in the milk and dairy products of lactating cows that were fed aflatoxin-contaminated feed. This metabolite is linked to hepatotoxic and carcinogenic health risks (Kourti et al., 2024). European Commission and Codex Elimentarious have prescribed a maximum permissible limit (MPL) for AFM1 at 50 ng/kg raw milk. The United States (US) and the Food Safety and Standards Authority of India (FSSAI) have set a maximum limit of 500 ng/kg raw milk (Hattimare et al., 2022). In the case of animal feed, maximum permissible limits of 5, 10, and 20 μg AFB1/kg have been set by the European Union (EU), China, and the US, respectively (Giovati et al., 2015).

Table 9.2 Mycotoxins, affected food commodities, and their maximum permissible limits

Mycotoxins	Forms	Fungus	Food commodity	Max. permissible limit by the EU	Max. permissible limit by the US	Reference
Aflatoxins	B1, B2, G1, G2, M1, M2	*Aspergillus bombycis*, *A. flavus*, *A. nomius*, *A. parasiticus*, *A. pseudotamarii*, *Penicillium* spp.	Cereals and grains, legumes, oilseeds, nuts, and coconut	Dried nuts—4 μg aflatoxin/kg. Dried grains—2 μg AFB1/kg. Milk—50 ng/kg.	Milk—500 ng/kg.	El-Sayed et al. (2022), Becker-Algeri et al. (2016), de Ruyck et al. (2015), Yu (2012), and Giovati et al. (2015)
Ochratoxin	A, B, C	*Aspergillus carbonarius*, *A. niger*, *A. ochraceus*, *Penicillium nordicum*, *P. verrucosum*	Cereals and grains, vegetables, oilseeds, wine, herbs, figs, beef, milk, eggs	Instant coffee—10 μg/kg. Roasted coffee—5 μg/kg. Juice and wine—2 μg/ kg. Processed cereal foods—3 μg/kg. Unprocessed cereal foods—5 μg/kg.	Cereal and coffee products—5 μg/kg. Dried fruits and nuts—10 μg/kg. Spices—15 μg/kg.	El-Sayed et al. (2022), Agriopoulou et al. (2020), and Heussner and Bingle (2015)
Trichothecenes	DON, T-2, HT-2,	*Fusarium crookwellense*, *F. culmorum*, *F. graminearum*	Cereals and grains	100 ng T-2/kg. 100 ng HT-2/kg.	1000 μg DON/kg.	Escrivá et al. (2015), Mostrom and Raisbeck (2012), and Wan et al. (2015)
Fumonisin	B1, B2, B3, etc.	*Fusarium culmorum*, *F. moniliform*, *F. verticilliode*	Maize	Maize—1000 μg/kg.	Maize—2–4 mg/kg. Popped maize—3 mg/kg.	El-Sayed et al. (2022), Becker-Algeri et al. (2016), Bryła et al. (2013), and Scott (2012)

(continued)

Table 9.2 (continued)

Mycotoxins	Forms	Fungus	Food commodity	Max. permissible limit by the EU	Max. permissible limit by the US	Reference
Zearalenone	F-2	*Fusarium crookwellense*, *F. culmorum*, *F. graminearum*	Maize, fodder, and grass	250 ng/kg body weight per day.	250 ng/kg body weight per day.	El-Sayed et al. (2022), Dänicke and Winkler (2015), and Zinedine et al. (2007)
Patulin		*Aspergillus clavatus*, *A. terreus*, *Penicillium careum*, *P. clavigerum*, *P. graseofulvum*	Wheat, fruits such as apples, grapes, pears, apricots, peaches, olives	Fruit juices—50 μg/kg. Other food products—10–50 μg/kg.	Fruits/apple juices—50 μg/kg.	El-Sayed et al. (2022), Bernal et al. (2019), and Eriksen et al. (2010)

4.1.2 Ochratoxin

Ochratoxins (OTs) are the secondary metabolites belonging to *Aspergillus* and *Penicillium* spp. (Bui-Klimke & Wu, 2015). These metabolites consist of phenylalanine and isocoumarin groups linked with amine bonding (Bayman & Baker, 2006). Cereals, pulses, coffee, nuts, spices, and fruits are the main sources for OTs, along with dairy, meat, and infant and baby foods (Wu et al., 2014). Biosynthesis of OTs varies at a wide range of pH from 3.0 to 10.0, in temperature ranges of 20–31 °C and above 0.8 A_w (Reddy & Bhoola, 2010). Its accumulation depends on protein consumption, mainly serum albumin (Heussner & Bingle, 2015). OTs are highly heat-resistant until 250 °C (Marin et al., 2013). OTs are primarily present in OTA, OTB, and OTC forms (Heussner & Bingle, 2015).

OTs are hepatotoxic, immunotoxic, neurotoxic, nephrotoxic, and teratogenic (Malir et al., 2013). OTA dose above 70 μg/kg daily consumption led to the endemic nephrotoxicity and carcinogenicity in humans and animals (Reddy & Bhoola, 2010; Bayman & Baker, 2006). A survey conducted in France and Italy reported the presence of OTA in blood samples in the range of 0.1–1.3 ng/mL and 0.12–2.84 ng/mL, respectively (Abouzied et al., 2002). A survey found that males had higher OTA levels than females. The EU has set MPL of OTA for imported foods, including cereal foods (at 3 μg/kg), grains (at 5 μg/kg), grape juice (at 2 μg/kg), instant coffee (at 10 μg/kg), roasted coffee (at 5 μg/kg), and wine (at 2 μg/kg) (Bayman & Baker, 2006). OTA concentration has been found to be higher than 80 mg/kg in slices of bread in European countries (Bui-Klimke & Wu, 2015).

4.1.3 Trichothecenes

Trichothecenes (TCs) are secondary metabolites of *Cephalosporium*, *Fusarium*, *Myrothecium*, *Spicellum*, *Stachybotrys*, *Trichoderma*, and *Verticimonosporium* spp. (McCormick et al., 2011; Wang et al., 2014). These are members of a mycotoxin group that contains more than 200 sesquiterpenoid compounds (Awuchi et al., 2021). These are low molecular compounds with both hydrophobic and hydrophilic groups, aiding those entering the skin and GI tract (Pestka, 2008; McCormick et al., 2011). Eggs and meat are the common sources of TCs. Deoxynivalenol (DON) is the most common TC, also known as vomitoxin (Wang et al., 2014). DON is associated with pro-inflammatory activities. It is highly heat-resistant, and remains stable at temperatures of up to 170 °C when fried for 30 minutes (Sobrova et al., 2010).

TCs cause carcinogenesis, diarrhoea, weak immunity, disturbed neurotransmissions, lower body mass, retarded growth, oral lesions, and vomiting (Li et al., 2011; Wan et al., 2015). TCs can impact the synthesis of nucleic acids and proteins, as well as influence processes such as apoptosis, cytokine activation, lipid peroxidation, and mitochondrial function (Mostrom & Raisbeck, 2012; Wan et al., 2015). A small exposure to DON can result in growth impairment in children (Wu et al., 2014). Additionally, it was associated with Kashin–Beck disease, a condition characterized by degenerative osteoarthritis that affects approximately 2.5 million people in China.

4.1.4 Fumonisins

Fumonisins (FuMs) are the carcinogenic mycotoxins produced by *Fusarium* genera (Marasas et al., 2004; Becker-Algeri et al., 2016). *Aspergillus niger and A. awamori* have also been reported with biosynthetic genes for FuMs (Bryła et al., 2013). FuMs resemble sphinganine structurally with hydroxylated hydrocarbon chains attached to tricarballylic acid, amino and methyl groups (Stockmann-Juvala & Savolainen, 2008; Marin et al., 2013). FuMs were produced in the A_w range of 0.95–0.99 at temperatures between 20 and 30 °C (Bryła et al., 2013). FuMs are categorized into four main groups, namely FuMA, FuMB, FuMC, and FuMP (Haschek et al., 2001; Scott, 2012). Beer, beans, maize, millet, mushroom, rice, sorghum, and soybeans are the highly FuM-contaminated foods (Scott, 2012). Generally, edibles have been found to be contaminated with B1, B2, and B3 FuMs, where B1 has been reported as the most toxic in nature (Soriano et al., 2005).

FuMs in maize grains were responsible for oesophageal cancer in Transkei, South Africa (Sydenham et al., 1990). It was found that 44.3% of breast milk samples of northern Tanzania had FuMB1 levels ranging between 6.57 and 471.05 ng/mL, while 10.3% of samples contained FuMB1 levels more than 200 ng/mL (EU MPL for FuMs in infant food) (Magoha et al., 2014). The EU has set the MPL of 1 mg/kg FuMB1 and FuMB2 for maize products and 800 μg/kg for breakfast cereals and snacks. USFDA (US Food and Drug Administration) has set the MPL of 2–4 mg/kg for corn and 3 mg/kg for popcorn (Scott, 2012).

4.1.5 Sterigmatocystins

Sterigmatocystins (STCs) are carcinogenic, hepatotoxic, mutagenic, and teratogenic metabolites of fungal genera, including *Aspergillus*, *Bipolaris*, *Chaetomium*, *Emiricella*, and *Penicillium* (Veršilovskis & De Saeger, 2009). Barley, corn, rice, peanuts, and wheat were found to be infected with STCs, which are classified as class 2B carcinogenic for humans (Gao et al., 2015).

About 25.6% of samples of Latvian grain (barley, buckwheat, oats, ryes, and wheat) have been found to contain STCs ranging between 0.7 and 83.0 μg/kg (Veršilovskis et al., 2008). More than 50% of samples contained STCs higher than 20 μg/kg (MPL set by Czech Republic and Slovakia).

4.1.6 Zearalenone

Zearalenone (ZEA) is a carcinogenic mycotoxin produced by *Fusarium* genera, formerly known as F-2 mycotoxin (de Ruyck et al., 2015; Dänicke & Winkler, 2015). ZEA has a similar structural morphology as 17β-estradiol and the ability to competitively bind oestrogen receptors in animals (Escrivá et al., 2015; de Ruyck et al., 2015). The major hosts of ZEA contamination are maize, barley, oats, rice, wheat, sorghum, rye, and soybean (Zinedine et al., 2007).

ZEA in food caused feminization in pigs by reducing testosterone levels and impairing spermatogenesis (D'Mello et al., 1999). Infertility, decreased milk production, and low oestrogen levels have been reported in cows. ZEA produced by *Fusarium* spp. along with FMs, could co-participate in contamination. ZEA has shown antagonistic effects (Fink-Gremmels & Malekinejad, 2007; de Ruyck et al., 2015). ZEA is also known for its anti-inflammatory activities.

4.1.7 Patulin

Patulin (PAT) is a neurotoxic and immunotoxic polyketide metabolite produced by *Penicillium* spp. (Vidal et al., 2019). Initially, PAT was utilized as an antibiotic; however, subsequent research revealed its potential to cause toxicity in humans. This toxicity can result in symptoms such as haemorrhages, nausea, ulceration, and vomiting, although it has not been classified as carcinogenic (Awuchi et al., 2021). Apples and their processed products are reported to be the primary sources of PAT contamination (Zhong et al., 2018).

The EU has set MPL for PAT at 50 μg/ kg in processed fruit products, ciders, nectars, juices, spirits, and fermented drinks. For apple puree, apple compote, and solid apple products, the MPL of 25 μg/kg has been set for adults, and 10 μg/kg for babies and infants (Iqbal et al., 2024).

4.1.8 Citrinin

It is a well-recognized nephrotoxin and has been reported to be associated with Japanese yellowed rice disease (Bennett & Klich, 2003). Citrinin synergistically works with Ochratoxin A and slows the RNA synthesis (Ashiq, 2015; Jeswal & Kumar, 2015). Citrinin is present in colour-producing fungus genera of *Monascus* spp., also banned by the European Union and United States because of its nephrotoxic and hepatotoxic nature (Dufossé et al., 2014).

In conclusion, stringent testing and quality control measures are crucial for minimizing the risks associated with mycotoxins and ensuring that fungal additives are free from these harmful substances. However, the complete elimination of mycotoxins remains challenging due to their persistence and the complexity of food supply chains (Magan & Aldred, 2007). To reduce the risk of contamination, adopting good agricultural practices such as proper harvesting, drying, and storage techniques is essential. Effectively addressing contamination and mycotoxin risks requires a combination of preventive strategies, advanced detection methods, and continuous monitoring to ensure food safety and quality. By implementing these measures, the food industry can benefit from FMs while minimizing associated risks.

4.2 Safety and Regulatory Issues

Safety and regulatory issues pose significant challenges in the use of food matrices (FMs) in the food industry. Regarding regulatory aspects, Quorn™, a mycoprotein derived from *Fusarium venenatum*, is classified as Generally Recognized as Safe (GRAS). However, this fungus produces a trichothecene known as 4,15-diacetoxyscirpenol (DAS), necessitating careful monitoring during processing (Dufossé et al., 2014). To ensure that FMs are free from mycotoxins, rigorous testing and quality control measures are required. These regulations are essential to confirm that metabolites are safe for consumption and do not pose health risks.

Various naturally synthesized antifungals, such as Luotonin-A, have been shown to inhibit many fungal phytopathogens, including azoxystrobin (Yang et al., 2020). Although these antifungals are naturally occurring, but further optimization in their composition, along with proper evaluation and registration procedures, is necessary.

Additionally, clear labelling of FMs is vital to inform consumers about the presence of these additives (Freeman, 2015). This transparency allows individuals with allergies or sensitivities to make informed choices. By developing robust testing methods, optimized production processes, and ensuring regulatory standards, a safe and effective use of FMs can be achieved in the food industry.

4.3 Consumer Acceptance

The use of fungal metabolites (FMs) in the food industry presents a significant challenge, primarily due to concerns regarding safety, naturalness, and a general unfamiliarity with fungal-derived ingredients (Stoev, 2015). Many consumers view FMs sceptically, especially those obtained from fermentation or bioengineering processes (Adekoya et al., 2017). A common misconception is that these metabolites are unsafe, largely due to a lack of understanding about their origins and benefits (Smith et al., 2016). This perception can impede their widespread acceptance and utilization in food products.

To overcome this challenge, it is essential to educate consumers about the safety and advantages of these metabolites (Anato et al., 2024). Transparent communication regarding production processes, the natural origins of these metabolites, and their health benefits can help establish trust (Adekoya et al., 2017). Adhering to strict regulatory standards and implementing quality control measures can reassure consumers about the safety of FMs (Nada et al., 2022). Furthermore, clear labelling and marketing strategies that emphasize the natural and sustainable aspects of FMs can attract health-conscious consumers who seek clean-label products (Mishra et al., 2022). Highlighting the roles of FMs in enhancing flavour, preservation, and nutritional value can help shift consumer perceptions. Additionally, proactively addressing any potential safety concerns will further foster acceptance.

4.4 *Production Constraints*

Economic and production constraints pose significant challenges in the use of FMs in the food industry (Kappes et al., 2023). High costs associated with the production, management, detection, control, and safety of FMs is a primary concern area (Mukhtar et al., 2023). About 25–50% of stored cereals are affected by fungal contamination globally, while 5–10% of cereals remain unconsumed, resulting in notable economic losses (Alizadeh et al., 2022). These various factors raise the produce cost, thus a specialized growth media, controlled cultivation conditions, and sophisticated bioreactor systems are essential. Scaling up FMs' production from laboratory to industrial levels also poses difficulties (Garcia-Ochoa & Gomez, 2008; Meyer et al., 2016). Factors such as bioreactor design, nutrient supply, aeration, and contamination control become increasingly challenging as production volume increases (Ganeshan et al., 2021).

Additionally, the need for substantial raw materials, energy, and water raises sustainability concerns (Dhiman et al., 2024; Onomu & Okuthe, 2024). Practices that minimize environmental impact while ensuring resource efficiency are essential (Blasi et al., 2023; Bilal et al., 2020; Boyd et al., 2020). Finally, meeting regulatory standards for safety and quality can be time-consuming and costly that can delay the introduction of new FMs to the market (Andarge & Lichtenberg, 2020).

5 Solutions and Innovations

5.1 *Advances in Biotechnology*

Since the 1990s, fungal identification has advanced to recognize spoilage strains, aided by biotechnology for enhancing the production and application of FMs (Filtenborg et al., 1996; Roth et al., 2023). Advance techniques aid in isolating new strains and FMs from unexplored environments and their identification. *Wallemia sebi*, a xerophilic fungus, was identified to produce various metabolites such as wallidimidione, 15-azasterol, 24, 28-dihydro-15-azasterol, and alkaloid cyclopentapyridine (Peng et al., 2011; Desroches et al., 2014; Jančič et al., 2016). An actinobacterium strain G10 from Algerian soils inhibits OTA by suppressing the growth of *Aspergillus carbonarium* (El Khoury et al., 2017). G10 may disrupt the OTA biosynthesis pathway by inhibiting the expression of acpks, acOTApks, and acOTAnrps genes.

RNA transfer from plants to pathogens has been observed, and inhibitory RNAs have been used to target fungal genes to inhibit *Botrytis cinereal*, *Fusarium graminearum*, and *Sclerotina sclerotiorum* (Wang et al., 2016b; Wang & Jin, 2017; McLoughlin et al., 2018). Natural antifungals like natamycin, produced by *Streptomyces natlensis*, are applied in cheese, juice, wine, yogurt, and sausages by inhibiting *Fusarium oxysporum* (Dalhoff & Levy, 2015). Genetic engineering has improved antifungal resistance in crops like bananas (Bubici et al., 2019) and enhanced the production of medicinal compounds (chloramaphenicol, kanamycin, and tetracycline) from strains like *Streptomyces aureofaciens*, *S. kanamyceticus*, and *S. venezuelae* (Odeyemi et al., 2020). This includes the introduction of genes that boost the synthesis of desired compounds or the suppression of pathways that produce unwanted by-products.

Quick identification techniques and transcriptomics can further enhance antifungal biotechnology (Mishra et al., 2022; Hefny et al., 2024). Sustainable substrates for fungal cultivation have also gained traction (Madusanka et al., 2024), supported by advanced methods like high-throughput screening and metabolomics for rapid detection and analysis of metabolites (Yang et al., 2019; Sarnaik et al., 2020; Bayona et al., 2022).

5.2 Role of Microbes in Metabolite Detoxification

Genetic engineering of fungi plays a crucial role in enhancing the production and application of FMs in the food industry. *Aspergillus niger* and *A. flavus* have a natural capability to detoxify AFB1 (Wu et al., 2008). *Bacillus subtilis* and *Micrococcus luteus* can bind FuMB1 and FuMB2 (Niderkorn et al., 2009). The peptidoglycan in the bacterial cell walls has a binding site for FuMBs with tricarballylic acid (Umesha et al., 2017). Bacterium *Sphingopyxis* spp. hydrolyzes FuMB1 to HFuMB1 (hydrolyzed FuMB1) by carboxylesterase, followed by deamination of HFuMB1 using aminotransferase (Heinl et al., 2010). The anti-aflatoxin effect of *A. flavus* in maize (Mauro et al., 2018; Camiletti et al., 2018), almonds (Picot et al., 2018), pepper (Ortega-Beltran et al., 2019; Ezekiel et al., 2019), fig and pistachio (Ortega-Beltran et al., 2019) has been determined. PAT was effectively reduced by *S. cerevisiae* during fermentation via physical adsorption (Mendieta et al., 2018). PAT binds with protein and N-H/O-N carbohydrate chains of yeast cell walls and resulted to adsorb (Zhang et al., 2019). Actinobacteria have produced various antifungal enzymes that degrade/detoxify mycotoxins (Campos-Avelar et al., 2020).

5.3 Safe Storage and Processing Conditions

Inappropriate storage, moisture levels, and insect damage aid in increasing the risk of fungal growth and mycotoxin biosynthesis (Amaike & Keller, 2011). Temperatures above 36 °C can inhibit the production of FMs (O'Brian et al., 2007), while the 28–37 °C range can affect *A. flavus* gene expression. Alkaline pH, antioxidants, and oxidized N_2 inhibit AFs biosynthesis (Georgianna & Payne, 2009). FuMs are more prevalent in the outer bran of maize and can be reduced through a proper debranning process (Bullerman & Bianchini, 2007). FuMs decrease above 150 °C, and reached up to 90% reduction at a treatment of 175 °C for 1 hour (Marin et al., 2013; Bullerman & Bianchini, 2007). Frying tortilla chips at 190 °C for 15 minutes resulted in a 67% reduction in FuMs, while baking corn muffins at 175 and 200 °C led to 16% and 28% reductions, respectively. An A_w level of 0.85–0.86 can inhibit fungal metabolism and reduce FuM production (Bryła et al., 2013).

PAT can be significantly reduced during fermentation with *S. cerevisiae* (10 g/L cells), achieving 53.97% and 85.88% reductions in a 20% gluconate solution (500 μg/L PAT) after 6 and 24 hours, respectively (Zhang et al., 2019). Sorghum fermentation with *Lactobacillus fermentum* led to an 84% reduction in TCs and a 98% reduction in AFB1 (Adebo et al., 2019). OTA is primarily stored in the outer seed coat, pericarp, and seed husk. This can be removed and destroyed at temperatures above 250 °C (Petzinger & Ziegler, 2000; Marin et al., 2013). Deoxynivalenol remains unaffected by frying, but leaching may effectively reduce it due to its water-soluble nature (Sobrova et al., 2010).

5.4 Emerging Technologies

Emerging technologies are transforming the methods used to reduce mycotoxins and enhance food safety and quality. Traditional approaches often do not completely eliminate pathogenic fungi and mycotoxins, but recent advancements demonstrate significantly improved efficacy. For instance, high-pressure processing (HPP) has proven effective in inactivating spores and reducing the levels of DON and ZEA in maize grains (Kalagatur et al., 2018). HPP treatment at 380 MPa and 60 °C for 30 minutes completely inhibited spore germination, while 550 MPa and 45 °C for 20 minutes resulted in zero colony-forming units (CFU). Additionally, HPP, when combined with moderate heat (600 MPa and 85 °C) has been shown to inactivate *Taaromyces macrosporus* and *Aspergillus fischeri* spores in strawberry puree in 13 and 7 minutes, respectively (Timmermans et al., 2020). All spores were inactivated at 90 °C and 600 MPa in 3–7 minutes.

Figs were treated with ultrasound at 40 kHz for 30 minutes, followed by soaking in a 60% glucose solution for 24 hours and drying at 60 °C (Villalobos et al., 2019). This treatment resulted in mold counts falling below detection limits and a significant reduction in mycotoxin production. Additionally, a 98% suppression of fungi was achieved using 30 minutes of ultrasound at 26 kHz and an intensity of 1 W/cm^2 in an anolyte solution (Rudik et al., 2020).

Pulsed electric field (PEF) treatment with pulse widths of 65, 51, and 20 microseconds resulted in a maximum reduction of AFB1 by 94.88%, 92.25%, and 91.93% at pH levels of 10.0, 7.0, and 4.0, respectively (Vijayalakshmi et al., 2018). Furthermore, a disinfection efficiency of 99.47–99.84% was attained against *Fusarium oxysporum* using PEF treatment at an electric field intensity of 8 kV/cm for 60 seconds (Zhong et al., 2019).

5.5 Miscellaneous Strategies

Ozone (O_3) is recognized as GRAS for its ability to degrade mycotoxins due to its effective oxidant and disinfectant properties (Piemontese et al., 2018). Aflatoxin B1 and G1 degrade more rapidly than B2 and G2 when exposed to a concentration of 13.5 ppm of ozone, with treatment times of 3–7 minutes and 3–4 minutes, respectively (Agriopoulou et al., 2016). A maximum degradation of DON by 78.66% in whole wheat kernels and flour was achieved at a moisture content of 20.10%, using an ozone concentration of 100 mg/L over treatment duration of 60 minutes (Wang et al., 2016a). A 12-hour treatment of durum wheat grain with ozone concentrations of 32.5 g and 48 g per hour resulted in a reduction of DON levels by 31.17% and 47.77%, respectively (Piemontese et al., 2018). While the 6-hour treatment using 55 g of ozone per hour achieved a 30.98% reduction in DON content.

Chitosan has been shown to decrease mycotoxin concentrations in palm kernel cake (Pirouz et al., 2020). Under optimized conditions of pH 4.0, temperature of 35 °C, and duration of 8 hours, reductions of 92.95%, 46.58%, 79.48%, and 83.11% were observed for aflatoxins B1, B2, G1, and G2, respectively, as well as reductions of 87.96%, 52.61%, 89.85%, and 88.73% for OTA, ZEA, FB1, and FB2, respectively.

Low-pressure cold plasma (LPCP) treatment of hazelnuts using air gases and sulphur hexafluoride (SF6) separately for 20 minutes achieved a reduction of 50.53% and 20.95% in total AFs, respectively (Basaran et al., 2008). A 10-minute treatment of maize with cold atmospheric-pressure plasma (CAP) resulted in a 65% reduction in AFB1 and a 64% reduction in FuMB1 (Wielogorska et al., 2019). Additionally, an 8-minute CAP treatment reduced 93% in AFs, 90% in TCs, and 93% in fusariotoxins, along with the complete elimination of ZEAs (Hojnik et al., 2019).

A nanocomposite made from Al_2O_3, bentonite, and activated carbon reduced mycotoxins in beer samples by 87%, with an adsorption capacity of 450 μg/g of nanocomposite (González-Jartín et al., 2019). Another composite combining alginate, activated carbon, pectin, or graphene oxide achieved a 70% reduction in mycotoxins, with an adsorption capacity of 598 ng/g of nanocomposite. Nanoparticles have been found to be effective in reducing various mycotoxins, including DON, 3-acetylDON, FuMB1, FuMB2, T-2, HT-2, Nivalenol, and ZEAs, as well as inhibiting fungal colonies (Tarazona et al., 2019). Silver nanoparticles (AgNPs) were found to be effective in degrading several strains of *Fusarium*, including *F. culmorum*, *F. oxysporum*, *F. proliferatum*, *F. poae*, and *F. verticillioides*. A dose of 10 μg AgNPs/mL inhibited the growth of *F. poae* after just 2 hours, while a concentration of 45 μg AgNPs/mL inhibited the other *Fusarium* species in less than 4 hours. Additionally, upconversion nanoparticles (UCNP@TiO_2), at a concentration of 6.5 mg and under a 50 W xenon lamp, when dispersed in a 20 ppm DON solution at pH 8.0, achieved photocatalytic degradation of DON to the extent of 96% in 90 minutes and a complete degradation in 120 minutes (Zhou et al., 2020).

5.6 *Regulatory Approaches and Standards*

Regulatory bodies, such as the USFDA and the European Food Safety Authority (EFSA), have established strict safety standards that manufacturers must meet before their products can be approved for market release (Zovi et al., 2025). These standards are designed to prevent harmful contaminants, such as mycotoxins, from entering the food supply. Some essential oils are used as antifungal agents in coating films to help preserve horticultural products (Bibow & Oleszek, 2024). European Regulation No. 1334/2008 requires rigorous testing and evaluation of metabolites for potential toxicity, allergenicity, and other health risks (Shephard, 2016).

Additionally, regulations establish quality control measures that producers must follow to ensure the consistency and purity of food materials (Jenkins & Grzywacz, 2000). These measures include guidelines for Good Manufacturing Practices (GMP), encompassing all aspects of production from raw material sourcing to final product packaging and ensuring control over mycotoxin levels (Piotrowska, 2021; Yu et al., 2023). Regulatory frameworks also mandate traceability and transparency, which involve meticulous documentation and record-keeping practices allowing tracking throughout the supply chain. These standards ensure that food materials meet required quality benchmarks, promote transparency, and instil confidence in manufacturers and consumers (King et al., 2017; Chatterjee et al., 2023).

6 Future Prospects

Researchers are concentrating on several key areas to unlock the full potential of FMs in the food industry, which has a promising future. Ongoing research and innovation are expected to lead to new applications, enhance production processes, and contribute to healthier, more sustainable food products. Next paragraph Genetic and Metabolic Engineering: Advances in genetic and metabolic engineering will improve the production of fungal metabolites. Fungi have the potential to serve as a sustainable protein source, offering a synthesis rate comparable to that of an omnivorous diet (Derbyshire & Delange, 2021). By optimizing metabolic pathways and introducing novel genes, scientists aim to increase yields, improve efficiency, and develop new metabolites with unique properties.

Sustainable Production: There is an increasing focus on sustainable production methods. FMs are being used at the industry level for their enzymes, which include amylase, cellulase, lipase, protease, and xylanase (Ferreira et al., 2015; Karimi et al., 2019). This includes utilizing renewable substrates, such as agricultural waste, and developing eco-friendly extraction and purification techniques. These approaches not only lower production costs but also reduce environmental impacts. Next paragraph Advanced Detection and Analysis: Integrating advanced detection and analytical methods, such as high-throughput screening, metabolomics, and multi-omics, will enable more precise identification and quantification of FMs

(Chen et al., 2019; Shankar & Sharma, 2022). These techniques will help ensure the quality, safety, and efficacy of the final products.

Exploration of Novel Applications: Researchers are investigating new applications of FMs in areas such as functional foods, nutraceuticals, and natural preservatives. In recent years, mycoprotein has gained attention due to the rapid growth of fungi (Majumder et al., 2024). Mycelium from species like *Aspergillus oryzae*, *Monascus purpureus*, *Neurospora intermedia*, and *Rhizopus oryzae* is being consumed for its fibrous texture as a meat substitute (Filho et al., 2019; Sar et al., 2020). These applications can potentially enhance the nutritional value and shelf life of food products.

7 Conclusion

FMs are pivotal in the food industry, offering a wealth of benefits that span flavour enhancement, preservation, colouring, nutritional supplementation, fermentation, and biocontrol. These natural compounds are produced by fungi such as *Aspergillus*, *Penicillium*, and *Saccharomyces* species, are integral to various food production processes, contributing to food products'sensory and nutritional qualities. However, their use is not without challenges. Ensuring consistent quality, scaling up production, regulatory compliance, stability, and consumer acceptance are significant hurdles that must be addressed. Contamination and mycotoxin risks further complicate their application, necessitating stringent safety and quality control measures. Advancements in genetic engineering, sustainable substrates, and advanced detection methods drive innovation in this field. These solutions improve production efficiency and cost-effectiveness, as well as enhance the safety and sustainability of fungal metabolite applications. Collaborative research and consumer education are crucial for increasing acceptance and fostering trust in these natural compounds.

The future of FMs in the food industry is promising. Continued research and technological advancements will unlock new applications and improve existing processes, contributing to healthier and more sustainable food products. Strengthening regulatory frameworks, promoting sustainable production practices, and educating consumers about the benefits and safety of FMs will be key to harnessing their full potential. As the industry evolves, FMs will undoubtedly play an increasingly vital role in shaping the future of food production.

References

Abdullah, R. R. H., El-Wahab, A. H. A., & El-Salam, S. A. A. (2024). Insecticidal activity and possible modes of action of secondary metabolites of some fungal strains and wild plants as natural pesticides against *Spodoptera frugiperda*. *Beni-Suef University Journal of Basic and Applied Sciences, 13*, 9. https://doi.org/10.1186/s43088-024-00467-z

Abouzied, M. M., Horvath, A. D., Podlesny, P. M., Regina, N. P., Metodiev, V. D., Kamenova-Tozeva, R. M., Niagolova, N. D., Stein, A. D., Petropoulos, E. A., & Ganev, V. S. (2002). Ochratoxin A concentrations in food and feed from a region with balkan endemic nephropathy. *Food Additives & Contaminants, 19*(8), 755–764. https://doi.org/10.1080/02652030210145036

Adebo, O. A., Kayitesi, E., & Njobeh, P. B. (2019). Reduction of mycotoxins during fermentation of whole grain sorghum to whole grain ting (a southern African food). *Toxins, 11*(3), 180. https://doi.org/10.3390/toxins11030180

Adekoya, I., Njobeh, P., Obadina, A., Chilaka, C., Okoth, S., de Boevre, M., & de Saeger, S. (2017). Awareness and prevalence of mycotoxin contamination in selected Nigerian fermented foods. *Toxins, 9*(11), 363. https://doi.org/10.3390/toxins9110363

Agriopoulou, S., Koliadima, A., Karaiskakis, G., & Kapolos, J. (2016). Kinetic study of aflatoxins degradation in the presence of ozone. *Food Control, 61*, 221–226. https://doi.org/10.1016/j.foodcont.2015.09.013

Agriopoulou, S., Stamatelopoulou, E., & Varzakas, T. (2020). Advances in occurrence, importance, and mycotoxin control strategies: Prevention and detoxification in foods. *Foods, 9*(2), 137. https://doi.org/10.3390/foods9020137

Alam, K., Mazumder, A., Sikdar, S., Zhao, Y., Hao, J., Song, C., Wang, Y., Sarkar, R., Islam, S., Zhang, Y., & Li, A. (2022). Streptomyces: The biofactory of secondary metabolites. *Frontiers in Microbiology, 13*. https://doi.org/10.3389/fmicb.2022.968053

Alberts, A. W. (1990). Lovastatin and simvastatin – Inhibitors of HMG COA reductase and cholesterol biosynthesis. *Cardiology, 77*(4), 14–21. https://doi.org/10.1159/000174688

Alizadeh, A. M., Golzan, S. A., Mahdavi, A., Dakhili, S., Torki, Z., & Hosseini, H. (2022). Recent advances on the efficacy of essential oils on mycotoxin secretion and their mode of action. *Critical Reviews in Food Science and Nutrition, 62*(17), 4726–4751. https://doi.org/10.1080/10408398.2021.1878102

Amaike, S., & Keller, N. P. (2011). Aspergillus flavus. *Annual Review of Phytopathology, 49*(1), 107–133. https://doi.org/10.1146/annurev-phyto-072910-095221

Amaral-Fonseca, M., Morellon-Sterling, R., Fernández-Lafuente, R., & Tardioli, P. W. (2021). Optimization of simultaneous saccharification and isomerization of dextrin to high fructose syrup using a mixture of immobilized amyloglucosidase and glucose isomerase. *Catalysis Today, 362*, 175–183. https://doi.org/10.1016/j.cattod.2020.03.021

Anato, A., Headey, D., Hirvonen, K., Pokharel, A., Tessema, M., Wu, F., & Baye, K. (2024). Feed handling practices, aflatoxin awareness and children's milk consumption in the Sidama region of Southern Ethiopia. *One Health, 18*, 100672. https://doi.org/10.1016/j.onehlt.2023.100672

Andarge, T., & Lichtenberg, E. (2020). Regulatory compliance under enforcement gaps. *Journal of Regulatory Economics, 57*(3), 181–202. https://doi.org/10.1007/s11149-020-09405-0

Arunthirumeni, M., Vinitha, G., & Shivakumar, M. S. (2023). Antifeedant and larvicidal activity of bioactive compounds isolated from entomopathogenic fungi *Penicillium* sp. for the control of agricultural and medically important insect pest (*Spodoptera litura* and *Culex quinquefasciatus*). *Parasitology International, 92*, 102688. https://doi.org/10.1016/j.parint.2022.102688

Ashiq, S. (2015). Natural occurrence of mycotoxins in food and feed: Pakistan perspective. *Comprehensive Reviews in Food Science and Food Safety, 14*(2), 159–175. https://doi.org/10.1111/1541-4337.12122

Awuchi, C. G., Ondari, E. N., Ogbonna, C. U., Upadhyay, A. K., Baran, K., Okpala, C. O. R., Korzeniowska, M., & Guiné, R. P. F. (2021). Mycotoxins affecting animals, foods, humans, and plants: Types, occurrence, toxicities, action mechanisms, prevention, and detoxification strategies—A revisit. *Foods, 10*(6), 1279. https://doi.org/10.3390/foods10061279

Bains, A., Chawla, P., Kaur, S., Najda, A., Fogarasi, M., & Fogarasi, S. (2021). Bioactives from mushroom: Health attributes and food industry applications. *Materials, 14*(23), 7640. https://doi.org/10.3390/ma14237640

Basaran, P., Basaran-Akgul, N., & Oksuz, L. (2008). Elimination of *Aspergillus parasiticus* from nut surface with low pressure cold plasma (LPCP) treatment. *Food Microbiology, 25*(4), 626–632. https://doi.org/10.1016/j.fm.2007.12.005

Bayman, P., & Baker, J. L. (2006). Ochratoxins: A global perspective. *Mycopathologia, 162*(3), 215–223. https://doi.org/10.1007/s11046-006-0055-4

Bayona, L. M., De Voogd, N. J., & Choi, Y. H. (2022). Metabolomics on the study of marine organisms. *Metabolomics, 18*(3). https://doi.org/10.1007/s11306-022-01874-y

Becker-Algeri, T. A., Castagnaro, D., de Bortoli, K., de Souza, C., Drunkler, D. A., & Badiale-Furlong, E. (2016). Mycotoxins in bovine milk and dairy products: A review. *Journal of Food Science, 81*(3). https://doi.org/10.1111/1750-3841.13204

Behera, B. C. (2020). Citric acid from *Aspergillus niger*: A comprehensive overview. *Critical Reviews in Microbiology, 46*(6), 727–749. https://doi.org/10.1080/1040841x.2020.1828815

Bendejacq-Seychelles, A., Gibot-Leclerc, S., Guillemin, J., Mouille, G., & Steinberg, C. (2023). Phytotoxic fungal secondary metabolites as herbicides. *Pest Management Science, 80*(1), 92–102. https://doi.org/10.1002/ps.7813

Bennett, J. W., & Klich, M. (2003). Mycotoxins. *Clinical Microbiology Reviews, 16*(3), 497–516. https://doi.org/10.1128/cmr.16.3.497-516.2003

Bensmail, S., Boudjema, K., Naimi-Fazouane, F., Bensmail, S., Djouahra-Fahem, D., Ferhoum, F., & Bourfis, N. (2023). Factors affecting acid protease production by *Mucor circinelloides* MG603064.1 through SmF process: Characterization and fromage frais making. *BioTechnologia, 104*(4), 333–349. https://doi.org/10.5114/bta.2023.132770

Bernal, Á. R. R., Reynoso, C. M., Londoño, V. A. G., Broggi, L. E., & Resnik, S. L. (2019). *Alternaria* toxins in Argentinean wheat, bran, and flour. *Food Additives & Contaminants: Part B, 12*(1), 24–30. https://doi.org/10.1080/19393210.2018.1509900

Bhardwaj, N., Kumar, B., Agrawal, K., & Verma, P. (2021). Current perspective on production and applications of microbial cellulases: A review. *Bioresources and Bioprocessing, 8*(1). https://doi.org/10.1186/s40643-021-00447-6

Bibow, A., & Oleszek, W. (2024). Essential oils as potential natural antioxidants, antimicrobial, and antifungal agents in active food packaging. *Antibiotics, 13*(12), 1168. https://doi.org/10.3390/antibiotics13121168

Bilal, M., Wang, Z., Cui, J., Ferreira, L. F. R., Bharagava, R. N., & Iqbal, H. M. (2020). Environmental impact of lignocellulosic wastes and their effective exploitation as smart carriers – A drive towards greener and eco-friendlier biocatalytic systems. *The Science of the Total Environment, 722*, 137903. https://doi.org/10.1016/j.scitotenv.2020.137903

Blasi, A., Verardi, A., Lopresto, C. G., Siciliano, S., & Sangiorgio, P. (2023). Lignocellulosic agricultural waste valorization to obtain valuable products: An overview. *Recycling, 8*(4), 61. https://doi.org/10.3390/recycling8040061

Boyd, C. E., D'Abramo, L. R., Glencross, B. D., Huyben, D. C., Juarez, L. M., Lockwood, G. S., McNevin, A. A., Tacon, A. G. J., Teletchea, F., Tomasso, J. R., Tucker, C. S., & Valenti, W. C. (2020). Achieving sustainable aquaculture: Historical and current perspectives and future needs and challenges. *Journal of the World Aquaculture Society, 51*(3), 578–633. https://doi.org/10.1111/jwas.12714

Bruno, A., Pandolfo, G., Crucitti, M., Troili, G. M., Spina, E., Zoccali, R. A., & Muscatello, M. R. A. (2018). Red yeast rice (RYR) supplementation in patients treated with second-generation antipsychotics. *Complementary Therapies in Medicine, 37*, 167–171. https://doi.org/10.1016/j.ctim.2018.03.007

Bryła, M., Roszko, M., Szymczyk, K., Jędrzejczak, R., Obiedziński, M. W., & Sękul, J. (2013). Fumonisins in plant-origin food and fodder – A review. *Food Additives & Contaminants Part A, 30*(9), 1626–1640. https://doi.org/10.1080/19440049.2013.809624

Bubici, G., Kaushal, M., Prigigallo, M. I., Cabanás, C. G., & Mercado-Blanco, J. (2019). Biological control agents against *Fusarium* wilt of banana. *Frontiers in Microbiology, 10*. https://doi.org/10.3389/fmicb.2019.00616

Bui-Klimke, T. R., & Wu, F. (2015). Ochratoxin A and human health risk: A review of the evidence. *Critical Reviews in Food Science and Nutrition, 55*(13), 1860–1869. https://doi.org/10.1080/10408398.2012.724480

Bullerman, L. B., & Bianchini, A. (2007). Stability of mycotoxins during food processing. *International Journal of Food Microbiology, 119*(1–2), 140–146. https://doi.org/10.1016/j.ijfoodmicro.2007.07.035

Burke, F. M. (2015). Red yeast rice for the treatment of dyslipidemia. *Current Atherosclerosis Reports, 17*, 1–6. https://doi.org/10.1007/s11883-015-0495-8

Camiletti, B. X., Moral, J., Asensio, C. M., Torrico, A. K., Lucini, E. I., De La Paz Giménez-Pecci, M., & Michailides, T. J. (2018). Characterization of Argentinian endemic *Aspergillus flavus* isolates and their potential use as biocontrol agents for mycotoxins in maize. *Phytopathology, 108*(7), 818–828. https://doi.org/10.1094/phyto-07-17-0255-r

Campos-Avelar, I., Colas De La Noue, A., Durand, N., Fay, B., Martinez, V., Fontana, A., Strub, C., & Schorr-Galindo, S. (2020). Minimizing ochratoxin a contamination through the use of actinobacteria and their active molecules. *Toxins, 12*(5), 296. https://doi.org/10.3390/toxins12050296

Chai, K. F., Ng, K. R., Samarasiri, M., & Chen, W. N. (2022). Precision fermentation to advance fungal food fermentations. *Current Opinion in Food Science, 47*, 100881. https://doi.org/10.1016/j.cofs.2022.100881

Chatterjee, S., Dhole, A., Krishnan, A. A., & Banerjee, K. (2023). Mycotoxin monitoring, regulation and analysis in India: A success story. *Foods, 12*(4), 705. https://doi.org/10.3390/foods12040705

Chaudhary, V., Katyal, P., Poonia, A. K., Kaur, J., Puniya, A. K., & Panwar, H. (2022). Natural pigment from *Monascus*: The production and therapeutic significance. *Journal of Applied Microbiology, 133*(1), 18–38. https://doi.org/10.1111/jam.15308

Chávez, R., Vaca, I., & García-Estrada, C. (2023). Secondary metabolites produced by the blue-cheese ripening mold *Penicillium roqueforti*; Biosynthesis and regulation mechanisms. *Journal of Fungi, 9*(4), 459. https://doi.org/10.3390/jof9040459

Chen, F., Ma, R., & Chen, X. (2019). Advances of metabolomics in fungal pathogen–plant interactions. *Metabolites, 9*(8), 169. https://doi.org/10.3390/metabo9080169

Chopra, H., Mishra, A. K., Baig, A. A., Mohanta, T. K., Mohanta, Y. K., & Baek, K. H. (2021). Narrative review: Bioactive potential of various mushrooms as the treasure of versatile therapeutic natural product. *Journal of Fungi, 7*(9), 728. https://doi.org/10.3390/jof7090728

Chugh, R. M., Mittal, P., Namratha, M. P., Arora, T., Bhattacharya, T., Chopra, H., Cavalu, S., & Gautam, R. K. (2022). Fungal mushrooms: a natural compound with therapeutic applications. *Frontiers in Pharmacology, 13*. https://doi.org/10.3389/fphar.2022.925387

Chugh, P., Kaur, J., Soni, R., Sharma, A., & Soni, S. K. (2023). A low-cost process for efficient hydrolysis of deoiled rice bran and ethanol production using an inhouse produced multi-enzyme preparation from *Aspergillus niger* P-19. *Journal of Material Cycles and Waste Management, 25*(1), 359–375. https://doi.org/10.1007/s10163-022-01538-y

Coban, H. B. (2019). Organic acids as antimicrobial food agents: Applications and microbial productions. *Bioprocess and Biosystems Engineering, 43*(4), 569–591. https://doi.org/10.1007/s00449-019-02256-w

D'Arrigo, P., Rossato, L. A. M., Strini, A., & Serra, S. (2024). From waste to value: Recent insights into producing vanillin from lignin. *Molecules, 29*(2), 442. https://doi.org/10.3390/molecules29020442

D'Mello, J., Placinta, C., & Macdonald, A. (1999). *Fusarium* mycotoxins: a review of global implications for animal health, welfare and productivity. *Animal Feed Science and Technology, 80*(3–4), 183–205. https://doi.org/10.1016/s0377-8401(99)00059-0

da Silva, R. R. (2017). Bacterial and fungal proteolytic enzymes: Production, catalysis and potential applications. *Applied Biochemistry and Biotechnology, 183*(1), 1–19. https://doi.org/10.1007/s12010-017-2427-2

Dalhoff, A. A., & Levy, S. B. (2015). Does use of the polyene natamycin as a food preservative jeopardise the clinical efficacy of amphotericin B? A word of concern. *International Journal of Antimicrobial Agents, 45*(6), 564–567. https://doi.org/10.1016/j.ijantimicag.2015.02.011

Dänicke, S., & Winkler, J. (2015). Invited review: Diagnosis of zearalenone (ZEN) exposure of farm animals and transfer of its residues into edible tissues (carry over). *Food and Chemical Toxicology, 84*, 225–249. https://doi.org/10.1016/j.fct.2015.08.009

Darwesh, O. M., Matter, I. A., Almoallim, H. S., Alharbi, S. A., & Oh, Y. K. (2020). Isolation and optimization *of Monascus ruber* OMNRC45 for red pigment production and evaluation of the pigment as a food colorant. *Applied Sciences, 10*(24), 8867. https://doi.org/10.3390/app10248867

de Ruyck, K., De Boevre, M., Huybrechts, I., & De Saeger, S. (2015). Dietary mycotoxins, co-exposure, and carcinogenesis in humans: Short review. *Mutation Research-Reviews in Mutation Research, 766*, 32–41. https://doi.org/10.1016/j.mrrev.2015.07.003

Derbyshire, E. J., & Delange, J. (2021). Fungal protein—What is it and what is the health evidence? A systematic review focusing on mycoprotein. *Frontiers in Sustainable Food Systems, 5*. https://doi.org/10.3389/fsufs.2021.581682

Deshmukh, R., Mathew, A., & Purohit, H. J. (2014). Characterization of antibacterial activity of bikaverin from *Fusarium* sp. HKF15. *Journal of Bioscience and Bioengineering, 117*(4), 443–448. https://doi.org/10.1016/j.jbiosc.2013.09.017

Deshmukh, S. K., Prakash, V., & Ranjan, N. (2018). Marine fungi: A source of potential anticancer compounds. *Frontiers in Microbiology, 8*, 2536. https://doi.org/10.3389/fmicb.2017.02536

Desroches, T. C., McMullin, D. R., & Miller, J. D. (2014). Extrolites of *Wallemia sebi*, a very common fungus in the built environment. *Indoor Air, 24*(5), 533–542. https://doi.org/10.1111/ina.12100

Dhiman, S., Kaur, P., Narang, J., Mukherjee, G., Thakur, B., Kaur, S., & Tripathi, M. (2024). Fungal bioprocessing for circular bioeconomy: Exploring lignocellulosic waste valorization. *Mycology: An International Journal on Fungal Biology, 15*(4), 538–563. https://doi.org/10.1080/21501203.2024.2316824

do Prado, F. G., Pagnoncelli, M. G. B., De Melo Pereira, G. V., Karp, S. G., & Soccol, C. R. (2022). Fermented soy products and their potential health benefits: A review. *Microorganisms, 10*(8), 1606. https://doi.org/10.3390/microorganisms10081606

Dufossé, L., Fouillaud, M., Caro, Y., Mapari, S. A., & Sutthiwong, N. (2014). Filamentous fungi are large-scale producers of pigments and colorants for the food industry. *Current Opinion in Biotechnology, 26*, 56–61. https://doi.org/10.1016/j.copbio.2013.09.007

Dufossé, L., Fouillaud, M., & Caro, Y. (2021). Fungi and fungal metabolites for the improvement of human and animal nutrition and health. *Journal of Fungi, 7*(4), 274. https://doi.org/10.3390/jof7040274

Dumas, E., Feurtey, A., de la Vega, R. C. R., Prieur, S. L., Snirc, A., Coton, M., Thierry, A., Coton, E., Piver, M. L., Roueyre, D., Ropars, J., Branca, A., & Giraud, T. (2020). Independent domestication events in the blue-cheese fungus *Penicillium roqueforti*. *Molecular Ecology, 29*(14), 2639–2660. https://doi.org/10.1111/mec.15359

El Khoury, R., Mathieu, F., Atoui, A., Kawtharani, H., El Khoury, A., Afif, C., Maroun, R., & El Khoury, A. (2017). Ability of soil isolated actinobacterial strains to prevent, bind and biodegrade ochratoxin A. *Toxins, 9*(7), 222. https://doi.org/10.3390/toxins9070222

Elbanhawy, A. A., Elsherbiny, E. A., El-Mageed, A. E. A., & Abdel-Fattah, G. M. (2019). Potential of fungal metabolites as a biocontrol agent against cotton aphid, *Aphis gossypii* Glover and the possible mechanisms of action. *Pesticide Biochemistry and Physiology, 159*, 34–40. https://doi.org/10.1016/j.pestbp.2019.05.013

El-Gendi, H., Saleh, A. K., Badierah, R., Redwan, E. M., El-Maradny, Y. A., & El-Fakharany, E. M. (2022). A comprehensive insight into fungal enzymes: Structure, classification, and their role in mankind's challenges. *Journal of Fungi, 8*(1), 23. https://doi.org/10.3390/jof8010023

El-Kady, I. A., Zohri, A. N. A., & Hamed, S. R. (2014). Kojic acid production from agro-industrial by-products using fungi. *Biotechnology Research International, 2014*(1), 642385. https://doi.org/10.1155/2014/642385

El-Sayed, R. A., Jebur, A. B., Kang, W., & El-Demerdash, F. M. (2022). An overview on the major mycotoxins in food products: Characteristics, toxicity, and analysis. *Journal of Future Foods, 2*(2), 91–102. https://doi.org/10.1016/j.jfutfo.2022.03.002

Eriksen, G. S., Jäderlund, K. H., Moldes-Anaya, A., Schönheit, J., Bernhoft, A., Jaeger, G., Rundberget, T., & Skaar, I. (2010). Poisoning of dogs with tremorgenic Penicillium toxins. *Medical Mycology, 48*(1), 188–196. https://doi.org/10.3109/13693780903225821

Escrivá, L., Font, G., & Manyes, L. (2015). In vivo toxicity studies of fusarium mycotoxins in the last decade: A review. *Food and Chemical Toxicology, 78*, 185–206. https://doi.org/10.1016/j.fct.2015.02.005

Ezati, P., & Rhim, J. W. (2021). Starch and agar-based color-indicator films integrated with shikonin for smart packaging application of shrimp. *ACS Food Science & Technology, 1*(10), 1963–1969. https://doi.org/10.1021/acsfoodscitech.1c00292

Ezekiel, C. N., Ortega-Beltran, A., Oyedeji, E. O., Atehnkeng, J., Kössler, P., Tairu, F., Hoeschle-Zeledon, I., Karlovsky, P., Cotty, P. J., & Bandyopadhyay, R. (2019). Aflatoxin in chili peppers in Nigeria: Extent of contamination and control using atoxigenic *Aspergillus flavus* genotypes as biocontrol agents. *Toxins, 11*(7), 429. https://doi.org/10.3390/toxins11070429

Ferreira, J. A., Lennartsson, P. R., & Taherzadeh, M. J. (2015). Production of ethanol and biomass from thin stillage by *Neurospora intermedia*: A pilot study for process diversification. *Engineering in Life Sciences, 15*(8), 751–759. https://doi.org/10.1002/elsc.201400213

Filho, P. F. S., Andersson, D., Ferreira, J. A., & Taherzadeh, M. J. (2019). Mycoprotein: Environmental impact and health aspects. *World Journal of Microbiology and Biotechnology, 35*(10). https://doi.org/10.1007/s11274-019-2723-9

Filtenborg, O., Frisvad, J., & Thrane, U. (1996). Moulds in food spoilage. *International Journal of Food Microbiology, 33*(1), 85–102. https://doi.org/10.1016/0168-1605(96)01153-1

Fink-Gremmels, J., & Malekinejad, H. (2007). Clinical effects and biochemical mechanisms associated with exposure to the mycoestrogen zearalenone. *Animal Feed Science and Technology, 137*(3–4), 326–341. https://doi.org/10.1016/j.anifeedsci.2007.06.008

Freeman, A. (2015). Transparency for food consumers: Nutrition labeling and food oppression. *American Journal of Law & Medicine, 41*(2–3), 315–330. https://doi.org/10.1177/0098858815591520

Ganeshan, S., Kim, S. H., & Vujanovic, V. (2021). Scaling-up production of plant endophytes in bioreactors: Concepts, challenges and perspectives. *Bioresources and Bioprocessing, 8*(1). https://doi.org/10.1186/s40643-021-00417-y

Gao, W., Jiang, L., Ge, L., Chen, M., Geng, C., Yang, G., Li, Q., Ji, F., Yan, Q., Zou, Y., Zhong, L., & Liu, X. (2015). Sterigmatocystin-induced oxidative DNA damage in human liver-derived cell line through lysosomal damage. *Toxicology In Vitro, 29*(1), 1–7. https://doi.org/10.1016/j.tiv.2014.08.007

Garcia-Ochoa, F., & Gomez, E. (2008). Bioreactor scale-up and oxygen transfer rate in microbial processes: An overview. *Biotechnology Advances, 27*(2), 153–176. https://doi.org/10.1016/j.biotechadv.2008.10.006

Gardiner, D. M., Kazan, K., Praud, S., Torney, F. J., Rusu, A., & Manners, J. M. (2010). Early activation of wheat polyamine biosynthesis during *Fusarium* head blight implicates putrescine as an inducer of trichothecene mycotoxin production. *BMC Plant Biology, 10*(1), 289. https://doi.org/10.1186/1471-2229-10-289

Georgianna, D. R., & Payne, G. A. (2009). Genetic regulation of aflatoxin biosynthesis: From gene to genome. *Fungal Genetics and Biology, 46*(2), 113–125. https://doi.org/10.1016/j.fgb.2008.10.011

Giovati, L., Magliani, W., Ciociola, T., Santinoli, C., Conti, S., & Polonelli, L. (2015). AFM1 in milk: Physical, biological, and prophylactic methods to mitigate contamination. *Toxins, 7*(10), 4330–4349. https://doi.org/10.3390/toxins7104330

González-Jartín, J. M., De Castro Alves, L., Alfonso, A., Piñeiro, Y., Vilar, S. Y., Gomez, M. G., Osorio, Z. V., Sainz, M. J., Vieytes, M. R., Rivas, J., & Botana, L. M. (2019). Detoxification agents based on magnetic nanostructured particles as a novel strategy for mycotoxin mitigation in food. *Food Chemistry, 294*, 60–66. https://doi.org/10.1016/j.foodchem.2019.05.013

Hameed, A., Hussain, S. A., Yang, J., Ijaz, M. U., Liu, Q., Suleria, H. A. R., & Zeng, X. (2017). Antioxidants potential of the filamentous fungi (*Mucor circinelloides*). *Nutrients, 9*(10), 1101. https://doi.org/10.3390/nu9101101

Haschek, W. M., Gumprecht, L. A., Smith, G., Tumbleson, M. E., & Constable, P. D. (2001). Fumonisin toxicosis in swine: An overview of porcine pulmonary edema and current perspectives. *Environmental Health Perspectives, 109*(suppl 2), 251–257. https://doi.org/10.1289/ehp.01109s2251

Hattimare, D., Shakya, S., Patyal, A., Chandrakar, C., & Kumar, A. (2022). Occurrence and exposure assessment of Aflatoxin M1 in milk and milk products in India. *Journal of Food Science and Technology, 59*(6), 2460–2468. https://doi.org/10.1007/s13197-021-05265-4

He, W. (2025). Recent advances in metabolomic research within food mycology. *Journal of Food Protection, 88*(5), 100494. https://doi.org/10.1016/j.jfp.2025.100494

Hefny, Z. A., Ji, B., Elsemman, I. E., Nielsen, J., & Van Dijck, P. (2024). Transcriptomic meta-analysis to identify potential antifungal targets in *Candida albicans*. *BMC Microbiology, 24*(1). https://doi.org/10.1186/s12866-024-03213-8

Heinl, S., Hartinger, D., Thamhesl, M., Vekiru, E., Krska, R., Schatzmayr, G., Moll, W., & Grabherr, R. (2010). Degradation of fumonisin B1 by the consecutive action of two bacterial enzymes. *Journal of Biotechnology, 145*(2), 120–129. https://doi.org/10.1016/j.jbiotec.2009.11.004

Heitmann, M., Zannini, E., & Arendt, E. (2018). Impact of *Saccharomyces cerevisiae* metabolites produced during fermentation on bread quality parameters: A review. *Critical Reviews in Food Science and Nutrition, 58*(7), 1152–1164. https://doi.org/10.1080/10408398.2016.1244153

Heussner, A., & Bingle, L. (2015). Comparative ochratoxin toxicity: A review of the available data. *Toxins, 7*(10), 4253–4282. https://doi.org/10.3390/toxins7104253

Hojnik, N., Modic, M., Tavčar-Kalcher, G., Babič, J., Walsh, J. L., & Cvelbar, U. (2019). Mycotoxin decontamination efficacy of atmospheric pressure air plasma. *Toxins, 11*(4), 219. https://doi.org/10.3390/toxins11040219

Hu, D., Yang, X., Hu, C., Feng, Z., Chen, W., & Shi, H. (2021). Comparison of ergosterol and vitamin D2 in mushrooms *Agaricus bisporus* and *Cordyceps militaris* using ultraviolet irradiation directly on dry powder or in ethanol suspension. *ACS Omega, 6*(44), 29506–29515. https://doi.org/10.1021/acsomega.1c03561

Iqbal, S. Z., Waseem, M., Razis, A. F. A., Bhatti, I. A., Khaneghah, A. M., Mohammed, O. A., Lakshminarayanan, S. P., & Iqbal, M. (2024). Mycotoxin patulin contamination in various fruits and estimating its dietary impact on the consumers: From orchard to table. *Heliyon, 10*(10), e30252. https://doi.org/10.1016/j.heliyon.2024.e30252

Jančič, S., Frisvad, J. C., Kocev, D., Gostinčar, C., Džeroski, S., & Gunde-Cimerman, N. (2016). Production of secondary metabolites in extreme environments: Food- and airborne *Wallemia* spp. produce toxic metabolites at hypersaline conditions. *PLoS One, 11*(12), e0169116. https://doi.org/10.1371/journal.pone.0169116

Jenkins, N. E., & Grzywacz, D. (2000). Quality control of fungal and viral biocontrol agents – assurance of product performance. *Biocontrol Science and Technology, 10*(6), 753–777. https://doi.org/10.1080/09583150020011717

Jeswal, P., & Kumar, D. (2015). Mycobiota and natural incidence of aflatoxins, ochratoxin A, and citrinin in Indian spices confirmed by LC-MS/MS. *International Journal of Microbiology, 2015*, 1–8. https://doi.org/10.1155/2015/242486

Kalagatur, N. K., Kamasani, J. R., Mudili, V., Krishna, K., Chauhan, O. P., & Sreepathi, M. H. (2018). Effect of high pressure processing on growth and mycotoxin production of *Fusarium graminearum* in maize. *Food Bioscience, 21*, 53–59. https://doi.org/10.1016/j.fbio.2017.11.005

Kalaivani, M., Sabitha, R., Kalaiselvan, V., & Rajasekaran, A. (2010). Health benefits and clinical impact of major nutrient, red yeast rice: A review. *Food and Bioprocess Technology, 3*(3), 333–339. https://doi.org/10.1007/s11947-009-0197-8

Kappes, A., Tozooneyi, T., Shakil, G., Railey, A. F., McIntyre, K. M., Mayberry, D. E., Rushton, J., Pendell, D. L., & Marsh, T. L. (2023). Livestock health and disease economics: A scoping review of selected literature. *Frontiers in Veterinary Science, 10*. https://doi.org/10.3389/fvets.2023.1168649

Kardos, N., & Demain, A. L. (2011). Penicillin: The medicine with the greatest impact on therapeutic outcomes. *Applied Microbiology and Biotechnology, 92*(4), 677–687. https://doi.org/10.1007/s00253-011-3587-6

Karimi, S., Soofiani, N. M., Lundh, T., Mahboubi, A., Kiessling, A., & Taherzadeh, M. J. (2019). Evaluation of filamentous fungal biomass cultivated on vinasse as an alternative nutrient source of fish feed: Protein, lipid, and mineral composition. *Fermentation, 5*(4), 99. https://doi.org/10.3390/fermentation5040099

Kebede, H., Abbas, H., Fisher, D., & Bellaloui, N. (2012). Relationship between aflatoxin contamination and physiological responses of corn plants under drought and heat stress. *Toxins, 4*(11), 1385–1403. https://doi.org/10.3390/toxins4111385

Keller, N. P. (2019). Fungal secondary metabolism: Regulation, function and drug discovery. *Nature Reviews Microbiology, 17*(3), 167–180. https://doi.org/10.1038/s41579-018-0121-1

Kew, M. C. (2013). Aflatoxins as a cause of hepatocellular carcinoma. *Journal of Gastrointestinal & Liver Diseases, 22*(3), 305–310.

Khan, R. (2024). Mycotoxins in food: Occurrence, health implications, and control strategies- A comprehensive review. *Toxicon, 248*, 108038. https://doi.org/10.1016/j.toxicon.2024.108038

Khan, R., Brishti, F. H., Arulrajah, B., Goh, Y. M., Rahim, M. H. A., Karim, R., Hajar-Azhari, S., Kit, S. K., Anwar, F., & Saari, N. (2024). Mycoprotein as a meat substitute: Production, functional properties, and current challenges-A review. *International Journal of Food Science & Technology, 59*(1), 522–544. https://doi.org/10.1111/ijfs.16791

Kim, J. H., Chang, P. K., Chan, K. L., Faria, N. C., Mahoney, N., Kim, Y. K., Martins, M. D. L., & Campbell, B. C. (2012). Enhancement of commercial antifungal agents by kojic acid. *International Journal of Molecular Sciences, 13*(11), 13867–13880. https://doi.org/10.3390/ijms131113867

King, T., Cole, M., Farber, J. M., Eisenbrand, G., Zabaras, D., Fox, E. M., & Hill, J. P. (2017). Food safety for food security: Relationship between global megatrends and developments in food safety. *Trends in Food Science & Technology, 68*, 160–175. https://doi.org/10.1016/j.tifs.2017.08.014

Klich, M. A. (2007). *Aspergillus flavus*: The major producer of aflatoxin. *Molecular Plant Pathology, 8*(6), 713–722. https://doi.org/10.1111/j.1364-3703.2007.00436.x

Kourti, D., Angelopoulou, M., Petrou, P., & Kakabakos, S. (2024). Sensitive aflatoxin M1 detection in milk by ELISA: Investigation of different assay configurations. *Toxins, 16*(12), 515. https://doi.org/10.3390/toxins16120515

Krupodorova, T., Barshteyn, V., Dzhagan, V., Pluzhnyk, A., Zaichenko, T., & Blume, Y. (2024). Enhancement of antioxidant activity and total phenolic content of *Fomitopsis pinicola* mycelium extract. *Fungal Biology and Biotechnology, 11*(1). https://doi.org/10.1186/s40694-024-00187-0

Kumar, V., Ahluwalia, V., Saran, S., Kumar, J., Patel, A. K., & Singhania, R. R. (2021). Recent developments on solid-state fermentation for production of microbial secondary metabolites: Challenges and solutions. *Bioresource Technology, 323*, 124566. https://doi.org/10.1016/j.biortech.2020.124566

Lahlali, R., Ezrari, S., Radouane, N., Kenfaoui, J., Esmaeel, Q., Hamss, H. E., Belabess, Z., & Barka, E. A. (2022). Biological control of plant pathogens: A global perspective. *Microorganisms, 10*(3), 596. https://doi.org/10.3390/microorganisms10030596

Li, Y., Wang, Z., Beier, R. C., Shen, J., De Smet, D., De Saeger, S., & Zhang, S. (2011). T-2 toxin, a trichothecene mycotoxin: Review of toxicity, metabolism, and analytical methods. *Journal of Agricultural and Food Chemistry, 59*(8), 3441–3453. https://doi.org/10.1021/jf200767q

Li, Y., Zhang, H., Chen, Z., Fan, J., Chen, T., Xiao, Y., Jie, J., Zeng, B., & Zhang, Z. (2022). Overexpression of a novel gene Aokap2 affects the growth and kojic acid production in

Aspergillus oryzae. *Molecular Biology Reports, 49*(4), 2745–2754. https://doi.org/10.1007/s11033-021-07084-4

Li, K., Qiao, K., Xiong, J., Guo, H., & Zhang, Y. (2023). Nutritional values and bio-functional properties of fungal proteins: Applications in foods as a sustainable source. *Foods, 12*(24), 4388. https://doi.org/10.3390/foods12244388

Lins, U. M., Mendonça, R. D. S., Dantas, S. S., de Souza, A. F., Montero-Rodríguez, D., Andrade, R. F. D. S., & Campos-Takaki, G. M. (2024). Bioconversion of alternative substrates for the biosynthesis of HMG-CoA reductase inhibitors by *Aspergillus* spp. strains with antimicrobial potential. *Fermentation, 10*(7). https://doi.org/10.3390/fermentation10070367

Liu, W., An, C., Shu, X., Meng, X., Yao, Y., Zhang, J., Chen, F., Xiang, H., Yang, S., Gao, X., & Gao, S. S. (2020). A dual-plasmid CRISPR/CAS system for mycotoxin elimination in polykaryotic industrial fungi. *ACS Synthetic Biology, 9*(8), 2087–2095. https://doi.org/10.1021/acssynbio.0c00178

Liu, J., Chen, Z., Yang, X., Sun, Z., & Mandlaa, N. (2024). Malt in storage: Changes in amylase, protease, and cellulase activities and their effects on the Maillard reaction. *Journal of Food Composition and Analysis, 132*, 106352. https://doi.org/10.1016/j.jfca.2024.106352

Lotfy, W. A., Ghanem, K. M., & El-Helow, E. R. (2007). Citric acid production by a novel *Aspergillus niger* isolate: II. Optimization of process parameters through statistical experimental designs. *Bioresource Technology, 98*(18), 3470–3477. https://doi.org/10.1016/j.biortech.2006.11.032

Lull, C., Wichers, H. J., & Savelkoul, H. F. J. (2005). Antiinflammatory and immunomodulating properties of fungal metabolites. *Mediators of Inflammation, 2005*, 63–80. https://doi.org/10.1155/MI.2005.63

Machiani, M. A., Javanmard, A., Machiani, R. H., & Sadeghpour, A. (2022). Arbuscular mycorrhizal fungi and changes in primary and secondary metabolites. *Plants, 11*(17), 2183. https://doi.org/10.3390/plants11172183

Madusanka, C., Udayanga, D., Nilmini, R., Rajapaksha, S., Hewawasam, C., Manamgoda, D., & Vasco-Correa, J. (2024). A review of recent advances in fungal mycelium based composites. *Discover Materials, 4*(1). https://doi.org/10.1007/s43939-024-00084-8

Mafe, A. N., & Büsselberg, D. (2024). Mycotoxins in food: Cancer risks and strategies for control. *Foods, 13*(21), 3502. https://doi.org/10.3390/foods13213502

Magan, N., & Aldred, D. (2007). Post-harvest control strategies: Minimizing mycotoxins in the food chain. *International Journal of Food Microbiology, 119*(1–2), 131–139. https://doi.org/10.1016/j.ijfoodmicro.2007.07.034

Magoha, H., De Meulenaer, B., Kimanya, M., Hipolite, D., Lachat, C., & Kolsteren, P. (2014). Fumonisin B1 contamination in breast milk and its exposure in infants under 6 months of age in Rombo, Northern Tanzania. *Food and Chemical Toxicology, 74*, 112–116. https://doi.org/10.1016/j.fct.2014.09.008

Majumder, R., Miatur, S., Saha, A., & Hossain, S. (2024). Mycoprotein: Production and nutritional aspects: A review. *Sustainable Food Technology, 2*(1), 81–91. https://doi.org/10.1039/d3fb00169e

Malir, F., Ostry, V., & Novotna, E. (2013). Toxicity of the mycotoxin ochratoxin A in the light of recent data. *Toxin Reviews, 32*(2), 19–33. https://doi.org/10.3109/15569543.2013.782504

Marasas, W. F., Riley, R. T., Hendricks, K. A., Stevens, V. L., Sadler, T. W., Waes, J. G., Missmer, S. A., Cabrera, J., Torres, O., Gelderblom, W. C., Allegood, J., Martínez, C., Maddox, J., Miller, J. D., Starr, L., Sullards, M. C., Roman, A. V., Voss, K. A., Wang, E., & Merrill, A. H. (2004). Fumonisins disrupt sphingolipid metabolism, folate transport, and neural tube development in embryo culture and in vivo: A potential risk factor for human neural tube defects among populations consuming fumonisin-contaminated maize. *The Journal of Nutrition, 134*(4), 711–716. https://doi.org/10.1093/jn/134.4.711

Marin, S., Ramos, A., Cano-Sancho, G., & Sanchis, V. (2013). Mycotoxins: Occurrence, toxicology, and exposure assessment. *Food and Chemical Toxicology, 60*, 218–237. https://doi.org/10.1016/j.fct.2013.07.047

Martău, G. A., Călinoiu, L., & Vodnar, D. C. (2021). Bio-vanillin: Towards a sustainable industrial production. *Trends in Food Science & Technology, 109*, 579–592. https://doi.org/10.1016/j.tifs.2021.01.059

Martínez-Rodríguez, P., Guerrero-Rubio, M. A., Henarejos-Escudero, P., García-Carmona, F., & Gandía-Herrero, F. (2022). Health-promoting potential of betalains in vivo and their relevance as functional ingredients: A review. *Trends in Food Science & Technology, 122*, 66–82. https://doi.org/10.1016/j.tifs.2022.02.020

Mauro, A., Garcia-Cela, E., Pietri, A., Cotty, P., & Battilani, P. (2018). Biological control products for aflatoxin prevention in Italy: Commercial field evaluation of atoxigenic *Aspergillus flavus* active ingredients. *Toxins, 10*(1), 30. https://doi.org/10.3390/toxins10010030

McCormick, S. P., Stanley, A. M., Stover, N. A., & Alexander, N. J. (2011). Trichothecenes: From simple to complex mycotoxins. *Toxins, 3*(7), 802–814. https://doi.org/10.3390/toxins3070802

McLoughlin, A. G., Wytinck, N., Walker, P. L., Girard, I. J., Rashid, K. Y., De Kievit, T., Fernando, W. G. D., Whyard, S., & Belmonte, M. F. (2018). Identification and application of exogenous dsRNA confers plant protection against *Sclerotinia sclerotiorum* and *Botrytis cinerea. Scientific Reports, 8*(1). https://doi.org/10.1038/s41598-018-25434-4

Meena, M., Prajapati, P., Ravichandran, C., & Sehrawat, R. (2021). Natamycin: A natural preservative for food applications—A review. *Food Science and Biotechnology, 30*(12), 1481–1496. https://doi.org/10.1007/s10068-021-00981-1

Mendieta, C. R., Gómez, G. V., Del Río, J. C. G., Cuevas, A. C., Arce, J. M., & Ávila, E. G. (2018). Effect of the addition of *Saccharomyces cerevisiae* yeast cell walls to diets with mycotoxins on the performance and immune responses of broilers. *The Journal of Poultry Science, 55*(1), 38–46. https://doi.org/10.2141/jpsa.0170019

Meyer, V., Andersen, M. R., Brakhage, A. A., Braus, G. H., Caddick, M. X., Cairns, T. C., De Vries, R. P., Haarmann, T., Hansen, K., Hertz-Fowler, C., Krappmann, S., Mortensen, U. H., Peñalva, M. A., Ram, A. F. J., & Head, R. M. (2016). Current challenges of research on filamentous fungi in relation to human welfare and a sustainable bio-economy: A white paper. *Fungal Biology and Biotechnology, 3*(6). https://doi.org/10.1186/s40694-016-0024-8

Miller, S. A., Ferreira, J. P., & LeJeune, J. T. (2022). Antimicrobial use and resistance in plant agriculture: A one health perspective. *Agriculture, 12*(2), 289. https://doi.org/10.3390/agriculture12020289

Mishra, B., Mishra, A. K., Kumar, S., Mandal, S. K., Nsv, L., Kumar, V., Baek, K., & Mohanta, Y. K. (2022). Antifungal metabolites as food bio-preservative: Innovation, outlook, and challenges. *Metabolites, 12*(1), 12. https://doi.org/10.3390/metabo12010012

Mohamed, H., Naz, T., Yang, J., Shah, A. M., Nazir, Y., & Song, Y. (2021). Recent molecular tools for the genetic manipulation of highly industrially important Mucoromycota fungi. *Journal of Fungi, 7*(12), 1061. https://doi.org/10.3390/jof7121061

Mohammed, Y. M. M., & Badawy, M. E. I. (2020). Potential of phytopathogenic fungal isolates as a biocontrol agent against some weeds. *Egyptian Journal of Biological Pest Control, 30*(1). https://doi.org/10.1186/s41938-020-00295-0

Mohankumari, H. P., Naidu, K. A., Narasimhamurthy, K., & Vijayalakshmi, G. (2021). Bioactive pigments of *Monascus purpureus* attributed to antioxidant, HMG-CoA reductase inhibition and anti-atherogenic functions. *Frontiers in Sustainable Food Systems, 5*, 590427. https://doi.org/10.3389/fsufs.2021.590427

Moreira, M., Feijoo, G., & Lema, J. (2003). Fungal bioreactors: Applications to white-rot fungi. *Reviews in Environmental Science and Bio/Technology, 2*(2–4), 247–259. https://doi.org/10.1023/b:resb.0000040463.80808.dc

Mostrom, M. S., & Raisbeck, M. F. (2012). Trichothecenes. In R. C. Gupta (Ed.), *Veterinary toxicology* (pp. 1239–1265). Academic Press. https://doi.org/10.1016/B978-0-12-385926-6.00108-3

Moura, M. A. F. E., Gomes, D. C., & Takahashi, J. A. (2023). Applications of fungi secondary metabolites in the food industry. In M. Carocho, S. A. Heleno, & L. Barros (Eds.), *Natural secondary metabolites* (pp. 739–776). Springer, Cham Springer eBooks. https://doi.org/10.1007/978-3-031-18587-8_24

Mukhtar, K., Nabi, B. G., Ansar, S., Bhat, Z. F., Aadil, R. M., & Khaneghah, A. M. (2023). Mycotoxins and consumer's awareness: Recent progress and future challenges. *Toxicon, 232*, 107227. https://doi.org/10.1016/j.toxicon.2023.107227

Murphy, E. J., Rezoagli, E., Major, I., Rowan, N. J., & Laffey, J. G. (2020). β-Glucan metabolic and immunomodulatory properties and potential for clinical application. *Journal of Fungi, 6*(4), 356. https://doi.org/10.3390/jof6040356

Nada, S., Nikola, T., Bozidar, U., Ilija, D., & Andreja, R. (2022). Prevention and practical strategies to control mycotoxins in the wheat and maize chain. *Food Control, 136*, 108855. https://doi.org/10.1016/j.foodcont.2022.108855

Naeem, M., Manzoor, S., Abid, M., Tareen, M. B. K., Asad, M., Mushtaq, S., Ehsan, N., Amna, D., Xu, B., & Hazafa, A. (2022). Fungal proteases as emerging biocatalysts to meet the current challenges and recent developments in biomedical therapies: An updated review. *Journal of Fungi, 8*(2), 109. https://doi.org/10.3390/jof8020109

Narvhus, J. A., & Gadaga, T. H. (2003). The role of interaction between yeasts and lactic acid bacteria in African fermented milks: A review. *International Journal of Food Microbiology, 86*(1–2), 51–60. https://doi.org/10.1016/S0168-1605(03)00247-2

Niderkorn, V., Morgavi, D., Aboab, B., Lemaire, M., & Boudra, H. (2009). Cell wall component and mycotoxin moieties involved in the binding of fumonisin B1 and B2 by lactic acid bacteria. *Journal of Applied Microbiology, 106*(3), 977–985. https://doi.org/10.1111/j.1365-2672.2008.04065.x

Nielsen, J. C., & Nielsen, J. (2017). Development of fungal cell factories for the production of secondary metabolites: Linking genomics and metabolism. *Synthetic and Systems Biotechnology, 2*(1), 5–12. https://doi.org/10.1016/j.synbio.2017.02.002

Nohynek, G. J., Kirkland, D., Marzin, D., Toutain, H., Leclerc-Ribaud, C., & Jinnai, H. (2004). An assessment of the genotoxicity and human health risk of topical use of kojic acid [5-hydroxy-2-(hydroxymethyl)-4H-pyran-4-one]. *Food and Chemical Toxicology, 42*(1), 93–105. https://doi.org/10.1016/j.fct.2003.08.008

O'Brian, G., Georgianna, D., Wilkinson, J., Yu, J., Abbas, H., Bhatnagar, D., Cleveland, T., Nierman, W., & Payne, G. (2007). The effect of elevated temperature on gene transcription and aflatoxin biosynthesis. *Mycologia, 99*(2), 232–239. https://doi.org/10.1080/15572536.2007.11832583

Odeyemi, O. A., Alegbeleye, O. O., Strateva, M., & Stratev, D. (2020). Understanding spoilage microbial community and spoilage mechanisms in foods of animal origin. *Comprehensive Reviews in Food Science and Food Safety, 19*(2), 311–331. https://doi.org/10.1111/1541-4337.12526

Onomu, A. J., & Okuthe, G. E. (2024). The application of fungi and their secondary metabolites in aquaculture. *Journal of Fungi, 10*(10), 711. https://doi.org/10.3390/jof10100711

Ortega-Beltran, A., Moral, J., Picot, A., Puckett, R. D., Cotty, P. J., & Michailides, T. J. (2019). Atoxigenic *Aspergillus flavusIsolates* endemic to almond, fig, and pistachio orchards in California with potential to reduce aflatoxin contamination in these crops. *Plant Disease, 103*(5), 905–912. https://doi.org/10.1094/pdis-08-18-1333-re

Papadaki, E., & Mantzouridou, F. T. (2023). Application of *Aspergillus niger* for extracellular tannase and gallic acid production in non-sterile table olive processing wastewaters. *Waste and Biomass Valorization, 15*(2), 1199–1212. https://doi.org/10.1007/s12649-023-02242-0

Parapouli, M., Vasileiadi, A., Afendra, A., & Hatziloukas, E. (2020). *Saccharomyces cerevisiae* and its industrial applications. *AIMS Microbiology, 6*(1), 1–32. https://doi.org/10.3934/microbiol.2020001

Peng, X., Wang, Y., Liu, P., Hong, K., Chen, H., Yin, X., & Zhu, W. (2011). Aromatic compounds from the halotolerant fungal strain of *Wallemia sebi* PXP-89 in a hypersaline medium. *Archives of Pharmacal Research, 34*(6), 907–912. https://doi.org/10.1007/s12272-011-0607-0

Pestka, J. (2008). Mechanisms of deoxynivalenol-induced gene expression and apoptosis. *Food Additives & Contaminants Part A, 25*(9), 1128–1140. https://doi.org/10.1080/02652030802056626

Petzinger, N., & Ziegler, N. (2000). Ochratoxin A from a toxicological perspective. *Journal of Veterinary Pharmacology and Therapeutics, 23*(2), 91–98. https://doi.org/10.1046/j.1365-2885.2000.00244.x

Picot, A., Doster, M., Islam, M., Callicott, K., Ortega-Beltran, A., Cotty, P., & Michailides, T. (2018). Distribution and incidence of atoxigenic Aspergillus flavus VCG in tree crop orchards in California: A strategy for identifying potential antagonists, the example of almonds. *International Journal of Food Microbiology, 265*, 55–64. https://doi.org/10.1016/j.ijfoodmicro.2017.10.023

Piemontese, L., Messia, M. C., Marconi, E., Falasca, L., Zivoli, R., Gambacorta, L., Perrone, G., & Solfrizzo, M. (2018). Effect of gaseous ozone treatments on DON, microbial contaminants and technological parameters of wheat and semolina. *Food Additives & Contaminants Part A, 35*(4), 761–772. https://doi.org/10.1080/19440049.2017.1419285

Piotrowska, M. (2021). Microbiological decontamination of mycotoxins: Opportunities and limitations. *Toxins, 13*(11), 819. https://doi.org/10.3390/toxins13110819

Pirouz, A. A., Selamat, J., Iqbal, S. Z., & Samsudin, N. I. P. (2020). Efficient and simultaneous chitosan-mediated removal of 11 mycotoxins from palm kernel cake. *Toxins, 12*(2), 115. https://doi.org/10.3390/toxins12020115

Pouris, J., Kolyva, F., Bratakou, S., Vogiatzi, C. A., Chaniotis, D., & Beloukas, A. (2024). The role of fungi in food production and processing. *Applied Sciences, 14*(12), 5046. https://doi.org/10.3390/app14125046

Raghuvanshi, V., & Katiyar, D. (2024). Terrein: Isolation, synthesis, and upscaling of a functional scaffold for multifaceted biomedical applications. *Medicinal Chemistry Research*, 1–16. https://doi.org/10.1007/s00044-024-03294-0

Rana, N., Walia, A., & Gaur, A. (2013). α-Amylases from microbial sources and its potential applications in various industries. *National Academy Science Letters, 36*(1), 9–17. https://doi.org/10.1007/s40009-012-0104-0

Rao, M. P. N., Xiao, M., & Li, W. (2017). Fungal and bacterial pigments: Secondary metabolites with wide applications. *Frontiers in Microbiology, 8*. https://doi.org/10.3389/fmicb.2017.01113

Reddy, L., & Bhoola, K. (2010). Ochratoxins—Food contaminants: Impact on human health. *Toxins, 2*(4), 771–779. https://doi.org/10.3390/toxins2040771

Redondo-Blanco, S., Fernández, J., López-Ibáñez, S., Miguélez, E. M., Villar, C. J., & Lombó, F. (2020). Plant phytochemicals in food preservation: Antifungal bioactivity: A review. *Journal of Food Protection, 83*(1), 163–171. https://doi.org/10.4315/0362-028x.jfp-19-163

Richard, J. L. (2007). Some major mycotoxins and their mycotoxicoses—An overview. *International Journal of Food Microbiology, 119*(1–2), 3–10. https://doi.org/10.1016/j.ijfoodmicro.2007.07.019

Roth, M. G., Westrick, N. M., & Baldwin, T. T. (2023). Fungal biotechnology: From yesterday `to tomorrow. *Frontiers in Fungal Biology, 4*. https://doi.org/10.3389/ffunb.2023.1135263

Rudik, F. Y., Morgunova, N. L., & Krasnikova, E. S. (2020). Decontamination of grain by ultrasound. *IOP Conference Series Earth and Environmental Science, 421*(2), 022022. https://doi.org/10.1088/1755-1315/421/2/022022

Salama, H. A., Badr, A. N., Elkhadragy, M. F., Hussein, A. M. S., Shaban, I. A. S., & Yehia, H. M. (2021). New antifungal microbial pigment applied to improve safety and quality of processed meat-products. *Microorganisms, 9*(5), 989. https://doi.org/10.3390/microorganisms9050989

Sar, T., Ferreira, J. A., & Taherzadeh, M. J. (2020). Bioprocessing strategies to increase the protein fraction of *Rhizopus oryzae* biomass using fish industry sidestreams. *Waste Management, 113*, 261–269. https://doi.org/10.1016/j.wasman.2020.06.005

Sarnaik, A., Liu, A., Nielsen, D., & Varman, A. M. (2020). High-throughput screening for efficient microbial biotechnology. *Current Opinion in Biotechnology, 64*, 141–150. https://doi.org/10.1016/j.copbio.2020.02.019

Scott, P. (2012). Recent research on fumonisins: A review. *Food Additives & Contaminants Part A, 29*(2), 242–248. https://doi.org/10.1080/19440049.2010.546000

Sevgili, A., & Erkmen, O. (2019). Improved lycopene production from different substrates by mated fermentation of *Blakeslea trispora*. *Foods, 8*(4), 120. https://doi.org/10.3390/foods8040120

Shankar, A., & Sharma, K. K. (2022). Fungal secondary metabolites in food and pharmaceuticals in the era of multi-omics. *Applied Microbiology and Biotechnology, 106*(9–10), 3465–3488. https://doi.org/10.1007/s00253-022-11945-8

Shanmuga, P. A. (2021). Screening of fungi for production and purification of omega-3 fatty acid. *Journal of Medical Pharmaceutical and Allied Sciences, 10*(4), 3089–3093. https://doi.org/10.22270/jmpas.v10i4.1098

Sharma, A., Arya, S. K., Singh, J., Kapoor, B., Bhatti, J. S., Suttee, A., & Singh, G. (2023). Prospects of chitinase in sustainable farming and modern biotechnology: An update on recent progress and challenges. *Biotechnology and Genetic Engineering Reviews*, 1–31. https://doi.org/10.1080/02648725.2023.2183593

Shephard, G. S. (2016). Current status of mycotoxin analysis: A critical review. *Journal of AOAC International, 99*(4), 842–848. https://doi.org/10.5740/jaoacint.16-0111

Shi, K., Tang, R., Huang, T., Wang, L., & Wu, Z. (2017). Pigment fingerprint profile during extractive fermentation with *Monascus anka* GIM 3.592. *BMC Biotechnology, 17*, 1–11. https://doi.org/10.1186/s12896-017-0366-1

Singh, V. K., & Kumar, A. (2023). Secondary metabolites from endophytic fungi: Production, methods of analysis, and diverse pharmaceutical potential. *Symbiosis, 90*(2), 111–125. https://doi.org/10.1007/s13199-023-00925-9

Smith, M., Madec, S., Coton, E., & Hymery, N. (2016). Natural co-occurrence of mycotoxins in foods and feeds and their in vitro combined toxicological effects. *Toxins, 8*(4), 94. https://doi.org/10.3390/toxins8040094

Sobrova, P., Adam, V., Vasatkova, A., Beklova, M., Zeman, L., & Kizek, R. (2010). Deoxynivalenol and its toxicity. *Interdisciplinary Toxicology, 3*(3). https://doi.org/10.2478/v10102-010-0019-x

Son, S. W., Kim, H. Y., Choi, G. J., Lim, H. K., Jang, K. S., Lee, S. O., Sung, N. D., & Kim, J. C. (2008). Bikaverin and fusaric acid from *Fusarium oxysporum* show antioomycete activity against *Phytophthora infestans*. *Journal of Applied Microbiology, 104*(3), 692–698. https://doi.org/10.1111/j.1365-2672.2007.03581.x

Soriano, J., González, L., & Catalá, A. (2005). Mechanism of action of sphingolipids and their metabolites in the toxicity of fumonisin B1. *Progress in Lipid Research, 44*(6), 345–356. https://doi.org/10.1016/j.plipres.2005.09.001

Stockmann-Juvala, H., & Savolainen, K. (2008). A review of the toxic effects and mechanisms of action of fumonisin B1. *Human & Experimental Toxicology, 27*(11), 799–809. https://doi.org/10.1177/0960327108099525

Stoev, S. D. (2015). Foodborne mycotoxicoses, risk assessment and underestimated hazard of masked mycotoxins and joint mycotoxin effects or interaction. *Environmental Toxicology and Pharmacology, 39*(2), 794–809. https://doi.org/10.1016/j.etap.2015.01.022

Stoev, S. D. (2024). Food security, underestimated hazard of joint mycotoxin exposure and management of the risk of mycotoxin contamination. *Food Control, 159*, 110235. https://doi.org/10.1016/j.foodcont.2023.110235

Sugiharto, S., Yudiarti, T., & Isroli, I. (2016). Assay of antioxidant potential of two filamentous fungi isolated from the Indonesian fermented dried cassava. *Antioxidants, 5*(1), 6. https://doi.org/10.3390/antiox5010006

Sun, Z., Wu, Y., Long, S., Feng, S., Jia, X., Hu, Y., Ma, M., Liu, J., & Zeng, B. (2024). *Aspergillus oryzae* as a cell factory: Research and applications in industrial production. *Journal of Fungi, 10*(4), 248. https://doi.org/10.3390/jof10040248

Svobodová, K., & Novotný, Č. (2018). Bioreactors based on immobilized fungi: Bioremediation under non-sterile conditions. *Applied Microbiology and Biotechnology, 102*(1), 39–46. https://doi.org/10.1007/s00253-017-8575-z

Sydenham, E. W., Thiel, P. G., Marasas, W. F. O., Shephard, G. S., Van Schalkwyk, D. J., & Koch, K. R. (1990). Natural occurrence of some *Fusarium* mycotoxins in corn from low and high esophageal cancer prevalence areas of the Transkei, Southern Africa. *Journal of Agricultural and Food Chemistry, 38*(10), 1900–1903. https://doi.org/10.1021/jf00100a004

Takahashi, J. A., Barbosa, B. V. R., Martins, B. d. A., Guirlanda, C. P., & Moura, M. A. F. (2020). Use of the versatility of fungal metabolism to meet modern demands for healthy aging, functional foods, and sustainability. *Journal of Fungi, 6*(4), 223. https://doi.org/10.3390/jof6040223

Tarazona, A., Gómez, J. V., Mateo, E. M., Jiménez, M., & Mateo, F. (2019). Antifungal effect of engineered silver nanoparticles on phytopathogenic and toxigenic *Fusarium* spp. and their impact on mycotoxin accumulation. *International Journal of Food Microbiology, 306*, 108259. https://doi.org/10.1016/j.ijfoodmicro.2019.108259

Timmermans, R., Hayrapetyan, H., Vollebregt, M., & Dijksterhuis, J. (2020). Comparing thermal inactivation to a combined process of moderate heat and high pressure: Effect on ascospores in strawberry puree. *International Journal of Food Microbiology, 325*, 108629. https://doi.org/10.1016/j.ijfoodmicro.2020.108629

Tiwari, P., & Bae, H. (2022). Endophytic fungi: Key insights, emerging prospects, and challenges in natural product drug discovery. *Microorganisms, 10*(2), 360. https://doi.org/10.3390/microorganisms10020360

Tong, Z., Zheng, X., Tong, Y., Shi, Y. C., & Sun, J. (2019). Systems metabolic engineering for citric acid production by *Aspergillus niger* in the post-genomic era. *Microbial Cell Factories, 18*, 1–15. https://doi.org/10.1186/s12934-019-1064-6

Turner, T. L., Kim, E., Hwang, C., Zhang, G. C., Liu, J. J., & Jin, Y. S. (2017). Conversion of lactose and whey into lactic acid by engineered yeast. *Journal of Dairy Science, 100*(1), 124–128. https://doi.org/10.3168/jds.2016-11784

Umesha, S., Manukumar, H. M. G., Chandrasekhar, B., Shivakumara, P., Kumar, J. S., Raghava, S., Avinash, P., Shirin, M., Bharathi, T. R., Rajini, S. B., Nandhini, M., Rani, G. G. V., Shobha, M., & Prakash, H. S. (2017). Aflatoxins and food pathogens: Impact of biologically active aflatoxins and their control strategies. *Journal of the Science of Food and Agriculture, 97*(6), 1698–1707. https://doi.org/10.1002/jsfa.8144

Uzma, F., Mohan, C. D., Hashem, A., Konappa, N. M., Rangappa, S., Kamath, P. V., Yaraguppi, D. A., Siddaiah, C. N., Ramesh, B. T., Satish, S., & Chowdappa, S. (2018). Endophytic fungi—Alternative sources of cytotoxic compounds: A review. *Frontiers in Pharmacology, 9*, 309. https://doi.org/10.3389/fphar.2018.00309

VanderMolen, K. M., Raja, H. A., El-Elimat, T., & Oberlies, N. H. (2013). Evaluation of culture media for the production of secondary metabolites in a natural products screening program. *AMB Express, 3*(1). https://doi.org/10.1186/2191-0855-3-71

Veršilovskis, A., & De Saeger, S. (2009). Sterigmatocystin: Occurrence in foodstuffs and analytical methods – An overview. *Molecular Nutrition & Food Research, 54*(1), 136–147. https://doi.org/10.1002/mnfr.200900345

Veršilovskis, A., Bartkevičs, V., & Miķelsone, V. (2008). Sterigmatocystin presence in typical Latvian grains. *Food Chemistry, 109*(1), 243–248. https://doi.org/10.1016/j.foodchem.2007.12.066

Vidal, A., Ouhibi, S., Ghali, R., Hedhili, A., De Saeger, S., & De Boevre, M. (2019). The mycotoxin patulin: An updated short review on occurrence, toxicity and analytical challenges. *Food and Chemical Toxicology, 129*, 249–256. https://doi.org/10.1016/j.fct.2019.04.048

Vijayalakshmi, S., Nadanasabhapathi, S., Kumar, R., & Kumar, S. S. (2018). Effect of pH and pulsed electric field process parameters on the aflatoxin reduction in model system using response surface methodology. *Journal of Food Science and Technology, 55*(3), 868–878. https://doi.org/10.1007/s13197-017-2939-3

Villalobos, M. C., Serradilla, M. J., Martín, A., Ruíz-Moyano, S., Casquete, R., Hernández, A., & Córdoba, M. G. (2019). Use of efficient drying methods to improve the safety and quality of dried fig. *Journal of Food Processing and Preservation*, e13853. https://doi.org/10.1111/jfpp.13853

Wadhwa, K., Kapoor, N., Kaur, H., Abu-Seer, E. A., Tariq, M., Siddiqui, S., Yadav, V. K., Niazi, P., Kumar, P., & Alghamdi, S. (2024). A comprehensive review of the diversity of fungal secondary metabolites and their emerging applications in healthcare and environment. *Mycobiology, 52*(6), 335–387. https://doi.org/10.1080/12298093.2024.2416736

Walker, G., & Stewart, G. (2016). *Saccharomyces cerevisiae* in the production of fermented beverages. *Beverages, 2*(4), 30. https://doi.org/10.3390/beverages2040030

Wan, Q., Wu, G., He, Q., Tang, H., & Wang, Y. (2015). The toxicity of acute exposure to T-2 toxin evaluated by the metabonomics technique. *Molecular BioSystems, 11*(3), 882–891. https://doi.org/10.1039/c4mb00622d

Wang, M., & Jin, H. (2017). Spray-induced gene silencing: a powerful innovative strategy for crop protection. *Trends in Microbiology, 25*(1), 4–6. https://doi.org/10.1016/j.tim.2016.11.011

Wang, Z., Wu, Q., Kuča, K., Dohnal, V., & Tian, Z. (2014). Deoxynivalenol: Signaling pathways and human exposure risk assessment—An update. *Archives of Toxicology, 88*(11), 1915–1928. https://doi.org/10.1007/s00204-014-1354-z

Wang, L., Luo, Y., Luo, X., Wang, R., Li, Y., Li, Y., Shao, H., & Chen, Z. (2016a). Effect of deoxynivalenol detoxification by ozone treatment in wheat grains. *Food Control, 66*, 137–144. https://doi.org/10.1016/j.foodcont.2016.01.038

Wang, M., Weiberg, A., Lin, F., Thomma, B. P. H. J., Huang, H., & Jin, H. (2016b). Bidirectional cross-kingdom RNAi and fungal uptake of external RNAs confer plant protection. *Nature Plants, 2*(10). https://doi.org/10.1038/nplants.2016.151

Wayllace, N. M., Martín, M., Busi, M. V., & Gomez-Casati, D. F. (2023). Microbial glucoamylases: Structural and functional properties and biotechnological uses. *World Journal of Microbiology and Biotechnology, 39*(11), 293. https://doi.org/10.1007/s11274-023-03731-z

Wielogorska, E., Ahmed, Y., Meneely, J., Graham, W. G., Elliott, C. T., & Gilmore, B. F. (2019). A holistic study to understand the detoxification of mycotoxins in maize and impact on its molecular integrity using cold atmospheric plasma treatment. *Food Chemistry, 301*, 125281. https://doi.org/10.1016/j.foodchem.2019.125281

Wu, F., Liu, Y., & Bhatnagar, D. (2008). Cost-effectiveness of aflatoxin control methods: Economic incentives. *Toxin Reviews, 27*(3–4), 203–225. https://doi.org/10.1080/15569540802393690

Wu, F., Groopman, J. D., & Pestka, J. J. (2014). Public health impacts of foodborne mycotoxins. *Annual Review of Food Science and Technology, 5*(1), 351–372. https://doi.org/10.1146/annurev-food-030713-092431

Yamada, R., Yoshie, T., Wakai, S., Asai-Nakashima, N., Okazaki, F., Ogino, C., Hisada, H., Tsutsumi, H., Hata, Y., & Kondo, A. (2014). Aspergillus oryzae-based cell factory for direct kojic acid production from cellulose. *Microbial Cell Factories, 13*, 71. https://doi.org/10.1186/1475-2859-13-71

Yang, Q., Zhang, A., Miao, J., Sun, H., Han, Y., Yan, G., Wu, F., & Wang, X. (2019). Metabolomics biotechnology, applications, and future trends: A systematic review. *RSC Advances, 9*(64), 37245–37257. https://doi.org/10.1039/c9ra06697g

Yang, G. Z., Zhang, J., Peng, J., Zhang, Z., Zhao, W., Wang, R., Ma, K., Li, J., Liu, Y., Zhao, Z., & Shang, X. F. (2020). Discovery of luotonin A analogues as potent fungicides and insecticides: Design, synthesis and biological evaluation inspired by natural alkaloid. *European Journal of Medicinal Chemistry, 194*, 112253. https://doi.org/10.1016/j.ejmech.2020.112253

Yin, G., Zhao, H., Pennerman, K. K., Jurick, W. M., Fu, M., Bu, L., Guo, A., & Bennett, J. W. (2021). Genomic analyses of *Penicillium* species have revealed patulin and citrinin gene clusters and novel loci involved in oxylipin production. *Journal of Fungi, 7*(9), 743. https://doi.org/10.3390/jof7090743

Yu, J. (2012). Current understanding on aflatoxin biosynthesis and future perspective in reducing aflatoxin contamination. *Toxins, 4*(11), 1024–1057. https://doi.org/10.3390/toxins4111024

Yu, W., Pei, R., Zhou, J., Zeng, B., Tu, Y., & He, B. (2023). Molecular regulation of fungal secondary metabolism. *World Journal of Microbiology and Biotechnology, 39*(8). https://doi.org/10.1007/s11274-023-03649-6

Zhang, Z., Li, M., Wu, C., & Peng, B. (2019). Physical adsorption of patulin by *Saccharomyces cerevisiae* during fermentation. *Journal of Food Science and Technology, 56*(4), 2326–2331. https://doi.org/10.1007/s13197-019-03681-1

Zhong, L., Carere, J., Lu, Z., Lu, F., & Zhou, T. (2018). Patulin in apples and apple-based food products: The burdens and the mitigation strategies. *Toxins, 10*(11), 475. https://doi.org/10.3390/toxins10110475

Zhong, C., Guan, X., Fan, Z., Song, W., Chen, R., Wang, Y., Sun, X., & He, S. (2019). Pulsed electric field disinfection treatment of *Fusarium oxysporum* in nutrient solution. *Water Science & Technology Water Supply, 19*(7), 2116–2122. https://doi.org/10.2166/ws.2019.090

Zhou, Y., Wu, S., Wang, F., Li, Q., He, C., Duan, N., & Wang, Z. (2020). Assessing the toxicity in vitro of degradation products from deoxynivalenol photocatalytic degradation by using upconversion nanoparticles@TiO2 composite. *Chemosphere, 238*, 124648. https://doi.org/10.1016/j.chemosphere.2019.124648

Zhu, G. Y., Shi, X. C., Wang, S. Y., Wang, B., & Laborda, P. (2022). Antifungal mechanism and efficacy of kojic acid for the control of *Sclerotinia sclerotiorum* in soybean. *Frontiers in Plant Science, 13*, 845698. https://doi.org/10.3389/fpls.2022.845698

Zinedine, A., Soriano, J. M., Moltó, J. C., & Mañes, J. (2007). Review on the toxicity, occurrence, metabolism, detoxification, regulations and intake of zearalenone: An oestrogenic mycotoxin. *Food and Chemical Toxicology, 45*(1), 1–18. https://doi.org/10.1016/j.fct.2006.07.030

Zovi, A., Vitiello, A., Sabbatucci, M., Musazzi, U. M., Sagratini, G., Cifani, C., & Vittori, S. (2025). Food supplements marketed worldwide: A comparative analysis between the European and the U.S. regulatory frameworks. *Journal of Dietary Supplements*, 1–16. https://doi.org/10.1080/19390211.2024.2389397

Chapter 10
Fungal Endophytes: A Potential Source of Novel Biosynthetic Gene Clusters (BGCs) Correlating with Specialized Metabolites and Their Use in Plant Disease Management

J. H. Ashwini, Vipin Verma, Ish Prakash, Basudev Basak, and Deeba Kamil

1 Introduction

Fungicides offer a quick and effective solution for controlling devastating crop diseases, but their use is often associated with environmental risks. Excessive application can drive the development of antifungal resistance and facilitate cross-kingdom pathogen transmission. Growing ecological concerns promote microbe-derived compounds as eco-friendly and sustainable alternatives. As a result, utilizing endophytes for biological control is increasingly recognized as a highly effective and sustainable alternative to conventional chemical methods for managing plant diseases (Ashwini et al., 2025).

Among many biocontrol agents, fungal endophytes stand out as promising candidates. These filamentous fungi reside within plant tissues and maintain a symbiotic relationship without inducing any visible symptoms or harm to the host plant (Hashem et al., 2023). In addition to enhancing plant growth, these fungi boost the plant's resilience by mitigating a range of abiotic stresses, such as drought, salinity, heavy metal toxicity, environmental pollutants, waterlogging, and temperature extremes, as well as biotic pressures including pathogens and herbivores, through multiple adaptive mechanisms (Su et al., 2021).

Beyond promoting plant health and stress tolerance, fungal endophytes have also gained considerable interest for their capacity to synthesize metabolites that resemble with their respective host plants (Singh, 2023). They serve as inexhaustible reservoirs of valuable metabolites with diverse biological functions. Endophytes house diverse bioactive compounds like alkaloids, terpenoids, quinones, steroids, phenolic acids, and tannins, making them promising candidates for managing plant pathogens. Their biotechnological relevance stems from their ability to synthesize

J. H. Ashwini · V. Verma · I. Prakash · B. Basak · D. Kamil (✉)
Division of Plant Pathology, ICAR-IARI, New Delhi, India

B. P. Singh et al. (eds.), *Fungal Additives and Bioactives in Food Processing Industries*, Fungal Biology, https://doi.org/10.1007/978-3-032-04519-5_10

secondary metabolites with potential applications as biocontrol agents, as well as in anticancer and immunosuppressive therapies. However, it is important to note that not all SMs directly inhibit pathogens, only some possess strong antibacterial or antifungal activity, making them especially relevant as candidates for novel drugs, toxins, or pigments (Gunatilaka, 2006). Additionally, these compounds are thought to influence fungal interactions with other organisms, further highlighting their ecological and applied significance.

The synthesis of these bioactive compounds is governed by specific genetic elements known as biosynthetic gene clusters (BGCs) which orchestrate their production, regulation and transport (Shen et al., 2022). However, only a limited subset of BGCs are involved in the production of the naturally derived antibiotics and pharmaceutical compounds that are currently in use. A major challenge in harnessing these clusters lies in their cryptic or silent nature through artificial cultivation techniques (Zhang et al., 2024). Therefore, this chapter aims to provide an overview of the current tools and BGC databases used to identify potential BGCs, classify their types, and highlight their significance in plant disease management. This information is intended to assist researchers in leveraging these genetic resources to develop novel therapeutics and bio-based products more effectively.

2 Biosynthetic Gene Clusters (BGCs) in Fungal Endophytes

Fungal endophytes produce numerous bioactive metabolites, many of which play crucial roles in managing plant diseases. For example, species within the order *Xylariales* produce an array of metabolites, including cytochalasins, glucosides, terpenoids, azaphilones, macrolide polyketides, non-ribosomal peptides, lactones, and benzenoids (Becker & Stadler, 2021).

3 Categories

Biosynthetic gene clusters (BGCs) exhibit considerable diversity and can be classified according to the types of compounds they generate.

1. *Polyketide Synthase (PKS) Cluster*: The *polyketide synthase* (PKS) cluster is involved in the biosynthesis of polyketides. These compounds are known for their structural complexity and include important pharmaceuticals like erythromycin and tetracycline. PKS clusters are further alienated into Type I, Type II and Type III, each with distinct enzymatic architectures and biosynthetic pathways. Type I PKS, for instance, operates in a modular fashion, with each module adding a specific building block to the growing polyketide chain
2. *Non-ribosomal Peptide Synthetase (NRPS) Cluster*: These clusters produce non-ribosomal peptides, which are recognized for their antimicrobial and

immunosuppressive effects. Unlike ribosomal peptides, NRPS products are assembled by large enzyme complexes that incorporate a wide variety of amino acids, often including nonstandard ones. This flexibility allows for the production of compounds like penicillin and vancomycin, which have been instrumental in medical treatments

3. *Terpenoid Biosynthetic Clusters*: Produce terpenes and terpenoids, the largest class of natural products. These compounds have roles ranging from ecological interactions to pharmaceutical applications. Terpenoid clusters often involve enzymes liketerpene *synthases* and cytochrome P450s, which play a key role in the basic compound diversity.

4 Techniques to Identify the BGCs

Traditionally, identifying potential SMs has relied on bioactivity screening, where bioactive compounds with specific properties are extracted and the corresponding genes are identified using biochemical methods. Subsequent research revealed that secondary metabolites (SMs) are generally encoded by genes grouped within biosynthetic gene clusters (BGCs), which not only contain the core biosynthetic genes but also incorporate regulatory elements such as promoters and transcription factors. BGCs may also include transporter genes that aid in the export of secondary metabolites, along with other genes that safeguard the producing organism from the toxic effects of its own metabolites (Luo et al., 2014).

Traditional methods for identifying SMs face significant challenges, including the difficulty of producing or extracting many compounds under laboratory conditions and the reliance on known pathways, which limits the discovery of novel compounds. For instance, no new antibiotic structural classes were introduced between 1962 and 2000, despite extensive efforts. In contrast, genomic approaches have enabled the prediction of over 33,000 BGCs across 1154 bacterial genomes (Cimermancic et al., 2014). This demonstrates that the particular bottleneck lies in exploration strategies rather than nature's capacity. Recent advancements in next generation sequencing and synthetic chemicals provide powerful solutions for identifying and characterizing previously unrecognized BGCs and their products.

4.1 Bioinformatics Pipelines for BGCs Mining

A range of bioinformatics pipelines have been established to mine biosynthetic gene clusters (BGCs) directly from available genomes (Table 10.1). Some of these tools are specifically tailored for fungal BGCs, including FunGeneClusterS (Vesth et al., 2016), Secondary Metabolite Unknown Regions Finder (SMURF) (Khaldi et al., 2010), TOUCAN (Almeida et al., 2020), fungiSMASH (Blin et al., 2017), and FunOrder (Vignolle et al., 2021). While these tools were originally designed for

Table 10.1 Available bioinformatics tools for fungal BGCs mining

Tools	Algorithm	Input	Structure prediction
SMURF	Hidden Markov Model	Protein sequences, chromosomal coordinates	No
FunGeneClusterS	Correlation values derived from gene expression coefficients	RNA seq data with annotations	No
TOUCAN	A supervised learning framework	Protein sequences	No
FunOrder	BLAST, Evolutionary information	Protein sequences	No
NPLinker	Input–Output Kernel Regression approach	Whole genome	Yes
antiSMASH6.0	BLAST, Hidden Markov Model, machine learning	Whole genome, metagenome	Yes
BiG-SCAPE and CORASON	antiSMASH-based	Genome, metagenome	Yes
NRPSpredictor2	Machine learning	Genome, metagenome	No

bacterial BGCs, they have been adapted for use with fungi. The identification of BGCs typically relies on signature-based algorithms that use profile hidden Markov models (pHMMs) or the Basic Local Alignment Search Tool (BLAST) to locate gene/s associated with the synthesis of specific natural products (NPs). Among these techniques, BLAST remains one of the most widely employed methods for BGC identification.

4.2 Phylogenetics Approaches

Phylogenetic information is a valuable feature for identifying biosynthetic gene clusters (BGCs). For example, NaPDoS (Natural Product Domain Seeker) is an evolution-based web-based identification pipeline to identify potential *ketosynthase* and condensation region of protein or nucleotide sequences (Ziemert et al., 2012). EvoMining, or evolutionarily guided genome mining, identifies functionally divergent paralogs of primary metabolic enzymes that have evolved to take part in specialized secondary metabolism (Schniete et al., 2021). A comprehensive computational workflow, combining CORASON and BiG-SCAPE, has been developed for large-scale BGC analysis. Core Analysis of Syntenic Orthologs (CORASON) applies phylogenomic methods to clarify evolutionary connections between BGCs by creating high-resolution, multi-locus phylogenies across gene cluster families (GCFs). On the other hand, the Biosynthetic Gene Similarity Clustering and Prospecting Engine (BiG-SCAPE) facilitates effective network study of BGCs and GCFs based on sequence similarity.

4.3 Metagenomics Data Analysis

Several computational tools have been established to identify natural product biosynthetic pathways from environmental genomic data. Environmental Surveyor of Natural Product Diversity version 2.0 (eSNaPD) is an e-platform that provides automated tools for analyzing biosynthetic gene cluster (BGC) diversity in crude metagenomic samples, metagenomic libraries, or culture collections (Reddy et al., 2014). eSNaPD estimates BGC diversity of the targeted sample and matches sequences to a curated database of known compounds to identify BGCs potentially linked to specific NPs. Another notable method is the CONKAT-seq, which detects biosynthetic domains that are physically clustered, a key feature of secondary metabolites of bacteria (Libis et al., 2019). CONKAT-seq identifies chromosomally clustered domains by analyzing the co-occurrence patterns of amplicons across multiple sub-pools within a library. This method facilitates large-scale exploration of biosynthetic potential in environmental samples and helps identify novel natural products, particularly from rare or uncultured microbiome members.

4.4 Next-Generation De Novo Fungal Genome Sequencing and Assembly Approaches

Once the genome of a fungus is sequenced and assembled into scaffolds, a key task is to identify those that may harbor genes for secondary metabolite (SM) production. Despite the wide structural diversity of fungal secondary metabolites, most belong to one of four major categories: non-ribosomal peptides, terpenoids, polyketides, and indole alkaloids—each distinguished by their specific carbon-based building blocks (Keller et al., 2005).

A practical method to detect gene clusters of interest involves searching for sequences that encode the hallmark enzymes specific to each SM category. This can be achieved using TBLASTN searches against a custom database derived from the fungal genome scaffolds, which can be built using the "makeblastdb" function in the NCBI BLAST suite.

Depending on the metabolite type, conserved functional domains within the core biosynthetic enzymes serve as reliable search targets. For example, the adenylation (A) domain is a signature of non-ribosomal peptide *synthetases* (NRPS), the *keto-synthase* (KS) domain signals polyketide *synthases* (PKS), terpene *synthase* domains indicate terpenoid production, and dimethylallyl tryptophan *synthase* (DMATS) domains are characteristic of genes involved in prenylated indole alkaloid biosynthesis.

The TBLASTN results provide a ranked list of scaffolds with potential SM clusters based on sequence similarity to the input domains. Additionally, this search process can be streamlined using user-friendly commercial platforms like CLC

Genomics Workbench or Geneious, which offer graphical user interfaces for conducting and interpreting BLAST searches.

4.5 Comparative Genomic Analysis-Based Biosynthesis Gene Cluster Mining

Next-generation sequencing (NGS) technologies present valuable opportunities to select appropriate reference genomes from related species for comparative studies. Since the genomes of even closely related species within the same genus often encode distinct sets of secondary metabolites (SMs), comparing the SM backbone biosynthesis genes of a target organism with those of a closely related species can help quickly identify novel secondary metabolites gene clusters. The above method is particularly effective with prior knowledge of the secondary metabolites repertoire in the test organism is available.

On the other hand, comparative analysis can also be explored to identify BCGs associated with a specific class of structurally similar secondary metabolites, even when produced by distantly related fungal species. This approach allows for the orthologous gene clusters study in organisms, which synthesize the same compound family. Additionally, comparative genomics plays a crucial role in defining the limits of secondary metabolite gene clusters that are frequently flanked by syntenic regions harboring highly conserved core genes (Machida et al., 2005). Syntenic regions across multiple genomes can be visualized using Mauve.

5 Enhancing Secondary Metabolite Production in Fungal Endophytes

Fungal strains often possess far more biosynthetic gene clusters (BGCs) than the number of secondary metabolites (SMs) they are known to produce. This mismatch is primarily due to the silent or low-level expression of many BGCs under conventional laboratory conditions (Schüller et al., 2020). To unlock this hidden biosynthetic potential, several techniques have been developed to induce or increase the activity of these otherwise inactive pathways.

5.1 One-Strain–Many-Compounds (OSMAC) Strategy

The OSMAC approach is based on the principle that changes in cultivation parameters can lead to the expression of different biosynthetic pathways within a single microbial strain. By systematically altering growth media, culture formats, oxygen availability, and adding enzyme inhibitors, researchers can prompt the production of a diverse array of secondary metabolites. In some cases, this strategy has enabled the identification of more than a dozen distinct compounds from a single fungal isolate (Bode et al., 2002).

5.2 Microbial Co-Culture Approach

Natural microbial communities exhibit intricate and dynamic interactions, ranging from mutualism to antagonism, which often evolve as adaptive strategies for defense. The triggering of cryptic biosynthetic gene clusters (BGCs) in certain microbes is shaped by ecological factors, particularly close associations with other microbial community and plant species. Co-cultivation or mixed fermentation involves cultivating multiple microbial strains together to mimic natural microbial interactions. This microbial "social networking" is facilitated by processes such as gene transfer and the release of small diffusible signaling molecules into the environment by individual strains [Moody, 2014]. Co-culture models are widely employed to trigger cryptic BGC expression in fungal endophytes.

5.3 Chemical Epigenetic Modification

The fungal secondary metabolism regulation is a highly multifaceted process controlled by a sophisticated interplay of cellular, chemical, and genetic elements. Chromatin-level mechanisms, such as gene silencing or activation, perform an important role in controlling the biosynthesis of fungal SMs. These regulatory challenges can be addressed using small chemical compounds known as epigenetic modifiers. These modifiers alter the chromatin structure by targeting post-translational modifications of histones and DNA, thereby promoting the SMs production through activation and suppression of the corresponding gene clusters. Chemical epigenetic modification involves growing test strains with epigenetic modifier compounds (histone *deacetylase* inhibitors or DNA *methyltransferase* inhibitors).

5.4 Molecule-Based Approaches

Activation of cryptic biosynthetic gene clusters (BGCs) through genetic engineering of test strains primarily relies on genome mining. This method targets individual biosynthetic gene clusters by performing a comprehensive genomic analysis of the selected strain, followed by the use of specialized bioinformatics software to pinpoint genes involved in secondary metabolite synthesis. Once the biosynthetic pathways are characterized, the strain of interest is genetically modified using appropriate synthetic biology techniques tailored to the identified pathway architecture.

6 Role of Fungal Endophytes and Their BGCs in Plant Disease Management

- The association of fungal endophytes with host plants in dynamic environments enhances the plant's inherent tolerance mechanisms against various stresses by modulating local and systemic responses. This interaction stimulates the secretion of antioxidants in sufficient quantities to neutralize reactive oxygen species (ROS), thereby preventing the overactivation of the plant's redox regulatory defense system under stress conditions. Additionally, these associations have been shown to help plants combat biotic stresses, boost phytohormone level, and position themselves as valuable biocontrol agents. They also serve as biofertilizers, aiding in pathogen resistance, crop enhancement, and soil quality improvement.
- Bioactive compounds from fungal endophytes can be used as natural pesticides, that is, biopesticide, an eco-friendly alternative to chemical fungicides (Table 10.2).

7 Fungal Endophytes in Disease Management: A Success Story

Researchers have made a significant stride in the battle to unravel the molecular mechanisms and biosynthetic pathways governing its antagonistic activity fungal endophytes. Present study reflects the first RNA-seq-based analysis of *Aspergillus niger*'s biocontrol mechanisms against *Fusarium oxysporum* f. sp. *psidii* (Fig. 10.1). Transcriptomic profiling revealed key upregulated DEGs linked to hydrolytic activity (AB *hydrolases*, cyanate *hydratase*, *esterase/lipase*, *endoglucanase* A), transporters (ATP-binding cassette transporter, MFS monosaccharide transporter, polyamine transporter 3), and secondary metabolism (*SidC*). These findings highlight the molecular basis of *A. niger*'s antifungal efficacy (Ganagaraj et al., 2024).

Aspergillus niger is a promising biocontrol agent against Fusarium wilt, yet the biosynthetic basis of its antifungal activity remains unexplored. This study utilized genome mining and metabolomic profiling to identify key secondary metabolites, revealing an average of 19 predicted compounds per genome (Fig. 10.2). A strong correlation between computational predictions and LC-MS/GC-MS validation confirmed the production of bioactive metabolites, including azanigerone A, aflavarin, pyranonigrin E, penicillin, and ochratoxin A. These findings highlight genome

Table 10.2 Some fungal endophytes that have been reported to produce bioactive compounds as their host plant

Host	Endophytic fungi	Bioactive compound	Target plant pathogen	Reference
Colletotrichum gloeosporioides and *Periconia* sp.	*Piper longum* and *Piper nigrum*	Piperine	Antimicrobial	Qiu et al. (2010)
Sordariomycetes sp.	*Eucommia ulmoides*	Chlorogenicacid	Antimicrobial	Chen et al. (2010)
Arisaema erubescens	*Phoma sp.*	Atrovenetinone and sclerodione	*Ustilago violacea*	Hussain et al. (2015)
Rhododendron tomentosum	*Fusarium tricinctum*	Trtesin	*Fusarium oxysporum*	Tejesvi et al. (2013)
Ginkgo biloba	*Chaetomium globosum*	Quercetin	Antimicrobial	Yuan et al. (2019)
Oryzae sativa	Talaromyces *pinophilus EF2*, Trichoderma longibrachiatum *EF5 and EF7*, Aspergillus terreus *EF6.*, *T. longibrachiatum*, Paecilomyces *tenuis EF1*, Nigrospora *sphaerica EF3*, and Nigrospora *oryzae EF4*	Bioactive compounds	*Macrophomina phaseolina*, *Fusarium oxysporum*, *Pythium* sp., *Magnaporthe grisea*, *Colletotrichum falcatum*, *Rhizoctonia solani*, bacterial pathogens including *Erwinia carotovora*, *Ralstonia solanacearum*, and *Xanthomonas oryzae*	Sornakili et al. (2020)
Capsicum annum	*Aspergillus niger*	Bioactive compounds	*Fusarium oxysporum* f.sp. *psidii*	Gangaraj et al. (2024) and Aziz et al. (2021)

mining as a reliable tool for discovering antifungal compounds and reinforce *A. niger*'s potential in sustainable plant disease management (Ish Prakash et al., 2025).

8 Challenges and Future Prospects

Identification of endophytic fungi capable of synthesizing plant-associated bioactive molecules offers a promising opportunity to utilize these microbes as alternative sources of valuable substances. However, a vast majority of endophytes remain

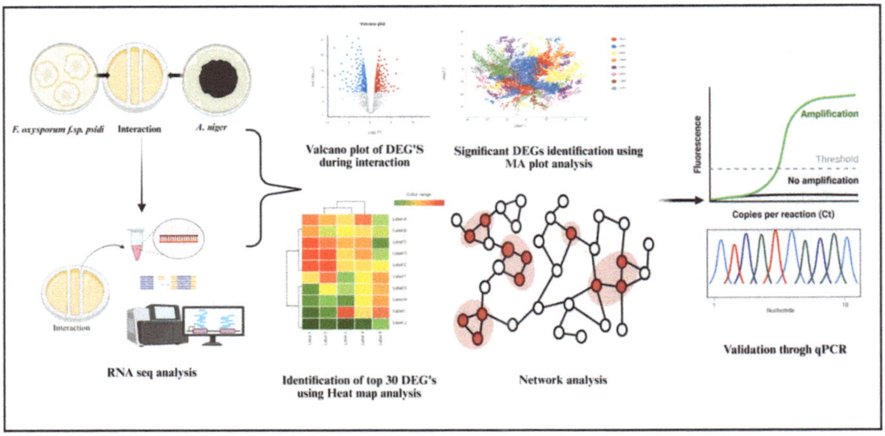

Fig. 10.1 Transcriptomic studies of the biocontrol strain *Aspergillus niger* AN-11 indicate altered expression levels of genes associated with hydrolytic enzyme activity and secondary metabolite biosynthesis pathways upon exposure to *Fusarium oxysporum* f. sp. *psidii*

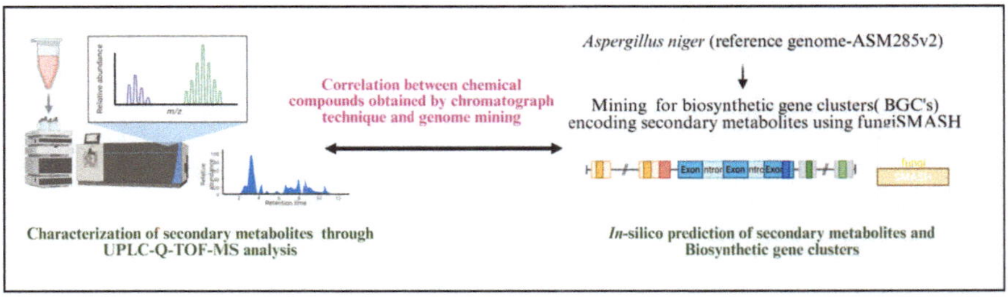

Fig. 10.2 Overlap observed between the chemical compounds predicted by BGC mining and those characterized through chromatographic techniques

unidentified due to the complexities associated with culturing them under laboratory conditions. Moreover, in-planta validation of their functional roles remains largely unexplored in plant pathology, creating a gap in our understanding of their true potential.

Advances in next-generation sequencing (NGS) and integrated omics approaches has significantly enhanced our capacity to investigate microbial symbionts, their unique metabolites, and the gene clusters responsible for their biosynthesis. While significant progress has been made in understanding BGCs in plants, many questions remain unanswered. One key area of interest is the extent of gene clustering in plant metabolism and its role in chemical defense mechanisms. Many plant-derived BGCs are known to produce compounds that help resist pests and pathogens, but their broader ecological functions are still not fully understood. Continued exploration of endophytes for their ecological roles and bioprospecting potential could lead to novel applications in healthcare, bioremediation, and eco-friendly agriculture.

When integrated with breakthroughs in genetic engineering and nanotechnology, these microorganisms offer substantial potential to address global challenges and promote sustainable progress.

9 Conclusions

Fungal endophytes are increasingly recognized as promising reservoirs of new bioactive molecules with diverse uses in medicine, agriculture, and environmental sustainability. Their ability to produce plant-associated metabolites, including clinically important compounds like paclitaxel, positions them as alternative and sustainable sources of valuable secondary metabolites. Advances in genomics and the characterization of biosynthetic gene clusters (BGCs) have begun to unravel the genetic basis of their metabolic diversity, paving the way for deeper insights into their biosynthetic capabilities. However, realizing their full potential requires overcoming key challenges, such as the unculturability of many strains, the loss of biosynthetic activity over time and the threat posed by habitat destruction to plant and fungal diversity. Future research must focus on exploring endophytes from understudied ecosystems, improving techniques for isolation and large-scale cultivation and leveraging genetic and metabolic engineering to enhance metabolite production. Interdisciplinary collaboration will be essential in transforming these findings into viable, eco-friendly biotechnological solutions, solidifying the role of endophytic fungi as vital contributors to future innovations in health, agriculture, and industry.

References

Almeida, H., Palys, S., Tsang, A., & Diallo, A. B. (2020). TOUCAN: A framework for fungal biosynthetic gene cluster discovery. *NAR Genomics and Bioinformatics, 2*(4), lqaa098.

Ashwini, J. H., Hiremath, S., Muttappagol, M., Nandan, M., Bharath, M., Shankarappa, K. S., Kumar, T. M., Basha, C. J., & Lakshminarayana Reddy, C. N. (2025). Screening of potential endophytes against fungal pathogens, Curvularia lunata, Bipolaris oryzae and pyricularia setariae using in vitro and computational approaches. *Journal of Phytopathology, 173*(2), e70050.

Aziz, L., Hamayun, M., Rauf, M., Iqbal, A., Arif, M., Husssin, A., & Khan, S. A. (2021). Endophytic Aspergillus niger reprograms the physicochemical traits of tomato under cadmium and chromium stress. *Environmental and Experimental Botany, 186*, 104456.

Becker, K., & Stadler, M. (2021). Recent progress in biodiversity research on the Xylariales and their secondary metabolism. *The Journal of Antibiotics, 74*, 1–23.

Blin, K., Wolf, T., Chevrette, M. G., Lu, X., Schwalen, C. J., Kautsar, S. A., & Medema, M. H. (2017). antiSMASH 4.0 – Improvements in chemistry prediction and gene cluster boundary identification. *Nucleic Acids Research, 45*(W1), W36–W41.

Bode, H. B., Bethe, B., Höfs, R., & Zeeck, A. (2002). Big effects from small changes: Possible ways to explore nature's chemical diversity. *Chem Bio Chem, 3*, 619–627.

Chen, X., Sang, X., Li, S., Zhang, S., & Bai, L. (2010). Studies on a chlorogenic acid-producing endophytic fungi isolated from Eucommia ulmoides Oliver. *Journal of Industrial Microbiology & Biotechnology, 37*, 447–454.

Cimermancic, P., Medema, M. H., Claesen, J., Kurita, K., Brown, L. C. W., Mavrommatis, K., Pati, A., Godfrey, P. A., Koehrsen, M., & Clardy, J. (2014). Insights into secondary metabolism from a global analysis of prokaryotic biosynthetic gene clusters. *Cell, 158*(2), 412–421.

Gangaraj, R., Nishmitha, K., Gupta, N. C., Akhtar, J., Shanmugaraj, C., & Kamil, D. (2024). Comparative transcriptome analysis of Aspergillus niger revealed its biocontrol mechanisms in response to the guava wilt pathogen *Fusarium oxysporum* f. sp. *psidii. Physiological and Molecular Plant Pathology, 134*, 102419.

Gunatilaka, A. A. L. (2006). Natural products from plant-associated microorganisms: Distribution, structural diversity, bioactivity, and implications of their occurrence. *Journal of Natural Products, 69*, 509–526. https://doi.org/10.1021/np058128n

Hashem, A. H., Attia, M., Kandil, E. K., Fawzi, M. M., Abdelrahman, A. S., Khader, M. S., Khodaira, M. A., Emam, A. E., Goma, M. A., & Abdelaziz, A. M. (2023). Bioactive compounds and biomedical applications of endophytic fungi: A recent review. *Microbial Cell Factories*. https://doi.org/10.1186/s12934-023-02118-x

Hussain, H., John, M., Al-Harrasi, A., Shah, A., Hassan, Z., Abbas, G., Rana, U. A., Green, I. R., Schulz, B., & Krohn, K. (2015). Phytochemical investigation and antimicrobial activity of an endophytic fungus Phoma sp. *Journal of King Saud University, Science, 27*(1), 92–95.

Keller, N. P., Turner, G., & Bennett, J. W. (2005). Fungal secondary metabolism – From biochemistry to genomics. *Nature Reviews Microbiology, 3*(12), 937–947.

Khaldi, N., Seifuddin, F. T., Turner, G., Haft, D., Nierman, W. C., Wolfe, K. H., & Fedorova, N. D. (2010). SMURF: Genomic mapping of fungal secondary metabolite clusters. *Fungal Genetics and Biology, 47*(9), 736–741.

Kim, C. K., Eo, J. K., & Eom, A. H. (2013). Diversity and seasonal variation of endophytic fungi isolated from three conifers in Mt. Taehwa, Korea. *Mycobiology, 41*, 82–85.

Libis, V., Antonovsky, N., Zhang, M., Shang, Z., Montiel, D., Maniko, J., Ternei, M. A., Calle, P. Y., Lemetre, C., Owen, J. G., et al. (2019). Uncovering the biosynthetic potential of rare metage nomic DNA using co-occurrence network analysis of targeted sequences. *Nature Communications, 10*(1), 3848.

Luo, Y., Cobb, R. E., & Zhao, H. (2014). Recent advances in natural product discovery. *Current Opinion in Biotechnology, 0*, 230–237.

Moody, S. C. (2014). Microbial co-culture: Harnessing intermicrobial signaling for the production of novel antimicrobials. *Future Microbiology, 9*, 575–578.

Prakash, I., Gangaraj, R., Ashwini, J. H., Vipin, K. V., & Kamil, D. (2025). *Genome mining of Aspergillus niger for secondary metabolite biosynthetic gene clusters and their real-time chromatographic validation with the potential A. niger AN11 strain* (p. 60). Proceedings of IPS Delhi Zonal Symposium.

Qiu, M., Xie, R. S., Shi, Y., Zhang, H., & Chen, H. M. (2010). Isolation and identification of two flavonoid-producing endophytic fungi from Ginkgo biloba L. *Annales de Microbiologie, 60*, 143–150.

Reddy, B. V. B., Milshteyn, A., Charlop-Powers, Z., & Brady, S. F. (2014). eSNaPD: A versatile, web-based bioinformatics platform for surveying and mining natural product biosynthetic diversity from metagenomes. *Chemistry and Biology, 21*(8), 1023–1033.

Schniete, J. K., Selem-Mojica, N., Birke, A. S., Cruz-Morales, P., Hunter, I. S., Barona-Gomez, F., & Hoskisson, P. A. (2021). ActDES – A curated actinobacterial database for evolutionary studies. *Microbial Genomics, 7*(1), 000498.

Shen, L., Roullier, C., Porée, F.-H., Gaslonde, T., Riffault-Valois, L., Grovel, O., Ruprich Robert, G., & Chapeland-Leclerc, F. (2022). Complementary strategies to unlock biosynthesis gene clusters encoding secondary metabolites in the filamentous fungus podospora anserina. *Journal of Fungi, 9*(1), 9–9.

Singh, K. (2023). Epigenetics. *Module Biomedical Science*. https://doi.org/10.1016/b9780-12-824315-2.00464-4

Sornakili, A., Thankappan, S., Sridharan, A. P., Nithya, P., & Uthandi, S. (2020). Antagonistic fungal endophytes and their metabolite-mediated interactions against phytopathogens in rice. *Physiological and Molecular Plant Pathology, 112*, 101525.

Su, Z., Zeng, Y., Li, X., Perumal, A. B., Zhu, J., Lu, X., Dai, M., Liu, X., & Lin, F. (2021). The endophytic fungus Piriformospora indica-assisted alleviation of cadmium in tobacco. *Journal of Fungi, 7*, 675.

Sudha, V., Govindaraj, R., Baskar, K., Al-Dhabi, & Duraipandiyan, V. (2016). Biological properties of endophytic fungi. *Brazilian Archives of Biology and Technology*, 59.

Tejesvi, M. V., Segura, D. R., Schnorr, K. M., Sandvang, D., Mattila, S., et al. (2013). An antimicrobial peptide from endophytic Fusarium tricinctum of Rhododendron tomentosum Harmaja. *Fungal Diversity, 60*, 153–159.

Vesth, T. C., Brandl, J., & Andersen, M. R. (2016). FunGeneClusterS: Predicting fungal gene clusters from genome and transcriptome data. *Synthetic and Systems Biotechnology, 1*(2), 122–129.

Vignolle, G. A., Schaffer, D., Zehetner, L., Mach, R. L., Mach-Aigner, A. R., & Derntl, C. (2021). FunOrder: A robust and semi-automated method for the identification of essential biosynthetic genes through computational molecular co-evolution. *PLoS Computational Biology,* **17**(9), e1009372.

Yuan, Z., Tian, Y., He, F., & Zhou, H. (2019). Endophytes from Ginkgo biloba and their secondary metabolites. *Chinese Medicine, 14*, 1–40.

Zhang, Y., Feng, L., Hemu, X., Tan, N., & Wang, Z. (2024). OSMAC strategy: A promising way to explore microbial cyclic peptides. *European Journal of Medicinal Chemistry*. https://doi.org/10.1016/j.ejmech.2024.116175

Ziemert, N., Podell, S., Penn, K., Badger, J. H., Allen, E., & Jensen, P. R. (2012). The natural product domain seeker NaPDoS: A phylogeny based bioinformatic tool to classify secondary metabolite gene diversity. *PLoS One, 7*(3), e34064.

Chapter 11
Applications of Fungi in Food Processing: Prospects and Challenges

Surabhi Pandey, Sweekriti Choudhry, and Anurag Singh

1 Introduction

The fascinating diversity of organisms found in the fungal kingdom has a significant influence on the health of ecosystems as well as animal and plant species. Fungi's development and metabolism can have a variety of effect (Bahram & Netherway, 2022). In addition to producing important materials like food through positive symbioses with plants, fungi are also the primary cause of changes like food deterioration, spoiling, and the generation of toxins. Despite these dual roles, humans have identified the importance of fungi in daily life and are employing them for a number of reasons. The area of food science has long made extensive use of fungus and their by-products. They are either eaten as-is or added to meals to help with other food processing methods which produce particular food items. Because of their rich nutritional profile and ability to reduce costs in the food industry, fungi are used in food biotechnology as filamentous fungi, fermentative yeasts, and mushrooms. In recent years, there has been a surge in cultivation to satisfy consumer needs because it is easier and less expensive to manufacture food and produce higher-quality items on a big scale using fungi (Pouris et al., 2024).

Fungal (microbial) metabolites, which are natural compounds produced by fungi and other microorganisms, include bioactive agents, colouring compounds, flavouring agents, and natural preservatives (Lin & Xu, 2022; Dufossé et al., 2021). This trend is a result of both growing consumer demand for clean-label, health-conscious products and growing awareness of the health and environmental risks associated with synthetic chemicals, which, although effective and cost-effective, often come with a host of issues. Artificial preservatives, such as sodium nitrite, BHA, and

S. Pandey · S. Choudhry · A. Singh (✉)
Department of Food Technology, Harcourt Butler Technical University, Kanpur, Uttar Pradesh, India
e-mail: a.singh@hbtu.ac.in

B. P. Singh et al. (eds.), *Fungal Additives and Bioactives in Food Processing Industries*, Fungal Biology, https://doi.org/10.1007/978-3-032-04519-5_11

BHT, have been linked to potential health risks, such as endocrine disruption and carcinogenicity when consumed over time (Mandal et al., 2018). Synthetic tastes and colourants can also lead to long-term metabolic problems, allergic responses, and hyperactivity in kids. Fungal metabolites, on the other hand, provide a naturally occurring, biocompatible substitute that is more in line with both environmental sustainability and human physiology. Secondary metabolites, which are substances that are not directly involved in growth or reproduction but frequently have strong biological activity, are produced by fungi in extraordinary quantities. Antioxidant, antibacterial, anti-inflammatory, and even anticancer effects have been discovered in several of these substances (Dufossé et al., 2021). For instance, the production of bioactive substances like statins and penicillin, which transformed contemporary medicine, is a well-known characteristic of Aspergillus and Penicillium species. These same fungus may be used in the food business to produce natural preservatives that prevent spoiling without sacrificing safety or nutritional value (Hyde et al., 2019). Carotenoids and polyketides—natural colours made from fungi—are also becoming more and more well-liked as substitutes for synthetic dyes. In addition to providing vivid colours, these colourants made from fungi also have extra health advantages including antioxidant activity. Natural pigments from microbial sources are biodegradable, safer to eat, and ecologically friendlier than manufactured colours, which frequently have little nutritional benefit and can be harmful (Afroz Toma et al., 2023).

Flavouring compounds are another promising frontier. Fungi can produce a wide range of aromatic metabolites that contribute to complex, authentic flavours in fermented foods and beverages. For instance, methyl ketones from *Penicillium* species are responsible for the distinctive flavour of blue cheese, while *Monascus purpureus*, used in traditional East Asian fermentation, imparts both colour and umami flavour to food. These natural flavour agents enhance taste without the need for synthetic chemicals, making them ideal for clean-label product development (Lindsay et al., 2022). Fungal metabolites are preferable because they are sustainable in addition to being safe and effective. The petrochemical synthesis of artificial additives is a wasteful and energy-intensive process compared to microbial fermentation. Agricultural waste materials can be used as substrates for fungus culture, converting waste into useful bioactive substances. This greatly lowers the carbon footprint of producing food and ingredients and is consistent with the ideas of the circular economy (Daniela et al., 2022). Fungal metabolites stand out as potent, sustainable substitutes for synthetic additives in the growing market for natural, health-conscious products. These microbial chemicals are changing the face of contemporary food research and business, from boosting food's nutritional profile to increasing safety, flavour, and appearance. An important step toward a more robust society is their integration (Chauhan & Rao, 2024).

Despite being readily accessible and convenient, processed foods present serious risks to the environment and human health. While providing little in the way of vital minerals or dietary fibre, these goods frequently include high levels of salt, saturated fats, processed sugars, and artificial additives. Frequent ingestion of these foods has been associated with a number of health complications, such as digestive disorders, obesity, cardiovascular disease, and type 2 diabetes (Weaver et al., 2014). Long-term chronic disorders may be exacerbated by the presence of chemical additives and preservatives, which may also disturb gut flora and increase inflammation (Zhou et al., 2023). Processed foods have a significant negative impact on the environment in addition to health. The production process uses a lot of water and energy, especially when making synthetic and meat-based products. Monoculture farming, which depletes soil nutrients and reduces biodiversity, is encouraged by the dependence on industrial agriculture to produce raw materials like corn, soy, and palm oil (Leite et al., 2022). Fungi are still strong friends in modern industrial processing and food science, especially when it comes to creating healthier and more environmentally friendly substitutes for traditional goods. In the large-scale synthesis of organic acids like citric acid, alcohol, and enzymes that improve the flavour or preserve food, fungi are now often employed as fermentation agents (Pouris et al., 2024). They are essential to the contemporary food business due to their biochemical diversity. The development of plant-based meat substitutes is among the most intriguing uses. Quorn and other products include mycoprotein, a protein-rich diet derived from filamentous fungus like Fusarium venenatum. In comparison to animal agriculture, it provides a full amino acid profile, a meat-like texture, and a far reduced environmental impact (Finnigan et al., 2019). Furthermore, Myceliophthora thermophilia (Renzetti et al., 2010) and other fungal enzymes are used to enhance tastes, soften dough, and extend the shelf life of processed foods, frequently taking the place of chemical additives. Bio-preservation is another area in which fungi are making contributions. Compounds produced from fungi are increasingly being used as important additions in processed foods since they are more sustainable and useful than traditional additives. These natural substances, which are produced by microbial fermentation, enhance nutrition, flavour, texture, and preservation while lowering reliance on artificial chemicals and resource-intensive procedures. One significant benefit of their multifunctionality is that fungal enzymes, such as amylases and proteases, improve texture, increase digestibility, and increase processing efficiency (Pouris et al., 2024).

1.1 Background on Fungi

Long before industrial food systems emerged, humans harnessed the natural properties of fungi to prepare and preserve food. Since ancient times, fungi have been used as food sources. They can grow as symbionts on trees or break down lignocellulosic woody biomasses. Fungi have been essential in traditional cuisines across the world, providing not only flavour and texture but also enhanced nutrition and shelf stability (Pouris et al., 2024). In East Asia, for instance, fermented soy products like tempeh and miso rely on beneficial fungi such as *Rhizopus* and *Aspergillus oryzae* (do Prado et al., 2022). These processes improve the digestibility of soybeans and enrich their nutrient profiles. In Europe, the mold *Penicillium* is responsible for the characteristic taste and appearance of blue cheeses and soft cheeses like Camembert

(Leclercq-Perlat, 2011). Fungi have also played a central role in bread-making and alcohol production for millennia. Yeasts such as *Saccharomyces cerevisiae* were used in ancient Egypt and Mesopotamia to ferment beer and leaven bread (Lahue et al., 2020). These traditional uses highlight fungi's longstanding role in enhancing food through natural fermentation, adding complexity and depth to both flavour and nutrition. Apart from these, Ascomycetes and basidiomycetes are fungi that produce edible fruiting bodies, which are considered to be a poor source of nutrients despite being high in proteins and low in fat (Amara & El-Baky, 2023). Since fungi bioremediate soil that has been contaminated by heavy metals and radioactive substances, moderate amounts of fungi should be consumed. Despite being regarded as delicacies, mushroom species including boletus, morels, and truffles are challenging to grow. Saprobic species like Pleurotus ostreatus, Lentinus edodes (shiitake), and Agaricus bisporus have been industrially produced in Asia for thousands of years. In Mexico, *Ustilago maydis*, a phytopathogenic fungus, produces black tumours over maize, a delicacy known as huitlacoche (Villagrán et al., 2023). Additionally, fungi are important to traditional and indigenous cultures, where they are used for their hallucinogenic qualities as well as to produce fermented foods and beverages like bread, cheese, rice, and beer (Singh et al., 2020).

1.2 Historical Perspectives of Fungi in Food

Despite the fact that just a small portion of fungus are edible and may be used to digest food, humans have relied on them for nourishment for thousands of years. Men have utilised fungi's inherent ability to ferment different grains and fruits from the beginning of agricultural techniques in order to improve the nutritional value and sensory appeal of particular foods, such bread and alcoholic beverages. Both fermented foods (such as bread, cheese, and rice, among other foods) and drinks (such as beer, cider, and wine) are produced using fungi. Because *Saccharomyces* yeasts can tolerate high levels of ethanol, unlike other microbes, they are primarily used to preserve the safety and quality of foods and beverages (Dupont et al., 2017).

The discovery of effective strains was first made possible by the unintentional fermentation and alcohol production by fungal species. Compared to their wild counterparts, domestic strains have improved and adapted. Around the world, edible mushrooms are grown and regarded as a part of a balanced diet (Naranjo-Ortiz & Gabaldón, 2019; Barzee et al., 2021). Because of *Saccharomyces cerevisiae*'s resistance to high ethanol levels, which guaranteed product stability and safety, early civilisations like those in Mesopotamia and Egypt relied on it for these procedures. Over 3000 years ago, *Aspergillus* species—specifically, *Aspergillus sojae* and *Aspergillus oryzae*—were used in China and Japan to produce sake and soy sauce, demonstrating the advanced nature of early fungal biotechnology. The discovery of *Penicillium* molds transformed the manufacture of cheese in Europe, allowing for the creation of famous types such as Roquefort and Camembert, which are renowned for their distinct flavours and textures (Khachatourians, 2003).

By the twentieth century, the potential of fungi had expanded beyond their traditional uses to include large-scale industrial applications, such as the commercial development of fungi like *Rhizopus oryzae*, *Aspergillus oryzae*, *Neurospora intermedia*, and *Monascus purpureus* as a source of mycoprotein, a sustainable meat substitute (Filho et al., 2019; Sar et al., 2020), and the cultivation of *Aspergillus niger* for the production of citric acid (Papagianni, 2007). More strains have been tailored for certain uses, such as increasing the synthesis of enzymes or increasing the efficiency of fermentation, because of developments in fungal genetics.

These achievements demonstrate how fungi have a dual function in preserving culinary traditions and fostering advancements in the field of food science. Fungi continue to play a crucial role in sustainability initiatives, promoting waste minimisation, nutrient recycling, and the creation of ecologically friendly food substitutes, in addition to the manufacture of fermented foods and beverages (Pouris et al., 2024; Singh et al., 2020).

1.3 *Importance of Fungi in Food Processing*

Fungi have been used to produce food and drinks, such as bread and alcoholic beverages, since ancient times. However, both molds and yeasts are important components of the industrial fermentation process used to produce food ingredients today. Through industrial fermentation, fungi provide a variety of substances used in food processing, including polyunsaturated fatty acids, enzymes, flavourings, vitamins, colourants, and acidulants. Fungi are used in fermentation operations for converting liquid or solid substrates into a variety of products. Although the substrates vary greatly, any substance that promotes microbial growth aims to create the required bioproduct as quickly as feasible, with the maximum rate and yield, consistently, and in the most straightforward and affordable manner (Copetti, 2019a).

Different types of ingredients and food products synthesised by fungi are shown in the Fig. 11.1 below.

2 Types of Fungi Used in Food Processing

2.1 *Yeasts*

Yeast is another kind of fungus that is frequently utilised in the food sector. To create wholesome, high-protein food for human and animal consumption, yeast cells and fungal mycelium are grown. Due to their capacity to perform fermentation operations, yeasts are preferred (Ludlow et al., 2016). They help produce staple foods like bread and cheese, which can improve food flavour and prolong its shelf life. One of the first methods of food preparation, fermentation has significant

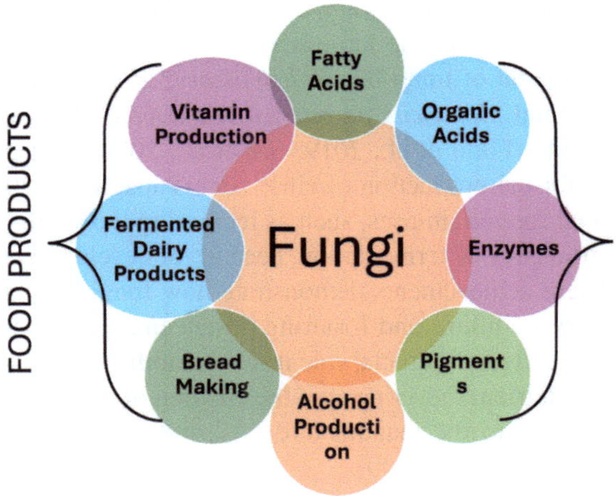

Fig. 11.1 Types of ingredients synthesised by fungi. (Adopted from Singh et al., 2020)

economic value. But only particular yeast strains are selected to carry out the fermentation process. There is a wealth of information about the existence, manufacturing, properties, and use of fermented foods (Pouris et al., 2024). The ability of particular yeast strains to withstand processing conditions, produce gas during fermentation, and provide a desirable flavour to the final product indicate the three main criteria that determine which strains are chosen (Suhaimi et al., 2021; Shurson, 2017).

2.1.1 Saccharomyces cerevisiae

As the most prevalent yeast species in bread and sourdoughs, *S. cerevisiae* is frequently referred to as baker's yeast or just "the yeast." Baker's yeasts, which were derived from the residues of the beer manufacturing process, have been utilised as a starting culture since the nineteenth century (Parapouli et al., 2020). For thousands of years, Saccharomyces cerevisiae has played a crucial role in the fermentation of food as well as beverages. For instance, S. cerevisiae is necessary for bread dough to effectively use the complex sugar maltose, withstand osmotic stress from the dough's semi-solid state, high salt content, and high sugar content, endure freezing and drying, and yield desired flavours and aromas (Lahue et al., 2020). Particularly noteworthy as a crucial participant in fermentation processes, Saccharomyces cerevisiae aids in the creation of wine, beer, and bread. This adaptable yeast species gives these delectable foods their distinctive flavours and textures by fermenting carbohydrates to produce carbon dioxide and alcohol (Pouris et al., 2024). Techniques have been created to meet the needs of the fermentation (beer, wine, cider, and distillates) and baking industries. In the health food industry, food grade yeasts are utilised as sources of high-nutritional-value proteins, enzymes, and vitamins. They are also used as nutritional supplements, food additives, conditioners, and flavouring agents, and they are used to produce microbiology media and animal feed (Singh et al., 2020). Many researchers have recently focused on vinification procedures that use fermentations using a variety of yeast inocula. This method is used for two key reasons: i) the endeavour to enhance the wines' organoleptic properties. ii) The demand for wines with reduced alcohol levels (Parapouli et al., 2020).

2.1.2 Non-Saccharomyces Yeasts

Non-Saccharomyces yeasts have historically been regarded as contaminants because many of the currently recognised species produce unwanted metabolites. Recent years have seen a shift in this area as a result of increased interest in specific strains that produce metabolites that enhance wine. Since most non-Saccharomyces yeasts have low fermentative power and low ethanol tolerance, particularly when SO_2 is present, using them necessitates using mixed fermentations (simultaneous or sequential) in conjunction with *Saccharomyces cerevisiae* to ensure a proper fermentation process. Numerous reviews have covered topics like the metabolic traits and key contributions of non-Saccharomyces to wine, how it improves wine qualities like acidity, and how it affects different oenological parameters. Among other things, they have included statistics regarding the number of commercial strains, more commercialised species, rules, patents, and the firms that offer the product. According to the study, there are about 42 commercial products based on non-Saccharomyces yeasts that may be used in different forms for winemaking. Of these products, 52% are from three species: *Torulaspora delbrueckii, Lachancea thermotolerans*, and *Metschnikowia pulcherrima* (Vejarano & Gil-Calderón, 2021). The scientific literature asserts that the use of non-Saccharomyces at the industrial level is still in its infancy because most applications have been conducted on an experimental basis. The oenological sector's increasing interest in creating wines with unique characteristics indicates that understanding of these yeasts is still an area in need of growth, according to Roudil et al., (2019), who described the progress of the supply of these commercial yeasts in recent years. A brief categorisation of yeasts has been given in Table 11.1.

2.2 *Molds*

In food processing, molds are important for fermentation, the synthesis of enzymes, and food cultivation (Pouris et al., 2024). Many fermented foods and drinks, including cheese, soy sauce, sake, and shoyu, are made with them. To turn substrates into desired products, molds and yeasts are employed as biotransformation agents (Copetti, 2019b), utilised to create enzymes that enhance food's flavour, texture, and fragrance. Fungal lipases, for instance, are used to reduce fat in meat and fish and enhance the flavour of dairy products. Fruit juices are extracted using pectinase, an enzyme that comes from the mold *Aspergillus niger* (Copetti, 2019a).

Table 11.1 Some Saccharomyces and Non-Saccharomyces Yeasts and their sensory impact

Type of yeast	Sensory impact with respect to *S. cerevisiae*	Reference
Non-Saccharomyces Yeast		
Torulaspora delbrueckii	Higher aroma intensity; complexity and fruity notes; reduces volatile activity	Azzolini et al. (2014) and Renault et al. (2009)
Metschnikowia pulcherrima	Higher smoky and floral/fruity notes; reduces ethanol content, enhances varietal aromas, increases glycerol production	Contreras et al. (2013) and González-Royo et al. (2014)
Hanseniaspora vineae	Higher preference, fruity, and floral scores	Lleixà et al. (2016)
Lachancea thermotolerans	Higher acidity; higher aroma intensity and quality, sensory acceptability; balances acidity improving wine complexity	Benito et al. (2017)
Saccharomyces Yeast		
Saccharomyces uvarum	Less acetic acid; acetaldehyde; higher aroma intensity	Česnik et al. (2023) and Varela et al. (2017)
Saccharomyces bayanus	Efficient sugar fermentation; fresh sensory profile; lower acetic acid and high glycerol	Česnik et al. (2023) Kelly et al. (2018)
Saccharomyces pastorianus	Subdued ester and higher alcohol levels, emphasising malt and hop flavours; fuller mouthfeel; higher malic acid degradation leading to reduced acidity	Česnik et al. (2023) Troianou et al. (2019)

2.2.1 *Aspergillus* spp.

One of the earliest fungal genera to be recognised is *Aspergillus*, a genus of molds that reproduces only asexually. This fungus is filamentous, global, and common; it is frequently isolated from soil, plant matter, and indoor air. In order to enhance its flavour, texture, digestibility, and nutritional value, food enzymes produced from this are used in the baking, in the production of fruit juice and cheese, wine, and in brewing. Pectinases, amylases, cellulases, glucanases, lipases, proteases, and other commercially available enzymes are utilised in the food industry. (Mojsov, 2016). Fungi are used to make a lot of fermented Asian cuisine. *Aspergillus oryzae* and *Aspergillus sojae* are important molds in some Asian food processing because they are used as the starting point for koji, which is a source of amylolytic, lipolytic, and proteolytic fungal enzymes that other microorganisms need to assimilate nutrients in subsequent fermentations. Additionally, the koji adds colour, flavour, and aroma—all of which are crucial for the overall character of fermented goods like sake and soy sauce. The health benefits of consuming koji and its derivative products have recently been investigated. These benefits include their antitumoural, anti-mutagenic, antioxidant, and stress-reduction properties, as well as their ability to reduce cholesterol and eliminate gluten allergies (Copetti, 2019b; Zhu & Tramper, 2013).

2.2.2 *Penicillium* spp.

With uses ranging from improving flavours and textures to acting as crucial ingredients in fermentation processes, *Penicillium* species are crucial to the preparation of food. The formation of blue cheese, white cheese, and meals fermented by mold depends on notable species such *Penicillium roqueforti*, *Penicillium camemberti*, and *Penicillium nalgiovense* (Meena et al., 2017). For instance, *P. roqueforti* develops in the veins of cheeses like Roquefort, Stilton, and Gorgonzola and is essential to the manufacture of blue cheese. It adds to the distinctively marbled look and tangy, earthy flavour. In similar way, *P. camemberti* is essential to white mold cheeses like Camembert and Brie, giving them a nutty, mushroom-like taste while also forming a white rind and softening the inside. Salami and other fermented meat products use *P. nalgiovense* as a surface mold to give them an appealing appearance and prevent unwanted microbial growth, through enzymatic activity, which breaks down proteins and lipids into smaller components that contribute to the sensory profile, improving flavour and aroma. Furthermore, *Penicillium* molds can withstand high salinity, which makes them perfect for fermenting and keeping food in such environments. *Penicillium* species are well-known for producing bioactive chemicals in addition to their conventional uses. Enzymes like lipases and proteases are released by *P. roqueforti* and *P. camemberti*, improving the flavour and texture of food. Additionally, they generate terpenes and methyl ketones, which are volatile substances essential to the scent profiles of cheeses ripened by mold (Mullan, 2014; Frisvad, 2014). Although *Penicillium* molds are useful for producing food, it is important to choose strains carefully to prevent contamination by species that create toxins, such as *P. expansum* or *P. nordicum*, which can produce dangerous mycotoxins such ochratoxin A and patulin. Quality control procedures guarantee the safe application of these molds, maintaining their advantages while reducing hazards. In conclusion, *Penicillium* molds are essential for producing fermented foods of superior quality, adding distinctive tastes, better textures, and better preservation. Their importance in both conventional and contemporary food processing methods is highlighted by their enzymatic and metabolic activity (Frisvad, 2014).

2.3 *Mushrooms*

A large woody or meaty fungus is called a mushroom. Ascomycetes (Morchella, Tuber, etc.) and Basidiomycetes (Agaricus, Auricularia, Tremella, etc.) are the two groups of higher fungi, which are referred to as mushrooms. Their heterotrophic eating style sets them apart. They are a great source of additional food because of their high protein content. Mushroom cultivation is very beneficial because of its ability to thrive on low-cost substrates, such as agricultural and industrial waste (rich in carbon and nitrogen). The great majority of mushrooms are rough, woody, bitter, and tasteless. Fresh mushrooms grown for commercial purposes are always safe to eat. Commercially produced mushrooms usually have a fleshy cap that is 2 to 10 cm diameter and grow to a height of 5 to 10 cm. Commercial production of edible mushrooms takes place in facilities with meticulous temperature and humidity controls. A particular bedding culture is supplemented with a pure culture of the fungal mycelium. Multiple mushroom crops are produced by each inoculation (Amara & El-Baky, 2023).

2.3.1 Edible Mushrooms

The uses of edible mushrooms are varied and include both medical and stand-alone uses. Certain mushrooms have grown in popularity and are now available in hypermarkets and local fresh food stores. The phylum Basidiomycota is home to the most widely grown and well-known mushrooms in the world. These include the following genera: Agaricus, which includes the common button mushroom, Agaricus bisporus; Pleurotus, which includes oyster mushrooms; Lentinus, which includes Lentinula edodes (shiitake); Auricularia, which includes Auricularia ssp. (wood ear or winter mushrooms); Flammulina, including Flammulina velutipes (enoki or winter mushrooms); Volvariella, which includes Volvariella volvacea (paddy straw mushroom); Tremella, which includes Tremella fuciformis (silver ear mushrooms); Agrocybe, which includes Agrocybe cylindracea; nameko (Pholiota nameko); and monkey head mushroom (Hypsizygus marmoreus). Compared to mushrooms grown in the wild, commercial mushrooms are safer. Furthermore, wild mushrooms are growing under specific circumstances and during specific periods of the year. Certain species, however, are better suited for natural collection (Amara & El-Baky, 2023).

2.3.2 Medicinal Mushrooms

For more than 2000 years, shiitake mushrooms have been used as food and as medicine throughout Asia. Polysaccharides, proteins, terpenoids, polyphenols, vitamins, and dietary fibres are all abundant in edible mushrooms. (Xu et al., 2014). Shiitake mushrooms have been used medicinally to treat a variety of conditions, including cancer, bronchial inflammation, heart disease, fungus and viral infections, blood cholesterol, diabetes, and immune-suppressive disorders. The possibility of growing and producing shiitake or golden oak mushrooms on a large scale is being investigated. The northeastern Indian region's economy depends on its output. Shiitake mushrooms are extremely valuable because of their distinct flavour as well as their health and nutritional advantages. Numerous bioactive substances, including lentinan, vitamins B1, B2, and C, dietary fibres, ergosterol, niacin, and minerals, are thought to contribute to its excellent nutritional value (Salwan et al., 2021). Proteins, lipids, carbs, vitamins, and minerals make up the remaining components of *L. edodes* fruit bodies, which are 88–92% water. The provitamin D2 (ergosterol) found in mushrooms is transformed into calciferol (vitamin D2) when exposed to heat and ultraviolet (UV) light. Additionally, it has the vitamin-B complex, which includes thiamine (B1), ribofavin (B2), and niacin (B12) (Dermiki et al., 2013; Poojary et al., 2017). Two other well-known therapeutic mushrooms are Reishi (*Ganoderma lucidum*) and Maitake (*Grifola frondosa*). *Ganoderma lucidum*, or reishi mushrooms, are prized for their therapeutic qualities, which include boosting the immune system, having anti-inflammatory and antioxidant benefits, and perhaps preventing cancer. They help preserve the liver, improve blood pressure and cholesterol, encourage relaxation and improved sleep, and support cardiovascular health. Reishi is a popular choice in complementary medicine because of these advantages, which are derived from its bioactive constituents such as polysaccharides and triterpenoids (De Carvalho Cristiane et al., 2015). *Grifola frondosa*, also referred to as maitake, has a number of therapeutic uses, mostly because of its bioactive substances, which include glycoproteins, polysaccharides, and β-glucans. These substances are well-known for their antiviral, antidiabetic, immunomodulatory, and anticancer effects. Maitake promotes cardiovascular health, helps control blood sugar, and strengthens the immune system. According to Wu et al., (2021), it also possesses antioxidant qualities that aid in lowering oxidative stress and may be useful in the treatment of metabolic diseases including diabetes and hyperlipidaemia.

3 Applications of Fungi in Food Processing

In the food processing industry, fungi are essential because they help produce a variety of substances, including organic acids (like citric and gluconic acids), enzymes (like amylases and proteases), vitamins, colours, and polyunsaturated fatty acids. These products made from fungi improve the flavour, texture, nutritional content, and preservation of food while promoting effective and sustainable industrial food processing. Additionally, it is used in the production of baked goods, dairy products, alcoholic drinks, and food preservation as mentioned in Table 11.2 (Barzee et al., 2021).

3.1 Fermentation Processes

3.1.1 Alcohol Production

Alcoholic beverages are made by fermenting the sugar found in fruits, grains, and berries. This procedure turns a liquid into a highly alcoholic beverage through distillation. Fermented by-products are utilised extensively and in large quantities by the population in several regions of India, Tibet, and Nepal. Fungi are used to make some of these alcoholic drinks (Walker & Stewart, 2016). Several people drink kombucha, an alcoholic beverage also known as tea fungus or tea mushroom, which has been proven to have several health advantages. The hue of this beverage might be either green or black. It is created via a bacterial and yeast symbiotic culture

Table 11.2 List of various fungi and their metabolites along with their applications

Ingredient	Compound	Producing Fungus	Application	Reference
Enzyme	Pectics	*Aspergillus* spp.	Fruit juice extraction, clarification and filtration; coffee mucilage removal	Copetti (2019a)
Enzyme	Proteases	*Aspergillus* spp.	Bread improvement, chill proofing of beer, milk coagulation; Dough conditioner, meat tenderisation	Ortiz et al., (2016) and Singh et al. (2020)
Organic Acid	Citric Acid	*Aspergillus niger, Candida lipolytica*	Emulsification of dairy products; Soft drinks, jams, jellies, candies, frozen fruits, wine	Copetti (2019a) and Singh et al. (2020)
Organic Acid	Gluconic Acid	*Aspergillus niger*	Acidity regulation in food products	Singh et al. (2020)
Fatty Acid	Ω-3 and Ω-6	*Mortierella alpine*	Added to improve functionality of food	Yang and Zhang (2016)
Flavouring	Blue Cheese Flavouring	*Penicillium roqueforti*	Flavouring agent	Singh et al. (2020)
Pigments and Vitamins	β-Carotene	*Blakeslea trispora*	Orange–red food colourants, vitamin A precursor and antioxidant.	Wang et al. (2013)

(SCOBY). In this, *Saccharomyces cerevisiae* is the yeast element and *Gluconacetobacter* is the bacterial element (Ayed et al., 2016). Drinking this beverage can have a number of health benefits, including lowering blood sugar, reducing constipation, and having anticancer properties (Kumari et al., 2021). Mostly through fermentation, fungi are essential to the manufacture of alcohol. *Saccharomyces cerevisiae*, a type of yeast, is the most often employed fungus in the production of alcoholic drinks, including wine, spirits, and beer. Yeast produces carbon dioxide and ethanol during fermentation by breaking down sugars from a variety of sources, such as grains, grapes, and molasses. A number of variables, including temperature, pH, and sugar concentration, can impact the process and have an impact on the amount of alcohol produced and yeast activity. Specialised fermentations can make use of fungi such as *Kluyveromyces* and *Pichia* in addition to *S. cerevisiae*. These fungi improve the efficiency and flavour of alcohol production. Fungal biotechnology advancements have produced improved strains with better fermentation properties, increased alcohol tolerance, and resistance to stressors like high sugar concentrations. According to Kumari et al., (2021) and Copetti (2019a), fungi play a crucial role in the alcohol sector by aiding in the fermentation process, flavour development, and production efficiency.

3.1.2 Bread Making

Bread is a staple food that has been around for a long time. It was first produced in ancient Egypt some 12,000 years ago, during the Neolithic period. The main components of bread are flour, water, and yeast—specifically, *Saccharomyces cerevisiae*—which ferments the flour's carbohydrates to produce ethanol and carbon dioxide (CO_2), which are necessary for the bread to leaven. This technique gives bread its distinct flavour and texture. To increase bread's shelf life, innovations such as the use of staling agents and emulsifiers have been introduced (Kumari et al., 2021). Bread is also made from a variety of grains, including rye, millet, sorghum, maize, and oats. Alternatives such as xanthan gum and guar gum are used to replicate the qualities of gluten in order to satisfy customers who are gluten-intolerant, improving the nutritional profile of bread (Cauvain, 2015; Lamacchia et al., 2014). Commercially available as dried cells, baker's yeast, mainly *Saccharomyces cerevisiae*, ferments the carbohydrates in wheat flour to produce ethanol and carbon dioxide, which is used as a leavening agent in bread manufacturing (Ali et al., 2012). In addition to environmental stressors including high sucrose concentration, air drying, and freeze-thaw cycles, which can induce osmotic stress to yeast cells, the kind of yeast and sugar present can also affect how well the fermentation process works. Yeast cells adjust to high sucrose levels in order to combat this. *Kazachstania exigua*, *Torulaspora delbrueckii*, and *Candida holmii* are osmotolerant and freeze-resistant substitutes for *S. cerevisiae* that can ferment dough in difficult circumstances such as high sucrose concentrations. These advancements in baking technology and alternative yeasts have significantly contributed to the evolution of the baking industry, meeting the growing demand for diverse and high-quality bread products (Kumari et al., 2021). There are several methods for manufacturing bread, such as the Chorleywood method, the sponge and dough method, and the straight dough method. Bulk fermentation and mechanical dough development are commonly used in industrial production. The latter entails letting the dough ferment gradually, which enhances its texture and adds gas bubbles, which are essential for the bread's ultimate shape and rise (Kumari et al., 2021).

3.1.3 Dairy Fermentation

Dairy fermentation is greatly aided by fungi, which help produce cheese, yoghurt, and other products with distinct flavours, textures, and scents. Certain fungi, notably *Penicillium* species, aid in the development of distinctive qualities in cheeses like Roquefort and Camembert throughout the ripening process. The fermentation of dairy-based drinks can also be improved by fungi, which will increase the probiotic content and improve the overall flavour and quality of fermented dairy products. Cheese, a solid or semisolid protein food made from milk, was historically the primary method of preserving milk before modern food processing techniques. Fungi such as *Penicillium camemberti* and *Penicillium roqueforti* are used in ripening cheeses like Camembert and Roquefort, respectively. These fungi produce methyl ketones, including 2-heptanone, contributing to flavour, and their proteases help soften the cheese by breaking down proteins. Mold ripening has been a part of cheese production for over 4000 years (Singh et al., 2021a, b). Fungi, including lactic acid bacteria and yeasts, are essential in the production of yogurt and other fermented dairy products. Specific strains such as *Lactobacillus delbrueckii* subsp. *bulgaricus* and *Streptococcus thermophilus*, along with yeasts like *Saccharomyces cerevisiae*, facilitate the conversion of lactose into lactic acid, which develops the distinctive flavours and textures of these products. These microorganisms significantly influence the acidity, aroma, and viscosity of fermented milk products (Pandit et al., 2024). Dairy fermentations frequently use fungal starter cultures to start and regulate the fermentation process. These cultures are made up of particular fungal strains that have been chosen for their capacity to enhance the texture and sensory qualities of fermented dairy products. To produce soft cheese, for instance, *Geotrichum candidum* is a popular fungal starting culture that helps create distinctive flavours and textures (Tamang et al., 2016). Additionally, fungus starting cultures help suppress unwanted bacteria, increasing product safety and shelf life (Pandit et al., 2024).

3.2 Food Preservation

3.2.1 Antimicrobial Properties

Numerous naturally occurring antimicrobial chemicals that have great potential for use in food processing are produced by fungi. These comprise peptidic natural products, terpenoids, and polyketides, which are examples of secondary metabolites. Penicillin and other β-lactam antibiotics, for instance, are produced by *Penicillium* species and prevent the synthesis of bacterial cell walls. By rupturing cell membranes, the depsipeptide beauvericin, which is found in *Beauveria bassiana*, exhibits antibacterial activity. Similarly, fungal respiration is inhibited by strobilurins, which are produced from *Strobilurus tenacellus*. These substances have two functions: they prolong food's shelf life and shield it from bacteria that cause spoiling. Mechanisms including rupture of cell membranes, reduction of microbial protein synthesis, and blockage of respiratory pathways are associated with the antibacterial characteristics (Jakubczyk & Dussart, 2020). Additionally, *Saccharomyces cerevisiae* demonstrated antioxidant and antibacterial properties (Fakruddin et al., 2017). Other beneficial qualities of *Candida tropicalis, Pichia sp.* and *Aureobasidium pullulans* include enzymatic or antibacterial capabilities against harmful bacteria, which make them potential probiotics for application in food and feed supplements (Syal & Vohra, 2013; Rousta et al., 2023).

3.2.2 Bio-preservation Techniques

Fungi contribute to environmental integrity and the preservation of food products' safety and quality in addition to producing the intended goods. The production of antibiotics by fungal strains aids in food preservation by limiting the growth of bacteria in food products. According to recent research, certain fungal strains contribute to food waste biodegradation, which is crucial for environmental sustainability. Additionally, they assist in turning food waste into forms that can be used (Nout, 2004). Fungi-induced post-harvest illnesses continue to be one of the main causes of food waste. The majority of current research efforts are concentrated on cutting-edge strategies to reduce these losses, such as the exploration of using different biocontrol agents made from fungus to treat crop illnesses that occur after harvest and extend the shelf life of agricultural products. Utilising antagonistic yeasts has emerged as a viable strategy to lessen fruit rotting brought on by fungi that produce mycotoxin. Because of their exceptional capacity to adsorb and biodegrade mycotoxins, the enzymes of several antagonistic yeasts have been the subject of extensive research in this area (Miranda et al., 2022).

3.3 Flavour and Aroma Enhancement

A variety of food products' taste development is greatly influenced by fungi. Alcohols, esters, and organic acids are among the volatile molecules that fungus make during fermentation processes and add to the distinctive flavours and scents of fermented foods. For instance, the manufacture of bread, beer, wine, and cheese depends on the development of particular taste compounds by yeast strains, which improves the sensory profiles and general acceptability among consumers (Pandit et al., 2024). Fungal enzymes have an impact on food products' scent (Copetti, 2019a). Since solid-state fermentation (SSF) provides a complex growth medium that closely resembles the natural habitat of filamentous fungi, it is a promising technique for producing natural flavours and aromas using these organisms. One of its defining characteristics is the fermentation that occurs on solid particles without the presence of free water. Usually, this is a controlled fermentation in which hyphae from filamentous fungi can pierce the solid particles, or bacteria or yeast can grow on the surface of the solid particles in a biofilm. More complex metabolites, such as volatile compounds that provide flavour and scent, are typically produced by bacteria under these situations. SSF has long been employed in food fermentation, including the ripening of cheese, the manufacture of mushrooms, and a portion of the manufacturing of soy sources (Lindsay et al., 2022).

3.4 Nutritional Enhancement

Food products can benefit from the addition of fungi to increase their nutritional value. Some fungal species, like filamentous fungus and yeasts, can produce minerals, other bioactive chemicals, and important vitamins, including B vitamins (Pandit et al., 2024).

3.4.1 Protein Fortification

Fungi are crucial to the fermentation processes of fermented grains and cereals, which are staple foods in many civilisations. For example, filamentous fungi such as Rhizopus and Aspergillus species are used as starter cultures in the preparation of traditional African fermented dishes including ogi, kenkey, and injera. These fungi aid in the hydrolysis of proteins, the breakdown of complex polysaccharides, and the creation of distinctive flavours and textures in finished goods. A food ingredient high in protein, mycoprotein is made from filamentous fungi, particularly the species Fusarium venenatum. It provides a different kind of protein for vegan and vegetarian diets. Additionally, fungi have the capacity to produce fungal biomass and single-cell protein (SCP), which can be used as substitute protein sources. To create biomass that is high in protein, some fungus species, as *Candidautilis* and *Fusarium venenatum*, are cultivated on a variety of substrates (Pandit et al., 2024).

3.4.2 Vitamin Production

Food products can benefit from the addition of fungi to increase their nutritional value. Minerals, other bioactive substances, and critical vitamins, particularly B vitamins, can be produced by several fungal species, including filamentous fungus and yeasts. For example, *Aspergillus oryzae* ferments soybeans to produce soy sauce, which increases the nutritional content of the finished food by synthesising vitamins and amino acids. Additionally, fungi can increase the nutritional advantages of food matrices by improving their digestibility and bioavailability of nutrients (Pandit et al., 2024).

4 Benefits of Fungal Applications in Food Processing

4.1 Economic Advantages

Through cost reduction and process simplification, fungi greatly enhance the food industry's economic efficiency. Large-scale production of food products with improved quality and longer shelf lives is made possible by their capacity to manufacture bioactive chemicals and enzymes. Because they take little energy and produce little waste, fungus fermentation processes—like those used to make cheese, bread, and alcoholic beverages—are economical. Furthermore, the bioconversion of agricultural by-products into valuable food ingredients by fungi lowers the cost of raw resources while generating marketable goods from otherwise wasted materials. Manufacturers benefit from the scalability and adaptability of fungal applications in food production (Pouris et al., 2024).

4.2 Sustainability and Environmental Impact

Due to their ability to facilitate ecologically friendly production methods, fungi are essential in advancing sustainability in the food industry. They lessen food waste and the environmental impact of food processing by participating in biodegradation and nutrient recycling. By transforming waste and agricultural residues into wholesome food products, fungi can support the circular economy paradigm. Using fungal enzymes as natural preservatives also lessens the environmental impact of synthetic compounds by eliminating the need for chemical additives. The use of fungi in biofertilisers and their ability to enhance soil health through mycorrhizal relationships both contribute to sustainable farming methods. A more sustainable and ecologically conscious food production system is facilitated by fungi through waste reduction, resource optimisation, and decreased reliance on non-renewable inputs (Pouris et al., 2024).

4.3 Health Benefits

Fungi play a variety of roles in food production and processing, providing numerous health benefits. They improve the development of flavours and add to the foods' sensory appeal. Fungi improve the overall nutritious profile of foods by generating vital vitamins, minerals, and bioactive substances. Their fermentation processes improve the nutrients' digestibility and bioavailability, which facilitates food absorption. Additionally, fungi have probiotic qualities that enhance digestive health and support a balanced gut microbiota. Their antibacterial, antioxidant, and anti-inflammatory properties also boost immunity and aid in the prevention of a number of illnesses. The importance of fungi in improving health and wellness is further highlighted by their role in the manufacture of functional foods and nutraceuticals (Pandit et al., 2024).

5 Challenges in the Use of Fungi in Food Processing

Fungi have long been utilised in food processing due to their beneficial properties in fermentation, enzyme production, and flavour development. However, their use also presents various challenges that must be addressed to ensure food safety, consumer acceptance, and regulatory compliance. The following sections outline the primary challenges in the use of fungi in food processing.

5.1 Safety Concerns

Safety is a major concern when incorporating fungi into food products. While many fungi are beneficial and safe, there are risks associated with contamination or improper use. These safety concerns primarily stem from the production of mycotoxins, allergens, and sensitivities among consumers.

5.1.1 Mycotoxin Production

Mycotoxins are toxic compounds produced by certain fungi, such as *Aspergillus*, *Penicillium*, and *Fusarium*. These toxins can contaminate food products during fungal growth, particularly under conditions of inadequate storage or improper handling. Aspergillus, Penicillium, and Fusarium are representative genera of mycotoxigenic fungi; Trichoderma, Trichothecium, and Alternaria have been discovered as important plant diseases or food pollutants (Ashiq, 2015). Mycotoxin production is influenced by certain environmental factors, although the level of contamination varies by region, farming methods, and the susceptibility of commodities to fungal penetration during storage and processing. Due to their carcinogenic, teratogenic, and mutagenic properties, mycotoxins—including aflatoxins, ochratoxin A, and fumonisins—pose a serious threat to food safety. In food processing, especially with fungi like *Aspergillus flavus* (commonly used in fermentation), ensuring that products do not contain harmful levels of mycotoxins is critical. To minimise the risks, strict monitoring of fungal strains, proper storage conditions, and post-harvest treatments are necessary. To protect the public from allergies and sensitivities, regulatory agencies including the U.S. Food and Drug Administration (FDA) and the European Food Safety Authority (EFSA) have established thresholds for mycotoxin contamination in foods (Awuchi et al., 2021; Pandit et al., 2024).

5.1.2 Allergens and Sensitivities

Another concern related to the use of fungi in food processing is the potential for allergic reactions and sensitivities. Fungal proteins and metabolites may trigger allergies in sensitive individuals. For instance, mold species commonly used in food fermentation, such as *Penicillium* and *Aspergillus*, have been identified as potential allergens. Symptoms of fungal allergies range from mild respiratory issues to more severe reactions like asthma. For consumers with fungal sensitivities, avoiding certain food products or ingredients may be necessary. The food industry must carefully assess the risk of cross-contamination, ensure proper labeling, and develop processes that reduce allergens' presence in the final product. More research into the identification of allergenic fungal proteins could improve safety and consumer confidence (Pandit et al., 2024).

5.2 Regulatory Issues

The majority of regions have regulations to guarantee the safety of human food and animal feed. These regulations vary based on whether the products are intended to be food (offering nutrients along with a pleasing taste and aroma), marketed as food additives (texture modifiers, preservatives, etc.), or used as feed and feed additives (Bagchi, 2006). The use of fungi in food processing is subject to stringent regulatory frameworks aimed at protecting public health. However, regulatory approval for novel fungal applications can be complex and varies by region. Different countries have different standards for the approval of new fungal species, strains, and their intended uses in food products. For example, in the European Union, novel food legislation requires that any new fungal strains used in food products be evaluated for safety before being introduced to the market. Similarly, the FDA in the U.S. evaluates the safety of new food ingredients, including fungi, through a Generally Recognised As Safe (GRAS) process. The primary focus of the Novel Food Regulation is on the safety and nutritional issues surrounding foods intended for human consumption. In the EU, three fungal strains (SCP producers) are approved for use in food. Prior to 1997, EU nations were consuming Saccharomyces cerevisiae, often known as Brewer's yeast or budding yeast. Quorn, a mycoprotein found in the microfungus Fusarium venenatum, is the second. The Novel Food Regulation ((EU) 2017/2470) has permitted the use of the third fungal strain, the yeast Yarrowia lipolytica; however, its use is limited in food supplements (Wiebe, 2004; Amara and El-Baky, 2023).

5.3 Consumer Acceptance

Consumer acceptance of fungi in food processing remains a significant challenge. While many consumers are familiar with traditional fermented foods such as cheese, soy sauce, and bread, the incorporation of fungi in new food products (such as fungi-based meat substitutes or fermented beverages) may evoke concerns about taste, texture, and safety. A major barrier to acceptance is the general public's perception of fungi. Many consumers associate fungi with mold growth or food spoilage, leading to hesitation in adopting fungal-based products. Therefore, educating consumers about the health benefits, safety measures, and traditional uses of fungi in food production is essential. Moreover, there may be resistance to consuming products made from genetically modified fungi or those developed using advanced fermentation technologies. Trust in the food industry and its ability to ensure product safety and quality will play a crucial role in overcoming these challenges.

5.4 Technical Limitations

The technical limitations in using fungi for food processing can also hinder its widespread adoption. One of the main challenges is the variability in fungal growth conditions. Fungi require specific environmental conditions (temperature, humidity, substrate) for optimal growth, which can be difficult to control at a large scale. Furthermore, fungal fermentation processes may require longer fermentation times compared to other microbial fermentation methods. This can increase production costs and reduce the efficiency of large-scale food manufacturing. Achieving consistent quality in the final product is another challenge, as fungal activity can vary with environmental fluctuations, leading to batch-to-batch inconsistencies. Additionally, the extraction and purification of fungal metabolites (such as enzymes or bioactive compounds) for use in food processing can be complex and resource-intensive. Improving fermentation technologies, optimising growth conditions, and developing more efficient methods of metabolite extraction will be essential to overcoming these technical challenges.

6 Practices to Address the Challenges

As the use of fungi in food processing grows, addressing the associated challenges becomes crucial for ensuring safety, improving consumer acceptance, and overcoming technical and regulatory hurdles. Several best practices and strategies can be employed by producers and regulatory bodies to mitigate risks and enhance the benefits of fungal applications in food. Addressing the challenges in the use of fungi in food processing is crucial to ensuring the safety, sustainability, and acceptance of

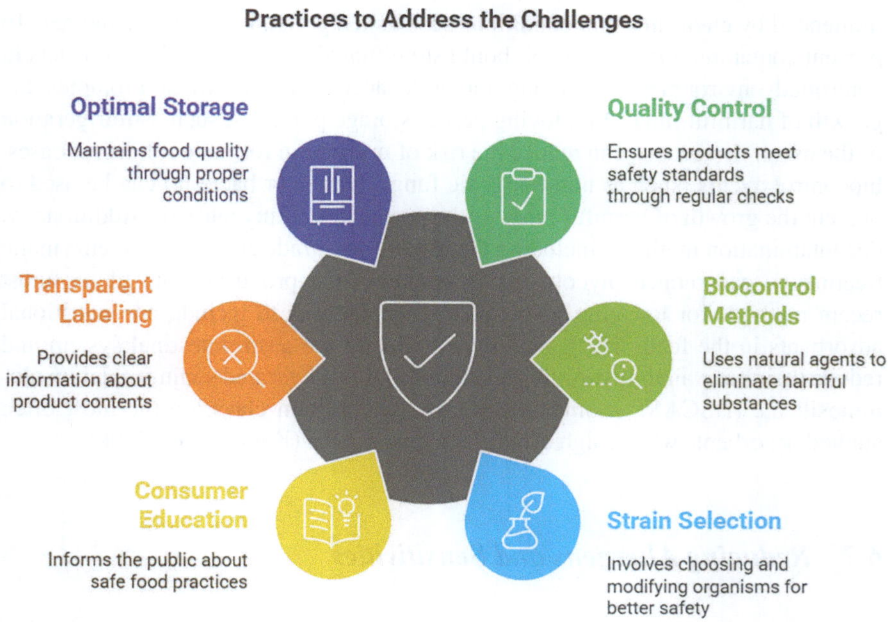

Fig. 11.2 Various methods to address the challenges associated with fungi

fungal-based food products. While fungi offer numerous benefits, such as enhancing nutritional content, improving fermentation, and providing alternative protein sources, their successful integration into the food industry depends on overcoming potential risks such as mycotoxin contamination, allergenic reactions, and regulatory barriers. Additionally, consumer acceptance is often hindered by perceptions of fungi as harmful or undesirable. By tackling these challenges, the food industry can ensure that fungi are used safely and effectively, fostering innovation, improving food security, and building trust among consumers. Brief methods to address these challenges are depicted in Fig. 11.2.

6.1 Safety Measures to Mitigate Mycotoxin Production

Regular monitoring of fungal strains used in food processing is critical to ensuring they do not produce harmful mycotoxins. This includes selecting non-toxigenic strains or strains with low mycotoxin production potential. Routine testing of raw materials and final food products for mycotoxin contamination is essential to meet regulatory standards and ensure consumer safety. For example, using analytical techniques such as high-performance liquid chromatography (HPLC) can help detect and quantify mycotoxins at low levels, thereby preventing contamination before products reach the market. Fungal growth and mycotoxin production are influenced by environmental conditions such as temperature, humidity, and pH. To prevent contamination, producers should store fungal cultures and final products in controlled environments, avoiding moisture accumulation, which promotes the growth of harmful fungi. Employing proper storage practices, such as refrigeration or the use of desiccants, can reduce the risk of mycotoxin formation. In some cases, biocontrol agents (such as non-toxigenic fungal strains or bacteria) can be used to prevent the growth of harmful fungi or the production of mycotoxins. Additionally, decontamination methods, including the use of food-grade adsorbents or enzymatic treatments, can reduce mycotoxins in contaminated products. One of the most recent methods for lowering mycotoxicosis in cattle is to include non-nutritional adsorbents in the feed, which bind mycotoxins in the gastrointestinal system and reduce their bioavailability. Activated carbons (ACs), hydrated sodium calcium aluminosilicate (HSCAS), zeolites, bentonites, and certain clays are the most often studied adsorbents with a high affinity for mycotoxins (Kabak et al., 2006).

6.2 Reducing Allergens and Sensitivities

To reduce the risk of allergenicity, selecting fungal strains that are less likely to produce allergenic proteins is essential. In some cases, genetic modification techniques can be applied to reduce the allergenic potential of fungal strains. This approach can help ensure that the use of fungi in food does not trigger adverse allergic reactions in consumers. Various food processing techniques, such as fermentation, can be optimised to break down allergenic compounds in fungi. For example, prolonged fermentation or heat treatment may reduce the levels of allergenic proteins. Additionally, modern filtration techniques can be employed to remove fungal spores or protein fragments from the final product, further decreasing allergenic risks. Clear labeling of food products containing fungi is crucial for individuals with known fungal allergies or sensitivities. Labels should include information about the presence of fungal ingredients and any allergens present. Educating consumers on how fungal allergens are processed, the safety measures in place, and potential alternatives can also help build trust in fungal-based food products.

6.3 Promoting Consumer Acceptance

One of the most effective ways to address consumer concerns about fungi in food is through education. Promoting the health benefits of fungal-based food products—such as their role in improving gut health, enhancing nutrient absorption, and providing alternative sources of protein—can increase consumer acceptance. Additionally, educating the public about the long history of safe fungal use in traditional foods like cheese, bread, and soy sauce can help shift negative perceptions. Consumers may be reluctant to try fungal-based food products due to concerns about taste, texture, or appearance. To address this, food manufacturers can invest in research and development to improve the sensory properties of fungal-based products. Techniques such as flavour masking, texture modification, and appearance enhancement can help ensure that fungal foods meet consumer preferences without compromising their nutritional or health benefits. Transparency in the production process is key to gaining consumer trust. Third-party certifications, such as organic, non-GMO, or allergen-free labels, can assure consumers of the safety and quality of fungal-based food products. Companies can also provide detailed information about sourcing, production methods, and safety testing to help consumers feel more confident about the foods they purchase.

7 Future Prospects for Fungi in Food Processing

The future of fungi in food processing holds tremendous promise as advancements in biotechnology, consumer preferences, and environmental sustainability converge to unlock new applications. As the global demand for sustainable food sources and healthier, more functional products increases, fungi are poised to play an even more prominent role. From enhancing nutrition and developing novel food products to contributing to waste reduction and resource efficiency, the potential applications of fungi in food processing are vast (Singh et al., 2021a, b). Fungal-based proteins are emerging as sustainable, nutritious alternatives to animal-derived proteins (Amara et al., 2023). Fungi, like Fusarium species and Saccharomyces cerevisiae, are rich in protein and can be tailored to meet nutritional needs. With their ability to grow on diverse substrates, including agricultural waste, and their lower environmental impact than livestock farming, fungi offer a promising solution for plant-based proteins. Products like Fusarium venenatum (Quorn) provide high-protein, low-fat meat substitutes, and the growing demand for ethical, sustainable food is expected to drive further innovation. Fungal proteins, rich in essential amino acids, are also highly bioavailable for human consumption (Wiebe, 2004). Fungi have long been used for their enzymes in food fermentation, such as in bread, cheese, and alcoholic beverages. Their enzymes (amylases, proteases, lipases) enhance flavour, texture, and digestibility and are being studied to improve food processing, reduce waste, and extract nutrients. Fungi are also a rich source of bioactive compounds like antioxidants, vitamins, and polyphenols, which offer health benefits. The use of fungal enzymes and bioactive compounds in functional foods is expected to grow, supporting gut health, immune function, and chronic disease prevention, thus expanding the functional food market (Rantasalo et al., 2018).

Sustainability is a key driving force in the future of fungi in food processing. Fungi's ability to grow on various organic waste materials, such as agricultural by-products, food scraps, and industrial waste, presents a valuable opportunity for reducing food waste and promoting circular economies (Leger et al., 2021). Fungal mycelium, the vegetative part of the fungus, is capable of breaking down complex organic matter, including lignocellulose, making it a potential tool for bioremediation. In the future, fungi could be employed in the valorisation of food waste, where fungal fermentation processes can convert waste materials into edible food products, such as mushroom-based snacks, or valuable by-products like biofuels and bioplastics. Additionally, fungi-based materials like mycelium could replace synthetic packaging materials, contributing to the reduction of plastic waste in the food industry. Mycelium is biodegradable, lightweight, and could serve as an eco-friendly alternative in product packaging, further supporting sustainability goals within the food sector.

Fungal fermentation is already a critical component of many traditional food products, but the future holds even more promise as new fermentation technologies emerge. Advanced bioreactor systems, precision fermentation, and genetic engineering of fungal strains are set to revolutionise the way fungi are used in food processing. For instance, precision fermentation allows for the tailored production of specific compounds, including flavours, proteins, and enzymes, through fungi. With the help of synthetic biology, fungal strains can be genetically engineered to produce novel compounds that enhance the nutritional profile of foods or create entirely new food products. As fermentation technology continues to evolve, the efficiency and scalability of fungal fermentation will improve, enabling the widespread use of fungi in food production. This can lead to the creation of more diverse and customised food products, offering consumers unique flavours, textures, and health benefits. The potential for the production of "designer" foods, where the characteristics of the food are controlled at the microbial level, is a key trend for the future of the food industry (Pouris et al., 2024).

As the awareness of fungi's potential in food processing grows, consumer acceptance is likely to follow. With increasing interest in plant-based, health-conscious, and sustainable foods, fungi-based products are increasingly positioned as viable and desirable alternatives in the marketplace. However, to further integrate fungi into mainstream food products, there will be a need for more education and outreach efforts that emphasise the safety, nutritional benefits, and versatility of fungi. Transparency in labeling, as well as clear communication about production methods and benefits, will be important to overcoming initial skepticism from consumers. In the future, consumer-focused innovation will help shape the demand for fungal foods. As the market for plant-based and alternative protein products expands, fungi could occupy a prominent space in offering novel, sustainable, and highly nutritious food options. With more consumer awareness, fungi-based products may become commonplace in households, restaurants, and food services (Mazac et al., 2022; Zurek et al., 2022).

8 Conclusions

The future of fungi in food processing is brimming with potential as advancements in biotechnology, sustainability, and consumer preferences align to unlock their diverse applications. Fungi, with their ability to enhance nutrition, create novel food products, and contribute to environmental sustainability, are set to play an increasingly prominent role in the food industry. The growing demand for sustainable, plant-based proteins positions fungal-based proteins as an alternative to animal-derived proteins, offering both nutritional benefits and a lower environmental footprint. Furthermore, fungi's use in fermentation, production of bioactive compounds, and waste reduction will continue to expand, making them a cornerstone of functional foods and circular economies. As innovative fermentation technologies, precision engineering, and consumer awareness evolve, fungi-based products are poised to become a significant part of the mainstream food market. With sustainability as a driving force and consumer preferences shifting toward more ethical, health-conscious choices, the future of fungi in food processing appears not only promising but transformative.

References

Afroz Toma, M., Rahman, M. H., Rahman, M. S., Arif, M., Nazir, K. H. M. N. H., & Dufossé, L. (2023). Fungal pigments: Carotenoids, riboflavin, and polyketides with diverse applications. *J Fungi (Basel).*, *9*(4), 454. https://doi.org/10.3390/jof9040454

Ali, A., Shehzad, A., Khan, M. R., Shabbir, M. A., & Amjid, M. R. (2012). Yeast, its types and role in fermentation during bread making process – A review. *Pakistan Journal of Food Sciences, 3*, 171–179. https://www.researchgate.net/publication/328052441

Amara, A. A., & El-Baky, N. A. (2023). Fungi as a source of edible proteins and animal feed. *Journal of Fungi, 9*(1), 73. https://doi.org/10.3390/jof9010073

Ashiq, S. (2015). Natural occurrence of mycotoxins in food and feed: Pakistan perspective. *Comprehensive Reviews in Food Science and Food Safety, 14*, 159–175.

Awuchi, C. G., Ondari, E. N., Ogbonna, C. U., Upadhyay, A. K., Baran, K., et al. (2021). Mycotoxins affecting animals, foods, humans, and plants: types, occurrence, toxicities, action mechanisms, prevention, and detoxification strategies – A revisit. *Foods (Basel, Switzerland), 10*, 1279.

Ayed, L., Abid, S. B., & Hamdi, M. (2016). Development of a beverage from red grape juice fermented with the Kombucha consortium. *Annals of Microbiology, 67*(1), 111–121. https://doi.org/10.1007/s13213-016-1242-2

Azzolini, M., Tosi, E., Lorenzini, M., Finato, F., & Zapparoli, G. (2014). Contribution to the aroma of white wines by controlled Torulaspora delbrueckii cultures in association with Saccharomyces cerevisiae. *World Journal of Microbiology and Biotechnology, 31*(2), 277–293. https://doi.org/10.1007/s11274-014-1774-1

Bagchi, D. (2006). Nutraceuticals and functional foods regulations in the United States and around the world. *Toxicology., 221*, 1–3. https://doi.org/10.1016/j.tox.2006.01.001

Bahram, M., & Netherway, T. (2022). Fungi as mediators linking organisms and ecosystems. *FEMS Microbiology Reviews, 46*(2), fuab058. https://doi.org/10.1093/femsre/fuab058

Barzee, T. J., Cao, L., Pan, Z., & Zhang, R. (2021). Fungi for future foods. *Journal of Future Foods, 1*(1), 25–37. https://doi.org/10.1016/j.jfutfo.2021.09.002

Benito, Á., Calderón, F., & Benito, S. (2017). The combined use of schizosaccharomyces pombe and Lachancea thermotolerans – Effect on the anthocyanin wine composition. *Molecules, 22*(5), 739. https://doi.org/10.3390/molecules22050739

Cauvain, S. (2015). Technology of breadmaking. *Springer eBooks*. https://doi.org/10.1007/978-3-319-14687-4

Česnik, U., Martelanc, M., Øvsthus, I., Vukajlović, T. R., Hosseini, A., Vodopivec, B. M., & Butinar, L. (2023). Functional characterization of saccharomyces yeasts from cider produced in hardanger. *Fermentation, 9*(9), 824. https://doi.org/10.3390/fermentation9090824

Chauhan, K., & Rao, A. (2024). Clean-label alternatives for food preservation: An emerging trend. *Heliyon., 10*(16), e35815. https://doi.org/10.1016/j.heliyon.2024.e35815

Contreras, A., Hidalgo, C., Henschke, P. A., Chambers, P. J., Curtin, C., & Varela, C. (2013). Evaluation of Non-Saccharomyces yeasts for the reduction of alcohol content in wine. *Applied and Environmental Microbiology, 80*(5), 1670–1678. https://doi.org/10.1128/aem.03780-13

Copetti, M. V. (2019a). Fungi as industrial producers of food ingredients. *Current Opinion in Food Science, 25*, 52–56. https://doi.org/10.1016/j.cofs.2019.02.006

Copetti, M. V. (2019b). Yeasts and molds in fermented food production: an ancient bioprocess. *Current Opinion in Food Science, 25*, 57–61. https://doi.org/10.1016/j.cofs.2019.02.014

Daniela, P., Cuadrado-Osorio, Ramírez-Mejía, J. M., Mejía-Avellaneda, L. F., Mesa, L., & Bautista, E. J. (2022). Agro-industrial residues for microbial bioproducts: A key booster for bioeconomy. *Bioresource Technology Reports, 20*, 101232., ISSN:2589-014X. https://doi.org/10.1016/j.biteb.2022.101232

De Carvalho Cristiane, S. M., Ceci, S., De Carvalho Leise, P., Marli T, D. A. M., Andre, L. M. S., Guilherme, P. A., & De Andrade Meire, C. N. (2015). Cultivation and bromatological analysis of the medicinal mushroom Ganoderma lucidum (Curt.: Fr.) P. Karst cultivated in agricultural waste. *African Journal of Biotechnology, 14*(5), 412–418. https://doi.org/10.5897/ajb2014.14022

Dermiki, M., Phanphensophon, N., Mottram, D. S., & Methven, L. (2013). Contributions of non-volatile and volatile compounds to the umami taste and overall flavour of shiitake mushroom extracts and their application as flavour enhancers in cooked minced meat. *Food Chemistry, 141*(1), 77–83. https://doi.org/10.1016/j.foodchem.2013.03.018

do Prado, F. G., Pagnoncelli, M. G. B., de Melo Pereira, G. V., Karp, S. G., & Soccol, C. R. (2022). Fermented soy products and their potential health benefits: A review. *Microorganisms., 10*(8), 1606. https://doi.org/10.3390/microorganisms10081606

Dufossé, L., Fouillaud, M., & Caro, Y. (2021). Fungi and fungal metabolites for the improvement of human and animal nutrition and health. *J Fungi (Basel)., 7*(4), 274. https://doi.org/10.3390/jof7040274

Dupont, J., Dequin, S., Giraud, T., Tacon, F. L., Marsit, S., Ropars, J., Richard, F., & Selosse, M. (2017). Fungi as a source of food. *Microbiology Spectrum, 5*(3). https://doi.org/10.1128/microbiolspec.funk-0030-2016

Fakruddin, M., Hossain, M. N., & Ahmed, M. M. (2017). Antimicrobial and antioxidant activities of Saccharomyces cerevisiae IFST062013, a potential probiotic. *BMC Complementary and Alternative Medicine, 17*(1). https://doi.org/10.1186/s12906-017-1591-9

Filho, P. F. S., Andersson, D., Ferreira, J. A., & Taherzadeh, M. J. (2019). Mycoprotein: Environmental impact and health aspects. *World Journal of Microbiology and Biotechnology, 35*(10). https://doi.org/10.1007/s11274-019-2723-9

Finnigan TJA, Wall BT, Wilde PJ, Stephens FB, Taylor SL, Freedman MR (2019) Mycoprotein: The future of nutritious nonmeat protein a symposium review. Curr Dev Nutr 3(6):nzz021. https://doi.org/10.1093/cdn/nzz021

Frisvad, J. (2014). PENICILLIUM | Penicillium/Penicillia in food production. *Elsevier eBooks*, 14–18. https://doi.org/10.1016/b978-0-12-384730-0.00249-4

González-Royo, E., Pascual, O., Kontoudakis, N., Esteruelas, M., Esteve-Zarzoso, B., Mas, A., Canals, J. M., & Zamora, F. (2014). Oenological consequences of sequential inoculation with non-Saccharomyces yeasts (Torulaspora delbrueckii or Metschnikowia pulcherrima) and Saccharomyces cerevisiae in base wine for sparkling wine production. *European Food Research and Technology, 240*(5), 999–1012. https://doi.org/10.1007/s00217-014-2404-8

Hyde, K. D., Xu, J., Rapior, S., et al. (2019). The amazing potential of fungi: 50 ways we can exploit fungi industrially. *Fungal Diversity, **97***, 1–136. https://doi.org/10.1007/s13225-019-00430-9

Jakubczyk, D., & Dussart, F. (2020). Selected fungal natural products with antimicrobial properties. *Molecules, 25*(4), 911. https://doi.org/10.3390/molecules25040911

Kabak, B., Dobson, A. D., & Var, I. I. L. (2006). Strategies to prevent mycotoxin contamination of food and animal feed: A review. *Critical Reviews in Food Science and Nutrition, 46*, 593–619.

Kelly, J., Yang, F., Dowling, L., Nurgel, C., Beh, A., Di Profio, F., Pickering, G., & Inglis, D. L. (2018). Characterization of Saccharomyces bayanus CN1 for fermenting partially dehydrated grapes grown in cool climate winemaking regions. *Fermentation, 4*(3), 77. https://doi.org/10.3390/fermentation4030077

Khachatourians, G. (2003). Fungi in food technology. *Mycology.* https://doi.org/10.1201/9780203913369.pt2

Kumari, R., Mishra, S., & Sachan, A. (2021). Fungi in food bioprocessing. *Fungal Biology*, 39–64. https://doi.org/10.1007/978-3-030-64406-2_4

Lahue, C., Madden, A. A., Dunn, R. R., & Heil, C. S. (2020). History and domestication of Saccharomyces cerevisiae in bread baking. *Frontiers in Genetics, 11*. https://doi.org/10.3389/fgene.2020.584718

Lamacchia, C., Camarca, A., Picascia, S., Di Luccia, A., & Gianfrani, C. (2014). Cereal-based gluten-free food: how to reconcile nutritional and technological properties of wheat proteins with safety for celiac disease patients. *Nutrients, 6*(2), 575–590. https://doi.org/10.3390/nu6020575

Leclercq-Perlat, M.-N. (2011). *Cheese | camembert, brie, and related varieties, encyclopedia of dairy sciences* (2nd ed., pp. 773–782., ISBN:9780123744074,). Academic. https://doi.org/10.1016/B978-0-12-374407-4.00091-1

Leger, D., Matassa, S., Noor, E., Shepon, A., Milo, R., & Bar-Even, A. (2021). Photovoltaic-driven microbial protein production can use land and sunlight more efficiently than conventional crops. *Proceedings of the National Academy of Sciences of the United States of America, 118*, e2015025118.

Leite, F. H. M., Khandpur, N., Andrade, G. C., Anastasiou, K., Baker, P., Lawrence, M., & Monteiro, C. A. (2022). Ultra-processed foods should be central to global food systems dialogue and action on biodiversity. *BMJ Glob Health., 7*(3), e008269. https://doi.org/10.1136/bmjgh-2021-008269

Lin, L., & Xu, J. (2022). Production of fungal pigments: molecular processes and their applications. *Journal of Fungi (Basel)., 9*(1), 44. https://doi.org/10.3390/jof9010044

Lindsay, M. A., Granucci, N., Greenwood, D. R., & Villas-Boas, S. G. (2022). Identification of new natural sources of flavour and aroma metabolites from solid-state fermentation of agro-industrial by-products. *Metabolites, 12*(2), 157. https://doi.org/10.3390/metabo12020157

Lleixà, J., Martín, V., Del Portillo, C. M., Carrau, F., Beltran, G., & Mas, A. (2016). Comparison of fermentation and wines produced by inoculation of hanseniaspora vineae and Saccharomyces cerevisiae. *Frontiers in Microbiology, 7*. https://doi.org/10.3389/fmicb.2016.00338

Ludlow, C. L., Cromie, G. A., Garmendia-Torres, C., Sirr, A., Hays, M., Field, C., Jeffery, E. W., Fay, J. C., & Dudley, A. M. (2016). Independent origins of yeast associated with coffee and cacao fermentation. *Current Biology, 26*(7), 965–971. https://doi.org/10.1016/j.cub.2016.02.012

Mandal, P., Rai, A., Mishra, S., Tripathi, A., & Das, M. (2018). *Mutagens in food* (pp. 133–160., ISBN:9780128092521). Academic. https://doi.org/10.1016/B978-0-12-809252-1.00007-9

Mazac, R., Meinila, J., Korkalo, L., Jarvio, N., Jalava, M., & Tuomisto, H. L. (2022). Incorporation of novel foods in European diets can reduce global warming potential, water use and land use by over 80. *Nature Food, 3*, 286–293.

Meena, M., Zehra, A., Dubey, M. K., Aamir, M., & Upadhyay, R. S. (2017). Penicillium enzymes for the food industries. *Elsevier eBooks*, 167–186. https://doi.org/10.1016/b978-0-444-63501-3.00009-0

Miranda, A. C., Leães, G. F., & Copetti, M. V. (2022). Fungal biofilms: Insights for the food industry. *Current Opinion in Food Science, 46*, 100846. https://doi.org/10.1016/j.cofs.2022.100846

Mojsov, K. (2016). Aspergillus enzymes for food industries. *Elsevier eBooks*, 215–222. https://doi.org/10.1016/b978-0-444-63505-1.00033-6

Mullan, W. (2014). STARTER CULTURES | Importance of selected genera. *Elsevier eBooks*, 515–521. https://doi.org/10.1016/b978-0-12-384730-0.00321-9

Naranjo-Ortiz, M. A., & Gabaldón, T. (2019). Fungal evolution: Major ecological adaptations and evolutionary transitions. *Biological Reviews/Biological Reviews of the Cambridge Philosophical Society, 94*(4), 1443–1476. https://doi.org/10.1111/brv.12510

Nout, M. J. R. (2004). *Useful role of fungi in food processing*. Wageningen University and Research.

Ortiz, G. E., Noseda, D. G., Mora, M. C. P., Recupero, M. N., Blasco, M., & Albertó, E. (2016). A comparative study of new Aspergillus strains for proteolytic enzymes production by solid state fermentation. *Enzyme Research, 2016*, 1–11. https://doi.org/10.1155/2016/3016149

Pandit, A., Darshane, P., & Dixit, M. (2024). Role of fungi in revolutionizing the food sector. *Indian Journal of Nutrition, 11*(1), 290.

Papagianni, M. (2007). Advances in citric acid fermentation by Aspergillus niger: Biochemical aspects, membrane transport and modeling. *Biotechnology Advances, 25*(3), 244–263. https://doi.org/10.1016/j.biotechadv.2007.01.002

Parapouli, M., Vasileiadi, A., Afendra, A., & Hatziloukas, E. (2020). Saccharomyces cerevisiae and its industrial applications. *AIMS Microbiology, 6*(1), 1–32. https://doi.org/10.3934/microbiol.2020001

Poojary, M. M., Orlien, V., Passamonti, P., & Olsen, K. (2017). Enzyme-assisted extraction enhancing the umami taste amino acids recovery from several cultivated mushrooms. *Food Chemistry, 234*, 236–244. https://doi.org/10.1016/j.foodchem.2017.04.157

Pouris, J., Kolyva, F., Bratakou, S., Vogiatzi, C. A., Chaniotis, D., & Beloukas, A. (2024). The role of fungi in food production and processing. *Applied Sciences, 14*(12), 5046. https://doi.org/10.3390/app14125046

Rantasalo, A., Landowski, C. P., Kuivanen, J., Korppoo, A., Reuter, L., Koivistoinen, O., Valkonen, M., Penttila, M., Jantti, J., & Mojzita, D. (**2018**). A universal gene expression system for fungi. *Nucleic Acids Research, 46*, e111.

Renault, P., Miot-Sertier, C., Marullo, P., Hernández-Orte, P., Lagarrigue, L., Lonvaud-Funel, A., & Bely, M. (2009). Genetic characterization and phenotypic variability in Torulaspora delbrueckii species: Potential applications in the wine industry. *International Journal of Food Microbiology, 134*(3), 201–210. https://doi.org/10.1016/j.ijfoodmicro.2009.06.008

Renzetti, S., Courtin, C., Delcour, J., & Arendt, E. (2010). Oxidative and proteolytic enzyme preparations as promising improvers for oat bread formulations: Rheological, biochemical and microstructural background. *Food Chemistry, 119*, 1465–1473. https://doi.org/10.1016/j.foodchem.2009.09.028

Roudil, L., Russo, P., Berbegal, C., Albertin, W., Spano, G., & Capozzi, V. (2019). Non-Saccharomyces commercial starter cultures: scientific trends, recent patents and innovation in the wine sector. *Recent Patents on Food Nutrition & Agriculture, 11*(1), 27–39. https://doi.org/10.2174/2212798410666190131103713

Rousta, N., Aslan, M., Akbas, M. Y., Ozcan, F., Sar, T., & Taherzadeh, M. J. (2023). Effects of fungal based bioactive compounds on human health: Review paper. *Critical Reviews in Food Science and Nutrition, 64*(20), 7004–7027. https://doi.org/10.1080/10408398.2023.2178379

Salwan, R., Katoch, S., & Sharma, V. (2021). Recent developments in shiitake mushrooms and their nutraceutical importance. *Fungal Biology*, 165–180. https://doi.org/10.1007/978-3-030-64406-2_10

Sar, T., Ferreira, J. A., & Taherzadeh, M. J. (2020). Bioprocessing strategies to increase the protein fraction of Rhizopus oryzae biomass using fish industry sidestreams. *Waste Management, 113*, 261–269. https://doi.org/10.1016/j.wasman.2020.06.005

Shurson, G. (2017). Yeast and yeast derivatives in feed additives and ingredients: Sources, characteristics, animal responses, and quantification methods. *Animal Feed Science and Technology, 235*, 60–76. https://doi.org/10.1016/j.anifeedsci.2017.11.010

Singh, S., Dhanjal, D. S., Thotapalli, S., Sonali, N., Sharma, P., & Singh, J. (2020). Importance and recent aspects of fungi-based food ingredients. *Elsevier eBooks*, 245–254. https://doi.org/10.1016/b978-0-12-821007-9.00019-x

Singh, A., Misra, M., Mishra, S., & Sachan, S. G. (2021a). Fungal production of food supplements. *Fungal Biology*, 129–142. https://doi.org/10.1007/978-3-030-64406-2_8

Singh, A., Misra, M., Mishra, S., & Sachan, S. G. (2021b). Fungal production of food supplements. In *Fungi in sustainable food production* (pp. 129–142). Springer.

Suhaimi, H., Dailin, D. J., Malek, R. A., Hanapi, S. Z., Ambehabati, K. K., Keat, H. C., Prakasham, S., Elsayed, E. A., Misson, M., & Enshasy, H. E. (2021). Fungal pectinases: Production and applications in food industries. *Fungal Biology*, 85–115. https://doi.org/10.1007/978-3-030-64406-2_6

Syal, P., & Vohra, A. (2013). Probiotic potential of yeasts isolated from traditional indian fermented foods. *International Journal of Microbiology Research, 5*(2), 390–398. https://doi.org/10.9735/0975-5276.5.2.390-398

Tamang, J. P., Watanabe, K., & Holzapfel, W. H. (2016). Review: Diversity of microorganisms in global fermented foods and beverages. *Frontiers in Microbiology, 7*. https://doi.org/10.3389/fmicb.2016.00377

Troianou, V., Toumpeki, C., Dorignac, E., Kogkou, C., Kallithraka, S., & Kotseridis, Y. (2019). Evaluation ofSaccharomyces pastorianusimpact to Sauvignon blanc chemical & sensory profile compared to different strains ofS. cerevisiae/bayanus. *BIO Web of Conferences, 12*, 02025. https://doi.org/10.1051/bioconf/20191202025

Varela, C., Barker, A., Tran, T., Borneman, A., & Curtin, C. (2017). Sensory profile and volatile aroma composition of reduced alcohol Merlot wines fermented with Metschnikowia pulcherrima and Saccharomyces uvarum. *International Journal of Food Microbiology, 252*, 1–9. https://doi.org/10.1016/j.ijfoodmicro.2017.04.002

Vejarano, R., & Gil-Calderón, A. (2021). Commercially available non-saccharomyces yeasts for winemaking: current market, advantages over saccharomyces, biocompatibility, and safety. *Fermentation, 7*(3), 171. https://doi.org/10.3390/fermentation7030171

Villagrán, Z., Martínez-Reyes, M., Gómez-Rodríguez, H., Ríos-García, U., Montalvo-González, E., Ortiz-Basurto, R. I., Anaya-Esparza, L. M., & Pérez-Moreno, J. (2023). Huitlacoche (*Ustilago maydis*), an iconic mexican fungal resource: Biocultural importance, nutritional content, bioactive compounds, and potential biotechnological applications. *Molecules, 28*(11), 4415. https://doi.org/10.3390/molecules28114415

Walker, G., & Stewart, G. (2016). Saccharomyces cerevisiae in the production of fermented beverages. *Beverages, 2*(4), 30. https://doi.org/10.3390/beverages2040030

Wang, Q., Luo, W., Gu, Q., Feng, L., Li, H., & Yu, X. (2013). Enhanced lycopene content in blakeslea trispora by effective mutation-screening method. *Applied Biochemistry and Biotechnology, 171*(7), 1692–1700. https://doi.org/10.1007/s12010-013-0468-8

Weaver, C. M., Dwyer, J., Fulgoni, V. L., 3rd, King, J. C., Leveille, G. A., MacDonald, R. S., Ordovas, J., & Schnakenberg, D. (2014). Processed foods: Contributions to nutrition. *The American Journal of Clinical Nutrition, 99*(6), 1525–1542. https://doi.org/10.3945/ajcn.114.089284

Wiebe, M. G. (2004). QuornTM Myco-protein – Overview of a successful fungal product. *Mycologist., 18*, 17–20. https://doi.org/10.1017/S0269915X04001089

Wu, J., Siu, K., & Geng, P. (2021). Bioactive ingredients and medicinal values of Grifola frondosa (Maitake). *Foods, 10*(1), 95. https://doi.org/10.3390/foods10010095

Xu, X., Yan, H., Tang, J., Chen, J., & Zhang, X. (2014). Polysaccharides in Lentinus edodes: Isolation, structure, immunomodulating activity and future prospective. *Critical Reviews in Food Science and Nutrition, 54*(4), 474–487. https://doi.org/10.1080/10408398.2011.587616

Yang, S., & Zhang, H. (2016). Enhanced polyunsaturated fatty acids production inMortierella alpinaby SSF and the enrichment in chicken breasts. *Food & Nutrition Research, 60*(1), 30842. https://doi.org/10.3402/fnr.v60.30842

Zhou, X., Qiao, K., Wu, H., & Zhang, Y. (2023). The impact of food additives on the abundance and composition of gut microbiota. *Molecules., 28*(2), 631. https://doi.org/10.3390/molecules28020631

Zhu, Y., & Tramper, J. (2013). Koji – Where East meets West in fermentation. *Biotechnology Advances, 31*(8), 1448–1457. https://doi.org/10.1016/j.biotechadv.2013.07.001

Zurek, M., Hebinck, A., & Selomane, O. (2022). Climate change and the urgency to transform food systems. *Science, 376*, 1416–1421.

Chapter 12
Fungal Additives in Food Packaging: A Sustainable Solution

Sahar Akhavan-Mahdavi

1 Introduction

The global food industry is under increasing pressure to adopt sustainable practices, particularly in food packaging. Conventional materials like polyethylene and polypropylene, derived from nonrenewable fossil fuels, contribute significantly to environmental degradation (Hussain et al., 2024). Over 8 million tons of plastic waste enter the oceans annually, with food packaging being a major contributor. These materials persist for centuries, accumulate in landfills and oceans, and release harmful chemicals, posing risks to ecosystems and human health. The lack of efficient recycling infrastructure further exacerbates these challenges, necessitating the exploration of sustainable alternatives (Santi et al., 2022).

In response, the packaging industry is shifting toward bio-based materials that offer biodegradability and reduced environmental impact. While synthetic biodegradable polymers, such as polylactic acid (PLA) and polyhydroxyalkanoates (PHA), have gained traction, their performance often falls short of conventional plastics in terms of mechanical strength, barrier properties, and cost-effectiveness. Additionally, some synthetic biodegradable materials still rely on energy-intensive production processes or nonrenewable resources, limiting their alignment with the principles of a circular economy. To overcome these limitations, researchers are exploring natural additives that enhance the functionality of bio-based packaging while maintaining sustainability (Al Mahmud et al., 2024; Das et al., 2024).

Fungal additives have emerged as a superior solution, offering a unique combination of biodegradability, biocompatibility, and enhanced performance. Derived from renewable fungal sources, these additives include polysaccharides (e.g., chitosan and glucans), enzymes, and secondary metabolites, which can be seamlessly

S. Akhavan-Mahdavi (✉)
Department of Food Materials and Process Design Engineering, Gorgan University of Agricultural Science and Natural Resources, Gorgan, Iran

B. P. Singh et al. (eds.), *Fungal Additives and Bioactives in Food Processing Industries*, Fungal Biology, https://doi.org/10.1007/978-3-032-04519-5_12

integrated into biodegradable polymer matrices. Chitosan, for instance, provides excellent antimicrobial and antioxidant properties, protecting packaged foods from microbial contamination and oxidative spoilage without the need for synthetic preservatives (Pellicer et al., 2017). Glucans enhance mechanical strength and barrier properties, improving the durability and protective capabilities of packaging films. Fungal enzymes and secondary metabolites further contribute to active packaging systems by extending shelf life and reducing food waste, aligning with consumer demand for clean-label products. Unlike synthetic additives, fungal-derived compounds are inherently biodegradable and sourced from renewable feedstocks, supporting a circular economy and minimizing environmental impact (Jafarzadeh et al., 2023).

This chapter provides a comprehensive overview of fungal additives in sustainable food packaging, detailing their functional properties, mechanisms of action, and integration into biodegradable polymers. It explores how these additives enhance packaging performance, offering antimicrobial activity, improved barrier properties, and extended shelf life, while addressing environmental concerns. Additionally, the chapter examines safety and regulatory considerations to ensure compliance with food safety standards, alongside recent advancements in extraction techniques and the development of fungal-based nanocomposites. By highlighting their practical applications and commercial potential, this chapter underscores the transformative role of fungal additives in revolutionizing the food packaging industry.

2 The Environmental Impact of Conventional Food Packaging Materials

The global food industry relies heavily on packaging to preserve food quality, ensure safety, and extend shelf life. However, conventional food packaging materials, such as polyethylene (PE), polypropylene (PP), and polystyrene (PS), are predominantly derived from nonrenewable fossil fuels and are nonbiodegradable (Mafe et al., 2024). These materials contribute significantly to environmental pollution, with millions of tons of plastic waste ending up in landfills and oceans annually. As shown in Fig. 12.1, food packaging is the largest contributor to global plastic waste, accounting for 40% of the total (Ncube et al., 2020). This alarming statistic underscores the urgent need for sustainable alternatives to conventional materials. As illustrated in Table 12.1, conventional materials like polyethylene and polypropylene have degradation times exceeding 500 years, contributing to long-term environmental pollution. Additionally, their production involves high carbon emissions, with polyethylene having a carbon footprint of 3.5 kg CO_2/kg (Mafe et al., 2024). Polystyrene, commonly used in disposable food containers, is particularly problematic due to its very low recyclability and high environmental persistence (Guillard et al., 2018). These materials not only persist in the environment but also break

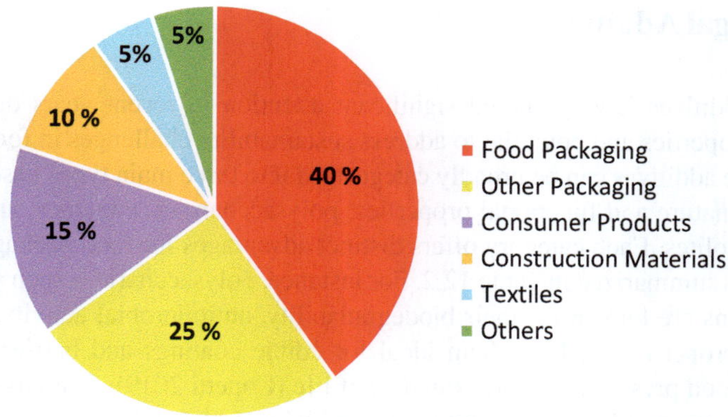

Fig. 12.1 Contribution of food packaging to global plastic waste

Table 12.1 Environmental impact of conventional food packaging materials

Material type	Degradation time	Carbon footprint (kg CO_2/kg)	Recyclability	Reference
Polyethylene (PE)	500+ years	3.5	Low	Yao et al. (2022)
Polypropylene (PP)	500+ years	3.2	Low	Zhang et al. (2020)
Polystyrene (PS)	500+ years	3.8	Very Low	Farrelly and Shaw (2017)
Biodegradable PLA	6–12 months	2.1	High	Taib et al. (2023)

down into microplastics, which contaminate soil, water, and marine ecosystems, ultimately entering the food chain and posing risks to human health and wildlife (Ncube et al., 2020). While biodegradable materials like polylactic acid (PLA) offer a more sustainable alternative, their adoption is still limited due to cost and scalability challenges. PLA, for instance, has a significantly lower carbon footprint (2.1 kg CO_2/kg) and degrades within 6–12 months, making it a promising option. However, its higher production costs and limited availability hinder its widespread use in the food industry. This gap in the market underscores the urgent need for innovative solutions like fungal additives, which are not only biodegradable but also derived from renewable sources (Dilkes-Hoffman et al., 2018).

Fungal additives, such as chitosan and glucans, present a viable alternative to conventional packaging materials. They are biodegradable, meaning they can be broken down by natural processes, reducing their environmental impact. Additionally, fungal additives exhibit antimicrobial properties, which enhance food safety and shelf life, and are derived from renewable sources, aligning with the principles of a circular economy. By replacing nonrenewable and nonbiodegradable materials, fungal additives offer a sustainable solution to the environmental challenges posed by conventional food packaging (Sid et al., 2021).

3 Fungal Additives

Fungal additives have garnered significant attention in recent years due to their unique properties and potential to address sustainability challenges in food packaging. These additives can be broadly categorized into three main types based on their chemical nature and functional properties: polysaccharides, enzymes, and secondary metabolites. Each category offers distinct advantages for food packaging applications, as summarized in Table 12.2. For instance, polysaccharides such as chitosan and glucans are known for their biodegradability, antimicrobial activity, and film-forming properties, making them ideal for edible coatings and barrier films that enhance food preservation and extend shelf life (Copetti 2019). Similarly, enzymes like lysozyme and laccase exhibit antimicrobial and biocatalytic properties, which are valuable for antimicrobial coatings and active packaging systems. Additionally, secondary metabolites, including phenolic compounds and terpenoids, provide antioxidant and antimicrobial effects, crucial for shelf-life extension and active packaging (Hernãndez et al., 2022). The production of fungal additives involves the cultivation, extraction, and purification of bioactive compounds from various fungal species, such as *Aspergillus, Penicillium*, and Trichoderma, as well as edible mushrooms like Agaricus and Pleurotus. As detailed in Table 12.3, different fungal species yield specific additives with varying levels of efficiency. For example, *Aspergillus niger* produces glucans through alkaline extraction, yielding high quantities with moderate purity, while *Trichoderma reesei* is a source of cellulase, an enzyme produced via fermentation with high yield and purity (Adeyeye 2016). *Pleurotus ostreatus* yields chitosan through deacetylation of chitin, offering

Table 12.2 Types of fungal additives and their functional properties

Type of additive	Examples	Key properties	Reference
Polysaccharides	Chitosan, glucans	Biodegradable, antimicrobial, film-forming	Mahmoud et al. (2019)
Enzymes	Lysozyme, laccase	Antimicrobial, biocatalytic	Sommer et al. (2018)
Secondary metabolites	Phenolic compounds, Terpenoids	Antioxidant, antimicrobial	Nawrot-Chorabik et al. (2022)

Table 12.3 Common fungal sources and their additives

Fungal species	Additive produced	Extraction method	Yield and purity	Reference
Aspergillus niger	Glucans	Alkaline extraction	High yield, moderate purity	Chuang et al. (2020)
Trichoderma reesei	Cellulase (enzyme)	Fermentation	High yield, high purity	Keshavarz and Khalesi (2016)
Pleurotus ostreatus	Chitosan	Deacetylation of chitin	Moderate yield, high purity	Golak-Siwulska et al. (2018)
Penicillium chrysogenum	Penicillin (secondary metabolite)	Fermentation	Low yield, high purity	Shaaban et al. (2023)

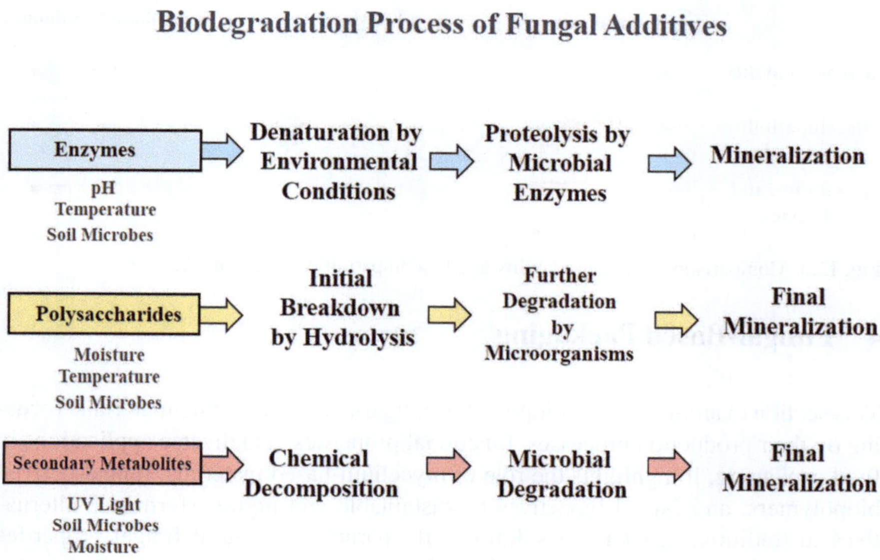

Fig. 12.2 Biodegradation process of fungal additives

moderate yield and high purity, and *Penicillium chrysogenum* produces penicillin, a secondary metabolite, through fermentation, albeit with lower yield but high purity. These production methods highlight the versatility and scalability of fungal additives for industrial applications (Nigam & Luke, 2016).

One of the most significant advantages of fungal additives is their biodegradability and biocompatibility, which make them ideal for sustainable food packaging. Unlike synthetic additives, fungal-derived materials can be broken down by natural processes, reducing environmental impact (Pundir et al., 2010).As illustrated in Fig. 12.2, fungal polysaccharides like chitosan and glucans are readily degraded by microorganisms in the environment, leading to the formation of nontoxic byproducts such as water, carbon dioxide, and biomass (Ye et al., 2018). Enzymes and secondary metabolites derived from fungi are also biodegradable, ensuring that they do not persist in the environment. The biodegradation process of fungal additives is further detailed in Fig. 12.2, showing the stepwise breakdown mechanism. Initially, fungal-based materials (such as chitosan and glucans) are attacked by microbial enzymes, leading to structural degradation into simpler compounds. These breakdown products are then metabolized by soil or aquatic microorganisms, eventually converting into CO_2, H_2O, and biomass, which reintegrate into the natural carbon cycle. This process emphasizes the environmental sustainability of fungal additives, making them an ideal alternative to synthetic food packaging materials. Furthermore, fungal additives are nontoxic and safe for human consumption, making them suitable for direct contact with food. Their biocompatibility extends to their use in medical and pharmaceutical applications, further validating their safety. A comparison of the biodegradability and biocompatibility of fungal additives, as shown in Fig. 12.3, highlights their low environmental impact and high safety profile, underscoring their potential as sustainable alternatives to conventional packaging materials (Adeyeye 2016).

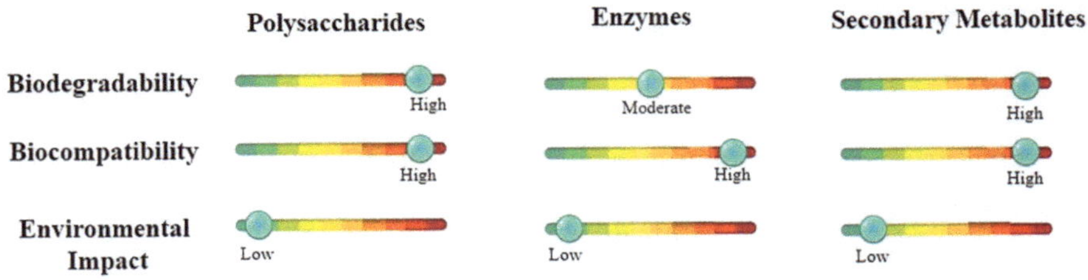

Fig. 12.3 Comparison of biodegradability and biocompatibility of fungal additives

4 Fungal-Based Packaging

This section examines the development of fungal-based packaging materials, focusing on their production processes, functional properties, and diverse applications in food packaging. It highlights the role of mycelium-based materials, fungal-derived biopolymers, and fungal bioactives as sustainable and high-performance alternatives to traditional packaging solutions. By harnessing natural fungal properties such as biodegradability, antimicrobial effects, and antioxidant activity, these materials offer an innovative approach to addressing food preservation, shelf-life extension, and environmental sustainability. The following subsections will explore the various stages of producing fungal-based packaging, their functional benefits, and their transformative potential in the food packaging industry (Manan et al., 2022).

Fungal-based packaging materials have garnered significant attention due to their sustainability, biodegradability, and potential for addressing the growing environmental concerns associated with plastic packaging. These materials are derived from either mycelium networks or fungal biopolymers such as chitosan and β-glucans, offering diverse functionalities that make them suitable for various food packaging applications. The production and processing of fungal-based packaging materials require a thorough understanding of the growth conditions, substrate selection, extraction processes, and processing techniques to optimize their performance as functional and eco-friendly packaging solutions (Rami et al., 2024).

4.1 Mycelium-Based Packaging Materials

Mushrooms are fungi with distinct reproductive and growth structures, including the vegetative mycelium (a network of hyphae) and the fruiting body that produces spores. Mycelium grows by attaching to a substrate and secreting enzymes that break down organic matter, enabling the fungus to absorb nutrients. This process, known as *colonization*, is crucial in producing mycelium-based biocomposites, which can be shaped in molds to form sustainable, biodegradable materials (Yang et al., 2021).

Mycelium biocomposites are created by growing the fungus around substrates like sawdust, resulting in a material that can replace environmentally harmful plastics, such as polystyrene, in packaging. These biocomposites can be arrested in growth through drying or heating, depending on the desired application. Mycelium-based materials are promising for packaging, as they are compostable and provide a sustainable alternative to petroleum-derived plastics. The growing demand for eco-friendly products among consumers, particularly the younger generation, has shifted market trends, making sustainability an important consideration for manufacturers. If cost-effective, mycelium biocomposites could revolutionize the packaging industry by reducing plastic waste and offering new opportunities for agricultural waste valorization (Sivaprasad et al., 2021; Verma et al., 2023). This shift aligns with growing corporate commitments to sustainability, exemplified by major companies like Coca-Cola and Unilever. The development of mycelium packaging is expected to minimize the environmental impact of nonbiodegradable materials, promoting local, renewable feedstocks and offering a green alternative in the global market.

Mycelium, the vegetative part of fungi, consists of a dense network of thread-like structures known as hyphae. Mycelium-based packaging materials are produced by cultivating fungal species on specific substrates, where the mycelium grows and forms a compact, interconnected network that can be harvested and processed into a durable material (Sivaprasad et al., 2021). The growth of mycelium is highly sensitive to various environmental factors, and optimizing these growth conditions is essential for achieving the desired properties in the final packaging material (Alemu et al., 2022). The optimization of growth conditions for mycelium-based packaging materials plays a crucial role in ensuring efficient fungal growth and high-quality production. The conditions such as temperature, humidity, pH, and nutrient sources significantly impact the growth and colonization of mycelium in the production process. As shown in Table 12.4, the optimal temperature range of 20–30 °C supports faster mycelium growth, while a humidity level of 80–90% enhances colonization and network density (Sydor et al., 2022). A neutral pH (6–7) is also essential for mycelial network formation, and the presence of nitrogen, carbon, and minerals in the nutrient source is critical for robust growth. The temperature range for most fungi species is between 20 and 30 °C, but some fungi, like *Pleurotus ostreatus* (oyster mushroom), may have optimal growth at slightly cooler temperatures. Moreover, the availability of nutrients such as nitrogen, carbon, and mineral salts directly impacts the growth rate and quality of the mycelium network. Optimizing these factors ensures the production of a robust, dense mycelial network that can be processed into strong and functional packaging materials (Manan et al., 2021).

Furthermore, various substrates are commonly used for mycelium growth in fungal packaging production. These substrates offer different nutritional components that affect the mycelium network's growth and the final properties of the material. For example, sawdust, as indicated in Table 12.5, contains cellulose and hemicellulose, which promote rapid colonization and a strong mycelial network, making it ideal for producing molds and sheets. In contrast, rice husks, which are rich in cellulose, lignin, and silica, provide high mechanical strength and water resistance, making them suitable for lightweight and strong materials for molds (Aiduang

Table 12.4 Optimization of growth conditions for mycelium-based packaging materials

Parameter	Optimal range	Effect on mycelium growth	Reference
Temperature	20–30 °C	Favors faster growth of mycelium	
Humidity	80–90%	Enhances colonization and network density	Verma et al. (2023)
pH	6–7	Neutral pH enhances mycelial network formation	Bagheriehnajjar et al. (2023)
Nutrient Source	Nitrogen, carbon, minerals	Essential for robust mycelial growth	Madusanka et al. (2024)

Table 12.5 Common substrates used for mycelium growth in packaging production

Substrate	Nutritional components	Effect on mycelium network	Common uses in packaging	Reference
Sawdust	Cellulose, hemicellulose	Rapid colonization, strong network	Molds and sheets	Schritt et al. (2021)
Straw	Cellulose, lignin	Dense, flexible network	Sheet-based packaging	Sreerag et al. (2024)
Rice husks	Cellulose, lignin, silica	High mechanical strength, water resistance	Lightweight, strong material for molds	Mbabali et al. (2024)
Corn stalks	Cellulose, hemicellulose	Dense, durable mycelium network	Molds and shapes for structural packaging	Meng et al. (2019)

et al., 2024). The selection of the substrate determines the speed of mycelium colonization, the density of the network, and the final material's properties. For example, substrates with higher cellulose content, such as wood sawdust, promote faster colonization and stronger mycelium growth. As the mycelium colonizes the substrate, it forms a dense, interconnected network that binds the particles together, creating a solid material that can be further processed into packaging films or molded shapes (Rajendran 2022).

Once mycelium has fully colonized the substrate, it can be harvested and processed into a variety of packaging materials. The transformation of mycelium into a functional biocomposite involves a series of key techniques aimed at enhancing its mechanical properties, durability, and overall performance for packaging applications. These techniques include compression, drying, and reinforcement, each of which plays a crucial role in ensuring the material meets the required standards for strength, flexibility, and stability. Compression is a critical step in the production of mycelium-based packaging (Aiduang et al., 2022). By compacting the mycelium network, the material becomes denser and gains increased mechanical strength, making it suitable for packaging applications that require a solid, robust structure. The degree of compression can be precisely controlled to achieve the desired balance between flexibility and rigidity. For instance, highly compressed mycelium is ideal for creating strong, rigid materials such as trays or containers, while a lighter compression might be used for more flexible applications like protective cushioning or sheets (Udayanga & Miriyagalla, 2021). Drying is another essential process that ensures the stability and longevity of mycelium-based packaging. After the mycelium has fully colonized its substrate, it undergoes a drying process to remove excess moisture, which could otherwise lead to fungal degradation or weakening of the material. Typically, air-drying or low-temperature oven drying (30–60 °C) is employed to achieve the optimal moisture content (Bitting et al., 2022). This step not only preserves the material but also enhances its resistance to microbial growth, making it stable and lightweight. The drying process is essential for preventing the growth of unwanted fungi and ensuring the material's shelf life. Reinforcement is often incorporated into mycelium-based packaging to further improve its mechanical properties, especially in terms of tensile strength and durability (Sivaprasad et al., 2021). This can be achieved by adding natural fibers or biopolymers to the mycelium network. For example, fibers from renewable plant sources like hemp, jute, or flax can be blended with the mycelium to increase its strength and enhance its performance in more demanding applications. Additionally, bio-based resins such as lignin, starch, or cellulose derivatives can be used to further strengthen the material and improve its resistance to wear and tear. This reinforcement not only enhances the material's functionality but also ensures that it maintains its eco-friendly and biodegradable properties. Together, these processes—compression, drying, and reinforcement—transform mycelium from a simple fungal network into a versatile and sustainable packaging material (Manan et al., 2022). By carefully controlling each step, manufacturers can create a wide range of mycelium-based biocomposites that offer an environmentally friendly alternative to traditional plastic packaging. These materials are not only biodegradable and compostable but also exhibit the potential to outperform conventional packaging solutions in certain applications, contributing to a more sustainable future (Lingam et al., 2023).

4.2 *Fungal-Derived Biopolymers for Packaging*

Fungal-derived biopolymers, particularly chitosan and β-glucans, are valuable materials for producing biodegradable and functional food packaging films. Chitosan, a polysaccharide derived from chitin (the primary component of fungal cell walls), has been extensively studied for its film-forming properties, antimicrobial activity, and biodegradability (Table 12.6). β-Glucans, on the other hand, are polysaccharides found in the cell walls of fungi that exhibit excellent film-forming properties, as well as beneficial health effects (Kandel 2022). The extraction of chitosan and β-glucans from fungal sources typically involves processes such as acid demineralization, deacetylation (for chitosan), or aqueous extraction (for β-glucans). In the case of chitosan, the process begins with the extraction of chitin from fungal biomass, followed by deacetylation using alkali treatment. For β-glucan extraction, fungi like *Saccharomyces cerevisiae* and *Aspergillus niger* are often used, where the polysaccharide is solubilized and then purified through filtration and precipitation (Muiruri et al., 2023).

Table 12.6 Properties of chitosan and β-glucan for packaging

Property	Chitosan	β-Glucan	Reference
Water solubility	Soluble in acidic conditions	Soluble in water, temperature dependent	Cazón and Vázquez (2019) and Chang et al. (2019)
Barrier properties	High resistance to oxygen and moisture	Good oxygen barrier but lower moisture resistance	Cazón and Vázquez (2020) and Razzaq et al. (2016)
Antimicrobial activity	Strong antimicrobial action against bacteria	Moderate antimicrobial effect	Bordenave et al. (2010) and Sherafatkhah Azari et al. (2021)
Film-forming capability	Excellent film-forming agent	Moderate film-forming ability	Akhavan-Mahdavi et al. (2023) and Sherafatkhah Azari et al. (2021)
Biodegradability	Biodegradable	Biodegradable	Peltzer et al. (2018) and Priyadarshi and Rhim (2020)

The functionality of fungal-derived biopolymers can be further enhanced by modifying their structures. For example, the incorporation of cross-linking agents can improve the mechanical strength and water resistance of chitosan-based packaging materials. Additionally, blending chitosan or β-glucans with other biodegradable polymers, such as poly(lactic acid) (PLA), can enhance their barrier properties, making them more effective at preventing the penetration of gases, moisture, and light, which are crucial for extending the shelf life of packaged food products. Fungal-derived biopolymers are primarily used to produce thin, flexible films that can be applied as coatings or used as standalone packaging materials. The most common technique for producing films from fungal biopolymers is casting, where a solution of the biopolymer is poured onto a flat surface and dried to form a film (Gaff et al., 2024). Other methods such as extrusion and electrospinning can also be employed to produce films with enhanced mechanical and barrier properties. Furthermore, the functionalization of fungal-based films through the incorporation of bioactive compounds can add additional functionalities, such as antimicrobial and antioxidant properties. For instance, chitosan films can be infused with essential oils, antimicrobial peptides, or plant-based extracts to further enhance their protective properties against microbial contamination and oxidation (Sreerag et al., 2024).

4.3 *Incorporation of Fungal Bioactives in Packaging Films*

The integration of fungal bioactives into packaging films has gained significant attention due to their potential to enhance food preservation. Fungal-derived compounds such as chitosan, β-glucan, phenolic compounds, terpenoids, and essential oils exhibit antimicrobial and antioxidant properties, making them excellent candidates for active packaging applications. These bioactive compounds can be

Table 12.7 Bioactive compounds incorporated into fungal packaging films

Bioactive compound	Source (Fungi)	Functionality	Common applications	Reference
Chitosan	*Aspergillus* spp., *Mucor* spp.	Antimicrobial, antioxidant	Active food packaging films	Akhavan-Mahdavi et al. (2023)
Phenolic compounds	*Penicillium* spp., *Fusarium* spp.	Antioxidant, antimicrobial	Food wraps, coatings	Ahmed et al. (2022)
Terpenoids	*Trichoderma* spp., *Ganoderma* spp.	Antioxidant, antimicrobial	Protective packaging	Stracquadanio et al. (2020)
Essential oils	*Ganoderma* spp., *Pleurotus* spp.	Antimicrobial, flavor preservation	Fresh produce packaging	Abou Fayssal et al. (2023)

embedded within the biopolymer matrix or applied as surface coatings to create functional packaging that enhances food safety and extends shelf life. Fungal secondary metabolites provide potent antimicrobial and antioxidant functionalities that can help mitigate food spoilage and contamination. Table 12.7 summarizes key bioactive compounds, their sources, functionalities, and common applications in packaging.

Several techniques are employed to incorporate fungal additives into biodegradable polymers, each offering unique advantages depending on the desired properties of the packaging material. Blending involves uniformly mixing fungal additives with the polymer matrix during the melt-processing stage, making it a simple, cost-effective, and scalable method suitable for films and rigid packaging. Coating applies fungal additives as a thin layer on the polymer surface, preserving the bulk properties while enhancing surface functionality, making it ideal for antimicrobial and antioxidant applications. Encapsulation incorporates fungal additives within micro- or nanoparticles before embedding them into the polymer matrix, allowing for controlled release of active compounds and improved stability, which is particularly beneficial for active packaging systems (Table 12.8).

Combining fungal additives with other natural additives can enhance the functional properties of biodegradable packaging materials, often resulting in improved antimicrobial, antioxidant, and mechanical performance. Chitosan combined with essential oils enhances antimicrobial activity due to the synergistic action of chitosan and essential oil components. Fungal enzymes paired with plant extracts contribute to improved antioxidant properties, helping to extend the shelf life of packaged food. Glucans integrated with nanocellulose enhance mechanical strength and barrier properties, making the packaging more durable and effective in food preservation. These synergistic combinations offer innovative solutions for developing high-performance, sustainable packaging materials (Delattre et al., 2023; Sharma et al., 2024).

Several case studies highlight the successful integration of fungal additives into food packaging, showcasing their benefits in real-world scenarios (Table 12.9). Kaya et al. (2022) demonstrated the potential of chitin-based films derived from white button mushrooms (Agaricus bisporus) incorporated with curcumin as a sustainable packaging solution. The study found that these films exhibited significant

Table 12.8 Summary of film-forming techniques for fungal-based films

Technique	Description	Effect on film properties	Common applications	Reference
Casting	Solution poured onto a flat surface and dried	Produces flexible, transparent films	Coatings, food wraps	Kaya et al. (2022)
Extrusion	Biopolymer extruded through a die	Creates continuous sheets and films	Packaging wraps, pouches	Pérez-Bassart et al. (2023)
Electrospinning	High-voltage electrospinning to form nanofibers	Produces thin, porous films	Active packaging, nanofiber films	Mirbagheri et al. (2023)
Freeze-drying	Freezing followed by sublimation	Lightweight, porous material	Lightweight food packaging	Varshan and Namasivayam (2024)

Table 12.9 Summary of successful applications of fungal additives in food packaging

Application	Fungal additive used	Key benefits	Outcome	Reference
Fresh Pistachio	Chitosan	Extended shelf life, reduced microbial load	Improved preservation	Akhavan-Mahdavi et al. (2024)
Cheese Packaging	. niger Mycelia (Chitin-Glucan Hydrogel)	Antioxidant effect, reduced moisture loss, improved shelf life	Increased pH, reduced peroxide number, and delayed microbial growth in Kashar cheese, sustainable preservation technique	
Meat Products	Chitin from *Agaricus bisporus*	Antimicrobial activity, antioxidant properties, biodegradability	Extended shelf life by 40%, biodegradable and eco-friendly packaging	Kaya et al. (2022)
Packaging	Mycelium from Pleurotus ostreatus	Biodegradability, cost-effective, sustainable alternative to EPS	Promising performance for packaging, thermal conductivity, and acoustic properties	Sivaprasad et al. (2021)
Packaging tools	Oyster mushroom spawn, rice straw	Biodegradable, eco-friendly, easy to degrade	Heat press drying produced the best biodegradable sheets, decomposing in 18 days, prototype model for future development	

antimicrobial properties, particularly against *E. coli*, and demonstrated excellent biodegradability in soil, with the films breaking down within 14 days. Additionally, the films improved the antioxidant activity and extended the shelf life of fresh chicken meat by approximately 40%. This research exemplifies the promising application of fungal-derived materials in the development of eco-friendly, active food packaging systems (Kaya et al., 2022). Another promising development is the

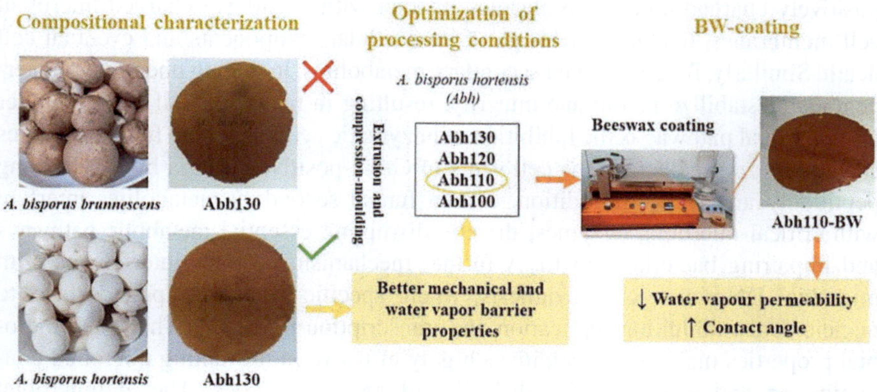

Fig. 12.4 Graphical abstract of Feasibility of *Agaricus bisporus* waste biomass to develop biodegradable food packaging materials (Pérez-Bassart et al., 2023, with copyright permission, Elsevier)

creation of mycelium-based bio-composites, which utilize natural fungal networks, such as *Pleurotus ostreatus*, to replace synthetic polystyrene in packaging applications. This material, developed by Sivaprasad et al. (2021), is biodegradable, cost-effective, and shows high production yields, offering an eco-friendly alternative to nonbiodegradable plastics (Sivaprasad et al., 2021). Pérez-Bassart et al. (2023) also explored Agaricus bisporus biomass residues to produce biodegradable food packaging films, comparing two varieties of *Agaricus bisporus*—brunnescens and hortensis. The study found that films made from *A. bisporus hortensis* exhibited superior mechanical and barrier performance, especially when coated with beeswax. These films biodegraded almost completely within 11 days under composting conditions, further advancing the potential of fungal waste biomass for eco-friendly packaging solutions (Fig. 12.4) (Pérez-Bassart et al., 2023).

5 Mechanisms of Action of Fungal Additives in Food Packaging

Fungal additives play a pivotal role in enhancing the functionality of food packaging materials. Their mechanisms of action are diverse and depend on the type of additive and its interaction with the food environment. The effectiveness of fungal-derived compounds in food packaging primarily stems from their antimicrobial and antioxidant properties, which contribute to extending shelf life and preserving food quality. Additionally, fungal additives interact with various food components, influencing their stability, sensory attributes, and overall functionality.

Fungal additives such as chitosan, glucans, and secondary metabolites exhibit potent antimicrobial activity by targeting various cellular components of microorganisms. One of the key mechanisms involves cell membrane disruption, where positively charged chitosan molecules interact with negatively charged microbial cell membranes, leading to leakage of intracellular components and eventual cell death. Similarly, fungal-derived secondary metabolites, including phenolics and terpenoids, destabilize membrane integrity, resulting in microbial cell lysis. Another antimicrobial pathway is the inhibition of enzymatic activity, where fungal enzymes such as lysozyme degrade the cell walls of Gram-positive bacteria by hydrolyzing peptidoglycan layers. In addition, certain fungal secondary metabolites interfere with critical microbial enzymes, thereby disrupting essential metabolic pathways and impairing bacterial growth. A further mechanism involves interference with microbial DNA and RNA synthesis, where specific fungal compounds bind to nucleic acids, inhibiting replication and transcription processes. These antimicrobial properties make fungal additives highly effective in preventing microbial contamination and extending the shelf life of packaged foods. The antimicrobial mechanisms of various fungal additives and their targeted microorganisms are summarized in Table 12.10.

Kaya et al. (2022) investigated the antibacterial properties of chitin-based films incorporated with curcumin (CGC) against *Escherichia coli* (Gram-negative) and *Staphylococcus aureus* (Gram-positive). The results demonstrated that the films exhibited notable antibacterial activity against *E. coli*, with the inhibition zones increasing proportionally with curcumin concentration. This effect is attributed to curcumin's ability to disrupt bacterial membrane potential and inhibit the polymerization of the FtsZ protein, which is crucial for bacterial cell division. However, the CGC films did not show significant inhibitory effects against *S. aureus*, likely due to strain-specific differences in curcumin susceptibility and the lower concentration of curcumin in the film-forming solution compared to the minimum inhibitory concentration reported for *S. aureus*. Additionally, the extraction method used for obtaining chitin from *Agaricus bisporus* might have influenced the antibacterial properties, as the absence of methanol in the process may have reduced the bioactivity of certain antimicrobial compounds. Overall, the study highlights the potential of CGC films as an effective antimicrobial packaging material, particularly for combating *E. coli* contamination in food products.

The antioxidant activity of fungal additives, particularly secondary metabolites like phenolic compounds and terpenoids, is essential in preventing oxidative spoilage and maintaining the nutritional and sensory quality of food. One of the primary antioxidant mechanisms is free radical scavenging, in which fungal antioxidants donate hydrogen atoms or electrons to neutralize reactive oxygen species (ROS) such as hydroxyl radicals and superoxide anions (Deshmukh & Gaikwad, 2024). This action helps to mitigate oxidative stress in food products, thereby preventing rancidity and discoloration. Another mechanism involves metal ion chelation, where fungal compounds bind to pro-oxidant metal ions such as iron and copper, preventing their catalytic role in lipid oxidation reactions. By inhibiting the activity of oxidative enzymes like lipoxygenase and polyphenol oxidase, fungal metabolites further reduce oxidative spoilage, thereby contributing to the stability of food during storage (Chuang et al., 2020). Kaya et al. (2022) also evaluated the antioxidant properties of curcumin-incorporated chitin-based films (CGC) and found that the

Table 12.10 Antimicrobial mechanisms of fungal additives

Fungal additive	Target microorganism	Mechanism of action	Reference
Chitosan	*Escherichia coli*	Cell membrane disruption	Ke et al. (2021)
Lysozyme	*Staphylococcus aureus*	Cell wall degradation	Ferraboschi et al. (2021)
Phenolic Compounds	*Salmonella* spp.	Enzyme inhibition, DNA binding	Takó et al. (2020)
Fungal Chitinases	*Listeria monocytogenes*	Breakdown of chitin in the cell wall, inhibition of biofilm formation	Bártová et al. (2019)
Glucans	*Aspergillus niger*	Activation of host immune response, inhibition of fungal spore germination	Briard et al. (2017)
Antifungal Proteins	*Candida albicans*	Disruption of fungal cell membrane integrity, pore formation	Zida et al. (2017)
Polysaccharides (e.g., Mannan)	*Saccharomyces cerevisiae*	Inhibition of spore germination, interference with cellular functions and structure	Faustino et al. (2021)
Secondary Metabolites (e.g., Griseofulvin)	*Fusarium* spp.	Disruption of microtubule formation, cell division inhibition	Xu et al. (2021)
Chitosan Derivatives	*Pseudomonas aeruginosa*	Disruption of outer membrane permeability, chelation of metal ions required for growth	Sahariah and Másson (2017)
Mycoproteins	*Campylobacter jejuni*	Interference with cell metabolism and protein synthesis, cell membrane disruption	Ma and Ma (2019)
Fungal Volatiles	*Bacillus subtilis*	Release of volatile organic compounds that inhibit bacterial growth and biofilm formation	Wang et al. (2021)
Polyketides (e.g., Penicillins)	*Helicobacter pylori*	Inhibition of cell wall synthesis through disruption of peptidoglycan cross-linking	Bugés Bascompte (2023)
Tannins	*Escherichia coli*	Protein precipitation, disruption of membrane integrity, inhibition of enzyme activity	Štumpf et al. (2020)
Terpenoids	*Streptococcus mutans*	Disruption of cell membrane function, inhibition of glucose uptake	Kamarehei et al. (2022)
Saponins	*Listeria monocytogenes*	Disruption of cell membrane integrity, alteration of permeability and ion leakage	Dong et al. (2020)
Alkaloids	*Candida tropicalis*	Disruption of fungal membrane, inhibition of cell proliferation and differentiation	Sulaiman et al. (2022)

addition of curcumin significantly enhanced the antioxidant activity of the films. This improvement is primarily attributed to the phenolic hydroxyl groups of curcumin, which are known for their ability to scavenge free radicals and inhibit oxidative degradation. The antioxidant activity of the films was assessed through various methods, including the measurement of total phenolic content, which was found to increase with higher curcumin concentrations. The films exhibited a strong capacity to protect against oxidative damage, thereby contributing to the preservation of food quality, especially for perishable items like meat. This enhanced antioxidant property suggests that CGC films could play an important role in active packaging applications, not only providing antimicrobial protection but also preserving the nutritional and sensory qualities of food by reducing oxidative stress during storage (Kaya et al., 2022).

Fungal additives interact with various food components, including proteins, lipids, and carbohydrates, influencing their structural and functional properties. Polysaccharides such as chitosan can form complexes with proteins, enhancing the mechanical strength of biodegradable packaging films. However, excessive interaction may lead to protein denaturation, potentially altering the texture and sensory attributes of food products. In the case of lipids, fungal antioxidants play a crucial role in preventing oxidative rancidity by scavenging free radicals and chelating metal ions (Friné et al., 2019). Additionally, fungal secondary metabolites may alter lipid stability in certain food matrices, affecting their shelf life and oxidative resistance. The interaction of fungal enzymes with carbohydrates is another critical aspect, as enzymes such as amylase can break down complex carbohydrates, influencing food texture and extending shelf life. Furthermore, fungal polysaccharides like glucans interact with starch molecules, improving the barrier properties of packaging films (Rami et al., 2024).

6 Sustainability and Environmental Impact

The shift toward sustainable packaging solutions has gained significant momentum due to growing concerns over plastic pollution, resource depletion, and environmental degradation. As the world increasingly prioritizes eco-friendly alternatives, fungal-based packaging materials have emerged as a promising solution. These materials, derived from fungi, offer an environmentally sustainable alternative to conventional petroleum-based plastics, with significant advantages in terms of biodegradability, compostability, and a lower carbon footprint (Moeini et al., 2021).

One of the most important attributes of fungal-based packaging is its biodegradability and compostability. These properties ensure that the packaging materials break down into nontoxic components when exposed to natural environments, unlike traditional plastics that persist for hundreds of years. Fungal packaging, typically made from mycelium-based biocomposites, decomposes within weeks to months under appropriate composting conditions. The biodegradation mechanism of these materials occurs through microbial enzymatic activity, which converts the packaging components into carbon dioxide, water, and biomass. The rate of this degradation depends on various environmental factors, such as temperature, humidity, and the microbial community present. For fungal packaging to be considered compostable, it must meet specific international standards. The European Standard EN 13432, for instance, requires materials to disintegrate within 12 weeks and biodegrade fully within six months. Similarly, ASTM D6400, a US standard, establishes criteria for compostability, ensuring that materials are nontoxic and can safely decompose into compost. These standards ensure that fungal packaging can seamlessly integrate into composting systems, contributing to waste reduction and supporting soil health.

In terms of environmental impact, fungal-based packaging presents clear advantages over traditional plastic packaging, particularly when evaluated through Life Cycle Assessment (LCA). LCA is a tool used to assess the environmental footprint of packaging materials from raw material extraction through to disposal. When comparing fungal-based packaging to conventional plastics, it becomes evident that fungal materials offer significant environmental benefits. They have a substantially lower carbon footprint, with values ranging from 0.5–1.5 kg CO_2 eq. per kg compared to 3–5 kg CO_2 eq. per kg for conventional plastics. Additionally, fungal packaging consumes considerably less energy and water in its production process.

Fungal-based packaging also contributes to the principles of a circular economy, which promotes the reuse and recycling of materials to minimize waste. By utilizing agricultural by-products such as wheat bran and sawdust as raw materials, fungal-based packaging reduces dependence on virgin resources. The mycelium growth process binds these materials naturally, eliminating the need for synthetic adhesives. At the end of its life cycle, fungal packaging can be composted, returning nutrients to the soil and supporting regenerative agriculture. This closed-loop system exemplifies the potential for fungal-based packaging to help transition toward more sustainable and circular production practices.

Despite the considerable environmental advantages, the large-scale adoption of fungal-based packaging faces several economic and industrial challenges. One of the primary obstacles is the higher cost of production compared to conventional plastic materials. The current production methods, such as solid-state fermentation, are still in the early stages of optimization, and the limited scalability of these processes makes fungal-based packaging more expensive. In addition to cost considerations, there are several industrial challenges to overcome. Scaling up production to meet the demands of global supply chains requires further optimization of production techniques. Furthermore, regulatory approvals are essential for market expansion. Fungal packaging must comply with standards set by organizations such as the FDA, EFSA, and ASTM for compostability before it can be widely adopted. Lastly, there are material performance concerns. While fungal packaging is biodegradable, enhancing its water resistance and mechanical strength is critical to making it suitable for a wider range of applications, particularly in food packaging.

7 Safety and Regulatory Considerations

The use of fungal additives in food packaging offers a promising alternative to traditional materials, but it requires careful consideration of safety, regulatory compliance, and commercialization challenges. This section discusses the consumer safety and toxicity studies, the regulatory frameworks governing fungal additives, and the hurdles that need to be overcome for their widespread adoption.

Ensuring the safety of fungal additives is essential for their application in food packaging. Comprehensive toxicity studies are conducted to evaluate their potential effects on human health and the environment. These studies typically focus on several key areas. Acute and chronic toxicity studies assess both the short-term and long-term health effects of fungal additives. For example, chitosan, a commonly used fungal derivative, has been extensively tested and is classified as Generally Recognized As Safe (GRAS) by the FDA. These studies help confirm that fungal additives do not pose significant health risks when used in packaging materials. Allergenicity is another important concern. The potential for allergenicity, especially from common fungal species, is carefully evaluated, with researchers investigating cross-reactivity with known allergens to minimize health risks. The environmental impact of fungal additives is also assessed to ensure their biodegradability and minimal harm to ecosystems. Fungal packaging is typically designed to break down into nontoxic components, unlike traditional plastics that persist in the environment for extended periods.

Regulatory frameworks ensure that fungal additives used in food packaging meet safety and quality standards. Compliance with these regulations is crucial for the market approval of fungal-based packaging solutions. Several key regulatory bodies play a role in the approval and safety evaluation of these materials. The Food and Drug Administration (FDA) provides guidelines for the use of food contact materials, including fungal additives. The FDA grants GRAS status to additives that meet safety criteria for human health. The European Food Safety Authority (EFSA) evaluates the safety of food additives and packaging materials within the European Union, issuing scientific opinions and approvals. Additionally, the Codex Alimentarius ensures harmonized food standards across nations, facilitating international trade and compliance. Despite the advantages of fungal-based packaging, several challenges must be addressed to enable its widespread commercialization. One significant challenge is the high production costs. The extraction and purification of fungal additives can be costly, which may affect the overall cost-effectiveness of fungal packaging. However, optimizing production processes can help reduce these costs and make the technology more viable for large-scale use. Scalability remains another challenge, as scaling up production while maintaining consistent quality is difficult. Investing in advanced manufacturing technologies and further research into production methods can help increase scalability and ensure that the materials can be produced efficiently at a larger scale. Consumer acceptance is also a key factor in commercialization. The perception of fungal-derived materials in food packaging can be a barrier to adoption, and educating consumers, as well as ensuring transparency about the safety and benefits of these materials, is essential for gaining market acceptance. Regulatory hurdles also present a challenge, as navigating the complex regulatory landscapes can be time-consuming and costly. Streamlining the regulatory approval process can help accelerate commercialization.

One of the primary concerns when using fungal materials in food packaging is the potential for mycotoxin contamination. Mycotoxins are toxic compounds produced by certain fungi, and their presence in packaging materials can pose a health risk. To ensure mycotoxin-free packaging, two main strategies are employed. First, the selection of non-mycotoxin-producing fungal strains is critical. GRAS fungal species such as *Ganoderma* and *Pleurotus* are preferred for packaging applications because they are less likely to produce harmful mycotoxins. Second, post-treatment sterilization is used to eliminate any residual fungal spores. Heat or chemical treatments are applied to fungal packaging materials to reduce the risk of mycotoxin contamination, ensuring that the final product is safe for use in food packaging.

For fungal packaging to be marketed as sustainable, it must comply with international compostability and biodegradability standards. Several regulations govern the environmental compliance of these materials. The EN 13432 standard in the European Union certifies compostability, ensuring that materials break down into nontoxic components within 12 weeks. Similarly, the ASTM D6400 standard in the United States certifies biodegradability and composting properties, while ISO 18606 provides a global standard for the environmental safety of bio-based materials.

8 Conclusions

The exploration of fungal-based additives in food packaging has revealed their immense potential to address critical challenges in sustainability, food preservation, and safety. Fungal-derived compounds, including polysaccharides such as chitosan and glucans, enzymes, and secondary metabolites, exhibit remarkable antimicrobial, antioxidant, and biodegradable properties, making them valuable components for active and intelligent packaging solutions. The incorporation of these compounds into biodegradable polymer matrices enhances packaging performance by improving mechanical strength, barrier properties, and the controlled release of bioactive agents. Additionally, they contribute to reducing plastic waste and environmental pollution, aligning with the global push for eco-friendly solutions.

The ability of fungal-based packaging to improve food preservation, extend shelf life, and ensure food safety positions it as a viable alternative to conventional synthetic materials. Advancements in fungal biotechnology, nanocomposite development, and controlled-release mechanisms are expected to expand the applicability of these materials within the food industry. Furthermore, emerging technologies such as edible packaging, nanotechnology, and active packaging systems offer exciting opportunities for innovation in this field. However, significant challenges remain, including the need for large-scale production, regulatory approvals, and cost-effective industrial implementation. Addressing these challenges is crucial for the widespread adoption of fungal-derived packaging materials. Continued research, technological advancements, and collaboration between academia, industry, and regulatory bodies will be essential in unlocking the full potential of fungal-based packaging for a more sustainable and circular economy in the food industry. Ultimately, fungal-based food packaging represents a promising step toward creating a greener and more resilient food system.

References

Abou Fayssal, S., El Sebaaly, Z., & Sassine, Y. N. (2023). Pleurotus ostreatus grown on agro-industrial residues: Studies on microbial contamination and shelf-life prediction under different packaging types and storage temperatures. *Foods, 12*(3), 524.

Adeyeye, S. A. (2016). Fungal mycotoxins in foods: A review. *Cogent Food & Agriculture, 2*(1), 1213127.

Ahmed, O. S., Tardif, C., Rouger, C., Atanasova, V., Richard-Forget, F., & Waffo-Téguo, P. (2022). Naturally occurring phenolic compounds as promising antimycotoxin agents: Where are we now? *Comprehensive Reviews in Food Science and Food Safety, 21*(2), 1161–1197.

Aiduang, W., Kumla, J., Srinuanpan, S., Thamjaree, W., Lumyong, S., & Suwannarach, N. (2022). Mechanical, physical, and chemical properties of mycelium-based composites produced from various lignocellulosic residues and fungal species. *Journal of Fungi, 8*(11), 1125.

Aiduang, W., Jatuwong, K., Luangharn, T., Jinanukul, P., Thamjaree, W., Teeraphantuvat, T., Waroonkun, T., & Lumyong, S. (2024). A review delving into the factors influencing mycelium-based green composites (MBCs) production and their properties for long-term sustainability targets. *Biomimetics, 9*(6), 337.

Akhavan-Mahdavi, S., Mirzazadeh, M., Alam, Z., & Solaimanimehr, S. (2023). The effect of chitosan coating combined with cold plasma on the quality and safety of pistachio during storage. *Food Science & Nutrition, 11*(7), 4296–4307.

Akhavan-Mahdavi, S., Mirbagheri, M. S., Assadpour, E., Sani, M. A., Zhang, F., & Jafari, S. M. (2024). Electrospun nanofiber-based sensors for the detection of chemical and biological contaminants/hazards in the food industries. *Advances in Colloid and Interface Science, 103111*.

Al Mahmud, M. Z., Mobarak, M. H., & Hossain, N. (2024). Emerging trends in biomaterials for sustainable food packaging: A comprehensive review. *Heliyon, 10*(1).

Alemu, D., Tafesse, M., & Mondal, A. K. (2022). Mycelium-based composite: The future sustainable biomaterial. *International Journal of Biomaterials, 2022*(1), 8401528.

Bagheriehnajjar, G., Yousefpour, H., & Rahimnejad, M. (2023). Multi-objective optimization of mycelium-based bio-composites based on mechanical and environmental considerations. *Construction and Building Materials, 407*, 133346.

Bártová, V., Bárta, J., & Jarošová, M. (2019). Antifungal and antimicrobial proteins and peptides of potato (Solanum tuberosum L.) tubers and their applications. *Applied Microbiology and Biotechnology, 103*, 5533–5547.

Bitting, S., Derme, T., Lee, J., Van Mele, T., Dillenburger, B., & Block, P. (2022). Challenges and opportunities in scaling up architectural applications of mycelium-based materials with digital fabrication. *Biomimetics, 7*(2), 44.

Bordenave, N., Grelier, S., & Coma, V. (2010). Hydrophobization and antimicrobial activity of chitosan and paper-based packaging material. *Biomacromolecules, 11*(1), 88–96.

Briard, B., Rasoldier, V., Bomme, P., ElAouad, N., Guerreiro, C., Chassagne, P., Muszkieta, L., Latgé, J.-P., Mulard, L., & Beauvais, A. (2017). Dirhamnolipids secreted from Pseudomonas aeruginosa modify anjpegungal susceptibility of Aspergillus fumigatus by inhibiting β1, 3 glucan synthase activity. *The ISME Journal, 11*(7), 1578–1591.

Bugés Bascompte, M. (2023). *Molecular mechanisms of antibiotic resistance in Helicobacter pylori.*

Cazón, P., & Vázquez, M. (2019). Applications of chitosan as food packaging materials. Sustainable agriculture reviews 36: Chitin and chitosan: applications in food, agriculture, pharmacy, medicine and wastewater treatment, 81–123.

Cazón, P., & Vázquez, M. (2020). Mechanical and barrier properties of chitosan combined with other components as food packaging film. *Environmental Chemistry Letters, 18*(2), 257–267.

Chang, J., Li, W., Liu, Q., Zhou, Y., Chen, X., Lyu, Q., & Liu, G. (2019). Preparation, properties, and structural characterization of β-glucan/pullulan blend films. *International Journal of Biological Macromolecules, 140*, 1269–1276.

Chuang, W. Y., Hsieh, Y. C., & Lee, T.-T. (2020). The effects of fungal feed additives in animals: A review. *Animals, 10*(5), 805.

Copetti, M. V. (2019). Fungi as industrial producers of food ingredients. *Current Opinion in Food Science, 25*, 52–56.

Das, P. P., Prathapan, R., & Ng, K. W. (2024). Advances in biomaterials based food packaging systems: Current status and the way forward. *Biomaterials Advances, 213988.*

Delattre, C., Cabrera-Barjas, G., Banerjee, A., Rodriguez-Llamazares, S., Pierre, G., Dubessay, P., Michaud, P., & Zamani, A. (2023). Production of fungal biopolymers and their advanced applications. In *Current Developments in Biotechnology and Bioengineering* (pp. 497–532). Elsevier.

Deshmukh, R. K., & Gaikwad, K. K. (2024). Natural antimicrobial and antioxidant compounds for active food packaging applications. *Biomass Conversion and Biorefinery, 14*(4), 4419–4440.

Dilkes-Hoffman, L. S., Lane, J. L., Grant, T., Pratt, S., Lant, P. A., & Laycock, B. (2018). Environmental impact of biodegradable food packaging when considering food waste. *Journal of Cleaner Production, 180*, 325–334.

Dong, S., Yang, X., Zhao, L., Zhang, F., Hou, Z., & Xue, P. (2020). Antibacterial activity and mechanism of action saponins from Chenopodium quinoa Willd. husks against foodborne pathogenic bacteria. *Industrial Crops and Products, 149*, 112350.

Farrelly, T. A., & Shaw, I. C. (2017). Polystyrene as hazardous household waste. *Household Hazardous Waste Management, 45.*

Faustino, M., Durão, J., Pereira, C. F., Pintado, M. E., & Carvalho, A. P. (2021). Mannans and mannan oligosaccharides (MOS) from Saccharomyces cerevisiae – A sustainable source of functional ingredients. *Carbohydrate Polymers, 272*, 118467.

Ferraboschi, P., Ciceri, S., & Grisenti, P. (2021). Applications of lysozyme, an innate immune defense factor, as an alternative antibiotic. *Antibiotics, 10*(12), 1534.

Friné, V.-C., Hector, A.-P., Manuel, N.-D. S., Estrella, N.-D., & Antonio, G. J. (2019). Development and characterization of a biodegradable PLA food packaging hold Monoterpene–Cyclodextrin Complexes against Alternaria alternata. *Polymers, 11*(10), 1720.

Gaff, M., Hosseini, S. B., Paulasová, D., Kamboj, G., Rezaei, F., & Paul, D. (2024). Functionality of production processes in mycelium-based composites. *Composite Structures, 118309.*

Golak-Siwulska, I., Kałużewicz, A., Spiżewski, T., Siwulski, M., & Sobieralski, K. (2018). Bioactive compounds and medicinal properties of Oyster mushrooms (Pleurotus sp.). *Folia Horticulturae, 30*(2), 191–201.

Guillard, V., Gaucel, S., Fornaciari, C., Angellier-Coussy, H., Buche, P., & Gontard, N. (2018). The next generation of sustainable food packaging to preserve our environment in a circular economy context. *Frontiers in Nutrition, 5*, 121.

Hernández, A., Rodríguez, A., Córdoba, M. G., Martín, A., & Ruiz-Moyano, S. (2022). Fungal control in foods through biopreservation. *Current Opinion in Food Science, 47*, 100904.

Hussain, S., Akhter, R., & Maktedar, S. S. (2024). Advancements in sustainable food packaging: From eco-friendly materials to innovative technologies. *Sustainable Food Technology, 2*(5), 1297–1364.

Jafarzadeh, S., Hadidi, M., Forough, M., Nafchi, A. M., & Mousavi Khaneghah, A. (2023). The control of fungi and mycotoxins by food active packaging: A review. *Critical Reviews in Food Science and Nutrition, 63*(23), 6393–6411.

Kamarehei, F., Mehdiabadi, M., & Naderi, F. (2022). Antibacterial effects of natural compounds on biofilm formation of Streptococcus mutans. *Clinical and Experimental Dental Research, 8*(6), 1426–1433.

Kandel, B. (2022). *Mycelium: Using mushrooms for the future packaging materials in Europe*.

Kaya, E., Kahyaoglu, L. N., & Sumnu, G. (2022). Development of curcumin incorporated composite films based on chitin and glucan complexes extracted from Agaricus bisporus for active packaging of chicken breast meat. *International Journal of Biological Macromolecules, 221*, 536–546. https://doi.org/10.1016/j.ijbiomac.2022.09.025

Ke, C.-L., Deng, F.-S., Chuang, C.-Y., & Lin, C.-H. (2021). Antimicrobial actions and applications of chitosan. *Polymers, 13*(6), 904.

Keshavarz, B., & Khalesi, M. (2016). Trichoderma reesei, a superior cellulase source for industrial applications. *Biofuels, 7*(6), 713–721.

Lingam, D., Narayan, S., Mamun, K., & Charan, D. (2023). Engineered mycelium-based composite materials: Comprehensive study of various properties and applications. *Construction and Building Materials, 391*, 131841.

Ma, N., & Ma, X. (2019). Dietary amino acids and the gut-microbiome-immune axis: Physiological metabolism and therapeutic prospects. *Comprehensive Reviews in Food Science and Food Safety, 18*(1), 221–242.

Madusanka, C., Udayanga, D., Nilmini, R., Rajapaksha, S., Hewawasam, C., Manamgoda, D., & Vasco-Correa, J. (2024). A review of recent advances in fungal mycelium based composites. *Discover Materials, 4*(1), 13.

Mafe, A. N., Edo, G. I., Akpoghelie, P. O., Joshua, O. A., Isoje, E. F., Igbuku, U. A., & Essaghah, A. E. A. (2024). Comparative analysis of the environmental impact of biopolymer-based and conventional plastic packaging in food engineering applications. *Al-Mustaqbal Journal of Sustainability in Engineering Sciences, 2*(2), 4.

Mahmoud, M. G., El Kady, E. M., & Asker, M. S. (2019). Chitin, chitosan and glucan, properties and applications. *World Journal of Agriculture and Soil Science, 3*(1), 1–19.

Manan, S., Ullah, M. W., Ul-Islam, M., Atta, O. M., & Yang, G. (2021). Synthesis and applications of fungal mycelium-based advanced functional materials. *Journal of Bioresources and Bioproducts, 6*(1), 1–10. https://doi.org/10.1016/j.jobab.2021.01.001

Manan, S., Atta, O. M., Shahzad, A., Ul-Islam, M., Ullah, M. W., & Yang, G. (2022). Applications of fungal mycelium-based functional biomaterials. In *Fungal biopolymers and biocomposites: prospects and avenues* (pp. 147–168). Springer.

Mbabali, H., Lubwama, M., Yiga, V. A., Were, E., & Kasedde, H. (2024). Development of rice husk and sawdust mycelium-based bio-composites: optimization of mechanical, physical and thermal properties. *Journal of The Institution of Engineers (India): Series D, 105*(1), 97–117.

Meng, L., Fu, Y., Li, D., Sun, X., Chen, Y., Li, X., Xu, S., Li, X., Li, C., & Song, B. (2019). Effects of corn stalk cultivation substrate on the growth of the slippery mushroom (Pholiota microspora). *RSC advances, 9*(10), 5347–5353.

Mirbagheri, M. S., Akhavan-Mahdavi, S., Hasan, A., Kharazmi, M. S., & Jafari, S. M. (2023). Chitosan-based electrospun nanofibers for diabetic foot ulcer management; recent advances. *Carbohydrate Polymers, 313*, 120512.

Moeini, A., Germann, N., Malinconico, M., & Santagata, G. (2021). Formulation of secondary compounds as additives of biopolymer-based food packaging: A review. *Trends in Food Science & Technology, 114*, 342–354.

Muiruri, J. K., Chuan Yeo, J. C., Zhu, Q., Ye, E., Loh, X. J., & Li, Z. (2023). Sustainable mycelium-bound biocomposites: Design strategies, materials properties, and emerging applications. *ACS Sustainable Chemistry & Engineering, 11*(18), 6801–6821.

Nawrot-Chorabik, K., Sułkowska, M., & Gumulak, N. (2022). Secondary metabolites produced by trees and fungi: Achievements so far and challenges remaining. *Forests, 13*(8), 1338.

Ncube, L. K., Ude, A. U., Ogunmuyiwa, E. N., Zulkifli, R., & Beas, I. N. (2020). Environmental impact of food packaging materials: A review of contemporary development from conventional plastics to polylactic acid based materials. *Materials, 13*(21), 4994.

Nigam, P. S., & Luke, J. S. (2016). Food additives: production of microbial pigments and their antioxidant properties. *Current Opinion in Food Science, 7*, 93–100.

Pellicer, E., Nikolic, D., Sort, J., Baró, M., Zivic, F., Grujovic, N., Grujic, R., & Pelemis, S. (2017). *Advances in applications of industrial biomaterials*. Springer.

Peltzer, M., Delgado, J. F., Salvay, A. G., & Wagner, J. R. (2018). β-Glucan, a promising polysaccharide for bio-based films developments for food contact materials and medical applications. *Current Organic Chemistry, 22*(12), 1249–1254.

Pérez-Bassart, Z., Reyes, A., Martínez-Abad, A., López-Rubio, A., & Fabra, M. J. (2023). Feasibility of Agaricus bisporus waste biomass to develop biodegradable food packaging materials. *Food Hydrocolloids, 142*, 108861.

Priyadarshi, R., & Rhim, J.-W. (2020). Chitosan-based biodegradable functional films for food packaging applications. *Innovative Food Science & Emerging Technologies, 62*, 102346.

Pundir, R. K., Jain, P., & Sharma, C. (2010). Antimicrobial activity of ethanolic extracts of Syzygium aromaticum and Allium sativum against food associated bacteria and fungi. *Ethnobotanical Leaflets, 2010*(3), 11.

Rajendran, R. C. (2022). Packaging applications of fungal mycelium-based biodegradable composites. In *Fungal biopolymers and biocomposites: Prospects and avenues* (pp. 189–208). Springer.

Rami, M. R., Forouzandehdel, S., & Aalizadeh, F. (2024). Enhancing biodegradable smart food packaging: Fungal-synthesized nanoparticles for stabilizing biopolymers. *Heliyon*.

Razzaq, H. A., Pezzuto, M., Santagata, G., Silvestre, C., Cimmino, S., Larsen, N., & Duraccio, D. (2016). Barley β-glucan-protein based bioplastic film with enhanced physicochemical properties for packaging. *Food Hydrocolloids, 58*, 276–283.

Sahariah, P., & Másson, M. (2017). Antimicrobial chitosan and chitosan derivatives: A review of the structure–activity relationship. *Biomacromolecules, 18*(11), 3846–3868.

Santi, R., Garrone, P., Iannantuoni, M., & Del Curto, B. (2022). Sustainable food Packaging: An integrative framework. *Sustainability, 14*(13), 8045.

Schritt, H., Vidi, S., & Pleissner, D. (2021). Spent mushroom substrate and sawdust to produce mycelium-based thermal insulation composites. *Journal of Cleaner Production, 313*, 127910.

Shaaban, R., Elnaggar, M. S., Khalil, N., & Singab, A. N. B. (2023). A comprehensive review on the medicinally valuable endosymbiotic fungi Penicillium chrysogenum. *Archives of Microbiology, 205*(6), 240.

Sharma, M., Tellili, N., Kacem, I., & Rouissi, T. (2024). Microbial biopolymers: From production to environmental applications – A review. *Applied Sciences, 14*(12), 5081.

Sherafatkhah Azari, S., Alizadeh, A., Roufegarinejad, L., Asefi, N., & Hamishehkar, H. (2021). Preparation and characterization of gelatin/β-glucan nanocomposite film incorporated with ZnO nanoparticles as an active food packaging system. *Journal of Polymers and the Environment, 29*, 1143–1152.

Sid, S., Mor, R. S., Kishore, A., & Sharanagat, V. S. (2021). Bio-sourced polymers as alternatives to conventional food packaging materials: A review. *Trends in Food Science & Technology, 115*, 87–104.

Sivaprasad, S., Byju, S. K., Prajith, C., Shaju, J., & Rejeesh, C. (2021). Development of a novel mycelium bio-composite material to substitute for polystyrene in packaging applications. *Materials Today: Proceedings, 47*, 5038–5044.

Sommer, S., Wegmann-Herr, P., Wacker, M., & Fischer, U. (2018). Influence of lysozyme addition on hydroxycinnamic acids and volatile phenols during wine fermentation. *Fermentation, 4*(1), 5.

Sreerag, N., Kashyap, P., Shilpa, V., Thakur, M., & Goksen, G. (2024). Recent advances on mycelium based biocomposites: synthesis, strains, lignocellulosic substrates, production parameters. *Polymer Reviews*, 1–30.

Stracquadanio, C., Quiles, J. M., Meca, G., & Cacciola, S. O. (2020). Antifungal activity of bioactive metabolites produced by Trichoderma asperellum and Trichoderma atroviride in liquid medium. *Journal of Fungi, 6*(4), 263.

Štumpf, S., Hostnik, G., Primožič, M., Leitgeb, M., Salminen, J.-P., & Bren, U. (2020). The effect of growth medium strength on minimum inhibitory concentrations of tannins and tannin extracts against E. coli. *Molecules, 25*(12), 2947.

Sulaiman, M., Jannat, K., Nissapatorn, V., Rahmatullah, M., Paul, A. K., de Lourdes Pereira, M., Rajagopal, M., Suleiman, M., Butler, M. S., & Break, M. K. B. (2022). Antibacterial and antifungal alkaloids from Asian angiosperms: Distribution, mechanisms of action, structure-activity, and clinical potentials. *Antibiotics, 11*(9), 1146.

Sydor, M., Cofta, G., Doczekalska, B., & Bonenberg, A. (2022). Fungi in mycelium-based composites: usage and recommendations. *Materials, 15*(18), 6283.

Taib, N.-A. A. B., Rahman, M. R., Huda, D., Kuok, K. K., Hamdan, S., Bakri, M. K. B., Julaihi, M. R. M. B., & Khan, A. (2023). A review on poly lactic acid (PLA) as a biodegradable polymer. *Polymer Bulletin, 80*(2), 1179–1213.

Takó, M., Kerekes, E. B., Zambrano, C., Kotogán, A., Papp, T., Krisch, J., & Vágvölgyi, C. (2020). Plant phenolics and phenolic-enriched extracts as antimicrobial agents against food-contaminating microorganisms. *Antioxidants, 9*(2), 165.

Udayanga, D., & Miriyagalla, S. D. (2021). Fungal mycelium-based biocomposites: An emerging source of renewable materials. *Microbial Technology for Sustainable Environment*, 529–550.

Varshan, G. A., & Namasivayam, S. K. R. (2024). A green chemistry principle for the biotransformation of fungal biomass derived chitosan into versatile Nano scale materials with high biocompatibility and potential biological activities – A review. *BioNanoScience*, 1–22.

Verma, N., Jujjavarapu, S. E., & Mahapatra, C. (2023). Green sustainable biocomposites: Substitute to plastics with innovative fungal mycelium based biomaterial. *Journal of Environmental Chemical Engineering, 11*(5), 110396.

Wang, K., Qin, Z., Wu, S., Zhao, P., Zhen, C., & Gao, H. (2021). Antifungal mechanism of volatile organic compounds produced by Bacillus subtilis CF-3 on Colletotrichum gloeosporioides assessed using omics technology. *Journal of Agricultural and Food Chemistry, 69*(17), 5267–5278.

Xu, K., Li, X.-Q., Zhao, D.-L., & Zhang, P. (2021). Antifungal secondary metabolites produced by the fungal endophytes: Chemical diversity and potential use in the development of biopesticides. *Frontiers in Microbiology, 12*, 689527.

Yang, L., Park, D., & Qin, Z. (2021). Material function of mycelium-based bio-composite: A review. *Frontiers in Materials, 8*, 737377.

Yao, Z., Seong, H. J., & Jang, Y.-S. (2022). Environmental toxicity and decomposition of polyethylene. *Ecotoxicology and Environmental Safety, 242*, 113933.

Ye, M., Liu, J., Ma, C., Li, Y.-Y., Zou, L., Qian, G., & Xu, Z. P. (2018). Improving the stability and efficiency of anaerobic digestion of food waste using additives: A critical review. *Journal of Cleaner Production, 192*, 316–326.

Zhang, Y., Li, F., Peng, N., & Peng, L. (2020). Environmental impact assessment of air-permeable plastic runway production in China. *Science of the Total Environment, 730*, 139073.

Zida, A., Bamba, S., Yacouba, A., Ouedraogo-Traore, R., & Guiguemdé, R. (2017). Anti-Candida albicans natural products, sources of new antifungal drugs: A review. *Journal de Mycologie Medicale, 27*(1), 1–19.

Chapter 13
Regulations on the Utilization of Fungal-Based Additives and Bio-Based Ingredients Used in Food Processing Industry

Heena Yadav, Anshul Dhawan, and Harinder Singh Oberoi

1 Introduction

Historically, natural-derived ingredients have been utilized for varied purposes. The use of synthetic ingredients and other additives over natural ones gained popularity only in past few decades. Although synthetic additives are cheaper and highly potent and produce more consistent effects in food products, their usage has been linked to several health issues and diseases (Carocho et al., 2014). Recent studies associating synthetic additives with cancer have led to increase in the popularity of additives derived from plants and microorganism (Martins et al., 2016; Sigurdson et al., 2017). One of the emerging sources of naturally derived additives is from microbes such as fungi, bacteria, algae, and actinomycetes, which are considered as more sustainable, reliable, and easily available alternative source (Rodriguez-Amaya, 2016). However, it is equally important to understand the regulations governing their use in food products to ensure consumer safety, maintain quality standards, and comply with legal requirements.

In today's time, fungal-based food additives and other bio-based ingredients play an essential role in the food processing sector, contributing to preservation, flavour enhancement, texture improvement, and nutritional enrichment (Fig. 13.1). Despite regaining popularity and acceptability, these ingredients face significant safety challenges. This includes potential mycotoxin contamination, allergenicity concerns

H. Yadav (✉)
Department of Food Science and Technology, National Institute of Food Technology Entrepreneurship and Management-Kundli, Sonipat, Haryana, India

A. Dhawan · H. S. Oberoi
Department Interdisciplinary Sciences, National Institute of Food Technology Entrepreneurship and Management (NIFTEM), Kundli, Sonepat, Haryana, India

B. P. Singh et al. (eds.), *Fungal Additives and Bioactives in Food Processing Industries*, Fungal Biology, https://doi.org/10.1007/978-3-032-04519-5_13

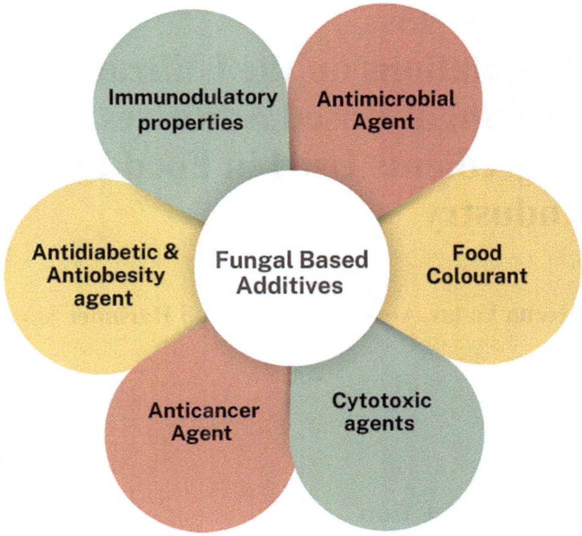

Fig. 13.1 Applications of fungal pigments in the food industry. (Adapted from Poorniammal et al., 2021)

particularly with enzyme preparations, strain-dependent metabolite variations affecting product consistency, and the risk of antimicrobial resistance gene transfer during production (Fernandes & Carvalho, 2017). The presence of secondary metabolites with unknown toxicological profiles poses additional challenges in safety assessment, especially for novel and genetically modified fungal strains (Magnuson et al., 2013).

Fungal-based food additives and bioactive compounds must meet strict criteria before approval to ensure safety and functionality. This includes toxicological evaluations to assess potential health risks such as toxicity, allergenicity, and genotoxicity. Additionally, labelling requirements ensure transparency by listing ingredients accurately and highlighting potential allergens, maintaining consumer awareness and regulatory compliance (Guerrand, 2018). This chapter provides an overview on the regulatory mechanism and approval process for fungal additives and other bio-based ingredients.

2 Global Regulatory Framework

Globally, the three main institutions regulating food additives (including the bio-based additives) globally are Joint Food and Agriculture Organization/World Health Organization (FAO/WHO) Expert Committee on Food Additives (JECFA); the Codex Alimentarius Commission (CAC); the European Food Safety Authority (EFSA) and the US Food and Drug Administration (FDA) (Cheeseman, 2014).

2.1 Role of JEFCA and Codex Alimentarius Commission

In 1956, FAO and WHO jointly founded JEFCA with an objective to gather and publicize knowledge about food additive and to establish their identity and purity specification. JECFA is responsible for developing detailed specifications for the identity and purity of additives including fungal-derived food additives, ensuring that these substances meet globally recognized standards for safety and functionality. The updated details on food additives are available in the compendium of food additives published on FAO website. The specifications cover various parameters, including chemical characterization, microbial purity criteria, permissible levels of heavy metals and other contaminants, and stringent controls on the conditions of production. It has generated numerous internationally recognized data and publications that are extensively utilized by various government agencies, industry, and research centres (FAO, 2011; FAO-JEFCA, 2025).

While the compendium does not explicitly categorize additives on the basis of the source, i.e. synthetic or bio-based, many are derived from natural sources such as plants, animals, or microorganisms. To identify bio-based food additives, individual monographs within the compendium may be referred, which provide information on the origin and production methods of each additive. Few examples of such additives included in the aforementioned compendium of food additives are mentioned in the Table 13.1 below:

The Codex Alimentarius Commission (CAC) is an international body jointly established by the FAO and WHO in the year 1962. CAC plays a crucial role in setting food standards that serve as a benchmark for governments, industry stakeholders, and consumers worldwide. These guidelines, developed to ensure food safety and fair trade practices, encompass a wide range of food ingredients, including fungal-derived additives and bioactive compounds (Codex, 2025).

Table 13.1 Example of fungi-based and bio-based additives included in JEFCA's Compendium of Food Additives

Food additive	Source
Fungi-based additives	
α-Amylase	*Aspergillus oryzae*
Lipase	*Rhizopus arrhizus*
Citric Acid	*Aspergillus niger*
Natamycin	*Streptomyces natalensis*
Glucose Oxidase	*Aspergillus niger*
β-Galactosidase (Lactase)	*Kluyveromyces lactis*
Other bio-based additives	
Xanthan Gum	*Xanthomonas campestris*
Monosodium Glutamate (MSG)	*Corynebacterium glutamicum*
Pectin	Citrus peel, Apple pomace
Carrageenan	*Kappaphycus alvarezii* (Red seaweed)
Alginate	Laminaria species

The Codex Committee on Food Additives (CCFA) is responsible for establishing global food additive standards under the Codex Alimentarius framework. General Standards for Food Additives (GFSA) recommends maximum permitted levels for food additives, assigns functional classes, and prepares priority lists for risk assessment by JECFA (Joint FAO/WHO Expert Committee on Food Additives). Moreover, CCFA recommends specifications of identity and purity of food additives and ensures that appropriate methods are available for analysis and labelling of the identified food additives (CCFA, 2025).

Through its work, CCFA harmonizes international food safety regulations, facilitating trade and consumer protection. It must be noted that CAC lacks regulatory jurisdiction; the enforcement of standards developed by Codex Alimentarius is entirely dependent on adoption into national regulatory frameworks.

2.2 Regulation and Approval Process for Food Additive in United States

In the United States (US), the US Food and Drug Administration (FDA) oversees the food regulations while FDA Science Board is the advisory scientific body. As a food regulator, the US FDA's fundamental goal is to safeguard the public health by ensuring the security, effectiveness, and safety of drugs for humans and animals, biological products, and medical devices, along with the safety of the national food supply, cosmetics, and radiation-emitting products (US FDA, 2025). In addition, the United States is a member nation of both the World Trade Organization (WTO) and the Codex Alimentarius Commission (CAC). The standards specified by CAC are suitably adopted in the food regulations of the nations (FDA, 2025).

The Federal Food, Drug, and Cosmetic Act (FD&C ACT) lists different types of food ingredients based upon their intended purpose and describes the FDA's authority over these food ingredients. Food additives and processing aids derived are regulated under the U.S. Code of Federal Regulations (CFR), Title 21. Furthermore, FDA also publishes guidance documents for consumers and food businesses. These guidance documents are not legally binding like CFR but serve as a detailed informative instrument for providing up to date information to the target audience (FD&C ACT, 2018; FDA, 2025).

CFR defines the food additives are defined as substances that are intended to become part of the food or alter food characteristics, either directly or indirectly. The directly added additives include the ingredients that are intentionally added to food for a specific purpose while indirect food additives include the ingredients or substances that unintentionally migrates into food from packaging or processing equipment (Rulis and Levitt, 2009). In addition, Section 201(s) of FD&C ACT exempts the use(s) of a substance that is GRAS from the definition of a food additive (FDA, 2010).

2.2.1 The Process of Approval of Food Additives

In 1960, FDA started market assessment of food additives as a process for implementation of FD&CA. For the same, a science-based pre-evaluation system with a clear objective was developed by independent scientists (Rulis & Levitt, 2009). Working within FD&C ACT and CFR, the **Office of Food Additive Safety (OFAS)** under FDA runs various programs for the use of substances that are food additives and Generally Recognized as Safe (GRAS). These food additives can be approved through two main processes. They are (a) the Generally Recognized as Safe (GRAS) determination or (b) the official Food Additive Petition (FAP) process (FDA, 2025; Magnuson et al., 2013).

Generally Recognized as Safe (GRAS) Ingredients
The definition of food additive in FD&CA Act mentions provisions for use of ingredients classified as GRAS and such ingredients do not require pre market review by FDA. Ingredients classified under GRAS must have comprehensive data (available publicly) to establish their safety and widely accepted by qualified experts for their safe use. In addition, GRAS uses must meet the same safety standard as for food additives, a reasonable certainty of no harm under the conditions of its intended use and have the same quantity and quality of information that would support the safety of a food additive. Examples of GRAS include canola oil, vinegar, and spices such as black pepper. (Source: FDA, 2025).

For fungal-derived additives and bio-based ingredients such as enzymes that do not qualify for GRAS status, a formal Food Additive Petition (FAP) must be submitted to the FDA for review and approval (Magnuson et al., 2013). The manufacturers are required to submit a detailed safety data package that includes acute and chronic toxicity studies, mutagenicity and genotoxicity assessments, reproductive toxicity evaluations, and, where applicable, long-term carcinogenicity studies (FDA, 2011).

Additionally, manufacturers must provide comprehensive chemical and manufacturing information, including detailed descriptions of the production process, validation of analytical methods, stability data, and impurity profiles. The regulatory framework ensures that all fungal-derived food ingredients introduced into the U.S. market meet the highest standards of safety and efficacy prior to approval for consumer use. Figure 13.2 illustrates the step-by-step process for the approval of food additives by the U.S. Food and Drug Administration (FDA, 2024).

The Food Additive Petition (FAP) process is rigorous, science-based, and time-consuming. The FDA typically requires approximately 24 months to reach a final decision on a petition submitted by manufacturers. Once a petition is accepted for review, the FDA publishes a notice in the Federal Register within 30 days. Subsequently, the FDA conducts a thorough risk assessment, which includes evaluations of toxicology, allergenicity, exposure levels, and cumulative dietary impact.

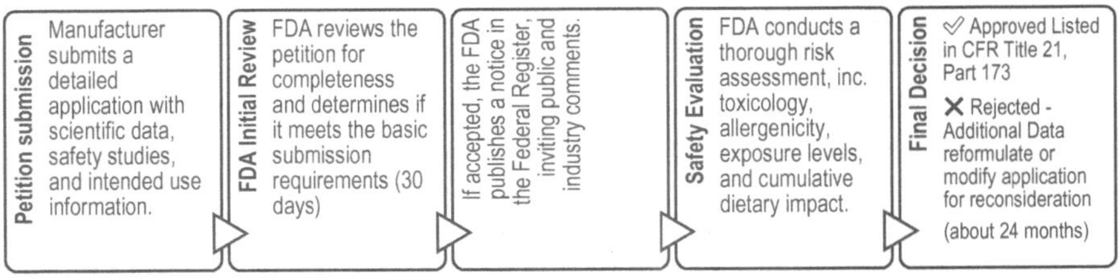

Fig. 13.2 Approval Process as per FAP for Food Additives

If the ingredient is determined to be safe—i.e. there is a "reasonable certainty of no harm" under its intended conditions of use—the petition is approved, and the additive is listed in Part 173 of Title 21 of the Code of Federal Regulations (CFR, 2024). If not, the petition is rejected, and the manufacturer must either submit additional data, reformulate the product, or revise and resubmit the application.

2.3 *Regulatory Policies in the European Union*

Until 2008, EU did not have underlaid specifications for food additives; many such ingredients, especially enzymes, were being regulated under processing aids' specifications of the member states. The variation in the standards of the member countries leading to complications in the implementation of the requirements led to the development of a comprehensive and uniform database for food additive. At present, EU has a well-defined regulatory system for food additives, ensuring their safety, transparency, and harmonized use by the member states.

Food Additives including those derived from plants and microorganisms are regulated under EU Regulation (EC) No 1333/2008, which outlines the authorization process, permissible use levels, and safety evaluation criteria for all food additives, including those derived from fungi. In addition, these regulations mandate clear declaration of fungal-based metabolites on the food product label. Figure 13.3 below illustrates how these additives are typically displayed on the label.

Since 1995, EU has been a member of World Trade Organisation (WTO) as well as the CAC. Similar to US and other member nations of CAC, EU also consider the food additives specifications prescribed by JEFCA. The food safety standards in EU are regulated by the European Commission (EC), Directorate General for Health and Food Safety (DG SANTE). Founded in 2002, European Food Safety Authority (EFSA) serves as an advising scientific organization. The primary goal of EFSA is to give independent scientific advice on food safety at all stages of food production and supply chain. These findings can provide a scientific basis for EU member states' legislation and policies affecting food and feed safety (Magnuson et al., 2013).

In contrast to US, EU has more compartmentalized approach, with separate regulations for food additives (Regulation EC No 1333/2008) and food enzymes (Regulation EC No 1332/2008) (EC-1332, 2008; EC-1333, 2008). The regulation

Fig. 13.3 Labelling requirements for fungal-based food additive

Ingredients:

XXXXX, XXXXXX, XXXXXX, XXXXX, XXXXXX, **E330, E260**

Allergen Statement:
Contains aspartame (E951), Azorubine (E122)

also specifies the community list of enzymes and their specific labelling requirements. In case of food additives, EU food additives database provides a comprehensive list of 411 food additives permitted in EU (EC, 2025). Unlike US, the definition of processing aids has not been included in the definition of food additives in EU regulations (EC-1333, 2008; EFSA, 2024).

Definitions

Food Additives: EC No 1333/2008 defines food additive as any substance not normally consumed as food but intentionally added to food for a technological function such as preservation, colour enhancement, emulsification, etc. Food Additives must be included in the Union List of Authorized Additives, which specifies permitted substances, their condition of use, and maximum limits. Fungal additives and other bio-based ingredients fall into the list if they perform such technological functions in the food.

Food Enzyme: As per EC No 1332/2008, "food enzyme is defined as a product obtained from plants, animals or microorganisms or products thereof, including a product obtained by a fermentation process using microorganisms that:

(i) contain one or more enzymes capable of catalysing a specific biochemical reaction; and
(ii) are added to food for a technological purpose during the manufacturing, processing, preparation, treatment, packaging, transport or storage of foods."

A 'food enzyme preparation' is defined as "a formulation consisting of one or more food enzymes in which substances such as food additives and/or other food ingredients are incorporated to facilitate their storage, sale, standardization, dilution or dissolution" (EC-1332, 2008; EFSA, 2024).

Further, Regulation EC No. 2015/2283 governs novel foods including plant-based and fungal metabolites. Any new fungal-based food additive must undergo pre-market approval, requiring scientific dossiers demonstrating safety and efficacy. Table 13.2 below depicts the examples of few novel fungal-based additives approved under EU.

Table 13.2 Novel fungal-derived additives and bioactives approved in the EU (2015/2283)

S. no.	Novel ingredient	Fungal source	Use
1.	Mycoprotein (Quorn)	*Fusarium venenatum*	Meat alternative, protein source
2.	Ergothioneine	*Aspergillus oryzae*, Tricholoma matsutake	Antioxidant, nutraceutical
3.	Beta-Glucans	*Saccharomyces cerevisiae*	Immune support, dietary fibre
4.	Xylooligosaccharides (XOS)	Fungal fermentation	Prebiotic, gut health enhancer
5.	Red Yeast Rice Extract	*Monascus purpureus*	Cholesterol-lowering, natural colourant

2.3.1 Approval Process

The EU classification of Qualified Presumption of Safety (QPS) serves a purpose similar to that of the Generally Recognized as Safe (GRAS) system in the United States. Microorganisms granted QPS status are not required to undergo a full safety assessment and are listed on the European Food Safety Authority (EFSA) website. Introduced in 2007, the QPS framework has been developed specifically for microbial and fungal-derived additives and provides a streamlined process for evaluating the safety of microorganisms based on well-defined taxonomic identity and established safety data (EFSA, 2024). The QPS process assesses microorganisms at the taxonomic unit (TU) level, taking into account their taxonomic classification, the existing body of scientific knowledge, and any potential safety concerns, such as antimicrobial resistance. Microorganisms with QPS status benefit from a simplified safety assessment process, which helps reduce regulatory burdens while maintaining high food safety standards. However, microorganisms with ambiguous taxonomy or known safety risks are excluded from QPS consideration. Figure 13.4 below provides an overview of the approval process under QPS framework for Food Additives.

Unlike the U.S. FDA system, which does not differentiate between Genetically Modified Organisms (GMOs) and non-GMOs in its evaluation, the EU has established a distinct and comprehensive regulatory pathway for Genetically Modified Microorganisms (GMMs). Food enzymes that fall under the scope of Regulation (EC) No 1829/2003 on genetically modified food and feed are also subject to Regulation (EC) No 1331/2008 concerning food enzymes (EC-1331, 2008). It is important to note that while QPS status facilitates the evaluation of food additives, it does not by itself guarantee market authorization (EC1829, 2003). Similar to other food additives and enzymes, EFSA is responsible for conducting a thorough scientific risk assessment before a product can be approved for market use. This includes a comprehensive evaluation of the GMM used in production, examining its genetic construct, modifications, intended functions, and the safety of the final product with respect to toxicity, allergenicity, and potential nutritional impact (EFSA, 2011; Aguilera et al., 2013; Schuler et al., 2019).

Fig. 13.4 Approval process under QPS for food additives. (Source: EFSA, 2024)

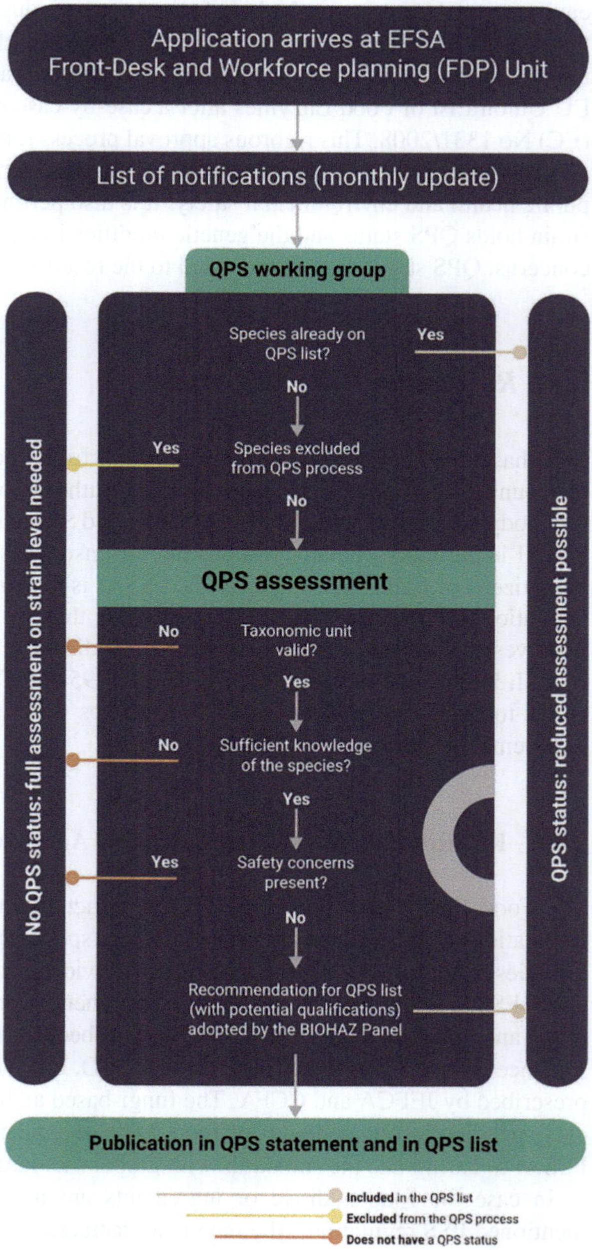

The production process plays a critical role in the regulatory assessment of GMM-based food additives. In most cases, food enzymes are categorized under Category 2, where GMMs are used during production but are not present in the final product. Under such conditions, the regulatory focus is on verifying the absence of residual DNA or harmful compounds in the end product (EFSA, 2011; Aguilera et al., 2013). Additionally, EFSA assesses dietary exposure levels and potential environmental impacts, particularly for large-scale industrial applications. Following the review, EFSA issues a scientific opinion that guides the regulatory decision-making process. Upon approval, GMM-derived food additives are included in the EU Union List of Food Enzymes after a case-by-case evaluation under Regulation (EC) No 1331/2008. This rigorous approval process reflects the EU's precautionary approach to genetically modified foods, balancing technological innovation with public health and environmental safety. It is also pertinent to note that if a parental strain holds QPS status and the genetic modification does not introduce new safety concerns, QPS status may be extended to the resulting GMM.

2.4 Regulatory Policies in India

India has a well-defined regulatory framework for implementation of food safety in the country. The Food Safety and Standards Authority of India (FSSAI) is the statutory body established under the Food Safety and Standards Act, 2006 (FSSA, 2006). FSSAI is the food regulator responsible for ensuring safe and wholesome food to the citizens of India. As an apex body, FSSAI is responsible for overseeing implementation of food safety regulations including those related to fungal-derived food additives, bioactives, and enzymes in the Indian market is the responsibility of FSSAI. India is the member of WTO since 1995 and CAC since 1963 and actively works to align with international best practices while addressing specific regional requirements and concerns (FSSAI, 2025).

2.4.1 Key Regulation with respect to Food Additives

The Food Safety and Standards (Food Products Standards and Food Additives) Regulations, 2011 specifies the standards and specifications for various food commodities. Annexure A of these regulations provides list of food additives permitted under FSSA, 2006. Though this list is comprehensive, no segregation for chemical, plant, and microorganism-based additives has been done (FSSAI, 2011).

Since India is the member of CAC and WTO, FSSAI aligns list of food additives prescribed by JEFCA and CCFA. The fungi-based additives and bio-based ingredients specified in compendium of food additives under JEFCA are adopted under FSSAI to harmonize the standards with global standards.

In case the food additive or ingredients are not specified under the above-mentioned FSS regulations, then the manufacturer needs to take approval from the food authority as per the procedure specified under Food Safety and Standards (Approval for Non-Specified Food and Food Ingredients) Regulations, 2017. These regulations underlay the requirements for approval of any novel food ingredient including food additives being derived from microorganisms to undergo through scrutiny before it is permitted to enter the Indian markets (FSSAI, 2017).

Further, the Food Safety and Standards (Labelling and Display) Regulation, 2020, specifies the information to be mentioned on the label of the food products including food additives. The food additives including fungal additives and plant-based additives included in the list of additives under Food Safety and Standards (Food Products Standards and Food Additives) Regulations, 2011, shall mention the FSSAI license number, name and address of the manufacturer, INS number as specified in Annexure A of the above-mentioned food regulation. In addition, every product shall mention the name/INS number of additive in the list of ingredients (FSSAI, 2020). The manufacturers obtaining a special approval under Food Safety and Standards (Approval for Non-Specified Food and Food Ingredients) Regulations, 2017, shall also comply with appropriate requirements specified under the labelling and display regulations (FSSAI, 2017).

2.4.2 Approval Process

FSSAI has established comprehensive guidelines for evaluating the safety and suitability of fungal-derived additives. The regulatory requirements include safety assessment protocols, compliance with purity criteria, and category-specific restrictions on permissible use levels. Additionally, manufacturers must provide detailed documentation on the manufacturing process, analytical testing methods, stability data, and relevant safety studies conducted within India.

FSSAI under Food Safety and Standards (Approval for Non-Specified Food and Food Ingredients) Regulations, 2017, also maintains separate guidelines for novel food products, including fungal-derived ingredients that do not have a history of traditional consumption in India. The pre-market approval process requires a scientific committee review, a thorough safety assessment, and, where applicable, clinical study data to validate the ingredient's health impact (FSSAI, 2017).

Fungal-derived enzymes are subject to specific regulatory requirements, including strain identification, documentation of non-pathogenicity, assessment of genetic modifications (if applicable), and standardization of enzyme activity units. Strict microbiological and heavy metal limits are also enforced to ensure the safety and quality of enzyme preparations used in food processing. Following approval, post-market surveillance is also carried out by the food authority for ensuring that any unforeseen risks are promptly identified and mitigated. Figure 13.5 below elaborates the step by step process for attaining approval for NSF food products and ingredient under FSSA, 2005.

3 Conclusion

This chapter presents a global overview on the regulatory framework and approval process for the Fungal-derived food additives across various regions of the globe. Fungal-derived food additives and their bio-based metabolites play a critical role in modern nutrition, with several compounds already permitted under Codex, US, and EU. JEFCA has developed individual monographs for select fungal-based

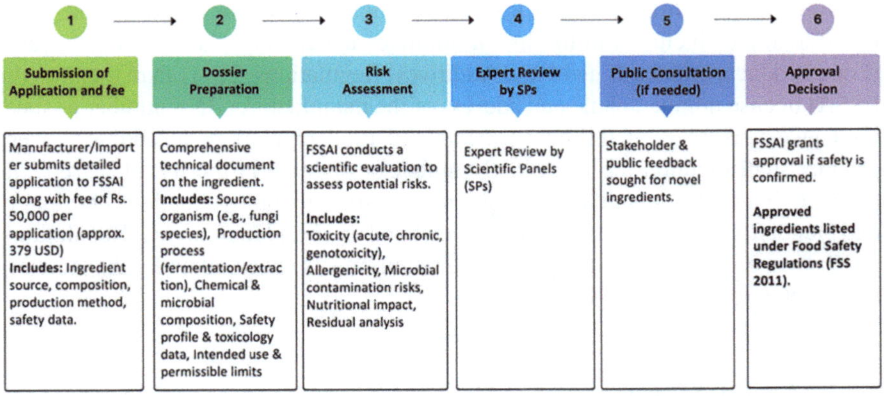

Fig. 13.5 Approval process for Non-specified Food Product or ingredients in India. (Source: FSSAI, 2017)

metabolites which is usually referred by Codex Alimentarius Commission. The USA, EU, and India require special approval process for use of these ingredients in the food sector. In the US, the approval mechanism is enabled through FAP while EU has implemented QPS framework for such ingredients. In India, a stepwise approval process for non-specified foods is required for these ingredients to enter the market. These comprehensive processes ensure through review of safety and efficacy of each fungal metabolites.

Although regulatory frameworks are designed to ensure both consumer safety and encourage innovation, several challenges still affect the approval and widespread use of fungal metabolites. These include the complexities involved in assessing risks, differences in standards across countries, and varying levels of consumer acceptance. To promote the safe and responsible use of fungal-based food additives, it's essential to enhance scientific evaluation processes and encourage greater international cooperation.

References

Aguilera, J., Gomes, A. R., & Olaru, I. (2013). Principles for the risk assessment of genetically modified microorganisms and their food products in the European Union. *International Journal of Food Microbiology, 167*(1), 2–7. https://doi.org/10.1016/j.ijfoodmicro.2013.03.013

Carocho, M., Barreiro, M. F., Morales, P., & Ferreira, I. C. (2014). Adding molecules to food, pros and cons: A review on synthetic and natural food additives. *Comprehensive Reviews in Food Science and Food Safety, 13*(4), 377–399. https://doi.org/10.1111/1541-4337.12065

CFR. (2024). Title 21 Part 171: Food additive petitions. https://www.accessdata.fda.gov/scripts/cdrh/cfdocs/cfcfr/CFRSearch.cfm?CFRPart=171. Accessed 10 Dec 2024.

Cheeseman, M. (2014). Global regulation of food additives. In *Food additives and packaging* (pp. 3–9). American Chemical Society. https://doi.org/10.1021/bk-2014-1162.ch001

Codex Alimentarius: International Food Standards. https://www.fao.org/fao-who-codexalimentarius/en/. Accessed 17 Jan 2025.

Codex Committee on Food Additives (CCFA). https://www.fao.org/fao-who-codexalimentarius/committees/committee/en/?committee=CCFA. Accessed 17 Jan 2025.

EC1331. Regulation (EC) No 1331/2008 of the European Parliament and of the Council of 16 December 2008 establishing a common authorisation procedure for food additives, food enzymes and food flavourings. 2008. https://eur-lex.europa.eu/legal-content/EN/TXT/?uri=CELEX:32008R1331. Accessed 10 Dec 2024.

EC1332. Regulation (EC) No 1332/2008 of the European Parliament and of the Council of 16 December 2008 on food enzymes. 2008. https://eur-lex.europa.eu/legal-content/EN/TXT/?uri=CELEX:32008R1332. Accessed 10 Dec 2024.

EC1829. Regulation (EC) No 1829/2003 of the European Parliament and of the Council of 22 September 2003. 2003. https://eur-lex.europa.eu/legal-content/en/ALL/?uri=CELEX%3A32003R1829. Accessed 15 Dec 2024.

EC-333. Regulation (EC) No 1333/2008 of the European Parliament and of the Council of 16 December 2008 on food additives. 2008. https://eur-lex.europa.eu/legal-content/EN/TXT/?uri=CELEX%3A32008R1333. Accessed 10 Dec 2024.

EFSA. Guidance on the risk assessment of genetically modified microorganisms and their products intended for food and feed use. 2011. https://efsa.onlinelibrary.wiley.com/doi/pdf/10.2903/j.efsa.2011.2193. Accessed 10 Jan 2025.

EFSA. Food enzymes. 2024. https://www.efsa.europa.eu/en/topics/topic/food-enzymes. Accessed 10 Dec 2024.

European Commission. Food additives database. EU Food and Feed Safety Portal. https://ec.europa.eu/food/food-feed-portal/screen/food-additives/search. Accessed 15 Jan 2025.

European Food Safety Authority. Qualified presumption of safety (QPS) assessment. https://www.efsa.europa.eu/en/applications/qps-assessment. Accessed 10 Dec 2024.

FDA. (1992). Statement of policy: Foods derived from new plant varieties; notice. *Fed Regist., 57*(104), 22983–23005. https://www.fda.gov/food/guidanceregulation/guidancedocuments-regulatoryinformation/biotechnology/ucm096095.htm. Accessed 15 Jan 2025

FDA. Guidance for industry, enzyme preparations: recommendations for submission of chemical and technological data for food additive petitions and GRAS notices. 2010. https://www.fda.gov/Food/GuidanceRegulation/GuidanceDocumentsRegulatoryInformation/ucm217685.htm. Accessed 15 Jan 2025.

FDA. Guidance for industry: questions and answers about the petition process. 2011. https://www.fda.gov/Food/GuidanceRegulation/ucm253328.htm#answers. Accessed 15 Jan 2025.

FDA. FDA Mission. 2025. https://www.fda.gov/aboutfda/whatwedo/default.html. Accessed 15 Jan 2025.

Fernandes, P., & Carvalho, F. (2017). Microbial enzymes for the food industry. In *Biotechnology of microbial enzymes* (pp. 513–544). Academic. https://doi.org/10.1016/B978-0-12-803725-6.00019-4

Food and Agriculture Organization of the United Nations. (2011). *Compendium of food additive specifications* (FAO JECFA Monographs 11). ISBN:978-92-5-107004-8. https://www.fao.org/3/a-i2358e.pdf

Food and Agriculture Organization of the United Nations. (n.d.). *Joint FAO/WHO Expert Committee on Food Additives (JECFA)*. Retrieved January 10, 2025, from https://www.fao.org/food-safety/scientific-advice/jecfa/en/

Food Safety and Standards Authority of India. (2006). *Food Safety and Standards Act*. https://www.fssai.gov.in/upload/uploadfiles/files/FOOD-ACT.pdf. Accessed 10 Dec 2024.

Food Safety and Standards Authority of India. (2011). *Food Safety and Standards (Food Products Standards and Food Additives) Regulations*. https://www.fssai.gov.in/upload/uploadfiles/files/Appendix%20A(2).pdf. Accessed 10 Dec 2024.

Food Safety and Standards Authority of India. (2017). *Food Safety and Standards (Approval for Non-Specified Food and Food Ingredients) Regulations*. https://www.fssai.gov.in/upload/uploadfiles/files/Compendium_FSS_NFS_FA_17_10_2022.pdf. Accessed 10 Dec 2024.

Food Safety and Standards Authority of India. (2020). *Food Safety and Standards (Licensing and Display) Regulations*. https://www.fssai.gov.in/upload/uploadfiles/files/Comp_Labelling.pdf. Accessed 10 Dec 2024.

Food Safety and Standards Authority of India. (n.d.-a). *Codex India – FSSAI*. https://www.fssai.gov.in/cms/codex.php. Accessed 10 Jan 2025.

Guerrand, D. (2018). Economics of food and feed enzymes: Status and prospectives. In *Enzymes in human and animal nutrition* (pp. 487–514). Academic. https://doi.org/10.1016/B978-0-12-805419-2.00026-5

Magnuson, B., Munro, I., Abbot, P., Baldwin, N., Lopez-Garcia, R., Ly, K., et al. (2013). Review of the regulation and safety assessment of food substances in various countries and jurisdictions. *Food Additives & Contaminants: Part A, 30*(7), 1147–1220. https://doi.org/10.1080/19440049.2013.795293

Martins, N., Roriz, C. L., Morales, P., Barros, L., & Ferreira, I. C. (2016). Food colorants: Challenges, opportunities and current desires of agro-industries to ensure consumer expectations and regulatory practices. *Trends in Food Science & Technology, 52*, 1–15. https://doi.org/10.1016/j.tifs.2016.03.009

Poorniammal, R., Prabhu, S., Dufossé, L., & Kannan, J. (2021). Safety evaluation of fungal pigments for food applications. *Journal of Fungi, 7*(9), 692. https://doi.org/10.3390/jof7090692

Rodriguez-Amaya, D. B. (2016). Natural food pigments and colorants. *Current Opinion in Food Science, 7*, 20–26. https://doi.org/10.1016/j.cofs.2015.08.004

Rulis, A. M., & Levitt, J. A. (2009). FDA'S food ingredient approval process: Safety assurance based on scientific assessment. *Regulatory Toxicology and Pharmacology, 53*(1), 20–31. https://doi.org/10.1016/j.yrtph.2008.10.003

Schuler, L., Zust, D., Vybiral, D., & Hau, P. (2019). *GM food regulations in the EU*. https://doi.org/10.1016/B978-0-08-100596-5.22609-7

Sigurdson, G. T., Tang, P., & Giusti, M. M. (2017). Natural colorants: Food colorants from natural sources. *Annual Review of Food Science and Technology, 8*(1), 261–280. https://doi.org/10.1146/annurev-food-030216-025923

U.S. Food and Drug Administration. (1992). New plant variety regulatory information. U.S. Food and Drug Administration. Notice. *Fed Regist, 57*(104), 22983–23005. https://www.fda.gov/food/food-new-plant-varieties/new-plant-variety-regulatory-information. Accessed 10 Dec 2024

U.S. Food and Drug Administration. (2025a). *FDA mission*. https://www.fda.gov/aboutfda/whatwedo/default.htm. Accessed 15 Jan 2025.

U.S. Food and Drug Administration. (2025b). Understanding how the FDA regulates food additives and GRAS ingredients. U.S. FDA. https://www.fda.gov/food/food-additives-and-gras-ingredients-information-consumers/understanding-how-fda-regulates-food-additives-and-gras-ingredients. Accessed 15 Jan 2025.

U.S. Food and Drug Administration. (n.d.). *Federal Food, Drug, and Cosmetic Act (FD&C Act)*. U.S. FDA. Retrieved from https://www.fda.gov/regulatory-information/laws-enforced-fda/federal-food-drug-and-cosmetic-act-fdc-act (Content current as of 29/03/2018)

Chapter 14
Application of Nanotechnology in Designing Fungal Additives-Based Food Ingredient

Pirthi Pal Singh, Purushottam Nagar, and Divakar Jaiswar

1 Introduction

Nanotechnology is defined as the study and application of techniques to produce materials at an extremely small scale. The word "nano" originates from the Greek word interpretate “dwarf” and in scientific terms, it relates to dimensions ranging from 1 to 100 nanometres. According to standard definitions, nanotechnology includes designing, fabricating characterizing, and utilizing materials with specific shape and size control at the nanoscale. At this size, materials demonstrate significantly altered chemical, physical, and biological properties compared to their bulk counterparts, leading to unique applications across various fields (Pathakoti et al., 2017).

The historical evidences of nanotechnology trace back centuries, with early examples including the use of gold nanoparticles in stained glass by the Flemish glassmaker John Utynam in 1449. Later, in the sixteenth century, Paracelsus, a Swiss physician also known as Theophrastus von Hohenheim, utilized gold nanoparticles for therapeutic applications. While these early applications demonstrate that the concept of manipulating matter at tiny scales is not new, the formal scientific foundation of nanotechnology emerged much later (Rafique et al., 2020). Richard Zsigmondy played a pivotal role by introducing the nanometre as a unit of measurement and developing techniques to analyse colloidal particles, later earned the Nobel Prize in Chemistry in 1925. Decades later, in 1974, Norio Taniguchi formally defined the word “nanotechnology” to describe precision engineering and material processing at the nanoscale, particularly in the fabrication of ultra-thin materials (Rafique et al., 2020).

P. P. Singh (✉) · P. Nagar · D. Jaiswar
Tirupati Innovation Centre (Tirupati Wellness Pvt Ltd.),
Paonta Sahib, Himachal Pradesh, India
e-mail: pirthipalsingh@yahoo.com

B. P. Singh et al. (eds.), *Fungal Additives and Bioactives in Food Processing Industries*, Fungal Biology, https://doi.org/10.1007/978-3-032-04519-5_14

Nanomaterials can be categorized by size, shape, structure, and properties. Their physicochemical properties, viz., solubility, diffusivity, optical and magnetic behaviour, toxicity, bioavailability, and thermodynamic stability exhibit distinct differences from their bulk counterparts (Sahoo et al., 2021). These unique properties endorse nanomaterials being highly valuable in numerous applications, including the food sector. Nanotechnology has enabled significant advancements in monitoring food quality and improving nutrient stability and bioavailability (Patra et al., 2018). It has also been evidenced to enhance texture and sensory properties of food materials while improving nutrient retention through encapsulation techniques (Anandharamakrishnan & Parthasarathi, 2019). Additionally, it has facilitated the reduction of artificial additives such as excess sugar, salt, fats, and synthetic preservatives, contributing to healthier food products (Singh et al., 2017). Beyond food characteristics enhancement, nanotechnology is stepping towards revolutionizing the entire agricultural food distribution network, influencing postharvest, food processing, packaging, shipping, logistics, and storage (Pathakoti et al., 2017). Innovations such as smart packaging materials help extend shelf life, improve safety, and prevent contamination by pathogens (Li & Xue, 2025). Lightweight yet durable nanomaterials ensure better protection during transit, while functional foods enriched with nano-encapsulated nutrients offer enhanced health benefits (Ashfaq et al., 2022; Ghosh et al., 2025). Given its broad potential, nanotechnology continues to drive innovation, providing sustainable, efficient, and health-conscious solutions that shape the future of food manufacturing and preservation (Singh et al., 2023a, b).

Food product packaging serves as a protective barrier against environmental variables such as temperature fluctuations, humidity, microbiological contamination, and exposure to ambient gases (Salgado et al., 2021). Nanotechnology enhances food safety by minimizing decomposition and nutrient loss, thereby extending the shelf life. Modern food packaging systems can be categorized in two types: active packaging and smart packaging. Smart packaging incorporates nano sensor that senses upon change in environment, providing real-time feedback to consumers regarding potential contamination or pathogen presence (Fuertes et al., 2016). These advanced systems utilize functional nanoparticles with specialized properties, such as antimicrobial activity or moisture regulation, achieved through agents like free radical scavengers, carbon dioxide scavengers, and emitters. Active packaging, in contrast, has a crucial function in food preservation by creating an inert shield against external environmental factors (Yildirim et al., 2018). This technology actively interacts with the food or its surrounding atmosphere to maintain freshness, inhibit microbial growth, and prevent oxidative degradation. Together, smart and active packaging systems leverage nanotechnology to assure quality, safety, and longevity of food during the supply chain.

Food additives are the materials that are incorporated into food to enhance its physical appearance, texture, taste, safety, freshness, and longevity as per World Health Organization (WHO) (Mwale, 2023). Today, food is not only intended to be consumed for the fulfilment of hunger, but its sensorial acceptance by the consumers is equally important. Hence, food additives played a vital role in achieving overall consumer acceptance. Food additives are classified into numerous categories: preservatives, e.g. sorbic acid, antioxidants, preservative like sodium benzoate, nutraceuticals like vitamins and minerals, chemicals (e.g. acids and digestive enzymes), colourants, and flavouring substances. Few additives in some cases show multiple benefits (Blekas, 2016). Since ages, fungal based additives have been used for the manufacturing of food additives and beverages, particularly alcoholic beverages. Currently, through fermentation process, both molds and yeasts have been used to obtain food ingredients. They are used as biotransformation agents for production scale manufacturing of food additives (Copetti, 2019). Selection of non-pathogenic fungal strains, advanced techniques of biosynthesis regulation, and genetic engineering helps to achieve higher rate of production of safe food-grade products. Fermentation-produced food additives aid to awareness onto clean label (Copetti, 2019).

This chapter makes a significant contribution by illustrating the new potential of nanotechnology in fungi-based food additive developments. By exploring the connection of nanoscience and food technology, it provides critical insights into how nanotechnology is revolutionizing in traditional food additives preparation. Food scientists, technologists and industry professionals are seeking to navigate the rapidly and newly evolving landscape of nano-enhanced food additives. By bridging scientific understanding with practical applications, this chapter contributes essential knowledge towards the development of more stable and functionally superior fungi-based food-additive products.

2 Basics and Size

In order to understand the scale of a nano-size range, the combined diameter of 5 atoms is about 10 Å (Å = angstrom), which in turn is equal to 1 nanometre (nm). Further, one nanometre is 10^{-9} metres, or about 10,000 times smaller than the diameter of a human hair. A strand of human DNA is approximately 2.5 nanometres in diameter, while the thickness of human hair ranges from 50,000 to 100,000 nanometres (Fig. 14.1).

3 Classification of Nanomaterials

Nanomaterials can be classified based on dimension in nanoscale which further characterized based on their nature like organic or inorganic composition (Fig. 14.2). Nanomaterial is synthesized from the various materials like polymer, phospholipids, proteins, metals, metal oxides, carbon-based, and quantum dots.

1. Zero-dimensional nanomaterials: These nanomaterials have all three axis (x,y,x) with size of more than 10 nm, e.g. quantum dots, fullerenes, and gold nanoparticles.

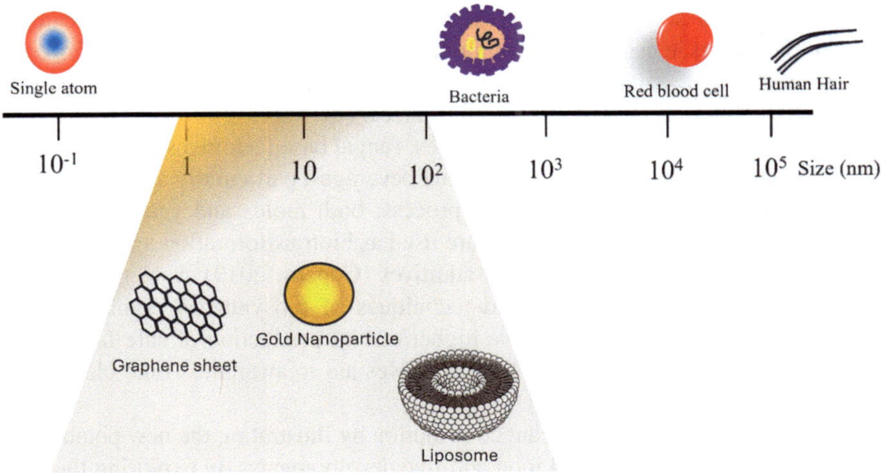

Fig. 14.1 Representation of nano scale size with various example

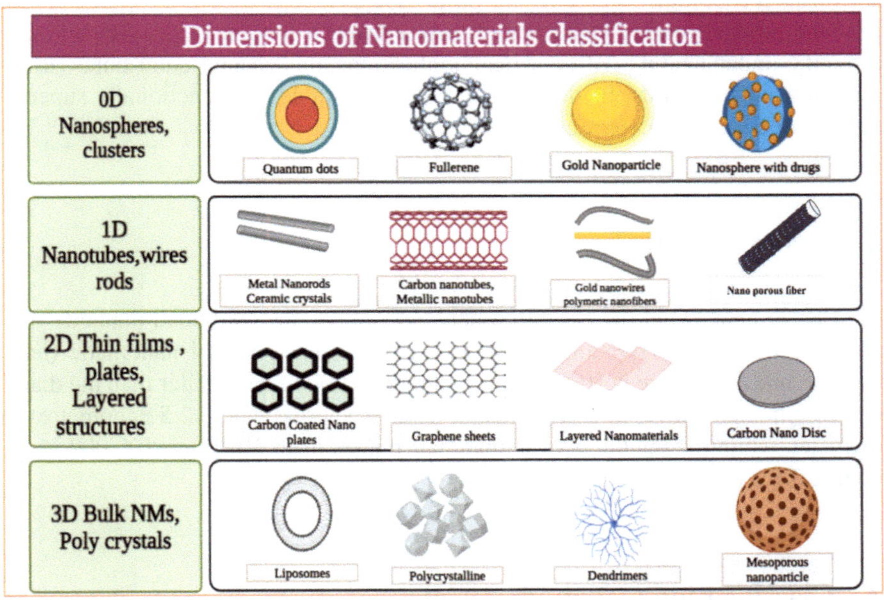

Fig. 14.2 Schematic representation of nanoparticle based on dimension along with example

2. One-dimensional nanomaterials: first and second dimensions (x,y) are within the range of nano size, but third axis dimension are outside of nanostructured. Eg nanofibres, nanotubes, nano rods, thin films and nanowires.
3. Two-dimensional nanomaterials(2D): 2 D nanomaterials have 2 dimensions outside the nanometre range but 1dimension is in the range of nano scale between 1 to 100 nm. It has plate like shape. Eg, Nano sheets or nano walls, nano coating layers and nano disc.

4. Three-dimensional nanomaterials(3D): 3-dimension material is outside of the nano scale range, (>100 nm), but its individual block is made up of nanometre materials. In three-dimensional nanomaterial, in which the 0D, 1D and 2D structural elements are together with each other. 3D nanocomposites form when more than two substance acts together to create unique composite with distinctive properties. Eg. bundles of nanowires and nanotubes and multi-nano layers, liposomes.

4 Manufacturing Methods of Nanomaterials

In general, there are two main procedures followed for manufacturing of nanomaterials ie. bottom-up and top-down methods, as shown in Fig. 14.3 (Mekuye & Abera, 2023).

1. The top-down method is destructive method, which involves a control process by which bulk materials converted into smaller materials, which ultimately transform into nanomaterials. Eg. Lithography, mechanical milling or ball milling, laser ablation, sputtering, electron explosion, arc discharge, and thermal decomposition.
2. Bottom-up method: The bottom-up is integrative method, which involves formation of material from atoms size to clusters and subsequently nanoparticles. Eg. Chemical vapor deposition (CVD), sol to gel formation and biological synthesis.

5 Application of Nanotechnology

In recent year the nanotechnology gained significant attention for its utilization in various industries of food and technology as mentioned in Fig. 14.4. Biosynthesis is an environmentally friendly and green approach to the synthesis of non-toxic and biological nanoparticles. Green synthesis nanoparticles have distinct and enhanced properties that make them suitable for biomedical applications. Microorganisms (bacteria, algae and fungi), biological templates and various plant parts are used in biosynthesis. To understand the application of nanotechnology on fungal-based food additives, fist we must understand the various food additive produced by fungi.

5.1 *Fungal Additives-Based Food Ingredients*

Food additive produced by fungi had been increased over years due to high demand across the globe due to its superior nutritional value and clean label. There is maximum requirement of fungal-based food additive those who belong to vegetarian category for process food or dietary supplements. Fungal-based food additive

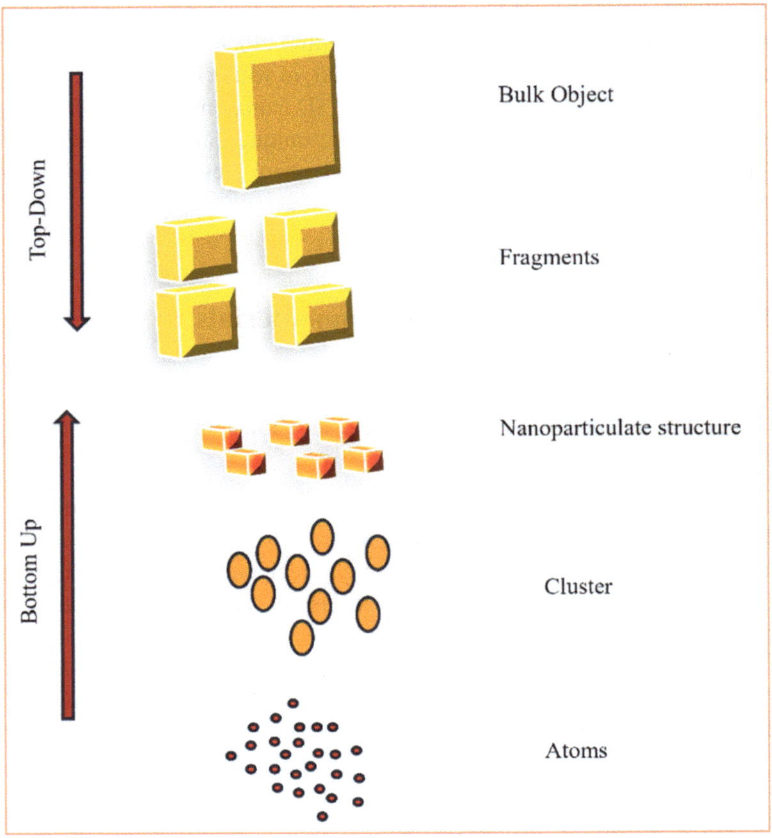

Fig. 14.3 Approach for manufacturing of nanoparticles: Top Down and Bottom Up

provides the clean label for the product (Choi et al., 2024). Food additive are substances added to enhance physical appearances, feel, taste and shelf life. Fungal-based protein can be good substitute of meat due to its filamentous structure mimic similar texture to the protein fibrils present in meat cell matrix. They are rich in protein and amino acid content, increasing dietary fibre intake and free of cholesterol (Derbyshire, 2022). The fungal-based protein sector showed significant growth, reaching a global valuation of $156.6 million in 2020. The market value expected to reach $386.6 million by 2030, representing a compound annual growth rate of 9.5% throughout the decade from 2021 to 2030 (Raju kale 2021). Naturally derived colour are safer in comparison with synthetic derived colours (Singh et al., 2023a, b). Worldwide the preference of food colourant sifted toward naturally obtained food colourants. The demand for naturally derived food colourant was estimated with global market value of USD 2.25 billion in 2025 and it is projected to grow 3.37 billion in 2030 (Mordor Intelligence 2025). Fungi produce vitamins as metabolite in their metabolic pathways. Selective modification of genome of fungi cell increases the production of vitamins safely (Dikkala et al., 2021). In

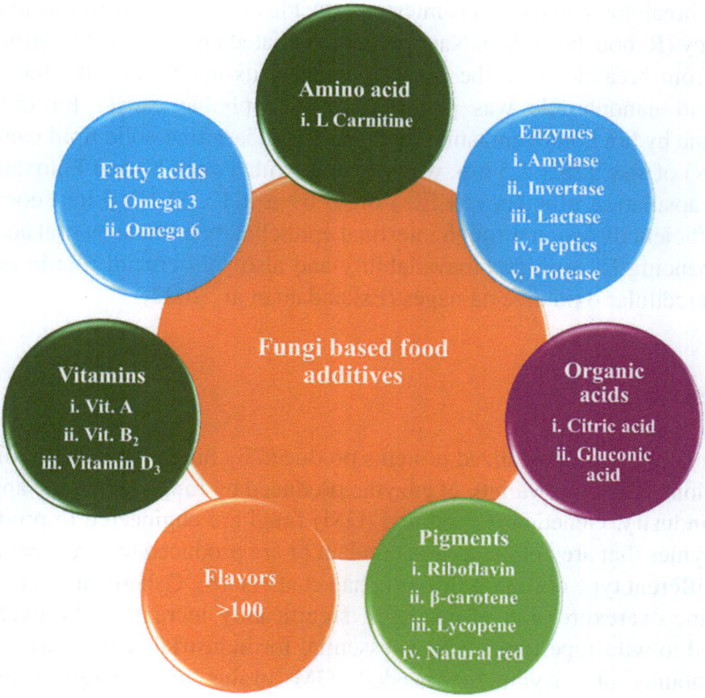

Fig. 14.4 Fungi-based food additives

industrially, the fungi are used through fermentation processes convert substrate into various food products. They are used for industrial scale biosynthesis of enzymes, proteins, organic acids, vitamins, amino acid, pigments and fatty acids (Fig. 14.4).

5.1.1 L-Carnitine

It is amino acid that facilitate fat metabolism converting it into energy. Crotonobetaine is converted into L carnitine through hydration by *Rhizomucor*, *Rhizopus* and *Neurospora* (Tian et al., 2009). few fungal strain such as *Saccharomyces cerevisiae*, *Penicillium*, *Actinomucor*, *Mucor*, *Neurospora*, *Rhizopus* and *Aspergillus* produce L-carnitine from γ-butyrobetaine via hydroxylation (Leung et al., 2010). L carnitine play pivotal role in energy production and metabolism (Longo et al., 2016). Particularly in sperm cells, facilitating the transport of fatty acids into mitochondria for energy production by generating ATP. Also, ATP is play very vital role for motility of sperm (Mateus et al., 2023). L-carnitine encapsulated in nanoparticle improve sperm quality by boosting viability, motility and provide protection from oxidative damages. The innovative nanoparticle-based delivery of L carnitine shows potential for advancing male infertility treatment (Rebouche, 1991). Dietary supplements L carnitine breakdown in the gastrointestinal tract leads to low absorption and reduces its efficacy (Rebouche, 1991). Nanoparticle mediated encapsulated L carnitine protects it from breakdown in the gut and enhance its uptake into the bloodstream. Solid lipid nanoparticle was prepared as hydrophobic carrier for delivery of L-carnitine by hot homogenization method. The L carnitine-solid lipid nanoparticle (LC-SLN) of size 128 ± 1.5 nm, was prepared with Tween 80 and Poloxamer 407® with encapsulation efficiency (EE%) of 69.09 ± 1.12. SLN's Lipid composition enable efficient diffusion through intestinal epithelial membrane on oral administration, enhancing L carnitine bioavailability and also paly crucial role in protecting from intercellular oxidative damages (Eskandani et al., 2022).

5.1.2 Enzymes

Fungal enzymes are specialized proteins produced by fungi that catalyse biochemical reactions. There are variety of enzyme produced by fungi having the application in food industry. Genetically modified (GM) fungi are engineered to produce specific enzymes that are not naturally abundant or are produced in insufficient quantities by different type of fungal strains (Zhao et al., 2018). Genetic modifications can lead to the overexpression of enzymes, significantly increasing the overall yield compared to wild-type fungi. This is essential for industrial-scale processes where large amounts of enzyme are needed. GM strains of *Aspergillus niger* and *Trichoderma reesei* are engineered to produce large quantities of cellulases and other enzymes, improving the efficiency of biomass degradation for biofuel production (Zhao et al., 2018). *Aspergillus niger* and other fungi can be genetically engineered to produce more stable enzymes with higher activity, such as xylanase and cellulase, which are essential for breakdown of biomass into biofuel (Prajapati et al., 2018; Zhao et al., 2018).

Enzyme immobilization in nanoscale was done to prevent loss of enzyme activity and to maintain their sustainability. There are various nano materials like nanoparticles, nanotubes, nanosheets gain attention for their as efficient immobilization matrices of enzyme as biocatalyst over chemical catalyst (Kim et al., 2008; Agustian et al., 2016). Commonly employed methods for enzyme immobilization involve covalent bonding and physical adsorption. There are few examples of enzyme immobilization on nano particles involve, α-amylase was immobilized on magnetic Fe_2O_3 nanoparticles (Fe_2O_3-NPs) to hydrolyse starch, The magnetic nanoparticle linked enzymes retained 94% of their original enzyme activity. The immobilization of enzymes was resistance to inactive, that immobilized enzymes was used for 8 cycle of batch production (Khan et al., 2012). In another study, Lipase immobilized on magnetic Fe_3O_4-chitosan nanoparticle was prepared by in situ co precipitation method. Lipase immobilized nanoparticle used to synthesize ascorbyl palmitate using ascorbic acid and palmitic acid. Lipase immobilization with 75% of immobilization efficiency help more than 50% of conversion of ascorbyl palmitate after 12 h of reaction. After ten cycle, immobilized lipase preserved

70% of its initial activity, this showed the suitable to reuse, easy to recover and enhance thermal stability of enzymes (Wang et al., 2015).

The enzyme invertase catalyse hydrolysis of sucrose into glucose and fructose producing invert sugar and it is used in food industries. This enzyme was immobilized over magnetic nanoparticle modified with silica and glycidyl methacrylate (GMA) by polymerization. The stability of enzyme activities of immobilized nanoparticles exhibited a 2.1 higher activity over the free form of the enzymes. Immobilized invertase lasted 6 weeks and showed a 21% decrease in activity after 12 reuse cycle (Bayramoglu et al., 2017).

Enzyme lactase (galactosidase) is produced from *Aspergillus oryzae* and *Kluyveromyces lactis*. Lactase is used to reduce lactose in whey and fluid milk, for production of oligosaccharide prebiotics in the food industry. Lactase enzymes conjugation on magnetic nanoparticles is obtained through covalently linked by carbodiimide chemistry to carboxylic acid activated magnetic particles. The conjugated lactase enzyme efficiency for lactose hydrolysis increased in comparison with free enzymes. The conjugated lactase enzymes were recovered after five time uses with 78% of activities (Talbert & Goddard, 2013).

A protease is an enzyme that facilitate breaking down proteins into smaller peptides or amino acids and enabling the formation of new protein. Chemically protease work by hydrolysing the peptide bonds within proteins by hydrolysis (Daggett, 2012). Fungal protease nanoparticles hold significant potential for various utilization in the food sector, includes meat tenderization, cheese manufacturing, protein hydrolysis and fruit juice clarification. The encapsulation of proteases in nanoparticles provides numerous advantages, such as increased enzyme stability, controlled release and more efficient use of the enzyme, all of which contribute to the improvement of food quality and the optimization of industrial food processing. These benefits make fungal protease nanoparticles a valuable tool in modern food manufacturing (Cabrera-Barjas et al., 2022).

5.1.3 Organic Acid

Organic acid like citric acid and gluconic acids are manufactured by the microbial fermentation process currently as flavouring agents, buffers, preservatives and technology adjuvants (Copetti, 2019). Citric acid is produced initially from the citrous fruits like lemons and limes. In industries the manufacturing of citric acid is produced by fermentation in bulk quantity. The most common strain for production of citric acid is used as filamentous fungi such as *Aspergillus niger*. There are different microbes utilized other than fungi such as yeast *Yarrowia lipolytica*, *Penicillium citrinum* and *Candida lipolytica* (Copetti, 2019; Wang et al., 2020). Citric acid production enhances from *Aspergillus niger* by using under-utilized cashew apple juice via improved strain and optimized process. The citric acid production achieved by supplementing the medium with 10% sucrose, metal ions and biogenic zinc oxide nanoparticles (ZnONPs)

These nanoparticles are stimulating the maturation of *Aspergillus niger* and enhance the citric acid production through interactions at the cellular level, improving nutrient uptake or enzyme activity. A Supplementation with sucrose metal ion and biogenic ZnONPs, combined with Taguchi Optimization, resulted in Citric acid Yield of 92.61% per day (Adeoye & Lateef, 2022).

Citric acid is used as co-polymer to prepare nano encapsulating agent. The copolymers are formed by linking citric acid, a naturally derived biodegradable material, with polyethylene glycol, a hydrophilic polymer known for its biocompatibility and non-toxicity. The copolymers have potential applications in drug delivery, gene therapy and other areas of nanomedicine. Their structure allows for the encapsulation of hydrophobic drugs within the hydrophilic PEG blocks, enhancing the bioavailability and targeted delivery of therapeutic agents (Naeini et al., 2010).

Nano-coating refers to the application of a fine, uniform layer of nanomaterials on the interface of fruits. This thin layer, composed of nanoparticles, serves as shield that can help preserve the quality of fruits by controlling moisture loss, oxidation and microbial contamination (Thakur et al., 2024). The nano-scale size of the coating allows for better penetration and adherence to the fruit surface, ensuring a more effective barrier against environmental factors that contribute to spoilage. Aloe gel is a natural substance with proven moisturizing, healing, and antimicrobial properties. It helps in reducing water loss and offers protective benefits for fruits (Suriati, 2020). Citric acid prevent oxidation by reducing browning and oxidative damage to fruits (Singh et al., 2022). It also has antimicrobial activity that inhibits the growth of spoilage-causing microorganisms (Su et al., 2014). When combined, aloe gel and citric acid form a biodegradable, eco-friendly coating that is both safe for consumers and effective in retaining the standards of the fruits and vegetables. Nano-coating with aloe gel and citric acid offers a novel and green approach to fruit preservation, providing benefits such as enhanced shelf life, better food quality and reduced environmental impact. This method holds promise for broader use in the food sector and it play a key role in the future of sustainable food packaging and preservation (Sánchez-Machado et al., 2017; Suriati, 2022).

Gluconic acid is an organic acid found in nature and it is derived from the oxidation of glucose. Gluconic acid is particularly manufactured via fermentation of glucose by microorganisms such as *Aspergillus niger*, a type of fungus, this biological process involves the oxidation of glucose yielding gluconic acid with the help of enzymes like glucose dehydrogenase (Roukas, 2000). In glucose oxidation, the activator facilitates the oxidative stress of glucose ($C_6H_{12}O_6$) through gold nanoparticles supported on Al_2O_3 whereby oxygen (O_2), resulting in the conversion of glucose into gluconic acid ($C_6H_{12}O_7$). it is a valuable product in various industries, including food, pharmaceutical and biotechnology. Methanobactin-mediated gold nanoparticle synthesis on alumina (Au/Al_2O_3) is a highly efficient and sustainable catalyst for glucose oxidation, leading to the formation of gluconic acid. This approach provides a green, biotechnological route for producing valuable chemicals like gluconic acid, contributing to cleaner, more sustainable industrial processes (Xin et al., 2014).

5.1.4 Pigments

Colour additives in food are substances used to impart or enhance colour, improve appearance, or make the food more appealing to consumers. These colourants can be either natural or synthetic and they are commonly utilized in the food manufacturing sector (Rana et al., 2021). Natural colours in food have been steadily increasing in recent years. Consumers' growing health consciousness has fuelled a demand for clean and natural ingredient containing products. Natural colours are perceived as healthier and safer, as they are derived from Microbial source, fruits, vegetables and other plant-based sources rather than synthetic chemicals (Barreto et al., 2023).

Microbial-origin pigments are an emerging class of natural colourants derived from microorganisms like bacteria, fungi, algae and yeast. These pigments offers multiple benefit over traditional plant and animal-based pigments, making them a promising alternative in the food sector, cosmeceuticals and pharmaceutical segments (Joshi et al., 2003). The advantages of microbial origin includes higher yield and cost-effectiveness, consistency and reliability, sustainability and environmental benefits, diversity and range of colours, faster production time, scalability and safety and ease of regulatory approval (Sen et al., 2019).

Natural pigments can be significantly sourced from filamentous fungi, which find application in diverse sectors includes food, cosmetics, pharmaceuticals and textiles. These fungi produce multiple range of vibrant colours, many of which are bioactive and offer additional functional benefits. The pigments produced by filamentous fungi can be extracted and used as natural colourants that are more sustainable and often have oxidative stress reducer, bactericidal and anti-inflammatory features (Kalra et al., 2020). The list of various fungi used for various colour production (Table 14.1). As consumer demand for natural and functional ingredients grows, fungal pigments are becoming an increasingly important alternative to traditional plant and animal-based colourants. With their high yield, low cost and environmentally friendly production processes, these pigments are poised to play a crucial role in multiple industrial sector, from food to cosmetics (Takahashi et al., 2020).

There is stability challenge of natural pigment from the external climatic condition like pH, light, moisture, heat and dietary matrices (Kalra et al., 2020). The stability of the pigment can be improved by the encapsulation by nanotechnology. Nano encapsulation involves enclosing the pigment inside nano-sized carriers (like nanoparticles, nano capsules, or nano emulsions), which provides a variety of protective and enhancing effects that can help address the common challenges related to pigment instability (Goiato et al., 2013). The photobleaching of *Monascus* produced pigments a group of natural colourants derived from the fungus *Monascus* is a significant challenge, as these pigments are sensitive to UV light and exposure to light during storage or processing can lead to colour degradation. *Monascus* pigments-encapsulated nanoparticles is prepared from hydroxypropyl cellulose (HPC) and poly (lactic-co-glycolic) acid (PLGA) crossed linked polymer by water-in-oil emulsion solvent diffusion method. HPC and PLGA-based nanoparticles were 67 nm and 155 nm respectively. PLGA nanoencapsulation is helping to protect

Table 14.1 Fungal pigments and their potential applications in food (Afroz Toma et al., 2023)

Fungi	Colour	Pigments	Application	Status
Ashbya gossypi	Yellow	Riboflavin	Food and Beverages	IP
Blakeslea trispora	Orange to yellow	β-carotene	Food colourant; anticancer and antioxidant activities	IP
Blakeslea trispora	Red	Lycopene	Food colourant, anticancer	IP
Cordyceps unilateralis	Deep, blood red	Naphtoquinone	Food colourant; anticancer and antibacterial activities	RP
Fusarium sporotrichioides	Yellow to orange/Red	β-carotene/Lycopene	Food colourants	RP
Fusarium verticillioides	Yellow	Napthoquinone	Antibacterial activity and food colourant	RP
Lactarius sp	Blue	Azulenes	Food and beverages	RP
Mucor circinelloides	Yellow to orange	β-carotene	Food colourant	DS
Monascus purpureus FTC 5357	Red	Monascorubramine	Food colourant	RP
Monascus ruber CCT 3802	Orange, yellow, and red	Monascorubrin	Food colourant	RP
Monascus purpureus	Yellow	Monascin	Food colourant	IP
Monascus sp.	Yellow	Ankaflavin (Azaphilone)	Food colourant, pharmaceutical, and antitumor and anti-inflammatory activities	IP
Monascus sp.	Orange	RubropuntatinA	Anticancer activity and food colourant	IP
Monascus sp.	Red	Monascorubramine	Antioxidant activity and food colourant	IP
Neurospora crassa	Yellow to orange	β-carotene	Food colourant	RP
Penicillium herquei	Yellow/blue	Atrovenetin	Atrovenetin	NK
Penicillium purpurogenum	*Penicillium purpurogenum*	Purpurogenone/Mitorubrin/Azaphilone/ Mitorubrinol	food, antioxidant, and pharmaceuticals	DS
Phycomyces blakesleeanus	*Yellow to orange*	β-carotene	Food colourant	RP
Talaromyces atroroseus	*Red*	Azaphilone	*Talaromyces atroroseus*	DP
Talaromyces sp.	*Red*	N-glutarylmonasc orubramine	*Food colourant*	IP

Note: *DS* development stage, *IP* industrial production, *RP* research project, *NK* not known photobleaching of *Monascus* pigments and improve photostability which is sensitive toward light exposure. The PLGA nanoparticles attribute the sustained release properties of the pigments (Matsuo et al., 2018).

Fusarium verticillioides is filamentous fungus that biosynthesizes natural pigments, which can range in colour and are of interest for their potential use as natural colourants. The optimal growth conditions for *Fusarium verticillioides* to maximize pigment production. Various factors, such as nutrient sources, temperature, pH levels and aeration conditions, are studied to determine the most effective conditions for pigment biosynthesis. The potential of nanoencapsulation as a method to protect and stabilize the pigment for use in various applications. Encapsulation helps to address issues related to pigment degradation due to light exposure, oxidation, or temperature sensitivity. Fusarium verticillioides-derived violet colourant was microencapsulated in a chitosan matrix cross-linked with sodium hydroxide and activated with glutaraldehyde. The natural violet colourant microsphere with chitosan prepared help to improve the stability and modulate the liberation of pigment (Mwaheb et al., 2024).

Fungal Carotenoids are natural ingredient that vary from yellow to orange-red in colour and it act as antioxidants. Carotenogenesis occurred upon exposure to light, particularly blue light (Dzurendova et al., 2022). β-carotene pigment is one of kind of carotenoids which also called as pro-vitamin A (Dufosse et al., 2014). β-carotene is biosynthesizes by few of the reported fungi like *Sclerotium rolfsii* and *Sclerotinia sclerotiorum* and the *ascomycetes Aspergillus giganteus*, *Cercospora nicotianae* and *Penicillium* sp. (Han et al., 2005). beta-carotene is pigment and used as nutraceutical ingredient, chemically it is low solubility, low bioavailable and also show chemical instability. Encapsulation in lipid base matrices like liposomes or nano emulsion, nanoparticle can improve stability and bioavailability (Moraes et al., 2013; Pezeshki et al., 2019). Beta carotene pro-liposomes particle prepared by spray drying mixture of Beta carotene, phospholipid, sucrose and anhydrous ethanol in specific ratio. These pro-liposomes used to prepare the liposomes of size 1500 nm by hydration of the with deionized water. The liposome dispersion is stabilized by using xanthan gum as thickening agent. The degradation of encapsulated beta-carotene is very minimal over period of 60 days upon storage under refrigeration (Moraes et al., 2013).

5.1.5 Vitamins

A vitamin is nutrients that is essential for normal growth and metabolism in living organisms. These essential compounds are needed in limited amounts by body because they are not produced in sufficient quantities, or at all, by the body itself. Vitamins are crucial for various physiological functions, such as immune support, energy production and cell repair (Combs Jr & McClung, 2016). Plant-based vitamin are being produced in very less quantity, hence there is need of vitamin production from different source. Microbial sources for vitamin production offer superior advantages over plant-based sources in terms of yield, cost-efficiency, scalability and consistency. The ability to genetically engineer microorganisms for enhanced production further enhances their potential in industrial applications. The customization of fungal genome cell factories produces vitamins with bio safe nature (Dikkala et al., 2021). Vitamin B1, also familiar as thiamine, is a hydrophilic in nature that support in retaining the health of nervous system, heart and muscles. Thiamine is involved in processes that help the body generate adenosine triphosphate (ATP), which is the primitive energy carrier at cellular level (Combs Jr & McClung, 2016). Vitamin B_1 (thiamine) is synthesized exogenous by *Aspergillus oryzae* due to its specific gene (Kubodera et al., 2003).

The nanoencapsulation of vitamin A was prepared to protect degradation in foodstuff by formation of solid lipid nanoparticle (SLN) and nano lipid carriers (NLCs). This nanoparticles with vitamin A are capable to protect breakdown upon storage and metabolism in stomach, to increase its oral bio accessibility. Nano encapsulation of vitamin A was fabricated by solvent free sonication method by using the total based of 195 mg of lipid in solid lipid Gelucire (GEL) and stearic acid (SA): liquid lipid mygliol (MYG) and oleic acid (OA) in the ratio of 3.3:1. The lipid phased contains Gelucire® (GEL) and Stearic acid melted above their melting points followed by addition of vitamin A. The tween 80 was added to the melted lipid and homogenized with a probe sonicator to prepare a nanoparticle suspension. The prepared nanoparticle were stable to temperature upto 70 °C for 15 min and also stable in different simulated gastric media (Resende et al., 2020). Vitamin B_2 (riboflavin), is another hydrophilic vitamin that is responsible for energy generation and several cellular activities. It helps to break down saccharides, fats and peptides into energy (Combs Jr & McClung, 2016). Riboflavin is also crucial for the maintenance of healthy skin, eyes and nervous system. *A. gossypii* is the main filamentous fungi that produces the vitamin B_2 in massive scale bioconversion. Vitamin B_2 is also being produced by yeast like *Candida guilliermondii* or *Debaryomyces subglobosus* (Copetti, 2019). Riboflavin is sensitive to light and can degrade, losing its effectiveness and potency when exposed to ultraviolet (UV) light, The degradation of riboflavin primarily occurs through exposure to light, heat, and oxidative environments (Sheraz et al., 2014). Encapsulation of vitamin B_2 protect from the external factors as humidity, air, oxygen, light and pH, enhancing the stability of the products. Riboflavin nano capsules using alginate and chitosan as wall material is developed by Nano Spray Dryer technology. Alginate-chitosan-riboflavin nanocapsules were in size range of 500–700 nm. The nanoencapsulation aid to rapid release vitamin B_2 (Mancilla Dávalos, 2014). Vitamin B_2 Nanoparticle are obtained with chitosan-coated poly (lactic-*co*-glycolic acid) (CS-PLGA(B_2) NP) by single emulsion solvent evaporation. Vitamin B_2 nanoparticle enhance the bioavailability for food fortification and nutraceutical applications. The prepared nanoparticles are of 200 nm display better stability. CS-PLGA(B_2) NP exhibited better photostability from UV light (Sathiensathaporn et al., 2024).

Vitamin D is a hydrophobic vitamin that responsible for healthy bones and supporting immunological activities in the body. It is unique because it can be synthesized by the body when the skin is in contact with sunlight, which is why it's often referred the "sunshine vitamin." Vitamin D helps the body to uptake calcium and phosphorus from meal, which are essential for supporting and retaining strong, healthy skeletal element and dental structure. It also helps in the regulation of calcium levels in the blood (OFFERS, 2017).

Cladonia rangiferina (reindeer lichen) produce a vitamin D precursor 7-dehydrocholesterol when exposed to ultraviolet (UV) light of the range of 282–310 nm and then, through a process of photochemical conversion, they produce vitamin D_3 (cholecalciferol) (Göring, 2018). *Saccharomyces cerevisiae* seems to be alternative of the production of cholecalciferol (Kessi-Pérez et al., 2022). Vitamin D_2 (ergocalciferol) and vitamin D_3 (cholecalciferol) are sensitive to heat, light, especially UV radiation. UV light can degrade vitamin D by breaking down its molecular structure, reducing its bioavailability (Grady & Thakker, 1980). Vitamin D_3 nanoparticle was developed with whey protein (WPI) by different calcium concentration and aggregation pH. The encapsulation of vitamin D_3 lead to delay in degradation of vitamin D_3 in presence of oxygen (Abbasi et al., 2014).

Vitamin C (VC) is very sensitive to get oxidated under light or heat and moisture. Liposomes containing VC were synthesized via the rapid expansion of supercritical solution (RESS) process. Vitamin C is loaded against phosphatidylcholine was successfully achieved till 75.38 ± 1.03%. Optimal Vitamin C loading was achieved under conditions of 25 MPa pressure and 48 °C temperature, with a feed ratio of 0.25 (Vitamin C to phosphatidylcholine) and resulting liposome size of 270.4 ± 5.2 nm. The other parametre like poly dispersity index and Zeta-potential was 0.254 ± 0.010 and −41.7 ± 0.9 mv, respectively (Jiao et al., 2019). In another study vitamin C stability was achieved by nano encapsulation technique. The vitamin C (VC) encapsulation was prepared by electrostatic interaction with glycidyl trimethylammonium chloride (GTMAC)-chitosan (GCh) subsequently cross-linking with phosphorylated-cellulose nanocrystals (PCNC) to form VC-GCh-PCNC nano capsule. Both VC loaded, phosphorylated-cellulose nanocrystals and Vitamin C-GTMAC-Chitosan with Sodium tripolyphosphate nano capsules in the range of 450 ± 8 and 428 ± 6 nm respectively with spherical shape. Encapsulation efficiency for VC-GCh-PCNC nanocrystal have 90.3 ± 0.42% and VC-GCh-TPP nanocapsules have 57.3 ± 0.28% in neutral pH. VC-GCh-PCNC nanoencapsulation showed slow release around 18% release over 14 days. PCNC nano crystal provided a better cross-linking potential with GCh creating a stable nano capsule that protect vitamin C from breakdown during extended storage (Baek et al., 2021).

5.1.6 Polyunsaturated Fatty Acid (PUFAs)

Polyunsaturated fatty acid contain more multiple double bond in their chemical structure, these are commonly referred to as vital fatty acids (Tarantino & Finelli, 2016). The α-linolenic acid (ALA) is a precursor of the longer chain ω-3 fatty acids eicosapentaenoic acid (EPA) and docosahexaenoic acid (DHA). linolenic acid (LA) is a progenitor of the longer chain ω-6 fatty acids gamma-linolenic acid (GLA), dihomogamma-linolenic acid (DGLA) and arachidonic acid (AA) (Papandreou, 2014). Fungi, like *P. irregulare*, *Aurantiochytrium* sp., *Mortierella alpine*, *Y.*

lipolytica, A. limacinum, S. limacinum, C. guilliermondii, generally are able to form longer polyunsaturated fatty acids like ω-3 fatty acid (Qin et al., 2023). ω-6 Arachidonic acid (ARA) are commercialized using the oleaginous fungus *Mortierella alpina* (Ochsenreither et al., 2016). EPA and DHA, the two primary ω-3 fatty acids, are essential for multiple health benefits, particularly for heart health, eye health, brain function, inflammatory control, skin health, and cognitive development (Swanson et al., 2012; Remila et al., 2021). ω-6 fatty acids are important fats that provide multiple benefits, such as supporting cell structure, brain function, immune health, hormonal balance, and skin health (Innes & Calder, 2018).

The stability of ω-3 and ω-6 fatty acids is a key challenge due to their high susceptibility to get oxidize. Oxidation can cause low of nutritional value, harmful compounds and unpleasant Flavours. To preserve the stability of these essential fatty acids, strategies such as using antioxidants, encapsulation, proper storage and careful handling are necessary. External factor like lipid self-oxidation, sunlight, metals, and heat have the impact on stability of the fatty acid (Nogueira et al., 2019). Corn starch with high content with amylose was utilized to synthesize nanosized complex with rich of ω-3 flax seed oil. This complex was converted in to powder by spray drying. ω-3 nano capsule is incorporated in bread formulation. Due to nano formulation, the sensory profile of the final food product did not change. It was demonstrated the that nanoencapsulation improve stability of ω-3 fatty acid by reducing thermal oxidation during baking and reduces the unpleasant taste in comparison with free oil (Gökmen et al., 2011). Various study has been done on nano formulation ω-3 poly unsaturated fatty acid containing in vitro or animal in the area of cardio vascular disease and cancer (Serini et al., 2019).

Arachidonic acid is an ω-6 polyunsaturated fatty acid (PUFA). Arachidonic acid-encapsulated microcapsules using coaxial electrospray offers a promising solution to enhance the stability, bioavailability and regulated release of this ω-6-fatty acid. By encapsulating arachidonic acid in a core-shell structure, the process helps safeguard it from oxidative damage, making it more effective and useful in various applications (Hu et al., 2021). The nanoencapsulation of eicosapentaenoic acid (EPA) and docosahexaenoic acid (DHA) in 6:1 ratio supports endothelial formation of nitric acid (NO) followed by it showed as anti-aggregatory effects. Nano emulsion was prepared by emulsification of ω-3 fatty acid (EPA:DHA in the ratio of 6:1) into solution containing surfactant like tween 80, span 80 and lecithin followed by mixing with gelatin and gummi arabicum solution (Remila et al., 2021).

Gamma linolenic acid is a type of ω6 polyunsaturated fatty acid (PUFA). The GLA- liposomes was prepared by ethanol injection method and homogenization method followed by added to 22% w/v poloxamer hydrogel. The prepared liposomes have particle size of 104.2 ± 2.15 nm, polydispersity index of 0.234 ± 0.0019 with encapsulation efficiency of 85.17 ± 2.05%. The release of GLA from the liposomes showed over 120 h *in vitro*. GLA liposomes showed anti-tumour activities in comparison to free GLA and liposomes with rat having solid tumour (Saad et al., 2023).

5.1.7 Flavours

Food flavour, comprised of volatile compounds, significantly influence human sensory perception and are widely used in diverse food products. It is a complex combination of multiple factors that involve taste, smell and mouthfeel (texture) (Premjit et al., 2022). More than 100 flavour are being produced by industries through fermentation process by using specific strain of microbes specifically fungi. As it is produced by fungi the same is classified as "natural flavour" (Copetti, 2019). Various strains of fungi are used to produces various flavour through fermentation process (Table 14.2). Flavour are sensitive to pH, heat and prone for oxidation, they required suitable means of protection so that they retain their nature for long time (Kim et al., 2019). Encapsulation of flavour enable stability bioavailability, reducing the impact of volatility and enabling a more sustained release profile in food products (Premjit et al., 2022).

Nanoencapsulation is a technique used to encapsulate bioactive compounds, such as flavours, in nanometre-sized carriers to protect, enhance, or control the release of these compounds in food. The process helps to strengthen the stability, solubility, bioavailability and sensory properties of food ingredients, particularly volatile substances like flavours that are prone to degradation or evaporation.

Below are some commonly used nanoencapsulation methods for flavour delivery in food (Burgos et al., 2017).

(a) Coacervation
(b) Spray drying
(c) Nanoprecipitation
(d) Nanoprecipitation
(e) Emulsification methods
(f) Emulsification methods
(g) Electrospray and electrospinning
(h) Ionic gelation

Table 14.2 List of flavour with fungi species

Flavour	Fungi strain	Flavour characteristics	Reference
2-alkanones, aldehydes and alkanols	*Penicillium roqueforti*	Blue cheese flavour	Bothast and Smiley (1977)
benzaldehyde	*Ischnoderma* spp.	Bitter almond flavour	Copetti (2019)
2-phenylethanol	*Saccharomyces* spp. *or Kluyveromyces* spp. *yeasts*	Alcohol with a roselike Odor	
Cinnamic acid esters	*Basidiomycetes*	Exotic fruit	
Cinnamaldehyde		Cinnamon	
Eugenol		Clove	
Methyl benzoate		Dry fruit	
Benzyl acetate		Jasmine	
4-alkanolide and 5-alkanolide, lactones	*C. lipolytica and Pichia ohmeri*	fruity, nutty, and fatty aroma	

Nano emulsion of clove essential oil (EO) and Oregano EO of diametre 180 to 250 nm embedded in methylcellulose matrix which help their preservative efficacy against microbes like yeast and Molds in bread for period of 15 days (Otoni et al., 2014). Eugenol encapsulation is done in nanoscale with diametre of 80–100 nm with chitosan, the same nanoparticle introduced into thermoplastic flour (TPF) by utilizing hot melt extrusion above 150 °C. Nanoparticle embedded thermoplastic flavour materials exhibits reduce oxidation activity and high-water vapor transmission rates. The possible use of eugenol-loaded chitosan nanoparticles as antioxidants in bioactive plastics for food packaging (Woranuch & Yoksan, 2013). Flavour nanoencapsulation in active food packaging system is an innovative approach designed to enhance the longevity, sustained flavour release and overall merits of food products.

6 Conclusions and Way Forward

The integration of nanotechnology with fungal-derived food additives represents a transformative advancement in food science, offering innovative solutions to existing challenges in additive functionality and application. This chapter has systematically demonstrated that nanoengineered fungal products exhibit superior stability, bioavailability and performance compared to conventional additives. Notably, nanoencapsulation techniques have proven particularly effective in enhancing the delivery and preservation of fungal metabolites, including enzymes, vitamins, pigments and fatty acids while overcoming the inherent limitations of using natural additives. Fungal-based nano additives presented compelling evidences for sustainable alternatives to synthetic ingredients, which aligns with growing consumer preferences towards clean-label products. Their nutritional superiority, coupled with the cost-effectiveness of fungal fermentation processes, highlight them as viable solutions for diverse food applications. Key technological breakthroughs, such as nanoparticle-immobilized enzymes with improved reusability, nano coatings for pigment stabilization and nano formulations for enhanced nutrient delivery validate the progress in this field. These innovations can potentially being translated into industrial applications, demonstrating scalability for commercial production, compatibility with existing processing systems and novel applications in active packaging and food preservation.

Looking forward, the convergence of fungal biotechnology and nanotechnology holds significant promise for addressing global needs in food innovation. The expanding market for natural ingredients will likely drive further development of these technologies, particularly in functional foods and nutraceuticals. However, realizing this potential requires continued research to optimize nano-delivery systems, rigorous safety evaluations and the establishment of robust regulatory frameworks. As the field evolves, maintaining a dual focus on technological advancement and responsible implementation, prioritizing consumer safety, regulatory compliance and public acceptance will be paramount. Ultimately, this interdisciplinary approach shall not only advance food science but will also contribute to a more sustainable and health-conscious food system.

References

Abbasi, A., et al. (2014). Stability of vitamin D3 encapsulated in nanoparticles of whey protein isolate. *Food Chemistry, 143*, 379–383.

Adeoye, A. O., & Lateef, A. (2022). Improving the yield of citric acid through valorization of cashew apple juice by Aspergillus niger: Mutation, nanoparticles supplementation and Taguchi technique. *Waste and Biomass Valorization, 13*(4), 2195–2206.

Afroz Toma, M., et al. (2023). Fungal pigments: Carotenoids, riboflavin, and polyketides with diverse applications. *Journal of Fungi, 9*(4), 454.

Agustian, J., et al. (2016). Enantio-conversion and-selectivity of racemic atenolol kinetic resolution using free Pseudomonas fluorescens lipase (Amano) conducted via transesterification reaction. *RSC Advances, 6*(31), 26077–26085.

Anandharamakrishnan, C., & Parthasarathi, S. (2019). *Food nanotechnology: Principles and applications*. CRC Press.

Ashfaq, A., et al. (2022). Application of nanotechnology in food packaging: Pros and Cons. *Journal of Agriculture and Food Research, 7*, 100270.

Baek, J., et al. (2021). Encapsulation and controlled release of vitamin C in modified cellulose nanocrystal/chitosan nanocapsules, *4*, 215–223.

Barreto, J. V. D. O., et al. (2023). Microbial pigments: major groups and industrial applications. *Microorganisms, 11*(12), 2920.

Bayramoglu, G., et al. (2017). Improvement stability and performance of invertase via immobilization on to silanized and polymer brush grafted magnetic nanoparticles. *Food Chemistry, 221*, 1442–1450.

Blekas, G. (2016). *Food additives: Classification, uses and regulation*. Academic/Elsevier.

Bothast, R., & Smiley, K. (1977). *Metabolites of Fungi used in food processing*.

Burgos, N., et al. (2017). *Nanoencapsulation of flavor and aromas in food packaging. Food Packaging* (pp. 567–601). Elsevier.

Cabrera-Barjas, G., et al. (2022). *Food biotechnology: Innovations and challenges. Future Foods* (pp. 697–719). Elsevier.

Choi, D., et al. (2024). Comprehensive review of clean-label antimicrobials used in dairy products, *23*(1), e13263.

Combs, G. F., Jr., & McClung, J. P. (2016). *The vitamins: fundamental aspects in nutrition and health*. Academic.

Copetti, M. V. (2019). Fungi as industrial producers of food ingredients. *Current Opinion in Food Science, 25*, 52–56.

Daggett, V. (2012). *The folding of proteins and nucleic acids*.

Derbyshire, E. (2022). Fungal-derived mycoprotein and health across the lifespan: A narrative review. *Journal of Fungi, 8*(7), 653.

Dikkala, P. K., et al. (2021). Fungal production of vitamins and their food industrial applications. In *Industrially important fungi for sustainable development: Volume 2: Bioprospecting for biomolecules* (pp. 563–596).

Dufosse, L., et al. (2014). Filamentous fungi are large-scale producers of pigments and colorants for the food industry. *Current Opinion in Biotechnology, 26*, 56–61.

Dzurendova, S., et al. (2022). Mucoromycota fungi as powerful cell factories for modern biorefinery. *Applied Microbiology and Biotechnology*, 1–15.

Eskandani, M., et al. (2022). Protective effect of l-carnitine-loaded solid lipid nanoparticles against H2O2-induced genotoxicity and apoptosis, *212*, 112365.

Fuertes, G., et al. (2016). Intelligent packaging systems: Sensors and nanosensors to monitor food quality and safety. *Journal of Sensors, 2016*(1), 4046061.

Ghosh, S., et al. (2025). Nanotechnology in the manufacturing of sustainable food packaging: A review. *Discover Nano, 20*(1), 36.

Goiato, M. C., et al. (2013). Effect of intrinsic nanoparticle pigmentation on the color stability of denture base acrylic resins. *The Journal of Prosthetic Dentistry, 110*(2), 101–106.

Gökmen, V., et al. (2011). Development of functional bread containing nanoencapsulated omega-3 fatty acids. *Journal of Food Engineering, 105*(4), 585–591.

Göring, H. (2018). Vitamin D in nature: A product of synthesis and/or degradation of cell membrane components. *Biochemistry (Moscow), 83*, 1350–1357.

Grady, L., & Thakker, K. (1980). Stability of solid drugs: degradation of ergocalciferol (vitamin D2) and cholecalciferol (vitamin D3) at high humidities and elevated temperatures. *Journal of Pharmaceutical Sciences, 69*(9), 1099–1102.

Han, J., et al. (2005). Effect of oxidative stress and exogenous β-carotene on sclerotial differentiation and carotenoid yield of Penicillium sp. PT95. *Letters in Applied Microbiology, 40*(6), 412–417.

Hu, M. X., et al. (2021). Arachidonic acid-encapsulated microcapsules with core-shell structure prepared by coaxial electrospray. *Journal of Applied Polymer Science, 138*(19), 50403.

Innes, J. K., & Calder, P. C. (2018). Omega-6 fatty acids and inflammation. *Prostaglandins, Leukotrienes and Essential Fatty Acids, 132*, 41–48.

Jiao, Z., et al. (2019). Preparation of vitamin C liposomes by rapid expansion of supercritical solution process: Experiments and optimization, *51*, 1–6.

Joshi, V., et al. (2003). "Microbial pigments."

Kalra, R., et al. (2020). Fungi as a potential source of pigments: Harnessing filamentous fungi. *Frontiers in Chemistry, 8*, 369.

Kessi-Pérez, E. I., et al. (2022). Yeast as a biological platform for vitamin D production: A promising alternative to help reduce vitamin D deficiency in humans. *Yeast, 39*(9), 482–492.

Khan, M. J., et al. (2012). Immobilization of porcine pancreatic α-amylase on magnetic Fe 2 O 3 nanoparticles: Applications to the hydrolysis of starch. *Biotechnology and Bioprocess Engineering, 17*, 377–384.

Kim, J., et al. (2008). Nanobiocatalysis and its potential applications. *Trends in Biotechnology, 26*(11), 639–646.

Kim, E. H.-J., et al. (2019). Dynamic flavor perception of encapsulated flavors in a soft chewable matrix. *Food Research International, 123*, 241–250.

Kubodera, T., et al. (2003). Thiamine-regulated gene expression of Aspergillus oryzae thiA requires splicing of the intron containing a riboswitch-like domain in the 5′-UTR. *FEBS Letters, 555*(3), 516–520.

Leung, I. K., et al. (2010). Structural and mechanistic studies on γ-butyrobetaine hydroxylase. *Chemistry & Biology, 17*(12), 1316–1324.

Li, D., & Xue, R. (2025). Nanostructured materials for smart food packaging: Integrating preservation and antimicrobial properties. *Alexandria Engineering Journal, 124*, 446–461.

Longo, N., et al. (2016). Carnitine transport and fatty acid oxidation. *Biochimica et Biophysica Acta (BBA)-Molecular Cell Research, 1863*(10), 2422–2435.

Mancilla Dávalos, V. (2014). *Nanoencapsulación de riboflavina en matrices poliméricas biodegradables empleando la técnica de secado por aspersión*.

Mateus, F. G., et al. (2023). L-carnitine and male fertility: Is supplementation beneficial? *Journal of Clinical Medicine, 12*(18), 5796.

Matsuo, T., et al. (2018). *Improvement of photobleaching of Monascus pigments using novel nanoencapsulation technology*, 琉球大学.

Mekuye, B., & Abera, B. (2023). Nanomaterials: An overview of synthesis, classification, characterization, and applications. *Nano Select, 4*(8), 486–501.

Moraes, M., et al. (2013). Liposomes encapsulating beta-carotene produced by the proliposomes method: Characterisation and shelf life of powders and phospholipid vesicles. *International Journal of Food Science & Technology, 48*(2), 274–282.

Mwaheb, M. A., et al. (2024). Fusarium verticillioides pigment: Production, response surface optimization, gamma irradiation and encapsulation studies. *BMC Biotechnology, 24*(1), 84.

Mwale, M. M. (2023). Food additives: Recent trends in the food sector. *Health Risks of Food Additives-Recent Developments and Trends in Food Sector.*

Naeini, A. T., et al. (2010). Poly (citric acid)-block-poly (ethylene glycol) copolymers – New biocompatible hybrid materials for nanomedicine. *Nanomedicine: Nanotechnology, Biology and Medicine, 6*(4), 556–562.

Nogueira, M. S., et al. (2019). Oxidation products from omega-3 and omega-6 fatty acids during a simulated shelf life of edible oils. *LWT, 101*, 113–122.

Ochsenreither, K., et al. (2016). Production strategies and applications of microbial single cell oils. *Frontiers in Microbiology, 7*, 1539.

OFFERS, J. (2017). *Vitamin D–fact sheet for health professionals.*

Otoni, C. G., et al. (2014). Edible films from methylcellulose and nanoemulsions of clove bud (Syzygium aromaticum) and oregano (Origanum vulgare) essential oils as shelf life extenders for sliced bread. *Journal of Agricultural and Food Chemistry, 62*(22), 5214–5219.

Papandreou, C. (2014). Polyunsaturated fatty acids in relation to sleep quality and depression in obstructive sleep apnea hypopnea syndrome. In *Omega-3 fatty acids in brain and neurological health* (pp. 337–347). Elsevier.

Pathakoti, K., et al. (2017). Nanostructures: Current uses and future applications in food science. *Journal of Food and Drug Analysis, 25*(2), 245–253.

Patra, J. K., et al. (2018). Application of nanotechnology in food science and food microbiology. *Frontiers in Microbiology, 9*, 714.

Pezeshki, A., et al. (2019). Nanostructured lipid carriers as a favorable delivery system for β-carotene. *Food Bioscience, 27*, 11–17.

Prajapati, A. S., et al. (2018). Review on cellulase and xylanase engineering for biofuel production, *14*(1), 38–44.

Premjit, Y., et al. (2022). Current trends in flavor encapsulation: A comprehensive review of emerging encapsulation techniques, flavour release, and mathematical modelling. *Food Research International, 151*, 110879.

Qin, J., et al. (2023). Biotechnological production of omega-3 fatty acids: Current status and future perspectives. *Frontiers in Microbiology, 14*, 1280296.

Rafique, M., et al. (2020). History and fundamentals of nanoscience and nanotechnology. In *Nanotechnology and photocatalysis for environmental applications* (pp. 1–25). Elsevier.

Rana, B., et al. (2021). The realm of microbial pigments in the food color market. *Frontiers in Sustainable Food Systems, 5*, 603892.

Rebouche, C. J. (1991). Quantitative estimation of absorption and degradation of a carnitine supplement by human adults. *Metabolism, 40*(12), 1305–1310.

Remila, L., et al. (2021). Nanoencapsulation of the omega-3 EPA: DHA 6: 1 formulation enhances and sustains NO-mediated endothelium-dependent relaxations in coronary artery rings and NO formation in endothelial cells. *Journal of Functional Foods, 87*, 104851.

Resende, D., et al. (2020). Nanoencapsulation approaches for oral delivery of vitamin A, *193*, 111121.

Roukas, T. (2000). Citric and gluconic acid production from fig by Aspergillus niger using solid-state fermentation. *Journal of Industrial Microbiology and Biotechnology, 25*(6), 298–304.

Saad, S., et al. (2023). Novel gamma linoleic acid encased in situ lipogel for augmented anti-tumor efficacy against solid tumor: In vitro and in vivo evaluation, *87*, 104768.

Sahoo, M., et al. (2021). Nanotechnology: Current applications and future scope in food. *Food Frontiers, 2*(1), 3–22.

Salgado, P. R., et al. (2021). Recent developments in smart food packaging focused on biobased and biodegradable polymers, *5*, 630393.

Sánchez-Machado, D. I., et al. (2017). Aloe vera: Ancient knowledge with new frontiers. *Trends in Food Science & Technology, 61*, 94–102.

Sathiensathaporn, S., et al. (2024). Nanoencapsulation of vitamin B2 using chitosan-modified poly (lactic-co-glycolic acid) nanoparticles: Synthesis, characterization, and in vitro studies on simulated gastrointestinal stability and delivery. *Journal of Food Science*.

Sen, T., et al. (2019). Microbial pigments in the food industry – Challenges and the way forward. *Frontiers in Nutrition, 6*, 7.

Serini, S., et al. (2019). Nanomedicine-based formulations containing ω-3 polyunsaturated fatty acids: Potential application in cardiovascular and neoplastic diseases. *International Journal of Nanomedicine*, 2809–2828.

Sheraz, M. A., et al. (2014). Photo, thermal and chemical degradation of riboflavin. *Beilstein Journal of Organic Chemistry, 10*(1), 1999–2012.

Singh, T., et al. (2017). Application of nanotechnology in food science: Perception and overview. *Frontiers in Microbiology, 8*, 1501.

Singh, S. K., et al. (2022). Citric acid, antioxidant effects in health. In *Antioxidants effects in health* (pp. 309–322). Elsevier.

Singh, R., et al. (2023a). Future of nanotechnology in food industry: Challenges in processing, packaging, and food safety. *Global Challenges, 7*(4), 2200209.

Singh, T., et al. (2023b). Natural bio-colorant and pigments: Sources and applications in food processing, *12*, 100628.

Sowah, D., et al. (2017). Vitamin D levels and deficiency with different occupations: A systematic review. *BMC Public Health, 17*, 1–25.

Su, L.-C., et al. (2014). Study on the antimicrobial properties of citrate-based biodegradable polymers, *2*, 23.

Suriati, L. (2020). Edible coating base on aloe gel with additives on strawberry fruit. *SEAS (Sustainable Environment Agricultural Science), 4*(1), 18–25.

Suriati, L. (2022). Nano coating of aloe-gel incorporation additives to maintain the quality of freshly cut fruits. *Frontiers in Sustainable Food Systems, 6*, 914254.

Swanson, D., et al. (2012). Omega-3 fatty acids EPA and DHA: Health benefits throughout life. *Advances in Nutrition, 3*(1), 1–7.

Takahashi, J. A., et al. (2020). Use of the versatility of fungal metabolism to meet modern demands for healthy aging, functional foods, and sustainability. *Journal of Fungi, 6*(4), 223.

Talbert, J. N., & Goddard, J. M. (2013). Characterization of lactase-conjugated magnetic nanoparticles. *Process Biochemistry, 48*(4), 656–662.

Tarantino, G., & Finelli, C. (2016). Lipids, Low-Grade Chronic Inflammation and NAFLD: A Ménage À Trois? In *Handbook of lipids in human function* (pp. 731–759). Elsevier.

Thakur, D., et al. (2024). Nanoemulsion edible coating for shelf-life improvement and quality control in perishable products. 100114.

Tian, J., et al. (2009). A novel strategy to improve the bioconversion of l-carnitine from crotonobetaine. *European Food Research and Technology, 229*, 721–724.

Wang, X.-Y., et al. (2015). Preparation Fe3O4@ chitosan magnetic particles for covalent immobilization of lipase from Thermomyces lanuginosus. *International Journal of Biological Macromolecules, 75*, 44–50.

Wang, B., et al. (2020). A novel green production process of citric acid on the pilot scale by directly recycling its extraction effluent. *Journal of Cleaner Production, 277*, 124068.

Woranuch, S., & Yoksan, R. (2013). Eugenol-loaded chitosan nanoparticles: I. Thermal stability improvement of eugenol through encapsulation. *Carbohydrate Polymers, 96*(2), 578–585.

Xin, J.-Y., et al. (2014). Methanobactin-mediated synthesis of gold nanoparticles supported over Al2O3 toward an efficient catalyst for glucose oxidation. *International Journal of Molecular Sciences, 15*(12), 21603–21620.

Yildirim, S., et al. (2018). Active packaging applications for food. *Comprehensive Reviews in Food Science and Food Safety, 17*(1), 165–199.

Zhao, C., et al. (2018). Mixed culture of recombinant Trichoderma reesei and Aspergillus niger for cellulase production to increase the cellulose degrading capability. *Biomass and Bioenergy, 112*, 93–98.

Chapter 15
Fungal Enzymes: Latest Developments in Production and Applications in Industry

Pirthi Pal Singh, Purushottam Nagar, Snigdha Chakraborty, Divakar Jaiswar, and Suresh Kumar Ravada

1 Introduction

Enzymes are nature's catalysts—biomolecules that accelerate biomolecular reaction within living organism. These protein molecules, and occasionally RNA molecules, are critical for biological process, ranging from digestion and metabolism to cellular communication. By catalyzing chemical reactions, enzymes enable these processes to occur more rapidly and efficiently, typically under mild conditions like ambient temperature and neutral pH, which would otherwise require harsh chemicals or extreme conditions in traditional chemical reactions (Wolfenden & Snider, 2001). This characteristic has made enzymes indispensable in a broad array of industries, ranging from pharmaceutical production to food processing. The importance of enzymes in industry is undeniable. In various sectors, they have replaced or enhanced traditional chemical processes, offering solutions that are not only more sustainable but also more efficient. The ever-growing demand for enzymatic processes in industrial applications is attributed to several factors where enzymes provide major benefits compared to traditional chemical methods, which include reduction in energy consumption, more selective reactions, and fewer by-products, making them a more sustainable option. Moreover, enzymes can operate in milder conditions, which not only improves energy efficiency but also leads to fewer negative environmental impacts. These advantages have made enzymes a central part of the shift toward green chemistry, which emphasizes the use of natural processes to create products with minimal waste and environmental harm (Aehle, 2007).

P. P. Singh (✉) · P. Nagar · S. Chakraborty · D. Jaiswar
Tirupati Innovation Centre (Tirupati Wellness Pvt Ltd.),
Paonta Sahib, Himachal Pradesh, India
e-mail: pirthipalsingh@yahoo.com

S. K. Ravada
Andhra University, Visakhapatnam, Andhra Pradesh, India

B. P. Singh et al. (eds.), *Fungal Additives and Bioactives in Food Processing Industries*, Fungal Biology, https://doi.org/10.1007/978-3-032-04519-5_15

Among the various types of enzymes available, fungal enzymes have gained significant prominence in recent years, especially in industries like pharmaceuticals and food processing. Traditionally, enzymes were sourced from plants and animals, such as rennet from animals for cheese-making or amylases from plants for starch breakdown. However, fungal enzymes offer several unique benefits that make them highly sought after in industrial applications (Dhevagi et al., 2021). One of the prime reasons fungal enzymes are recognized as increasingly popular is due to their ability to function under a diverse condition. Several fungal species produce enzymes that are not only effective at neutral pH but can also thrive in slightly acidic or alkaline environments, making them versatile for various industrial processes. Moreover, fungi, particularly Aspergillus, Trichoderma, and Penicillium, produce a broad spectrum of enzymes, including amylases, proteases, lipases, cellulases, and xylanases, which are critical to various applications including food production to biofuel generation. Fungi are also more robust than bacteria, capable of tolerating higher temperatures and diverse substrate conditions, which makes them ideal for industrial-scale enzyme production (Østergaard & Olsen, 2011).

Fungal enzymes are also gaining attention because they are produced through fermentation, a well-established and scalable method for industrial enzyme production. Solid-state fermentation (SSF) and submerged fermentation (SmF) are the primary methods used to cultivate fungi for enzyme production. SSF, in particular, is highly efficient in the production of extracellular enzymes in a solid substrate environment, and it is often preferred for the synthesis of enzymes that require complex substrate interactions (Doriya et al., 2016). Fungal enzymes are therefore not only a sustainable and cost-effective alternative to traditional methods, but also enable more efficient production through optimized fermentation processes. This chapter explores the role of enzymes in industrial applications that cannot be overstated. As biocatalysts, they provide an eco-friendly and efficient alternative to conventional chemical processes, offering numerous benefits, including sustainability, cost-effectiveness, and selectivity. Fungal enzymes, with their robustness, versatility, and high production yields, have positioned themselves at the forefront of industrial biotechnology. The continued interest in and demand for fungal enzymes across various sectors, particularly in pharmaceuticals and food production, reflects the growing recognition of their immense potential and advantages over other enzyme sources (Burns & Dick, 2002).

2 The Production of Fungal Enzyme

The synthesis of fungal enzyme involves a number of intricate steps, beginning with the careful selection of suitable fungal strains, followed by fermentation technique optimization, and ending with the formulation and purification of enzymes for a range of industrial uses. The unique qualities of fungal enzymes, including their stability in a wide range of environmental conditions and their ability to catalyze a wide variety of biochemical reactions, make it invaluable in pharmaceuticals, food,

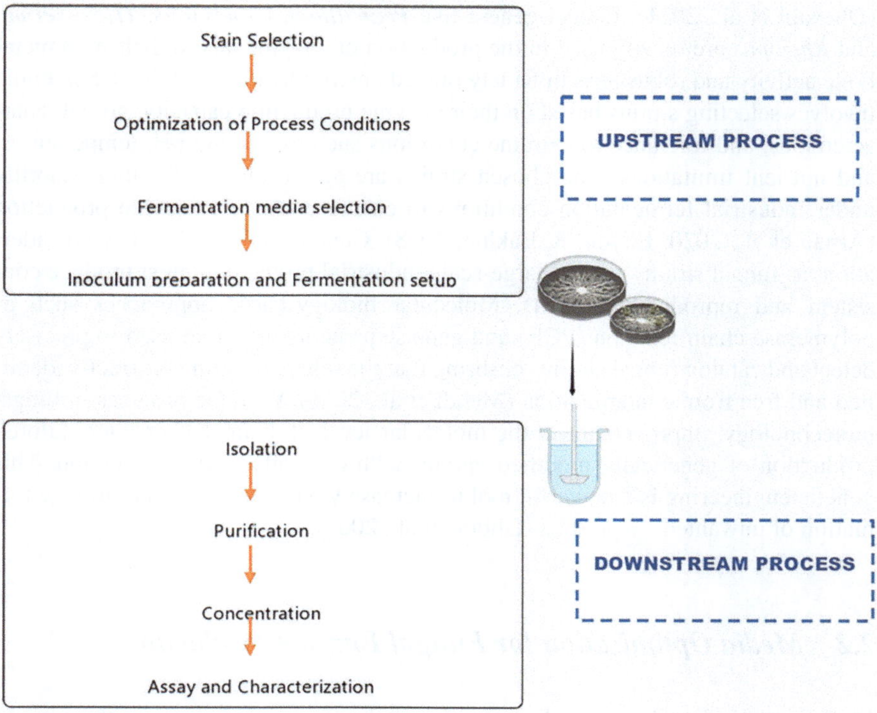

Fig. 15.1 The primary process in the production of fungal enzyme, transitioning between upstream and downstream process

and nutrition industry (Dhevagi et al., 2021). This section explores the crucial elements of producing fungal enzymes, such as the choice of fungal strains, the different fermentation techniques used, and the variables affecting enzyme yield, all of which enhance the productivity and scalability of enzyme synthesis. Figure 15.1 gives an overview of the production of fungal enzyme. Figure 15.1 represents a summarized view of the production of fungal enzyme.

2.1 Selection and Screening of Fungal Strain

The preliminary step in the upstream process is the selection and screening of fungal strains with high enzyme-producing potential. Fungal strains are typically isolated from diverse environments, such as soil, decaying organic matter, plant material, and fermented foods. Strain diversity is essential because different fungi may engineer enzyme with unique biochemical property, suitable for specific commercial applications. The choice of the fungal strain is rooted in its ability to produce the desired enzymes in high quantities, often under industrial-scale conditions

(Dhevagi et al., 2021). Fungal genera like *Penicillium*, *Aspergillus*, *Trichoderma*, and *Rhizopus* are mostly used in the production of enzyme due to their high metabolic activity and robustness in widely ranged environmental conditions. Screening involves selecting strains based on their enzyme production capacity, growth characteristics, and tolerance to extreme conditions such as varying pH, temperatures, and nutrient limitations. The chosen strains are also evaluated for their stability under industrial fermentation conditions to ensure consistent enzyme production (Arnau et al., 2020; Elrsoul & Bakhie, 2018). Genetic stability is a key consideration, as fungal strains used in large-scale industrial production must produce consistent and reproducible results. Molecular biology-based approaches such as polymerase chain reaction (PCR) and gene sequencing are often used to precisely detect and catalog fungal strains, ensuring that the selected strain is correctly identified and free from contamination (Mehdi et al., 2011). With the progress in fungal biotechnology, improvement in the molecular level of stains has enabled tailored production of genetically modified variant with controlled gene expression. This genetic engineering is a powerful tool to increase yield capacity and minimize formation of unwanted by-product (Ghorai et al., 2009).

2.2 Media Optimization for Fungal Enzyme Production

Media optimization is a crucial aspect of fungal enzyme production, as the nutrient component mix in the growth media directly affects the yield and efficiency of enzymes produced by fungal strains, since the media is the prime source of all required nutrient for fungal growth and biosynthesis. The optimization process involves adjusting various components in the culture media designed to escalate the yield of specific enzymes, ensuring maximum productivity under industrial-scale conditions (Sethi & Gupta, 2014). Carbon sources are the primary energy and substrate sources for fungal growth and enzyme synthesis. The choice of carbon source influences both rate of fungal growth and the specificity and level of enzyme biosynthesis. The most common carbon sources include simple sugars like glucose, sucrose etc, polysaccharides like cellulose,starch etc, and lignocellulosic materials like agricultural waste. The optimization of carbon concentration and its source is crucial as excessive carbon concentrations may repress enzyme biosynthesis due to catabolite repression and the nature of carbon source influences the type of enzyme being produced. For instance, cellulose and lignocellulosic materials are often favored for cellulase and xylanase production, while glucose is commonly used for amylase production (Elrsoul & Bakhie, 2018; Gautam et al., 2011). Second, nitrogen is the another indispensable nutrient required for biosynthesis of proteins, enzymes, and several other cellular constituents wherein the optimization of both the nitrogen source and the carbon-to-nitrogen ratio are Critical Material Attributes (CMA) for maximizing the enzyme biosynthesis and maintain optimal fungal growth rate. For nitrogen sources, organic compounds like yeast extract, peptone, or urea are widely employed, since they supply essential amino acids and peptides required for fungal metabolism and nitrogen sources from nonorganic sources such as ammonium salts are also frequently included in the medium to support fungal growth (Chen et al., 2008). In a study by S. P. Gautam et al., impact of various nitrogen and carbon sources was studied in the manufacture of cellulase enzyme by cellulolytic fungi, *A. niger* and *Trichoderma* sp. carbon sources like cellulose, carboxymethyl cellulose (CMC), sucrose, glucose, and maltose were used at a range of 0.5 to 3.0% (w/v) and nitrogen source like peptone, beef extract, sodium nitrate, ammonium nitrate, and yeast extract were employed, ranging from 0.5 to 3.0%. Where CMC and cellulose were revealed as the best carbon source to maximize production of enzyme by *A. niger* followed by CMC 1%, sucrose, glucose, and maltose, and 1.0% peptone as the prime nitrogen source, followed by ammonium nitrate and beef extract, whereas yeast and sodium nitrate produced the least cellulase enzyme. Upon employing *Trichoderma* sp., the superior carbon source was identified to be sucrose 1.0%, followed by glucose, cellulose, maltose, and CMC, whereas 1% yeasr extract brought about maximum enzyme activity in producing maximum cellulase component, followed by peptone (Gautam et al., 2011). S. P. Gautam et al.,has demonstrated the impact of various nitrogen and carbon sources during the manufacturing of cellulase enzyme by using cellulolytic fungi, A. niger and Trichoderma sp. Carbon sources like cellulose, carboxymethyl cellulose (CMC), sucrose, glucose, and maltose were used at a range of 0.5 to 3.0% (w/v) and nitrogen source like peptone, beef extract, sodium nitrate, ammonium nitrate, and yeast extract were in the range of 0.5 to 3.0%. The study concluded that the CMC and cellulose were the best carbon source, and 1.0% peptone was the best nitrogen source to maximize the production of enzyme by A. niger followed by CMC 1%, sucrose, glucose, maltose, and ammonium nitrate, beef extract, as the alternate carbon and nitrogen sources respectively. The study further illustrated that the yeast and sodium nitrate resulted in the least cellulase enzyme production. However, in case of the Trichoderma sp., glucose, cellulose, maltose, CMC and 1% yeast extract were found to be the preferred carbon and nitrogen source respectively (Gautam et al., 2011). Further, Stamm et al. reported that amylase production by Rhizopus oligosporus-ML-10, enhanced by 1.28-times upon addition of 25% maltose as carbon source and 0.25% yeast extract as nitrogen source in supplementing media (Stamm et al., 1976). In certain cases, by using inducer-repression mechanism, specific inducers are included to the media for triggering or eliminating the production of particular enzymes. For example, cellulose and lignocellulosic materials serves as inducers for cellulase production, while protein substrates induce protease synthesis. On the other hand, certain carbon sources such as glucose suppress enzyme production via catabolite repression (Wilson et al., 2007). Nona Mikiashvili et al. investigated the impact of the sources of nitrogen and carbon on enzymatic activity of *pleurotus ostreatus,* upon studying role of varied carbon sources in enzyme production capacity of *P. ostreatus* 98, media containing CMC reported laccase production, 2-fold higher compared to sodium gluconate. Compared to fungal cultivation, laccase level in *P. ostreatus* 98 with sodium gluconate in media reported 38-fold increase. In contrast, grapevine cutting sawdust, and insoluble crystalline cellulose reported poor growth substrate for laccase secretion. In presence of cellobiose and glucose, *Trametes pubescens* demonstrated considerable laccase synthesis in contrast low laccase level were caused by inadequate utilization of lactulose and α-cellulose. The highest manganese dependent-peroxidase (MnP) and peroxidase activity was reported in media with media with polysaccharide substrate, xylan. While comparatively elevated MnP activity was noticed in mannitol and crystalline cellulose, but lactosewas the appropriate substrate for peroxidase concentrate. In *P. ostreatus* 108, xylan substituted by sodium gluconate and cryatalline cellulose reported greatest activity of laccase and mannitol, glucose, maltose although provide good mycelia growth but reported lower level of enzyme activity. The maximum laccase and peroxidase activity was observed in 5-days with inorganic nitrogen source, and in 8-days with organic source. The catalytic activity of *P. ostreatus* enzyme was observed to decrease when organic nitrogen source was added to medium and the superior nitrogen source for laccase accumulation by both fungi stain was peptone and casein hydrosylate (Mikiashvili et al., 2006).

Furthermore, minerals like magnesium, potassium, calcium, and iron are essential for enzyme stability and activity, as they serve as cofactors in enzymatic reactions. For effective biosynthesis of enzymes like laccases or peroxidases trace elements, such as manganese or zinc are additionally incorporated. The use of vitamins and growth factors like biotin, thiamine, and pantothenic acid further boost the fungal metabolism and enzyme yield. Fungal growth and enzyme biosynthesis is highly susceptible to environmental conditions such as pH, temperature and oxygen level. The pH of the culture medium affects enzyme secretion, growth rate, and substrate utilization. Where most fungi prefer slightly acidic to neutral pH, typically ranging between 4.5 and 7.0, this can vary depending on the species. The optimal growth temperatures for fungi typically range from 25 °C to 30 °C, it must be optimized to ensure the highest possible enzyme activity. R. Sivakumar et al. tailored the production media for laccase production by Ganoderma sp. by subjecting to varied incubation condition like different pH such as 4.5, 5.0, 5.5, 6.0, and 6.5, various solvents like methanol, ethanol, propanol, 2-methyl 1-propanol, and isopropanol and varied copper concentration like 10 μM, 20 μM, 27 μM, 40 μM and 50 μM and nutrition media. Upon analyzing source of carbon like mannitol, lactose, glucose, maltose, and sucrose at 2% concentration and organic and inorganic nitrogen source at 0.2%, it was identified that 0.2% yeast extract as nitrogen source and 2% starch as carbon source with 10-day incubation period, 4% iso propanol, 6.0 pH, and 27 μM copper concentration maximized the laccase yield (Tamilvanan, 2020).

Additionally, fungal cultures, particularly filamentous fungus and SmF process requires adequate oxygen levels for the aerobic metabolism of fungi, the insufficiency of which can lead to reduced fungal growth and limited enzyme production. Media optimization also involves ensuring sufficient aeration, which requires optimizing the agitation speed and the volume of the fermentation vessel to ensure proper oxygen transfer. In large-scale production, this involves the use of bioreactors fitted with aeration and agitation systems to maintain the level of dissolved oxygen. Since production of fungal enzyme is a time-dependent process, with enzyme yields increasing during the growth phase and peaking at a specific time point, the fermentation time should be optimized based on the strain's growth curve and the time at which maximum enzyme activity is observed. Harvesting the enzymes at the right moment ensures that the maximum enzyme activity is retained (Gomes et al., 2023).

2.3 Inoculum Preparation and Fermentation Setup

Inoculum preparation is a crucial step that ensures uniform growth of an active and viable starter fungal culture in a controlled environment during the fermentation process. Inoculum is typically a conidia (spores) or mycelial fragments of the selected fungal strain, the source of which is selected based on the strain's growth characteristics, spore viability, and the enzyme to be produced. The pre-cultivation of inoculum includes small-scale growth of fungal spores or mycelia in a pre-culture nutrient-rich medium optimized to support the initial growth of the culture. This phase is usually carried out in a shake flask or small fermenter under controlled physical conditions. In a 12 to 48-hour incubation time, the growth of an optimal inoculum size of 1–5% v/v is determined to ensure a sufficient quantity of healthy inoculum for subsequent fermentation. Upon sufficient growth of the pre-culture, the inoculum is expanded in progressively larger fermenters or bioreactors in a series of steps to avoid shock or nutrient limitation under similar conditions to those of the main fermentation phase. Prior to the inoculation of the fermentation medium, the inoculum is assessed for spore count and viability by a hemocytometer or automated cell counter for determining spore concentration, ensuring that a sufficient number of viable spores are introduced to the fermentation media (Ballaminut et al., 2014; Schmidt et al., 2008).

Fermentation setup is the process of creating an optimal environment for the growth of the fungal culture obtaining the target enzymes. Based on scale of enzyme production, size of bioreactor is selected with volumes ranging from liters to thousands of liters. The vessel must also allow for precise control over environmental specifications such as, temperature, aeration, pH, and agitation, that are critical for fungal growth and enzyme production (Glenn & Goldman, 1976). To prevent contamination, both the fermentation medium and the fermentation vessel undergo sterilization, typically by autoclaving, ensuring that the fungal culture is the dominant microorganism in the system. Upon preparation and sterilization of the fermentation media, under sterile and controlled condition the inoculum is introduced into the fermenter, at a ratio of 5–10% of the total fermentation volume to ensure rapid colonization (Junker et al., 2006).

Chief parameters like temperature, agitation speed, pH, and oxygen supply are monitored closely and regulated according to requirement throughout the fermentation process. The fermentation process typically lasts from several hours to days, depending on the fungal strain and the required enzyme production. During fermentation, periodic sampling is required to monitor fungal growth (biomass), substrate consumption, and enzyme activity, ensuring that the culture remains in the optimal phase of growth and enzyme production. The production of enzymes can be evaluated by employing biochemical assays, such as colorimetric or fluorometric methods, to assess the extracellular enzyme activity over time. After the fermentation period, the fungal biomass is separated from the culture broth, and the enzymes are harvested by filtration or centrifugation to remove the fungal cells and debris, leaving the enzyme-rich supernatant for downstream processing. The harvested enzymes then are concentrated, purified, and formulated for their intended industrial applications (Persson et al., 1991).

3 Upstream Process for Fungal Enzyme Production

The upstream process in fungal enzyme production refers to the initial stages of cultivation, where key factors such as strain selection, media optimization, and fermentation conditions are meticulously controlled to ensure high yields of target enzymes. This process encompasses several crucial steps, including the selection of the fungal strain, optimization of culture conditions, inoculum preparation, and the establishment of ideal growth environments, which all hold a significant role in maximizing enzyme production and activity. Fermentation processes are pivotal in the commercial-scale for producing fungal enzymes, which serve essential roles in numerous sectors such as food processing, biofuel production, and pharmaceutical applications. The general goal of fermentation for fungal enzyme production is to optimize the metabolic conditions to maximize the yield, activity, stabilizing the enzyme. The techniques employed can vary widely based on the particular enzyme requirements, including the metabolic pathways involved in enzyme synthesis and secretion (Jonathan et al., 2021).

3.1 Solid-State Fermentation (SSF)

Primarily a biotechnological proceeding, SSF is based on the proliferation of microorganisms on solid substrate that provides both moisture as well as nutrients necessary for microbial growth, and do not need free-flowing liquid. The substrate in SSF can be an agricultural residue, such as wheat bran, rice husk, or sugarcane bagasse, which are rich in complex organic compounds like proteins, carbohydrates, and lipids. During SSF, fungal hyphae penetrate and colonize the solid substrate. The fungi secrete extracellular enzymes, such as amylases, proteases, cellulases, and lipases, which decompose down complex polymers into simpler molecules, which the fungi can then metabolize for growth. This enzymatic breakdown of the solid substrate also results in the production of valuable metabolites, including bioactive compounds, enzymes, and other secondary metabolites. The process typically occurs under aerobic conditions, and the absence of free water limits the risk of contamination and overgrowth by unwanted microorganisms. SSF is economically advantageous as it utilizes inexpensive and abundant agro-based industrial residues such as substrates, that would eles be discarded. Employing low-cost materials such as wheat bran or rice husk and limited aeration requirement reduces the overall production cost of enzymes. Fungi in SSF typically produce higher enzyme yields because the solid-state environment more closely mimics the natural habitat of many fungi, encouraging the production of extracellular enzymes such as in high quantities (Pandey et al., 1999). Moreover, production of enzyme in SSF tend to exhibit greater stability, which is crucial for their industrial applications. This stability is explained by higher hydrophobicity and enhanced rigidity cell wall. The solid matrix protects the enzymes from degradation, ensuring prolonged activity over extended periods (Verduzco-Oliva & Gutierrez-Uribe, 2020). Manufacturing of organic acid is chiefly found during SSF culture. *A. niger* has reported to produce abundant quantity of citric acid, and oxalic acid production is reported by *A. oryzae*, the mold used to saccharify rice, barley and ferment soya bean for producing soy sauce and miso (Biesebeke et al., 2002; Rodríguez Couto & Sanromán, 2005). Aeration is a significant factor to control several functions in SSF culture that includes carbon dioxide removal, oxygenation, heat dissipation and distribution of water vapor and volatile compounds that directly affects the enzyme production. Yield of α-amylase by *A. oryzae* was seen to significantly reduced in low oxygen concentration, thus proving impact of inadequate oxygen in enzyme expression (Rahardjo et al., 2005). A study investigated the production of cellulolytic, lipolytic, and proteolytic enzyme by employing zygomycetes fungi like *Rhizomucor miehei*, *Mucor corticolus*, *Gilbertella persicaria*, and *Rhizopus niveus* by employing agro-based industrial waste as substrate in two varied SSF system, one with untreated corn residue and other with a blend of wheat bran and the untreated corn residue. Rapid endoglucanase (CMCase) production was observed with peak at 48 h, while cellobiohydrolase (CBH) and β-glucosidase reached maximum activity after 72 h. Further, *R. Niveus* reported highest protease activity and incorporation of wheat bran enhanced enzyme production due to its rich nutritional profile. The tested fungi demonstrated potential for rapid cellulase production that would meet the demand of industrial enzyme using SSF method (Takó et al., 2015).

3.2 Submerged Fermentation

SmF, a widely used biotechnological approach that involves microorganisms, such as fungi, yeasts, or bacteria, are cultured in a liquid nutrient media. In this process, microorganisms are suspended in a liquid growth medium, typically composed of water, carbon, nitrogen sources, vitamins, and minerals, which provide optimal growth conditions for the organism. The submerged environment facilitates the easy distribution of nutrients, maintaining moisture content and ensuring uniform temperature and pH. Fungal enzymes produced through SmF, such as cellulases, amylases, proteases, and lipases, are of particular industrial importance due to their versatility and efficiency in several applications (Fazenda et al., 2008). SmF offers several advantages over SSF, particularly in terms of scalability and ease of control.

The liquid medium allows the precise regulation of factors such as pH, temperature, and oxygen supply, ensuring that the growing culture can achieve maximum productivity. The process is also easier to automate, providing more consistency in the production of enzymes and allowing for continuous or batch fermentation (Pleissner et al., 2014). In a study by M. Tako et al., production of cellulosic, lipolytic and proteolytic enzyme by *Gilbertella persicaria*, *Mucor corticolus*, *Rhizomucor miehei,* and *Rhizopus niveus* were investigated by SSF using agro-industrial waste that includes corn residue and wheat bran as substrate. Exploring to enhance the yield of wheat bran, it was reported that production of endoglucanase (CMCase) was rapid along with maximum activity after 48-h fermentation and cellobiohydrolase and β-glucosidase enzyme after 72-h by SSF method (Takó et al., 2015). A major challenge in SmF method is the decreased enzyme activity of the obtained fungal enzyme caused due to sheer agitation stress on mycelial cells and carbon catabolic repression (Gamarra et al., 2010). The production of cellulase by filamentous fungi is a natural procedure occurring in the absence of simple sugar. But the presence of degradation material in fungal fermentation system that breaks down to simple sugar results in catabolic repression (Suto & Tomita, 2001). The agitation in SmF results in formation of heavy foams which are controlled by anti-foaming agents in the media. But use of chemical components disrupts the viscosity of the culture media and reduce oxygen concentration the media affecting enzyme production. To uniformly distribute oxygen in the media, innovative impeller design like Rushton impeller are used specially for fungal Smf to reduce sheer stress on fungal morphology (Korsa et al., 2023; Yang et al., 2012).

While both SSF and SmF are employed for fungal enzyme manufacture, they differ significantly in their operational mechanics, scalability, and applications. One of the main differences is the type of growth environment. SSF involves growing fungi on solid substrates with limited moisture, whereas SmF requires a liquid medium where the fungi are suspended. This distinction results in varying enzyme production profiles and different technical challenges. In SSF, fungi are generally more exposed to the substrate, which can result in more concentrated enzyme production on the surface of the solid material, whereas in SmF, enzymes are more evenly distributed throughout the culture medium due to the liquid environment. SSF is typically favored when cost-effective, low-moisture substrates are available, and the product requires higher stability and longer shelf life. Additionally, SSF tends to result in a higher enzyme yield per unit of substrate due to the more extensive interaction between fungi and solid materials (Subramaniyam & Vimala, 2012). However, it requires more stringent control of humidity, temperature, and aeration to maintain optimal growth conditions. In contrast, SmF excels in scenarios requiring high scalability, ease of control, and consistent product quality. The use of liquid media in SmF allows for continuous or batch fermentation processes that are easier to automate, making it a more suitable option for large-scale industrial enzyme production. The increased oxygen availability in liquid cultures supports the rapid growth of filamentous fungi, which can result in faster enzyme production. However, SmF can be more prone to contamination and may require more sophisticated filtration and sterilization steps to maintain aseptic conditions. In terms of applications,

SSF is often preferred in industries where the substrate itself is crucial, such as in producing solid biofuels, certain fermented foods, and enzymes for specialized applications. SmF, on the other hand, is more widely used for the mass production of enzymes in pharmaceutical, nutraceutical, and food industries, where liquid fermentation processes can be easily optimized and controlled. Both methods, however, are critical in the production of fungal enzymes, with the preference of method depending on the specific requirements of the industrial application, the type of enzyme, and the scale of production (Singhania et al., 2010).

3.3 Fed-Batch and Continuous Fermentation Technique

Fermentation processes are essential mass production of fungal enzymes, and two widely employed techniques—fed-batch fermentation and continuous fermentation—offer distinct advantages focused on the desired output and the features of the producing enzyme.

The on-going fermentation research is focused in cost-efficient production of enzymatic products that allows for high yield of bioproducts. Fed-batch fermentation combines the elements associated with batch and continuous fermentation, where an initial volume of nutrient medium is inoculated with fungal spores to initiate fermentation and nutrients are added in controlled amounts during the fermentation process. This fermentation technique minimizes or rule out the substrate-associated growth inhibition or the production of metabolic by-products by the control of nutrient availability. Considering the detrimental effect of overfeeding and underfeeding upon cellular growth, optimized feeding strategy is crucial in fed batch fermentation method. The prime advantage of fed-batch fermentation in fungal enzyme production is that it allows for the precisely controlling key factors such as pH, substrate concentration, and dissolved oxygen, which are crucial for growing filamentous fungi and enzyme synthesis. With feeding nutrients over time, fungal cells maintain an exponential growth phase, maximizing enzyme production (Lee et al., 1999).

In fed control strategies, simple loop feedback method is the preliminary strategy. Also known as single-loop method, this scheme control feeding rate of nutrient by assessing pH or oxygen concentration and maintain them at a set point. Predetermined feeding strategy maintains cellular specific growth rates not by controlling nutrient feeding but by balancing yield (Suzuki et al., 1990). Furthermore, nutrient feeding is also controlled by assessing glucose uptake rate or demand. Rate of glucose feed and glucose utilization rate are observed to prevent glucose accumulation or depletion (Lee, 1996). Also, nutrient conversion rate is assessed to determine feed rate and the need of nutrients. Block diagrams with variables like nutrient and oxygen concentration are designed that allows the estimation of glucose utilization rate and nutrient conversion rate. The model paired with proportional-integral deviation controller or ON/OFF model maintains the set variables by manipulating feed ate. Another proposed method by Lin et al., mentions the estimation of glucose concentration by nutrient quantity added to fermenter and carbon dioxide evolution rate (Liu et al., 1993). Furthermore, to counter with real time uncertainty, the "Fuzzy set theory" was proposed by Zadeh in 1965 that deals with industrial complexity and nonlinear system behavior. This theory optimizes enzyme production by controlling basic parameters like temperature, pH, and bioreactor substrate concentration was monitored. Applying fuzzy control, despite the presence of several disturbances, the high biomass concentration proved robustness and stability of fuzzy control system in baker's yeast fed batch bioprocess (Mahjoub et al., 1994).

Unlike fed-batch fermentation, the continuous fermentation technique involves constant input of nutrients and the transferring of products and waste during fermentation process, maintaining a steady state where the microorganism exist in a balanced growth phase, providing a constant output of enzyme over time. In this system, continuous pump of fresh media is incorporated into the bioreactor and equal quantity of culture is withdrawn out simultaneously. It allows for the uninterrupted production of enzymes with the system operating continuously for days or even weeks, depending on the production rate and the reactor setup. Continuous fermentation has certain striking advantages, including consistent enzyme quality for the state processing, a reduced risk of contamination, significant reduction in processing time. Since the microorganism is kept in a steady growth phase, there is less fluctuation in enzyme yield compared to batch processes, and the system can be operated for extended periods, making it highly efficient for large-scale production. Additionally, because the culture is always fresh, nutrient limitations and waste buildup are minimized, helping to maintain a constant and optimal fermentation environment. This method is particularly beneficial when high volumes of fungal enzymes are required on a continuous basis. The major drawback of this method lies in its intrinsic nature. All the fermentation steps are executed in continuous manner and failure of one step will result in complete collapse of the batch (Maxon, 1955). In a study by C. Tamerler et al., producing lipase and esterase by *Ophiostoma piceae* and *Fusarium oxysprorum* was analyzed by fed batch cultivation and continuous batch method. The study reported, that fed batch strategy enhanced the production of lipolytic enzyme from both the fungal stains. Upon presence of only nitrogen and carbon source in the media, specific lipase activity increased by 80% and 23% in Ophiostoma piceae and Fusarium oxysprorum respectively and specific esterase activity increased by 20% and 15% (Tamerler & Keshavarz, 2000).

4 Downstream Process: Isolation and Purification of Fungal Enzyme

Downstream processing involves the recovering, purifying, and isolating of desired enzyme from the fermentation broth or culture medium after the fermentation or bioprocessing phase. The goal of downstream processing is to obtain a purified and concentrated product that meets the required quality standards for industrial or commercial use, while also removing impurities, by-products, and microbial cells (Jonathan et al., 2021).

4.1 Isolation Method of Fungal Enzyme

4.1.1 Centrifugation Method

Centrifugation is a widely utilized separation technique in the downstream processing of fungal enzyme production. It exploits the differences in the density of particles in a liquid medium to separate solid materials, such as fungal cells, cellular debris, and other particulate matter, from the liquid phase, which contains the dissolved enzyme of interest. The process is governed by centrifugal force, which accelerates the sedimentation of particles on the basis of size, shape, density and force causes particles suspended in the liquid to experience an outward acceleration. For extracellular enzyme, after fermentation the culture broth is spun in a centrifuge to remove fungal cells and other solid impurities, leaving the enzyme-rich supernatant behind. Whereas, for intracellular enzyme, cells must first be disrupted by either mechanical disruption or chemical lysis to release the enzyme, subsequently, centrifugation is used to remove debris from the enzyme solution (Heinonsalo et al., 2012).

The two primary types of centrifugation techniques commonly used for enzyme isolation are differential centrifugation and density gradient centrifugation. Differential centrifugation method involves subjecting the culture broth to a series of increasingly higher speeds for separating particles on the basis of size as well as density. Initially, centrifugation at lower speed is used for eliminating larger, heavy components such as fungal mycelium or cellular debris. As the speed increases, smaller particles such as cell organelles and enzymes are gradually separated. This method is highly effective for separating fungal cells from the enzyme-rich culture broth in a time-efficient manner. In density gradient centrifugation, a gradient of a high-density substance (such as sucrose, cesium chloride, or Ficoll) is formed in a test tube. The fermentation broth is layered on top of the gradient, and the centrifuge creates a centrifugal force that forces the particles in the broth to move through the gradient. Components in the broth will settle at different levels within the gradient, depending on their density. This method is useful when a more refined separation is required, particularly when isolating intracellular enzymes from complex mixtures, as it allows for higher purity of the target enzyme (Verdín et al., 2015).

4.1.2 Filteration Technique

Filtration separation technique is primarily employed to remove solid particles, including fungal cells, cell debris, and other suspended impurities from the liquid culture medium, leaving the desired enzyme or metabolites in the supernatant. Filtration relies on the selective passage of liquids through a semipermeable membrane or porous material that captures particulate matter based on size, shape, and other physical properties. It is a primary clarification step in fungal enzyme isolation before more refined purification techniques like chromatography or ultrafiltration are used. The process is generally faster, more straightforward, and less resource-intensive compared to other separation techniques like centrifugation.

Additionally, filtration can be scaled up for industrial processes, idealizing it for bulk production of fungal enzymes (Fong et al., 2000).

Microfiltration is a low-pressure filtration technique used to separate larger particles (ranging from 0.1 to 10 microns in size) from the fermentation broth and is commonly used to remove fungal cells, large cell debris, and other suspended solid particles without significantly affecting the soluble enzyme in the supernatant (Naja et al., 2006). Ultrafiltration also called size-exclusion filtration operates under higher pressure conditions than microfiltration and uses membranes with smaller pore sizes (ranging from 1 to 100 nanometers). Ultrafiltration is typically used for the concentration and partial purification of fungal enzymes (Pant & Adholeya, 2009). Nanofiltration is a filtration process that operates between microfiltration and reverse osmosis in terms of pore size (ranging from 1 to 10 nanometers) and pressure. Nanofiltration membranes are particularly effective at removing small organic molecules and salts while retaining larger molecules like proteins and enzymes (Mondal et al., 2019).

4.2 Purification Technique for Fungal Enzymes

Purifying fungal enzymes is an enormously crucial process to achieve desired purity and activity levels for industrial applications. The aim of purification is to remove impurities, contaminants, and unwanted metabolites while retaining the enzyme's functional properties. Several techniques are employed, often in combination, to achieve high yields and purity of the target enzyme. This section deals with some common purification methods for fungal enzymes.

4.2.1 Precipitation Technique

Precipitation stands as a simple and most cost-efficient technique for enzyme purification. This process involves adding salts or organic solvents into the enzyme solution to cause the target enzyme and other proteins to form aggregate and settle out of solution. Ammonium sulfate the extensively used salt for protein precipitation, known as salting out method. This increase the ionic strength of the solution, which reduces the solubility of proteins, causing them to precipitate. The process is typically done in steps, where increasing concentrations of ammonium sulfate are introduced, and the precipitate is collected by centrifugation. This selective precipitation can help concentrate the enzyme and remove a significant portion of impurities. Although cost-effective, scalable, and relatively easy to perform, some enzymes may lose activity during precipitation due to the high salt concentrations in this precipitation technique. Study by Sunil. S. More et al. carried out the purification of fungal laccase enzyme by ammonium sulphate precipitation after filtration and centrifugation of culture filtrate at 5000RPM (More et al., 2011). Organic solvent method of precipitation includes solvents like ethanol, acetone, or isopropanol that precipitate proteins from the solution. The organic solvent decrease solubility of the enzyme, leading to the precipitation. Blanca Hernández-Rodríguez et al. studied reaction of organic solvent on the stability as well as activity of lipase enzyme produced by thermotolerant fungi stain, where solvents were categorized based on dipole moment and functional group. In nonpolar solvents like toluene, n-hexane and n-decane, lipase activity ranged from 70–147%, but stability was attributed to nonpolar solvents preserving the enzymes hydration shell. In protic solvents, ethanol and isopropanol, most biocatalysts showed reduced enzyme activity, as low as 12%. While enzyme in aprotic polar solvent showed uncommon stability as aprotic solvents often cause enzyme denaturation through dehydration and structural disruption (Hernández-Rodríguez et al., 2009). Selective precipitation is a commonly used technique in biochemistry and molecular biology that leverages the differences in solubility of proteins under varying conditions of ionic strength, pH or the presence of specific chemicals. When purifying fungal enzymes, this method can help isolate the enzyme of interest from a fungal culture medium or crude extract, allowing for further analysis and application. The principle behind selective precipitation relies on the idea that each protein has its own solubility profile and under specified conditions, proteins can be made to aggregate and precipitate out of solution, leaving the remaining soluble proteins behind. In a study by María Julia Boggione et al. the precipitation of endoglucanases from *Aspergillus niger* was performed by selective protein precipitation by the addition of polyelectrolytes. The study focused on investigating the polyelectrolyte-endoglucanase complex under optimal precipitation condition to design an appropriate downstream process. Polyvinylsulfonoic acid and chitosan were the primary polymers added, where both reported increased enzymatic activity post-precipitation and chitosan reporting greater stability. The precipitation curves indicated that enzyme-polymer interaction was mainly electrostatic and is temperature dependent. Further, polyvinylsulfonoic acid-enzyme complex was formed within 10 min as compared to chitosan-enzyme complex that was formed in 40 min. Based on purification factor and recovery of endoglucanase, high purification efficiency was obtained in commercial endoglucanase, and even 9-fold higher in fungal enzyme, indicating broad applicability (Anderson & Slotkin, 1975).

4.2.2 Dialysis Method

Dialysis method eliminates low molecular weight contaminants such as salts, small metabolites, or solvents from the enzyme solution, relying on the semipermeable nature of a dialysis membrane The process is driven by the concentration gradient, when there is a difference in concentration of solute molecule inside and outside the dialysis bag and will naturally diffuse from higher concentration to lower concentration until equilibrium is attained. The continuous exchange of the dialysis buffer helps removes the unwanted smaller components, leaving the enzyme in the concentrated solution inside the dialysis bag. The dialysis method is also used to remove salts, particularly after processes like ion exchange chromatography or salting out, where high concentrations of salts have the probability to inhibit enzyme activity.

Dialysis method is also employed to concentrate enzyme solutions. The very small pore size of dialysis membrane limits the flow of water into it, effectively concentrating the enzyme by removing the dissolved small molecules. This is particularly useful when large volumes of fermentation broth or enzyme solution need to be concentrated prior to further purification. Dialysis, being a relatively gentle process, allows for the exchange of buffer solutions without causing denaturation or destabilization of the enzyme. Dialysis is less selective than other purification methods, such as chromatography. While it efficiently removes small molecules, it may not provide high specificity in separating the enzyme from other similar-sized proteins or contaminants, thus is often combined with other methods for effective purification (Volesky et al., 1984). P. Palanivel et al. studied twenty three fungal stains belonging to five diverse genera was isolated from alkaline soil of Kolli Hills of Eastern Ghats, and screened for its protease activity. Under optimized condition of pH 9 and temperature 25–30 °C and with 1% starch as carbon source and 3% soyabean meal as nitrogen source, maximum protease enzyme was obtained by submerged fermentation. Crude enzyme was precipitated and purified by a two-step method involving ammonium precipitation followed by dialysis method that resulted in a 6.7-fold increase in specific gravity per mg protein, compared to crude enzyme. Further, a 1.4-fold increase in enzyme concentrate as compared to single-step purification, thus indicating dialysis as an effective step in purity protocol (Palanivel et al., 2013).

4.2.3 Chromatographic Methods

Chromatography, a sophisticated and extensively used technique for separation and purification of proteins, including fungal enzymes that takes advantage of the differences in the physical as well as chemical characteristics of proteins, that includes size, charge, hydrophobicity, and affinity for certain ligands. Chromatographic methods are typically more precise than precipitation techniques, offering a higher degree of purity and the ability to handle small to large quantities of sample. Ion exchange chromatography is a popular method for protein purification relying on their charge and involves the separation of proteins by their interaction with charged stationary phases in the presence of a mobile phase. In ion-exchange chromatography, the enzyme solution is moved through ion exchange resin packed column. The proteins bind to the resin depending on their net charge (positive or negative) and are then eluted using a gradient of salt or changes in pH that alter the ionic interactions. It is a highly selective and efficient method but careful optimization of conditions like salt gradients, pH is necessary since some enzyme activities may be sensitive to changes in ionic strength or pH. A study aimed to design an integrated bioseperation approach for purifying of cellulase-free xylanase from *Aspergillus niger* NRRL3, with a particular emphasis on achieving high purity suitability. The researchers employed a two-stage aqueous two-phase system as the primary extraction method, subsequently with ion exchange chromatography for final purification. Among the tested system, a combination of UCON/Citrate and PEG1450/Citrate phase reported to be most effective, yielding 80% enzymatic activity recovery and a purification factor of 4. However, the most the most significant improvement in purity was achieved through the application of ion exchange chromatography using a Q-Sepharose XL column. This step not only refined the extract but also resulted in a final purification factor of 11 and an overall activity recovery of 79.9%. The ion exchange process played a critical role in separating xylanase from other proteins and impurities, especially cellulolytic enzymes, thus producing a highly purified enzyme preparation. The effectiveness of the chromatography step was further validated using nano-LC–MS/MS, which confirmed the selective enrichment of xylanase and the absence of contaminating cellulases (Taddia et al., 2021), Gel filtration chromatography, or size-exclusion chromatography (SEC), separates out proteins on the basis of size, molecular weight and is employed in removing low molecular weight contaminants or for molecular weight determination. Passing the enzyme solution through a column that is packed with porous beads, the smaller molecules enters into pores and have a lengthy path, while bigger molecules traverse more quickly. As a result, proteins are separated by size, with larger enzymes eluting first. SEC is a gentle technique that does not require changes in pH or salt concentrations, which makes it ideal for sensitive enzymes. A study by L.Ngo et al. studied the purification of extracellular endo-polygalacturonase (endo-PG) produced by *Aspergillus awamori* L1 through submerged fermentation. A two-step purification process was employed consisting of precipitation followed by gel filtration chromatography. After precipitating of enzyme using ammonium sulfate, the enzyme was exposed to gel filteration employing Sephadex G-75 coloum, which significantly enhanced enzyme's purity and a 30.4-fold increase in purification factor with a recovery yield of 68.6% was resulted using this combined method (Ngo et al., 2008). Affinity chromatography is a highly specific and predominant technique which explores the specific binding interaction among an enzyme and a ligand or substrate. The enzyme binds to a ligand such as an antibody, substrate analog, or small molecule that is immobilized on a solid foundation such as a column matrix. Other nonspecific proteins are carried away, and the enzyme is eluted employing a buffer containing elevated concentration of free ligand or a condition that disrupts the enzyme-ligand binding. The highly selective method that allows for the purification of a target enzyme with minimal contaminants, often achieving a high degree of purity in a single step (Cao et al., 2011). Reverse phase chromatography (RPC) is a variation of liquid chromatography that uses a nonpolar stationary phase and a polar mobile phase. It is used to separate proteins based on their varying interactions with a hydrophobic stationary phase. Greater hydrophobic proteins interact rather strongly with the stationary phase and elute later as compared to lesser hydrophobic proteins. RPC is primarily used for the separation of hydrophobic enzymes or those that contain hydrophobic patches on their surface. This method provides high resolution and is effective for purifying proteins with varying levels of hydrophobicity. But RPC requires high concentrations of organic solvents that can denature sensitive enzymes (Soliday & Kolattukudy, 1983). Figure 15.2 represents the crucial parameters in producing fungal enzyme

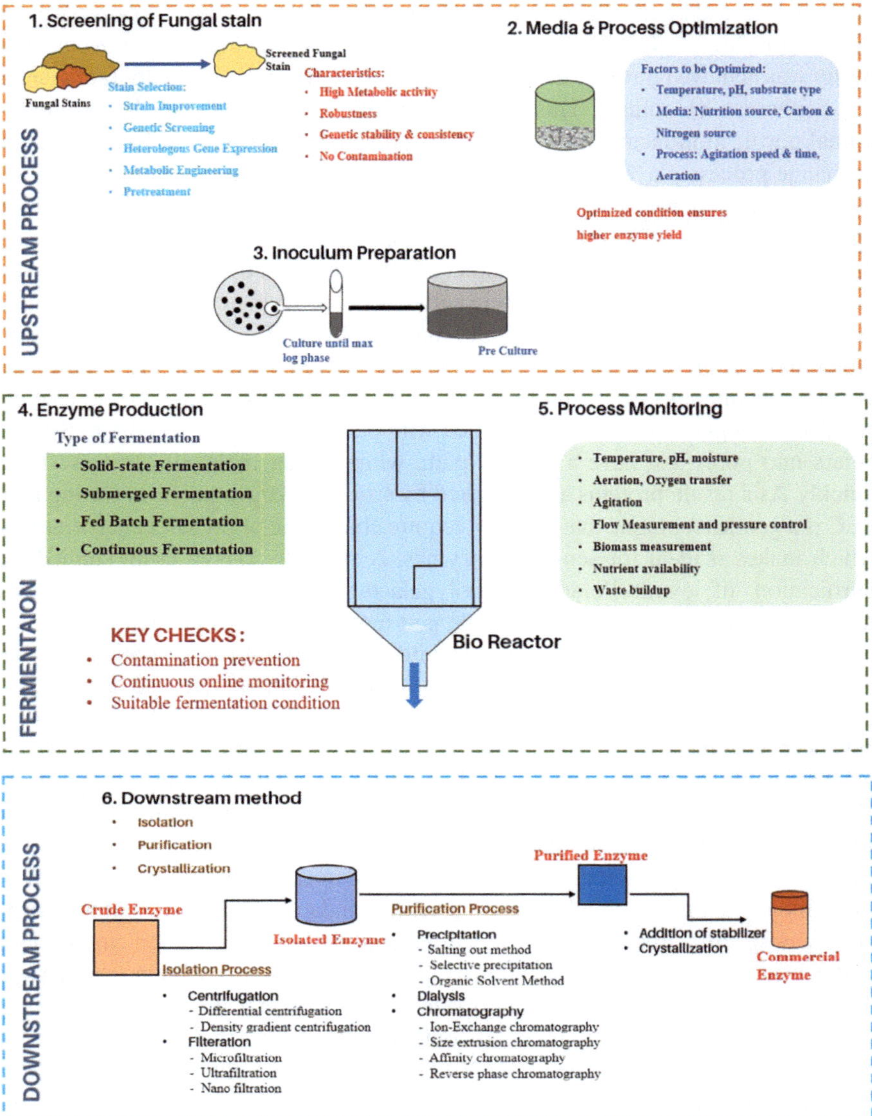

Fig. 15.2 Representation of critical parameters in the production of fungal enzyme

5 Assay and Characterization of Fungal Enzymes

Characterizing fungal enzymes after production is crucial for understanding their properties, optimizing their activity, and determining their potential applications in various industries. This process involves assessing factors such as substrate specificity, optimal conditions (temperature, pH), kinetic properties, and stability under different conditions. Techniques like SDS-PAGE, mass spectrometry, and enzyme activity assays elucidate enzyme's purity, molecular weight, and functional efficiency. Understanding these properties allows for the identification of the enzyme's industrial viability, including its use in sectors such as food processing, and pharmaceuticals. Moreover, enzyme characterization helps in improving enzyme production, stability, and performance, leading to more sustainable and cost-effective industrial processes.

5.1 *Enzyme Activity Assay*

Enzyme Activity Assay, is a fundamental step in characterizing a fungal enzyme, as it provides insight into the enzyme's functional capacity and efficiency. Enzyme activity is typically determined by measuring the rate at which the enzyme converts its substrate into a product. This can be done using various assays, dependent upon the variability of enzyme and the reaction it catalyzes (King et al., 2009). For example, in the case of amylases, the breakdown of starch into sugars is monitored, often through colorimetric methods that detect the formation of reducing sugars. Substrate specificity is also assessed, where the enzyme is tested against a range of potential substrates to determine which ones it can act upon efficiently. This allows researchers to identify the enzyme's preferred substrates and reaction mechanisms (McCleary, 1991). Additionally, enzyme activity can be quantified by monitoring changes in absorbance, fluorescence, or through the generation of a measurable product over time, offering a clear picture of the enzyme's catalytic power. Understanding enzyme activity is crucial not only for optimizing conditions for maximum performance but also for determining its industrial applicability in processes such as bioconversion, waste management, or drug production (Alcalde et al., 2002).

5.2 *Thermostability and pH Stability of Fungal Enzyme*

Optimal Condition optimization determines the specific environmental parameters that maximize the activity and stability of the fungal enzyme. These conditions include factors such as temperature, pH, ionic strength, and the occurrence cofactors, all of which can significantly influence enzyme performance. Enzymes, including fungal enzymes possess a favorable thermal range at which their activity is maximized. At temperatures below this range, enzyme activity is generally slow due to lower kinetic energy, while at greater temperatures denature or lose of enzyme's functional structure might occur. Understanding the thermal stability of the enzyme is crucial for industrial applications, as enzymes used in processes like food production or waste treatment may need to function under elevated temperatures (Maheshwari et al., 2000). Along with temperature, enzymes also possess a favorable pH range, within which the enzyme displays maximized potential. The ionization state of the enzyme's active site and the substrate is highly pH-dependent and deviations from the optimal pH leads to a reduction in enzyme activity or complete inactivation. For instance, proteases often work best in acidic conditions, while other enzymes, like amylases, may be more active in neutral or slightly alkaline environments, thus optimizing pH range for a particular enzyme (Roth et al., 2019). Also, certain enzymes require particular ions such as magnesium, calcium, or manganese, for optimal activity that act as cofactors, that stabilize the structure of enzyme or modulate the catalytic process. Testing the enzyme's activity with varying concentrations of metal ions or salts helps identify which ions are essential for its function and whether they enhance or inhibit activity. Additionally, ionic strength can influence enzyme-substrate interactions and enzyme stability, making it an important factor to test (Gao et al., 2015).

5.3 Kinetic Study

Studying enzyme's reaction rates under different substrate concentrations gives an overview to its efficiency and affinity. The key parameters determined in this step are Km (Michaelis constant) and Vmax (maximum velocity), Km represents the substrate concentration at which enzyme reaches half of its maximum velocity, thereby reflecting its affinity for the substrate. Whereas, Vmax represents maximum reaction velocity attained when enzyme is saturated with substrate, reflecting the enzyme's catalytic capacity (Chauve et al., 2010). A lower Km points towards high affinity, which means that enzyme is efficient in binding to the substrate at lower concentrations, and a higher Km points to lower affinity. Additionally, inhibition studies are also conducted to examine the effect on enzyme by specific inhibitors. The inhibition study provides valuable information about the enzyme's active site and how it interacts with its substrate, which is essential for understanding its regulation and potential for therapeutic applications (Mishra & Kumar, 2009).

6 Application of Fungal Enzyme

6.1 Pharmaceutical Application

Fungal enzymes have crucial impact pharmaceutical industry, in drug delivery systems and therapeutic benefits. These enzymes, derived from various species of fungi, offer unique advantages due to their ability to break down specific substrates, their stability under diverse conditions, and their bioactivity.

Fungal enzymes are increasingly being explored for their potential to enhance the efficacy of drug delivery systems. Incorporated into drug delivery matrices, fungal enzymes aid for drug release in a controlled manner. Employing enzymes such as cellulases, amylases, and proteases, a venture can be undertaken to modulate the drug release reacting to physiological alterations such as pH, temperature. These enzymes can break down the polymers in the delivery systems, releasing the drug at a specific site or over a sustained period (Grigoras, 2019). Fungal enzymes are also utilized in targeted drug delivery. Enzymes like glucanases or chitinases can specifically degrade components of the extracellular matrix in certain diseases, allowing drugs to reach deeper tissue layers. This method is precisely targets cancerous cell or infected tissues, ensuring targeted delivery of therapeutic agent while minimizing side effects. Fungal enzymes like proteases, can provoke drug release when they come into contact with specific biological components. This approach improves precision of drug delivery, allowing the drugs to be released only when needed, such as when an enzyme encounters a particular substrate (Wu et al., 2023).

Further, fungal enzymes are employed in several therapeutic applications, benefiting both the manufacturing of drugs and their treatment capabilities. Enzymes like amylases, lipases, and proteases are often used to aid digestion. They are included in enzyme replacement therapies for patients with conditions like cystic fibrosis, pancreatic insufficiency, or digestive enzyme deficiencies. Fungal-derived enzymes can help break down carbohydrates, fats, and proteins, aiding in better nutrient absorption (Hector et al., 2016). Some fungal enzymes have demonstrated anti-inflammatory properties, which can be used in therapies for inflammatory diseases like arthritis or inflammatory bowel disease (IBD) (Spagnuolo et al., 2017). Certain fungal proteases have shown anticancer activities by breaking down proteins involved in tumor progression and metastasis. For example, chitinases and glucanases may have a role in limiting the growth of cancer cells (Kzhyshkowska et al., 2007). Fungal enzymes such as glucanases, chitinases, and proteases can act as antimicrobial agents. By breaking down the cellular envelop of other fungi, bacteria, or even viruses, these enzymes can enhance the efficacy of antibiotics and antifungal drugs. They may also help to reduce microbial resistance by targeting the structural components of pathogens (Maria et al., 2005). Fungal-derived enzymes like collagenase and other proteases are used to treat wounds and burns by debriding necrotic tissue. These enzymes help break down damaged tissue, promoting the growth of healthy tissue. Fungal enzymes highlight an indispensable role in synthesizing bioactive molecules. Enzymes like cytochrome P450 monooxygenases and other oxidative enzymes are employed in pharmaceutical sector for biotransformation procedures, producing drugs or metabolites with high specificity and less toxic by-products (Van Bogaert et al., 2011).

6.2 Application in Nutraceutical Industry

Fungal enzymes have gained considerable attention in the nutraceutical industry for their potential to improve bioavailability, functionality, and stability of active components in various products. Their applications in drug delivery and therapeutic benefits are broad and can significantly improve both the formulation and efficacy of nutraceuticals. Fungal enzymes are integral to the development of nutraceutical products, helping in the processing, extraction, and enhancement of bioactive compounds. Fungal enzymes can improve the bioavailability of nutrients and bioactive compounds, such as vitamins, minerals, antioxidants, and polyphenols. For instance, enzymes like cellulases and xylanases break down plant fibers, releasing these bioactive molecules in a more bioavailable form. This is crucial for products such as functional foods, dietary supplements, and herbal formulations where increased absorption in the digestive tract is essential for therapeutic effects (Tharifkhan et al., 2021).

Fungal enzymes are involved in fermentation techniques to enhance the production of functional ingredients like probiotics, peptides, and secondary metabolites. Enzymes like amylases, lipases, and proteases are used to break down complex food matrices, releasing nutrients that can promote health benefits such as improving gut health, boosting immune function, or lowering cholesterol (Ranghar et al., 2019).

Enzymatic processing using fungal enzymes aids in breaking down plant cell walls, improving the texture and nutritional content of plant-based foods. For example, fungal enzymes like cellulase and hemicellulase are employed in fragmenting polysaccharides in plant materials, thereby optimizing nutrient absorption from food ingredients. This is particularly beneficial in plant-based protein isolates, such as soy, peas, and quinoa, which are commonly used in functional foods and supplements. Fungal enzymes are also used to fortify food products with essential vitamins, minerals, and other nutrients that might otherwise be lost in the processing phase. For example, enzymes are used to extract nutrients from grains, legumes, and fruits, making them easier to incorporate into fortified foods and beverages (De Souza & Kawaguti, 2021).

Fungal enzymes are gaining importance in drug delivery systems, particularly for nutraceuticals and natural therapeutic products. Their use in drug delivery system allows for more precise, efficient, and controlled delivery of bioactive compounds. Fungal enzymes can help in the microencapsulation of sensitive bioactive compounds like vitamins, minerals, and polyphenols, protecting from decayingin extreme environmental factors like light, heat, or oxidation. Fungal enzymes can degrade the outer shell of microencapsulated compounds when they reach specific areas in the digestive tract or targeted tissues, facilitating controlled release. This method is widely used for delivering probiotics, herbal extracts, or vitamins (Zabot et al., 2022). Fungal enzymes are used to improve the stability and shelf-life of nutraceutical products. Enzymes like amylases, proteases, and lipases can be applied in food and pharmaceutical formulations to reduce the degradation of active compounds during processing or storage. This is particularly important for nutraceuticals containing delicate compounds, such as omega-3 fatty acids or polyphenols, which are susceptible to oxidation (De La Cruz Quiroz et al., 2019).

6.3 Application in Food Processing Industry

Fungal enzymes have enormous applications in the food-based industry in enhancing food processing, escalate nutritional benefits, and optimize texture, flavor, and overall quality of food products.

Fungal enzymes, such as amylases and lipases are involved to improve dough quality, fermentation, and bread texture and other bakery items. Amylases breaks starches present in flour into simpler form of sugars, that enhance the fermentation process by providing more sugar for yeast activity. Lipases improve the dough's texture and shelf life, while proteases help break down gluten, resulting in softer bread. Fungal proteases are widely used in cheese production to aid in curd formation and improve the texture and yield of cheese (Rosell & Dura, 2015). These enzymes help break down proteins in milk, leading to more efficient coagulation during the cheese-making process. Fungal enzymes like lipases are also used to enhance the flavor and characteristics of aged cheeses (El-Fadaly et al., 2015). Fungal glucanases and pectinases are employed in producing alcoholic beverages like beer and wine to break down complex polysaccharides like glucans and pectin in the raw materials. This enhances the clarity, filtration, and overall quality of the final product. These enzymes also aid in the extraction of fermentable sugars, which can increase the efficiency of fermentation (Vitolo, 2020).

Fungal enzymes such as pectinases and cellulases are used in the eliciting of juice from fruits and vegetables. These enzymes penetrate across the cell walls and pectin in fruit, resulting in higher juice yields, better clarity, and improved flavor profiles. They also reduce the viscosity of juices and increase their stability during storage (van Oort & Whitehurst, 2010). Fungal proteases such as papain or bromelain are used as meat tenderizers (Mohd Azmi et al., 2023). These enzymes break down muscle fibers and connective tissues present in meat, resulting tender texture. Fungal proteases are often incorporated into marinades and processing formulations for meat products. Fungal enzymes are increasingly used to improve digestibility and nutritional value of animal feed. Enzymes like xylanases, cellulases, and proteases help break down complex carbohydrates and proteins in feed ingredients like grains and oilseeds, making nutrients more accessible to animals (Ojha et al., 2019). This can improve the efficiency of animal growth, feed conversion, and overall health. Fungal enzymes play a role in improving the digestibility of fortifying ingredients such as vitamins, minerals, and antioxidants. For example, the use of enzymes in the breakdown of fibers in grains allows for better absorption of vitamins and minerals in the digestive system, making the fortification process more effective (Espinosa-Páez et al., 2017). Fungal enzymes can be used in the production of prebiotics and probiotics by breaking down complex carbohydrates like inulin or oligosaccharides in plants. This supports the growth of beneficial Fungal enzymes such as lipases and proteases are used to control the levels of certain fats and proteins in processed foods. By modifying these components, they can help improve the storage stability and shelf life of products such as baked goods, dairy, and meat products (Ojwach et al., 2022).

Fungal enzymes can also prevent rancidity and off-flavors in fats and oils, helping to extend the freshness of packaged food products. Gut bacteria and enhances the health benefits of dietary products, such as fermented foods or supplements (Shahani et al., 1976). Fungal enzymes are also introduced for the fermentation of non-dairy foods such as soy, tofu, and tempeh. These enzymes help break down complex carbohydrates and proteins, making the foods more digestible and enhancing the flavor profile of plant-based products (Ahsan et al., 2020). Fungal lipases are used in the modification of fats and oils for the production of healthier food options, such as low-fat or trans-fat-free products. By altering the structure of fatty acids, these enzymes can help produce oils with improved nutritional profiles, including omega-3 and omega-6 fatty acids, which have health benefits for cardiovascular health (Alessa Esperón-Rojas et al., 2022). Fungal enzymes contribute to sustainability efforts by reducing food waste in the processing of plant materials. Enzymatic treatments can make better use of food by-products like fruit peels, seeds, and stems, converting them into useful products like biofuels, animal feed, or dietary fibers. This helps reduce waste and enhances the value of food by-products (Sharma et al., 2022) (Fig. 15.3). Represents the majorly employed fungal enzymes in pharmaceutical, nutraceutical and food industry.

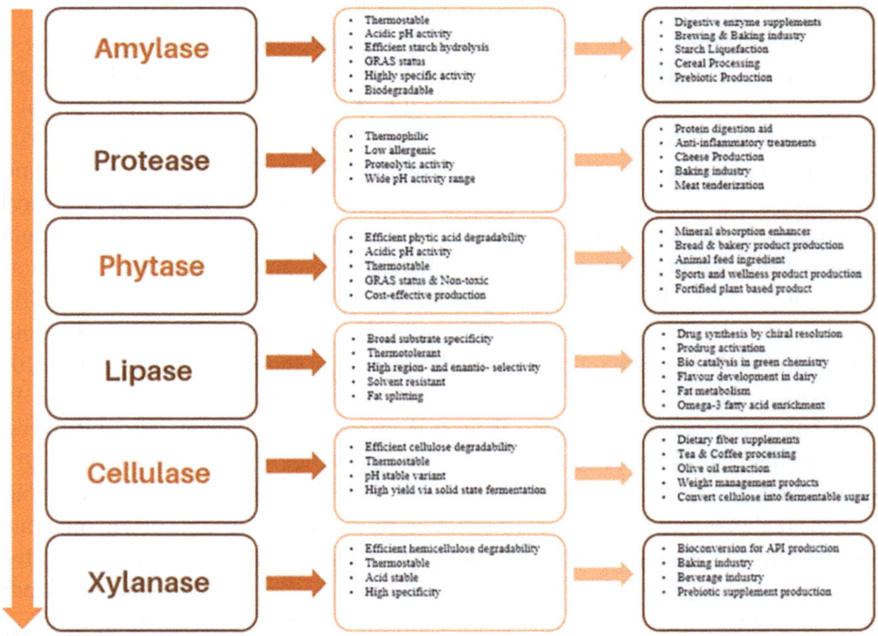

Fig. 15.3 Represents the majorly employed fungal enzymes in pharmaceutical, nutraceutical and food-based industry

7 Future Perspective

Using genetic engineering in fungal enzyme production has become an integral tool to enhance the productivity and specificity of the enzymes produced. Altering the genetic makeup of fungal strains, finds a scope in optimizing their metabolic pathways for greater enzyme secretion, increase the resistance to stress factors, or even introduce entirely new enzymatic functions. This is typically achieved through the use of recombinant DNA technology, where genes encoding for target enzyme are isolated, cloned, and inserted into the fungal genome using vectors. The engineered strain is then cultured under controlled conditions to produce the desired enzyme at elevated levels. Promoter engineering is another key strategy used to drive high-level expression of the enzyme-producing genes. Strong promoters, such as those from highly expressed genes, are incorporated to ensure robust transcription and enzyme production. Additionally, strain optimization through techniques such as site-directed mutagenesis or CRISPR-Cas9 enables the precise alteration of metabolic pathways to maximize enzyme yield or improve the functional properties of the enzyme, such as stability or substrate specificity. High-throughput screening methods are employed to evaluate large libraries of genetically modified fungal strains, quickly identifying those that exhibit superior enzyme production capabilities. Through genetic modification, the efficiency of the upstream process is significantly enhanced, leading to improved productivity and more cost-effective enzyme production on an industrial scale.

References

Aehle, W. (2007). *Enzymes in industry: production and applications* (3rd Completely Rev ed.). Wiley VCH.

Ahsan, S., Khaliq, A., Chughtai, M. F. J., Nadeem, M., Din, A. A., Hlebová, M., et al. (2020). Functional exploration of bioactive moieties of fermented and non-fermented soy milk with reference to nutritional attributes. *Journal of Microbiology, Biotechnology and Food Sciences, 10*(1), 145–149.

Alcalde, M., Bulter, T., & Arnold, F. H. (2002). colorimetric assays for biodegradation of polycyclic aromatic hydrocarbons by fungal laccases. *SLAS Discovery, 7*(6), 547–553.

Alessa Esperón-Rojas, A., Gabriela Mendoza-Sánchez, L., & Sergio, G. H. (2022). Structured acylglycerides emulsifiers with bioactive fatty acids as food ingredients. In *Value-addition in food products and processing through enzyme technology [Internet]* (pp. 221–233). Elsevier. Available from: https://linkinghub.elsevier.com/retrieve/pii/B9780323899291000044

Anderson, T. R., & Slotkin, T. A. (1975). Maturation of the adrenal medulla – IV. Effects of morphine. *Biochemical Pharmacology, 24*(16), 1469–1474.

Arnau, J., Yaver, D., & Hjort, C. M. (2020). Strategies and challenges for the development of industrial enzymes using fungal cell factories. In H. Nevalainen (Ed.), *Grand challenges in fungal biotechnology [Internet]* (pp. 179–210). Springer. Available from: http://link.springer.com/10.1007/978-3-030-29541-7_7

Ballaminut, N., Machado, K. M. G., Oliveira, L. H. D. S., & Matheus, D. R. (2014). Physiological characterization of fungal inoculum for biotechnological remediation of soils. *Brazilian Archives of Biology and Technology, 57*(4), 561–570.

Biesebeke, R., Ruijter, G., Rahardjo, Y. S. P., Hoogschagen, M. J., Heerikhuisen, M., Levin, A., et al. (2002). *Aspergillus oryzae* in solid-state and submerged fermentations. *FEMS Yeast Research, 2*(2), 245–248.

Burns, R. G., & Dick, R. P. (Eds.). (2002). *Enzymes in the environment: Activity, ecology, and applications*. Marcel Dekker.

Cao, H., Liu, D., Mo, X., Xie, C., & Yao, D. (2011). A fungal enzyme with the ability of aflatoxin B1 conversion: Purification and ESI-MS/MS identification. *Microbiological Research, 166*(6), 475–483.

Chauve, M., Mathis, H., Huc, D., Casanave, D., Monot, F., & Lopes, F. N. (2010). Comparative kinetic analysis of two fungal β-glucosidases. *Biotechnology for Biofuels, 3*(1), 3.

Chen, X.-E., Fang, X.-B., & Xia, W.-S. (2008). Strain improvement and optimization of the media composition of chitosanase-producing fungus Aspergillus sp. CJ 22-326. *African Journal of Biotechnology, 7*(14), 2501–2508.

De La Cruz Quiroz, R., Cruz Maldonado, J. J., Rostro Alanis, M. D. J., Torres, J. A., & Parra, S. R. (2019). Fungi-based biopesticides: Shelf-life preservation technologies used in commercial products. *Journal of Pest Science, 92*(3), 1003–1015.

De Souza, T. S. P., & Kawaguti, H. Y. (2021). Cellulases, hemicellulases, and pectinases: Applications in the food and beverage industry. *Food and Bioprocess Technology, 14*(8), 1446–1477.

Dhevagi, P., Ramya, A., Priyatharshini, S., Geetha Thanuja, K., Ambreetha, S., & Nivetha, A. (2021). Industrially important fungal enzymes: Productions and applications. In A. N. Yadav (Ed.), *Recent trends in mycological research [Internet]* (pp. 263–309). Springer. Available from: https://link.springer.com/10.1007/978-3-030-68260-6_11

Doriya, K., Jose, N., Gowda, M., & Kumar, D. S. (2016). Solid-state fermentation vs submerged fermentation for the production of l-asparaginase. In *Advances in food and nutrition research [Internet]* (pp. 115–135). Elsevier. Available from: https://linkinghub.elsevier.com/retrieve/pii/S104345261630016X

El-Fadaly, H. A., El-Kadi, S. M., Hamad, M. N., & Habib, A. A. (2015). Role of Fungal Enzymes in the Biochemistry of Egyptian Ras Cheese during Ripening Period. *OALib., 02*(08), 1–13.

Elrsoul, R. M. M. A. A., & Bakhie, S. E. A. (2018). Optimization of factors influencing cellulase production by some indigenous isolated fungal species. *JJBS., 11*(1), 31–36.

Espinosa-Páez, E., Alanis-Guzmán, M. A., Hernández-Luna, C., Báez-González, J., Amaya-Guerra, C., & Andrés-Grau, A. (2017). Increasing antioxidant activity and protein digestibility in Phaseolus vulgaris and Avena sativa by fermentation with the Pleurotus ostreatus Fungus. *Molecules, 22*(12), 2275.

Fazenda, M. L., Seviour, R., McNeil, B., & Harvey, L. M. (2008). Submerged culture fermentation of "higher fungi": The macrofungi. In *Advances in applied microbiology [Internet]* (pp. 33–103). Elsevier.

Fong, Y. K., Anuar, S., Lim, H. P., Tham, F. Y., & Sanderson, F. R. (2000). A modified filter paper technique for long-term preservation of some fungal cultures. *Mycologist, 14*(3), 127–130.

Gamarra, N. N., Villena, G. K., & Gutiérrez-Correa, M. (2010). Cellulase production by Aspergillus niger in biofilm, solid-state, and submerged fermentations. *Applied Microbiology and Biotechnology, 87*(2), 545–551.

Gao, W.-W., Zhang, F.-X., Zhang, G.-X., & Zhou, C.-H. (2015). Key factors affecting the activity and stability of enzymes in ionic liquids and novel applications in biocatalysis. *Biochemical Engineering Journal, 99*, 67–84.

Gautam, S. P., Bundela, P. S., Pandey, A. K., Khan, J., Awasthi, M. K., & Sarsaiya, S. (2011). Optimization for the production of cellulase enzyme from municipal solid waste residue by two novel cellulolytic fungi. *Biotechnology Research International, 2011*, 1–8.

Ghorai, S., Banik, S. P., Verma, D., Chowdhury, S., Mukherjee, S., & Khowala, S. (2009). Fungal biotechnology in food and feed processing. *Foodservice Research International, 42*(5–6), 577–587.

Glenn, J. K., & Goldman, J. (1976). Task delegation to physician extenders – some comparisons. *American Journal of Public Health, 66*(1), 64–66.

Gomes, D. G., Coelho, E., Silva, R., Domingues, L., & Teixeira, J. A. (2023). Bioreactors and engineering of filamentous fungi cultivation. In *Current developments in biotechnology and bioengineering [Internet]* (pp. 219–250). Elsevier. Available from: https://linkinghub.elsevier.com/retrieve/pii/B9780323918725000181

Grigoras, A. G. (2019). Drug delivery systems using pullulan, a biocompatible polysaccharide produced by fungal fermentation of starch. *Environmental Chemistry Letters, 17*(3), 1209–1223.

Hector, A., Chotirmall, S. H., Lavelle, G. M., Mirković, B., Horan, D., Eichler, L., et al. (2016). Chitinase activation in patients with fungus-associated cystic fibrosis lung disease. *The Journal of Allergy and Clinical Immunology, 138*(4), 1183–1189.e4.

Heinonsalo, J., Kabiersch, G., Niemi, R. M., Simpanen, S., Ilvesniemi, H., Hofrichter, M., et al. (2012). Filter centrifugation as a sampling method for miniaturization of extracellular fungal enzyme activity measurements in solid media. *Fungal Ecology, 5*(2), 261–269.

Hernández-Rodríguez, B., Córdova, J., Bárzana, E., & Favela-Torres, E. (2009). Effects of organic solvents on activity and stability of lipases produced by thermotolerant fungi in solid-state fermentation. *Journal of Molecular Catalysis B: Enzymatic, 61*(3–4), 136–142.

Jonathan, J., Tania, V., & Tanjaya, J. C. (2021). Recent advancements of fungal xylanase upstream production and downstream processing. *IJLS, 3*(1), 37–58.

Junker, B., Lester, M., Leporati, J., Schmitt, J., Kovatch, M., Borysewicz, S., et al. (2006). Sustainable reduction of bioreactor contamination in an industrial fermentation pilot plant. *Journal of Bioscience and Bioengineering, 102*(4), 251–268.

King, B. C., Donnelly, M. K., Bergstrom, G. C., Walker, L. P., & Gibson, D. M. (2009). An optimized microplate assay system for quantitative evaluation of plant cell wall–degrading enzyme activity of fungal culture extracts. *Biotechnology and Bioengineering, 102*(4), 1033–1044.

Korsa, G., Konwarh, R., Masi, C., Ayele, A., & Haile, S. (2023). Microbial cellulase production and its potential application for textile industries. *Annales de Microbiologie, 73*(1), 13.

Kzhyshkowska, J., Gratchev, A., & Goerdt, S. (2007). Human chitinases and chitinase-like proteins as indicators for inflammation and cancer. *Biomarker Insights, 2*, 117727190700200.

Lee, S. Y. (1996). High cell-density culture of Escherichia coli. *Trends in Biotechnology, 14*(3), 98–105.

Lee, J., Lee, S. Y., Park, S., & Middelberg, A. P. J. (1999). Control of fed-batch fermentations. *Biotechnology Advances, 17*(1), 29–48.

Liu, Y.-C., Wu, W.-T., & Tsao, J.-H. (1993). Fed-batch culture for L-lysine production via on-line state estimation and control. *Bioprocess Engineering, 9*(4), 135–139.

Maheshwari, R., Bharadwaj, G., & Bhat, M. K. (2000). Thermophilic fungi: their physiology and enzymes. *Microbiology and Molecular Biology Reviews, 64*(3), 461–488.

Mahjoub, M., Mosrati, R., Lamotte, M., Fonteix, C., & Marc, I. (1994). Fuzzy control of baker's yeast fed-batch bioprocess: A robustness study. *Foodservice Research International, 27*(2), 145–153.

Maria, G. L., Sridhar, K. R., & Raviraja, N. S. (2005). Antimicrobial and enzyme activity of mangrove endophytic fungi of southwest coast of India. *Journal of Agricultural Technology.*

Maxon, W. D. (1955). Continuous fermentation: A discussion of its principles and applications. *Applied Microbiology, 3*(2), 110–122.

McCleary, B. (1991). Measurement of polysaccharide degrading enzymes using chromogenic and colorimetric. *Chemistry in Australia*, 398–401.

Mehdi, M. K., Hamid, R., & Mohammad, F. (2011). Genetic manipulation of fungal strains for the improvement of heterologous genes expression (a mini-review). *African Journal of Biotechnology, 10*(41), 7939–7948.

Mikiashvili, N., Wasser, S. P., Nevo, E., & Elisashvili, V. (2006). Effects of carbon and nitrogen sources on Pleurotus ostreatus ligninolytic enzyme activity. *World Journal of Microbiology and Biotechnology, 22*(9), 999–1002.

Mishra, A., & Kumar, S. (2009). Kinetic studies of laccase enzyme of Coriolus versicolor MTCC 138 in an inexpensive culture medium. *Biochemical Engineering Journal, 46*(3), 252–256.

Mohd Azmi, S., Kumar, P., Sharma, N., Sazili, A., Lee, S.-J., & Ismail-Fitry, M. (2023). Application of plant proteases in meat tenderization: Recent trends and future prospects. *Food, 12*(6), 1336.

Mondal, S., Berihun, M., Kovacs, M., Vatai, G., & Koris, A. (2019). Partial refinement of fungal chitinase (Beauveria bassiana) with multistage membrane filtration. *Acta Alimentaria, 48*(3), 277–287.

Sunil S More, Renuka P S, Pruthvi K, Swetha M, S Malini, Veena S M Purification, and characterization of fungal laccase from *Pleurotus* sp. Enzyme Research 2011;2011:1–7.

Naja, G., Volesky, B., & Schnell, A. (2006). Comparative testing of tangential microfiltration for microbial cultures. *Biotechnology and Bioengineering, 95*(4), 584–598.

Ngo, L. T. A., Pham, T. L., & Le, V. V. M. (2008). Purification of endopolygalacturonase from submerged culture of Aspergillus awamori L1 using a two-step procedure: Enzyme precipitation and gel filtration. *International Food Research Journal, 15*(2), 135–140.

Ojha, B. K., Singh, P. K., & Shrivastava, N. (2019). Enzymes in the animal feed industry. In *Enzymes in food biotechnology [Internet]* (pp. 93–109). Elsevier. Available from: https://linkinghub.elsevier.com/retrieve/pii/B9780128132807000074

Ojwach, J., Adetunji, A. I., Mutanda, T., & Mukaratirwa, S. (2022). Oligosaccharides production from coprophilous fungi: An emerging functional food with potential health-promoting properties. *Biotechnology Reports, 33*, e00702.

Østergaard, L. H., & Olsen, H. S. (2011). Industrial applications of fungal enzymes. In M. Hofrichter (Ed.), *Industrial applications [Internet]* (pp. 269–290). Springer. Available from: http://link.springer.com/10.1007/978-3-642-11458-8_13

Palanivel, P., Ashokkumar, L., & Balagurunathan, R. (2013). Production, purification and fibrinolytic characterization of alkaline protease from extremophilic soil fungi. *International Journal of Pharma and Bio Sciences, 4*(2), 101–110.

Pandey, A., Selvakumar, P., Soccol, C., & Poonam, N. N. (1999). Solid state fermentation for the production of industrial enzymes. *Indian Academy of Sciences, 77*(1), 149–162.

Pant, D., & Adholeya, A. (2009). Concentration of fungal ligninolytic enzymes by ultrafiltration and their use in distillery effluent decolorization. *World Journal of Microbiology and Biotechnology, 25*(10), 1793–1800.

Persson, I., Tjerneld, F., & Hahn-Hägerdal, B. (1991). Fungal cellulolytic enzyme production: A review. *Process Biochemistry, 26*(2), 65–74.

Pleissner, D., Kwan, T. H., & Lin, C. S. K. (2014). Fungal hydrolysis in submerged fermentation for food waste treatment and fermentation feedstock preparation. *Bioresource Technology, 158*, 48–54.

Rahardjo, Y. S. P., Sie, S., Weber, F. J., Tramper, J., & Rinzema, A. (2005). Effect of low oxygen concentrations on growth and α-amylase production of Aspergillus oryzae in model solid-state fermentation systems. *Biomolecular Engineering, 21*(6), 163–172.

Ranghar, S., Agrawal, S., & Agrawal, P. K. (2019). Microbial products: Protein, enzyme, secondary metabolites and chemicals. In D. P. Singh & R. Prabha (Eds.), *Microbial interventions in agriculture and environment [Internet]* (pp. 347–384). Springer. Available from: http://link.springer.com/10.1007/978-981-32-9084-6_17

Rodríguez Couto, S., & Sanromán, M. A. (2005). Application of solid-state fermentation to ligninolytic enzyme production. *Biochemical Engineering Journal, 22*(3), 211–219.

Rosell, C. M., & Dura, A. (2015). Enzymes in bakeries. *Enzymes in Food and Beverage Processing, 1*, 34.

Roth, C., Moroz, O. V., Turkenburg, J. P., Blagova, E., Waterman, J., Ariza, A., et al. (2019). Structural and functional characterization of three novel fungal amylases with enhanced stability and pH tolerance. *IJMS., 20*(19), 4902.

Schmidt, K. R., Chand, S., Gostomski, P. A. H., Boyd-Wilson, K. S., Ford, C., & Walter, M. (2008). Fungal inoculum properties and its effect on growth and enzyme activity of trametes versicolor in soil. *Biotechnology Progress, 21*(2), 377–385.

Sethi, S., & Gupta, S. (2014). Optimization of cultural parameters for cellulase enzyme production from fungi. *BioLife., 2*(3), 989–996.

Shahani, K. M., Arnold, R. G., Kilara, A., & Dwivedi, B. K. (1976). Role of microbial enzymes in flavor development in foods. *Biotechnology and Bioengineering, 18*(7), 891–907.

Sharma, V., Tsai, M.-L., Nargotra, P., Chen, C.-W., Kuo, C.-H., Sun, P.-P., et al. (2022). Agro-industrial food waste as a low-cost substrate for sustainable production of industrial enzymes: A critical review. *Catalysts, 12*(11), 1373.

Singhania, R. R., Sukumaran, R. K., Patel, A. K., Larroche, C., & Pandey, A. (2010). Advancement and comparative profiles in the production technologies using solid-state and submerged fermentation for microbial cellulases. *Enzyme and Microbial Technology, 46*(7), 541–549.

Soliday, C. L., & Kolattukudy, P. E. (1983). Primary structure of the active site region of fungal cutinase, an enzyme involved in phytopathogenesis. *Biochemical and Biophysical Research Communications, 114*(3), 1017–1022.

Spagnuolo, R., Cosco, C., Mancina, R. M., Ruggiero, G., Garieri, P., Cosco, V., et al. (2017). Beta-glucan, inositol and digestive enzymes improve quality of life of patients with inflammatory bowel disease and irritable bowel syndrome. *European Review for Medical and Pharmacological Sciences, 21*(2), 102–107.

Stamm, O., Latscha, U., Janecek, P., & Campana, A. (1976). Development of a special electrode for continuous subcutaneous pH measurement in the infant scalp. *American Journal of Obstetrics and Gynecology, 124*(2), 193–195.

Subramaniyam, R., & Vimala, R. (2012). Solid state and submerged fermentation for the production of bioactive substances: A comparative study. *International Journal of Natural Sciences, 3*(3), 480–486.

Suto, M., & Tomita, F. (2001). Induction and catabolite repression mechanisms of cellulase in fungi. *Journal of Bioscience and Bioengineering, 92*(4), 305–311.

Suzuki, T., Yamane, T., & Shimizu, S. (1990). Phenomenological background and some preliminary trials of automated substrate supply in pH-Stat modal fed-batch culture using a setpoint of high limit. *Journal of Fermentation and Bioengineering, 69*(5), 292–297.

Taddia, A., Rito-Palomares, M., Mayolo-Deloisa, K., & Tubio, G. (2021). Purification of xylanase from Aspergillus niger NRRL3 extract by an integrated strategy based on aqueous two-phase systems followed by ion exchange chromatography. *Separation and Purification Technology, 255*, 117699.

Takó, M., Kotogán, A., Krisch, J., Vágvölgyi, C., Mondal, K. C., & Papp, T. (2015). Enhanced production of industrial enzymes in Mucoromycotina fungi during solid-state fermentation of agricultural wastes/by-products. *Acta Biologica Hungarica, 66*(3), 348–360.

Tamerler, C., & Keshavarz, T. (2000). Lipolytic enzyme production in batch and fed-batch cultures of Ophiostoma piceae andFusarium oxysporum. *Journal of Chemical Technology and Biotechnology, 75*(9), 785–790.

Tamilvanan, N. (2020). Optimization, production, purification of laccase enzyme from Bacillus sp. Yuan F. *JNDC., 2*(4), 35–43.

Tharifkhan, S. A., Perumal, A. B., Elumalai, A., Moses, J. A., & Anandharamakrishnan, C. (2021). Improvement of nutrient bioavailability in millets: Emphasis on the application of enzymes. *Journal of the Science of Food and Agriculture, 101*(12), 4869–4878.

Van Bogaert, I. N. A., Groeneboer, S., Saerens, K., & Soetaert, W. (2011). The role of cytochrome P450 monooxygenases in microbial fatty acid metabolism. *The FEBS Journal., 278*(2), 206–221.

van Oort, M., & Whitehurst, R. J. (Eds.). (2010). *Enzymes in food technology* (2nd ed.). Wiley-Blackwell.

Verdín, J., Sánchez-León, E., Fajardo, R., Leal-Morales, C. A., Bartnicki-Garcia, S., & Riquelme, M. (2015). Density gradient centrifugation for enrichment and identification of GFP-tagged chitosomal microvesicles of filamentous fungi. *BIO-PROTOCOL [Internet], 5*(19) Available from: http://www.bio-protocol.org/e1611

Verduzco-Oliva, R., & Gutierrez-Uribe, J. A. (2020). Beyond enzyme production: Solid State Fermentation (SSF) as an alternative approach to produce antioxidant polysaccharides. *Sustainability., 12*(2), 495.

Vitolo, M. (2020). Enzymes in the production of juices and beverages. *World Journal of Pharmacy and Pharmaceutical Sciences., 9*(3), 504–517.

Volesky, B., Luong, J. H. T., & Aunstrup, K. (1984). Microbial enzymes: Production, purification, and isolation. *Critical Reviews in Biotechnology, 2*(2), 119–146.

Wilson, C. J., Zhan, H., Swint-Kruse, L., & Matthews, K. S. (2007). The lactose repressor system: Paradigms for regulation, allosteric behavior and protein folding. *Cellular and Molecular Life Sciences, 64*(1), 3–16.

Wolfenden, R., & Snider, M. J. (2001). The depth of chemical time and the power of enzymes as catalysts. *Accounts of Chemical Research, 34*(12), 938–945.

Wu, S., Song, R., Liu, T., & Li, C. (2023). Antifungal therapy: Novel drug delivery strategies driven by new targets. *Advanced Drug Delivery Reviews, 199*, 114967.

Yang, Y., Xia, J., Li, J., Chu, J., Li, L., Wang, Y., et al. (2012). A novel impeller configuration to improve fungal physiology performance and energy conservation for cephalosporin C production. *Journal of Biotechnology, 161*(3), 250–256.

Zabot, G. L., Schaefer Rodrigues, F., Polano Ody, L., Vinícius Tres, M., Herrera, E., Palacin, H., et al. (2022). Encapsulation of bioactive compounds for food and agricultural applications. *Polymers, 14*(19), 4194.

Correction to: Novel Extraction Methods for Bioactive Compounds from Fungi: An Overview

Charu Agnihotri, Shubham Singh Patel, Zothanpuia, and Bhim Pratap Singh

Correction to:
Chapter 1 in: B. P. Singh et al. (eds.), *Fungal Additives and Bioactives in Food Processing Industries*, Fungal Biology, https://doi.org/10.1007/978-3-032-04519-5_1

The book was inadvertently published with an incorrect name information for one of the chapter authors as "Shubham Kumar Patel," instead of the correct name "Shubham Singh Patel" in the Front Matter and Chapter 1.

The updated version of this chapter can be found at
https://doi.org/10.1007/978-3-032-04519-5_1

B. P. Singh et al. (eds.), *Fungal Additives and Bioactives in Food Processing Industries*, Fungal Biology, https://doi.org/10.1007/978-3-032-04519-5_16

Index

B. P. Singh et al. (eds.), *Fungal Additives and Bioactives in Food Processing Industries*, Fungal Biology, https://doi.org/10.1007/978-3-032-04519-5

MIX
Papier aus verantwortungsvollen Quellen
Paper from responsible sources
FSC® C105338
FSC
www.fsc.org

If you have any concerns about our products,
you can contact us on
ProductSafety@springernature.com

In case Publisher is established outside the EU,
the EU authorized representative is:
Springer Nature Customer Service Center GmbH
Europaplatz 3, 69115 Heidelberg, Germany

Printed by Libri Plureos GmbH
in Hamburg, Germany